Georgia Land Surveying History and Law

FARRIS W. CADLE

The University of Georgia Press
Athens and London

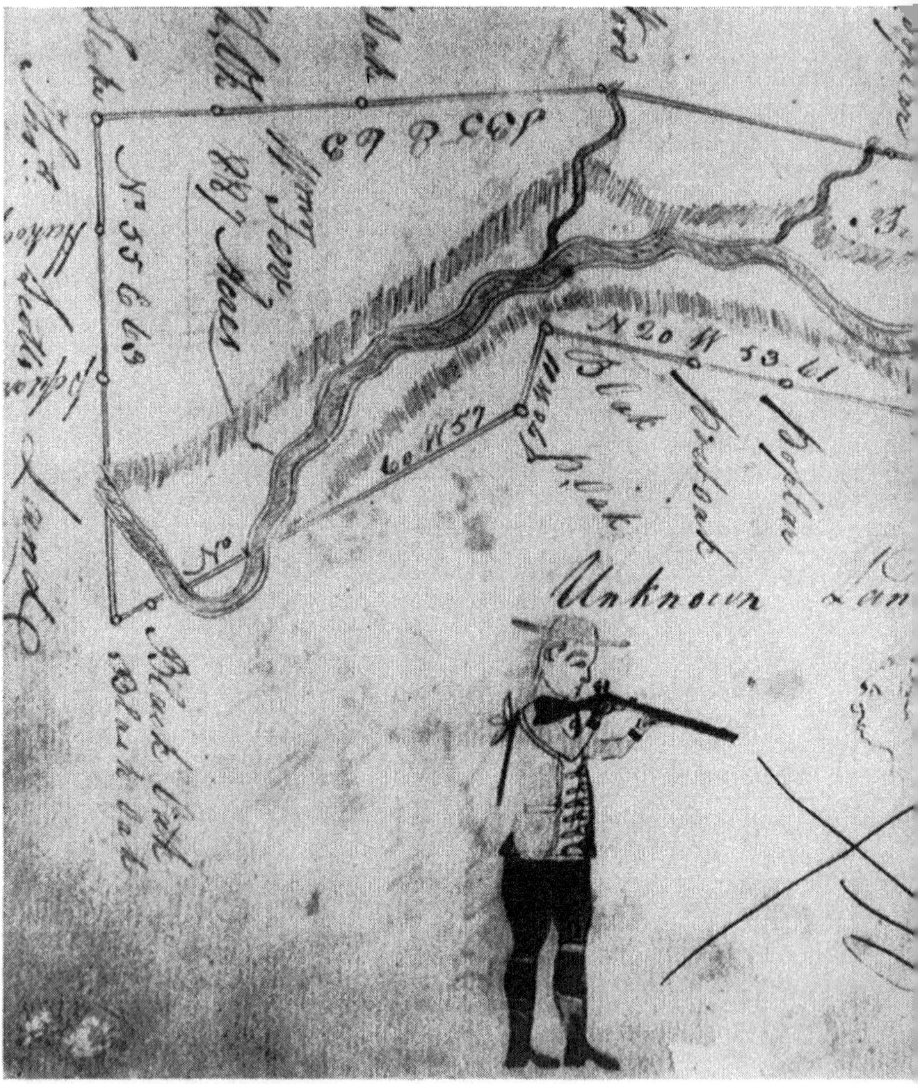

Georgia Land Surveying History and Law

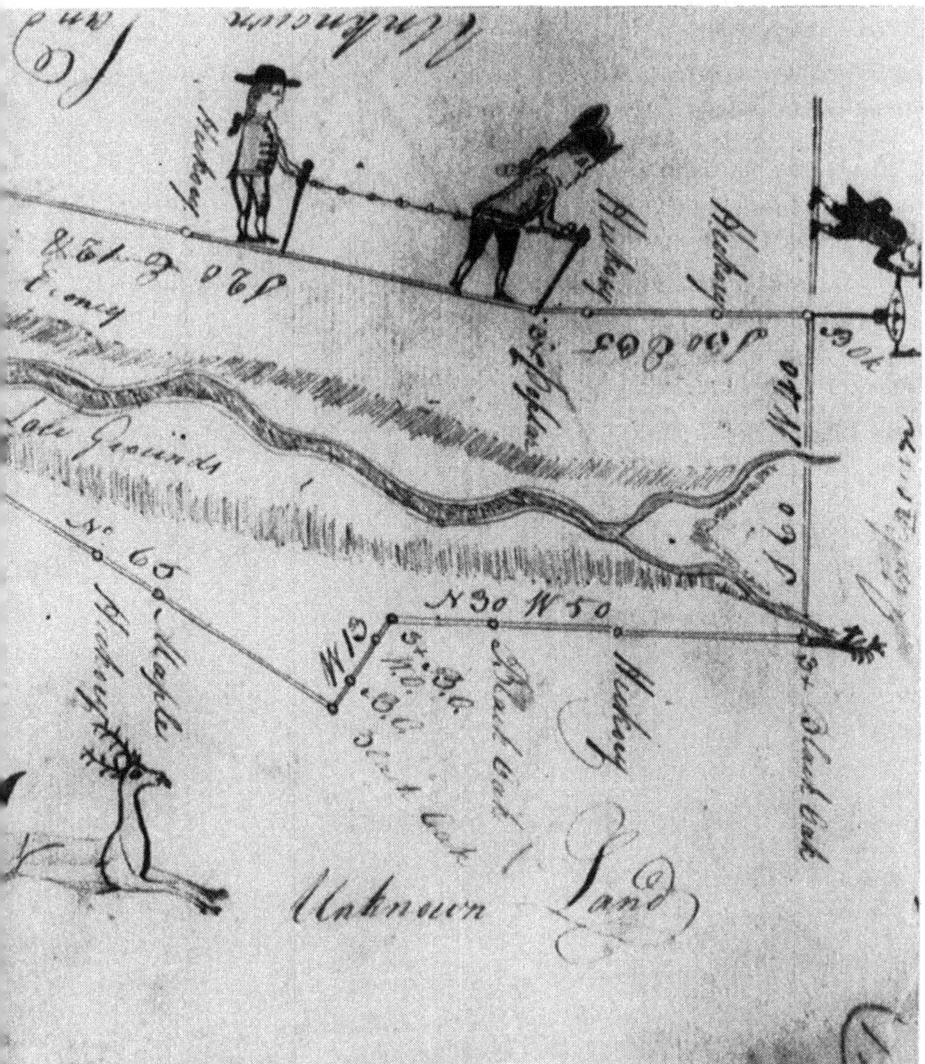

Athens, Georgia 30602

Set in Linotron 202 Baskerville by Tseng Information Systems
Printed and bound by Thomson-Shore
The paper in this book meets the guidelines
for permanence and durability of the Committee on
Production Guidelines for Book Longevity of
the Council on Library Resources.

Printed in the United States of America

95 94 93 92 91 5 4 3 2 1

Library of Congress Cataloging in Publication Data

Cadle, Farris W.
Georgia land surveying history and law / Farris W. Cadle.
p. cm.
Includes bibliographical references and index.
ISBN 0-8203-1257-6 (alk. paper)
1. Surveying—Law and legislation—Georgia—History.
2. Surveying—Georgia—History. I. Title.
KFG127.5.C33 1991
346.75804'32—dc20
[347.5806432] 90-40785
CIP

British Library Cataloging in Publication Data available

Contents

Illustrations

PART I

PART II

Preface

The land surveyor's most fundamental role is to measure property lines on the ground and then to describe these lines (usually in the form of a plat) from his measurements. It is manifestly evident that before such lines can be measured they must first be found, and it is here that the greatest perplexities in surveying occur and where surveyors often show their greatest incapacity. Rapidly rising land values and the increasing incidence and expense of boundary litigation demand that more serious consideration be given this matter.

The purpose of this book is to present, in lay terms, a comprehensive treatment of the legal and historical aspects of land surveying in Georgia, with emphasis on their practical import to the profession. Although written for land surveyors, this text may also prove useful to attorneys, realtors, title abstractors, historians, and anyone concerned with real property in Georgia. Every Georgia case and statute dealing with boundaries has been considered for discussion. Liberal use has been made of general legal works and the decisions of other states to supply points not decided in Georgia and, in a few instances, to compare the Georgia law to that of other states.

While the value of the legal perspectives presented are probably apparent, it is hoped that the historical perspectives provided will have just as useful an application. We are all inexorably caught up in the past. As E. Merton Coulter once stated, "The present is not something that appeared without a background. The present is nothing more than an outgrowth of the past." By knowing the past we have a better understanding of the present. In few aspects of human endeavor does this dictum play a more decisive role than with regard to the borders surrounding our freeholds, for it is one of the most basic concepts of real property that boundaries, once established, are fixed and unalterable. Yet most land surveyors go casually about their work, scarcely aware of the rich and complex panoply of historical factors that have shaped the boundaries they routinely measure and restore. These considerations have been brought home time and again to the author during his years of land surveying experience. He has been accorded the opportunity of retracing original Garden Lot and Farm Lot lines of Savannah Township, still intact and still serving their essential purpose of demarcating adjoining properties two and a half centuries after they were first laid down on the ground, probably by Noble Jones or Joseph Avery. His investigations led to the discovery that a century and a half ago his great-great-grandfather and great-great-uncle, perhaps newly arrived in Georgia and landless, acted as chain carriers in a survey performed for the purpose of obtaining a headright grant from the state. The lines they ran are still clearly discernable, marking the peripheries of a tract of land that was passed down through the generations until today it is in the possession of the author. The very warrant that authorized their action, the obtaining of which was the initial step in the granting process, along with the resulting plat, are on file in the Georgia Department of Archives and History. The author's ramblings carried him to a remote corner of Carroll County where he reconnoitered the land lot line that James A. Rogers and his crew were running on a damp, wintry day in 1827 when they were suddenly confronted by an angry party of Creek Indians, thus sparking a nationwide controversy.

One of the joys of writing a book is the opportunity of working with many wonderful people. It would be impossible to acknowledge individually all those who offered their time and expertise. Particular appreciation is due Eloise S. Bailey, Kevin B. Buice, Wiley J. Busbin, Gertrude Calvert, Kenneth Coleman, Raymond E. Davis, Robert S. Davis, Jr., Louis De Vorsey, Jr., Marion R. Hemperley, Nancy Grayson Holmes, Frank Parker Hudson, Margaret A. Johnsen, Elizabeth Fitzpatrick Johnson, David H. Robertson, José F. Rodriguez, Mrs. Albert B. Saye, the late Albert B. Saye, Sidney B. Shepherd, Ingrid Shields, Gordon B. Smith, Erwin C. Surrency, Durwin Talon, Philip H. L. Yonge, and Katrina DeFoor Wingate. The author is highly appreciative for the graciousness and thoroughly competent treatment he received from librarians and archivists at Duke University, Emory University, the Georgia Department of Archives and History, the Georgia Historical Society, the Library of Congress, the Museum of Early Southern Decorative Arts, the National Archives, the New York Public Library, the British Public Record Office, the University of Georgia, the University of North Carolina at Chapel Hill, and the University of Virginia. Through the persistent efforts of the personnel at these venerable institutions, the author has gained a tremendous appreciation for the invaluable and often ignored service that the custodians of "the memory of humanity" perform for society.

Part I. A History of the Surveying of the Public Domain of Georgia

Chapter 1. The Colonial Period

Because surveys performed today are either partially or wholly retracement surveys, whose primary object is to ascertain the lines as originally marked on the ground, some knowledge of the development of the land system in Georgia and the problems its early surveyors faced will be helpful to present-day land surveyors. In a larger sense the original surveys of Georgia are worthy of detailed investigation if only because they form the foundation for the settlement and development of the state.

Broadly speaking, the state of Georgia is laid out under three survey systems: (1) the "headright system," which covers generally the area northeastward of the Appalachee, Oconee, and Altamaha rivers, as well as original Glynn and Camden counties; (2) the "land lot and district system," which covers most of the remaining portion of the state; and (3) several small, isolated tracts, most of them Indian reserves or town reserves, which are within the land lot and district area but are not laid out under this system. To this enumeration a fourth could perhaps be added: the land system under the Trustees. Such further breakdown would serve no useful purpose, however, for when the headright system was inaugurated in 1755 grantees under the Trustees took out new grants to their lands under that system. Accordingly, land grants under the Trustees will be classified within the headright system.[1]

As far as future generations were concerned, the fundamental difference between the headright and the land lot and district systems is the way they were originally laid out. Under the headright system the individual tracts were surveyed piecemeal over a period of more than 170 years, and only the broadest stipulations were laid down in advance as to their locations and shapes. In contrast, under the land lot and district system, the individual surveys extended over a fairly large expanse of territory and were usually made by the same surveyor within a matter of a few weeks. Furthermore, the acts under which the system was created provided that the tracts were to be laid out in a more or less uniform

1. As this narrative progresses, frequent reference to the map on page 4 will help the reader to understand how Georgia's land system evolved.

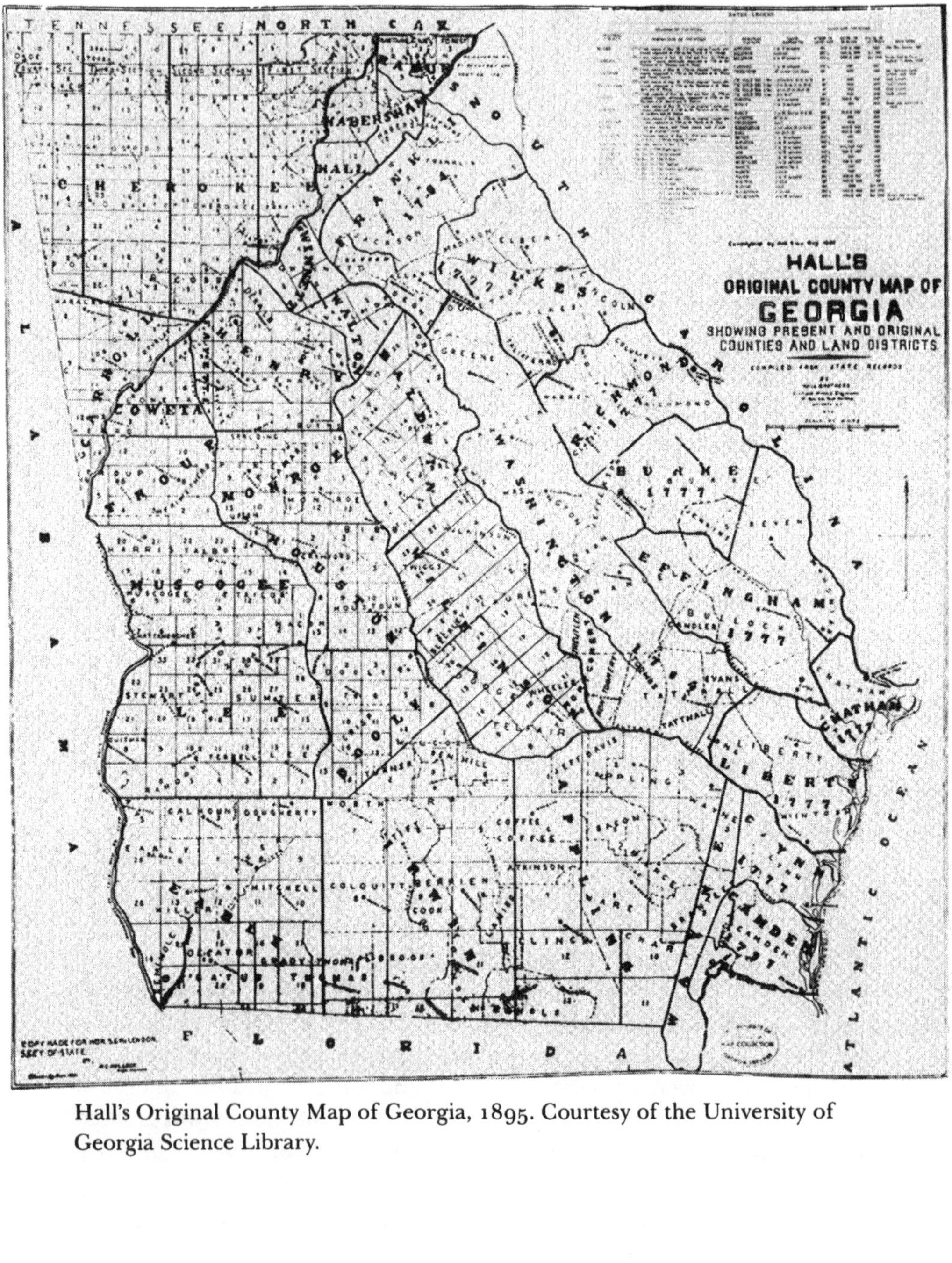

Hall's Original County Map of Georgia, 1895. Courtesy of the University of Georgia Science Library.

manner. The differing methods by which the two systems were originally surveyed has a profound impact on resurvey procedures.

With the notable exceptions of the Yazoo Fraud and a few other less spectacular (though no less shabby) land fraud schemes, one policy was firmly adhered to throughout the period of Georgia's settlement. The necessity of having a large reserve of able-bodied men for defense purposes and the desire to populate and develop the state as quickly as possible made it extremely undesirable that large tracts fall into the hands of a few nonresident speculators who would let the lands lay idle until they could be sold for exorbitant profits. Therefore, each settler was granted only as much land as he and his dependents could actually occupy and improve.

Land System under the Trustees

The rights by which the English claimed the North American continent date back to the explorations of John Cabot in 1497.[2] For more than a century and a half afterward, however, the English made little effort at colonization south of Jamestown. In 1663 and 1665 King Charles II granted to Edward Earl of Clarendon, George Duke of Albemarle, William Lord Craven, John Lord Berkley, Anthony Lord Ashley, Sir George Carteret, Sir William Berkley, and Sir John Colleton (afterward called the eight Lords Proprietors) a tract of land that embraced all the territory between twenty-nine degrees north latitude (a latitude passing through present-day central Florida) and thirty-six degrees, thirty minutes north latitude (about the latitude of the present north boundary of North Carolina) and from the Atlantic Ocean to the Pacific Ocean.[3] This territory was named Carolina. Because the Spaniards were already well established in Florida and the present southwestern United States and the French were gaining a foothold along the Mississippi River and Gulf Coast, the Carolina grant overlapped dominions claimed by both Spain and France.

Over the next several decades a number of successful English settlements were made in Carolina, but all attempts to settle the lands south

2. [B. Martyn], An Impartial Inquiry into the State and Utility of the Province of Georgia 56 (London 1741), *reprinted in* 1 Coll. Ga. Hist. Soc. 153, 183 (Savannah 1840); Johnson v. McIntosh, 21 U.S. (8 Wheat.) 543, 5 L. Ed. 681 (1823).

3. 1 Col. Rec. N.C. 20–33, 102–14; Fletcher v. Peck, 10 U.S. (6 Cranch) 87, 3 L. Ed. 162 (1810).

of the Savannah River failed. On July 25, 1729, the heirs of seven of the eight Lords Proprietors (all except John Carteret, sole heir of Lord Carteret) surrendered their rights and interests in Carolina to the British crown.[4] Shortly thereafter the province of Carolina was divided into North Carolina and South Carolina.

On June 9, 1732, King George II granted a charter incorporating twenty-one noblemen and gentlemen as "The Trustees for establishing the Colony of Georgia in America."[5] The Trustees were granted the crown's seven-eights interest in all the lands between the Savannah and Altamaha rivers and westerly from the heads of these two rivers respectively to the Pacific Ocean, along with all islands within twenty leagues of the eastern seacoast of the territory.[6] The Trustees were also able to secure, in trust, John Carteret's one-eighth interest in the lands.[7]

The initial contingent of colonists, led by James Edward Oglethorpe, embarked from England and arrived at Beaufort, South Carolina, in January 1732/3, having come by way of Charleston and Port Royal. The colonists were left at Beaufort while Oglethorpe, accompanied by several South Carolinians, reconnoitered the Savannah River for the purpose of selecting a site on which to build a town. Fifteen miles from the ocean they found what appeared a promising location—a steep, sandy bluff studded with tall, moss-laden longleaf pine and live oak. The Creek Indians who lived in the area called it Yamacraw. A preliminary agreement was expeditiously concluded with these Indians and the colonists fetched from Beaufort. They arrived at the site on February 1, 1732/3.[8]

A more comprehensive treaty was drawn up and signed between Oglethorpe and the Creek Nation on May 21, 1733.[9] This treaty was followed by two more treaties of cession—one at Coweta, the principal town of the

4. An Act for Establishing an Agreement with Seven of the Lords Proprietors of Carolina, 2 Geo. II c. 34 (1729); Am. St. Papers, 1 Public Lands 39; 3 Col. Rec. N.C. 32; Crittenden, *The Surrender of the Charter of Carolina,* 1 North Carolina Historical Review 383 (1924). All dates mentioned in this text before 1752 are Old Style, and all dates thereafter are New Style.

5. 1 Col. Rec. Ga. 11.

6. Id. 18.

7. J. McCain, Georgia as a Proprietary Province 134, 226, 250, 252–54, 256 (1917); A. Saye, New Viewpoints in Georgia History 48, 54 (1943).

8. A. Johnson, Georgia as Colony and State 48–49 (1938).

9. Gentleman's Magazine (London), July 1733, at 384; C. Jones, Historical Sketch of Tomo-chi-chi, Mico of the Yamacraws 27–37 (Albany, N.Y. 1868); H. McCall, The History of Georgia 248 (1909 reprint).

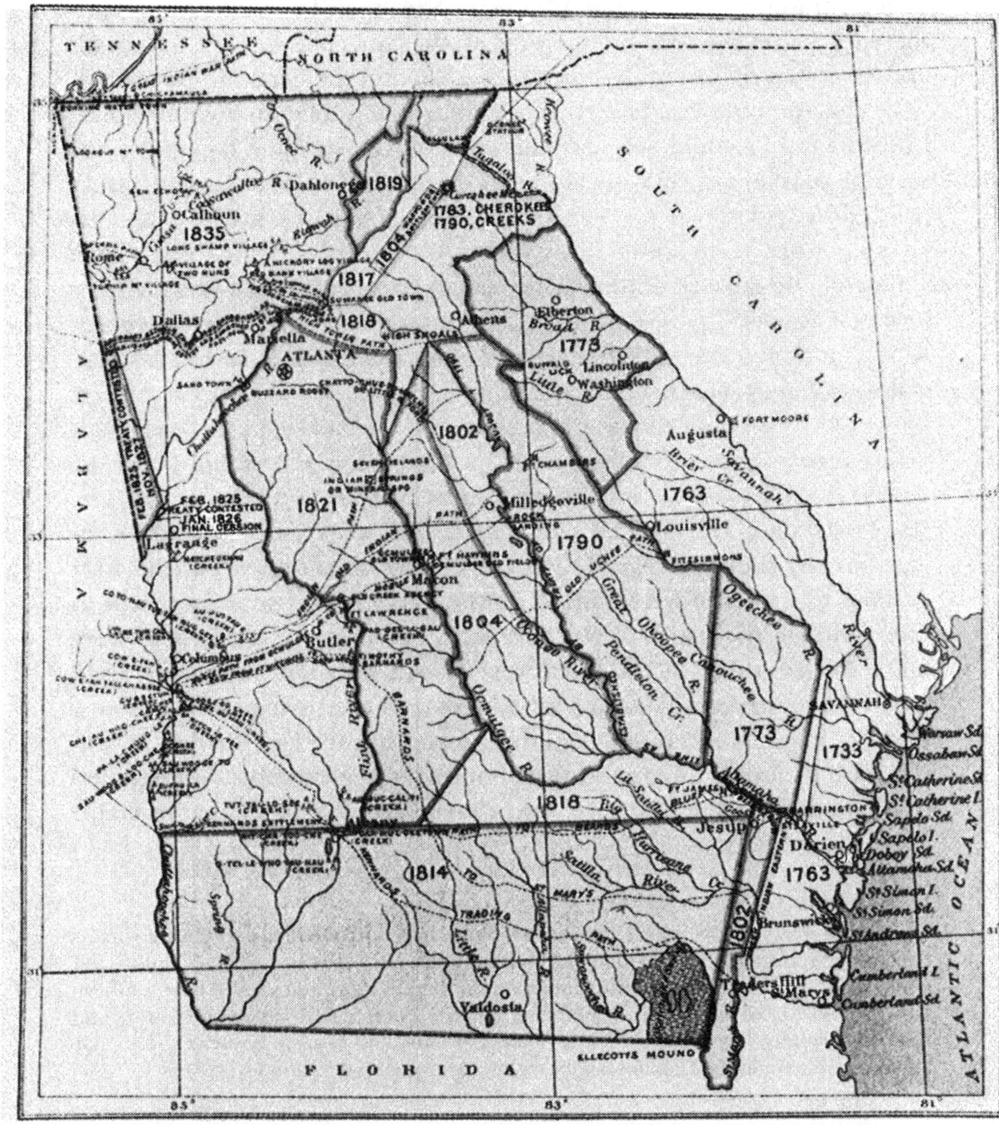

Indian land cessions in Georgia, 1733–1835. From Ulrich B. Phillips, *Georgia and State Rights* (1902).

Creek Nation, in 1739,[10] and another at Savannah in 1758.[11] Altogether these three treaties ceded to the English the lands between the Savannah and St. Johns rivers and between the ocean and the head of tidewater.[12]

Its charter authorized Georgia to employ the public surveyor of South Carolina until it could procure a surveyor of its own,[13] and it engaged the services of William Bull, the surveyor general of South Carolina. Bull, assisted by Oglethorpe, supervised the first survey task in Georgia—the laying out of Savannah.[14]

One of the most striking features about Savannah is the unusual pattern of its layout—a subject that has been a source of much lively debate.[15] The plan almost certainly had been devised before the colonists left England, and may have been conceived by Oglethorpe.[16] The town proper, physically and politically, was arranged into a series of wards, each subdivided into squares, tythings, and lots. Each ward was planned around a central square, which was flanked on its eastern and western sides by four "trust lots," reserved for public uses, and on its northern and southern ends by four tythings. The colonists had been organized into four tythings (a group of ten families) on the passage over, and they slept in four tents on the banks of the Savannah River before the houses were built. Each tything in the town was divided into ten house lots, each fronted by a street and backed by an alley and measuring sixty by ninety feet. Each family received one of these lots. Thus the first ward of Savannah with its four tythings and forty homesites may have been designated to accommodate the first forty-odd charity families that came to Geor-

10. 26 Col. Rec. Ga. 485; H. McCall, The History of Georgia 252 (1909 reprint).

11. 7 Col. Rec. Ga. 657–67; 28 Col. Rec. Ga. (pt. 1) 266.

12. The Creeks and Cherokees were the two principal Indian nations occupying what is now Georgia. It was they with whom the land-hungry whites were to deal in obtaining land cessions. At the time of the colony's founding the Creeks inhabited what is now southern Georgia and Alabama, while the Cherokees inhabited present-day northern Georgia and Alabama, western North Carolina and South Carolina, and eastern Tennessee. The map on page 7 illustrates the Indian cessions within the present bounds of Georgia.

13. 1 Col. Rec. Ga. 23.

14. 3 Col. Rec. Ga. 380; P. Gordon, Journal of Peter Gordon 37 (E. Coulter ed. 1963); South Carolina Gazette (Charleston), Mar. 31, 1733, at 3, col. 1. The central thoroughfare in Savannah is named for Bull.

15. See Bell, *A New Theory on the Plan of Savannah,* 48 Ga. Hist. Q. 147 (1964); W. Harden, A Suggestion as to the Origin of the Plan of Savannah (1885); Reps, *$C^2 + L^2 = S^2$? Another Look at the Origins of Savannah's Town Plan,* in Forty Years of Diversity 101 (H. Jackson ed. 1984); Savannah Morning News, July 4, 1976, at 10AA, col. 1.

16. 20 Col. Rec. Ga. 17.

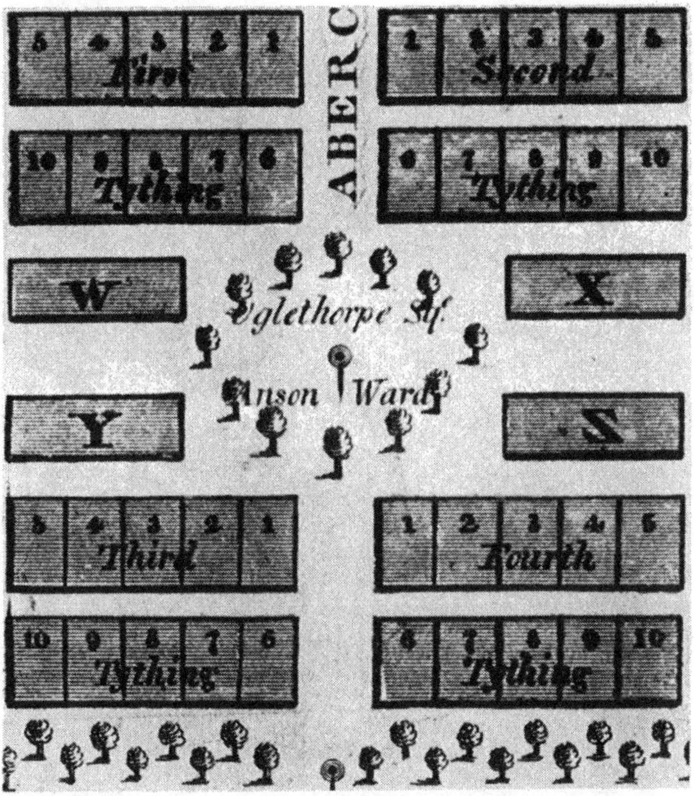

Map of Anson Ward in Savannah, showing configuration of house lots, tythings, trust lots, and square. Detail from map of Savannah by John McKinnon and C. C. Wright, 1820. Courtesy of the Georgia Historical Society.

gia. At the head of each tything was a tythingman who reported to a constable, the officer in charge of the ward.[17]

Surrounding this layout was the city common over which Oglethorpe's plan was faithfully followed and extended for more than a century until by 1855 no more common land was left.[18]

Georgia was founded largely as a refuge for the destitute and impoverished people of England. The charter for the province provided that

17. K. Coleman, Colonial Georgia 29 (1976); M. Lane, Savannah Revisited: A Pictoral History 11–27 (2d ed. 1973); 1 W. Stevens, A History of Georgia 98–101 (New York 1847).

18. M. Lane, Savannah Revisited: A Pictoral History 17 (2d ed. 1973).

Savannah. Map by John McKinnon and C. C. Wright, 1820. Courtesy of the Georgia Historical Society.

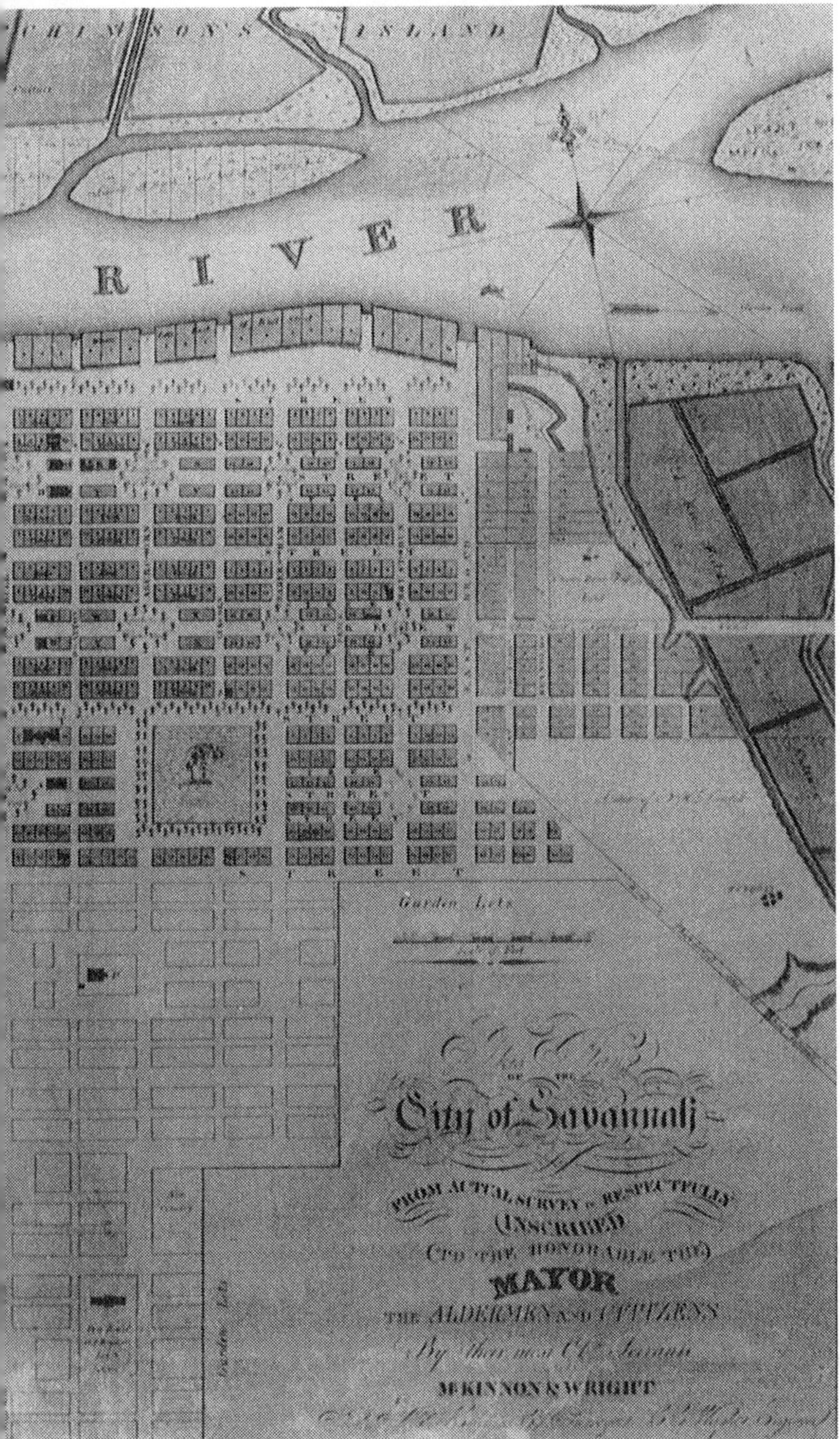
ISLAND
RIVER
Garden Lots
City of Savannah
FROM ACTUAL SURVEY & RESPECTFULLY
(INSCRIBED)
(TO THE HONORABLE THE)
MAYOR
THE ALDERMEN AND CITIZENS
M'KINNON & WRIGHT

no one in the colony could be granted more than five hundred acres of land, but the Trustees reduced this amount to fifty acres for colonists coming over at Trust expense (called "charity colonists"). Those who came at their own expense (called "adventurers") could be granted up to the maximum amount of five hundred acres.[19]

Initially lands were granted through the Trustees in London. Because it was cumbersome to grant numerous small parcels three thousand miles from where the land was located, the Trustees adopted the procedure of granting several thousand acres to a single colonist or to a small group of colonists to be held in trust and subdivided among the charity colonists when they should apply for land.[20] Savannah was originally laid off on a tract containing twenty-four square miles or 15,360 acres. Of this land, 12,000 acres were used for actual grants of 240 freeholds of fifty acres each, while the remainder was used for roadways, streets, squares, trust lots, and the town common.[21]

As mentioned, each charity family received a sixty-by-ninety-foot lot in the town. Each family also received a triangular-shaped "garden lot" of four and seven-eighths acres on the periphery of the town as well as a rectangular-shaped forty-five-acre "farm lot" a small distance from the town, for a total of approximately fifty acres.

In a manner remotely resembling the town layout, the farm lots of Savannah were arranged into units of ten, each unit surrounding two trust lots reserved for the use of the public generally. Like the farm lots, these trust lots contained forty-five acres each. The garden, farm, and trust lots were laid out contiguous to one another in a compact fashion, with regular road rights-of-way in between. These lands as well as the town lots were distributed to the original colonists by a lottery.[22]

Savannah, with its unique plan of the town proper and accompanying garden and farm lots, served as a model for most of the other villages laid

19. 1 Col. Rec. Ga. 22; J. McCain, Georgia as a Proprietary Province 232–34, 239–40 (1917).

20. 2 Col. Rec. Ga. 118, 123; 3 Col. Rec. Ga. 379; 32 Col. Rec. Ga. passim; Journal of the Earl of Egmont 102 (R. McPherson ed. 1962); 1 C. Jones, The History of Georgia 114, 155–61 (1883); J. McCain, Georgia as a Proprietary Province 257–58 (1917).

21. J. McCain, Georgia as a Proprietary Province 272–73 (1917).

22. Id.; H. Anderson, D. Douglas, & P. Tailfer, A True and Historical Narrative of the Colony of Georgia in America 30 (London 1741), *reprinted in* 2 Coll. Ga. Hist. Soc. 163, 199 (Savannah 1842); F. Moore, A Voyage to Georgia Begun in the Year 1735, at 7, 28–29 (London 1744), *reprinted in* 1 Coll. Ga. Hist. Soc. 79, 82, 97–98 (Savannah 1840).

out in Georgia under Trustee administration. All these outlying villages, however, with the exception of Augusta and Darien, soon dwindled and no longer exist.[23]

There were some decided disadvantages in laying out lots and farms so regularly. Much of the land in Georgia was not well suited for cultivation, and it was impossible to find large tracts that would be uniformly desirable. As a result, many settlers got tracts of poor land. There was also the inconvenience of having three separate and scattered parcels of land to tend.[24]

Adventurers received individual grants directly through the Trustees in London. Usually their lands were laid off outside the limits of the tracts assigned to charity colonists, though they were often allowed a building lot within the town. A reasonable effort seems to have been made to allow them to select locations best suited to their purposes. Care was also taken to mix the lands within each grant so some was high and some low for crop variation, but these details depended largely upon the surveyor in charge.[25]

When an adventurer brought indentured servants to the colony, a reserved tract was often laid out adjoining the adventurer's land to be granted to the indentured servants upon the expiration of their terms of service. In 1735, for example, the Trustees instructed Noble Jones to lay out a five-hundred-acre tract for the Moravian Count Zinzendorf and a two-hundred-acre tract adjoining it to be equally divided among Zinzendorf's ten indentured servants when their terms of service had ended.[26] The amount of land reserved for indentured servants was at first fixed at twenty-five acres, later at twenty acres, and in 1741 at fifty acres.[27] In 1741 the government in Georgia was reorganized and thereafter all land grants were received directly through officials in the colony.[28]

Although some irregularly shaped tracts were laid out under the Trustees, it is ironic that these surveys were, on the whole, the most regular

23. C. Jones, The Dead Towns of Georgia passim (1878).

24. 20 Col. Rec. Ga. 142; 21 Col. Rec. Ga. 322–24; 22 Col. Rec. Ga. (pt. 2) 100; J. McCain, Georgia as a Proprietary Province 274 (1917).

25. J. McCain, Georgia as a Proprietary Province 258, 273–74 (1917).

26. 29 Col. Rec. Ga. 44; Letter from Benjamin Martyn to Thomas Causton (Jan. 25, 1734/5) (Egmont Mss. in Phillipps Collection, Ser. No. 14207, at 292, UGL).

27. J. McCain, Georgia as a Proprietary Province 267 (1917).

28. Id. 260.

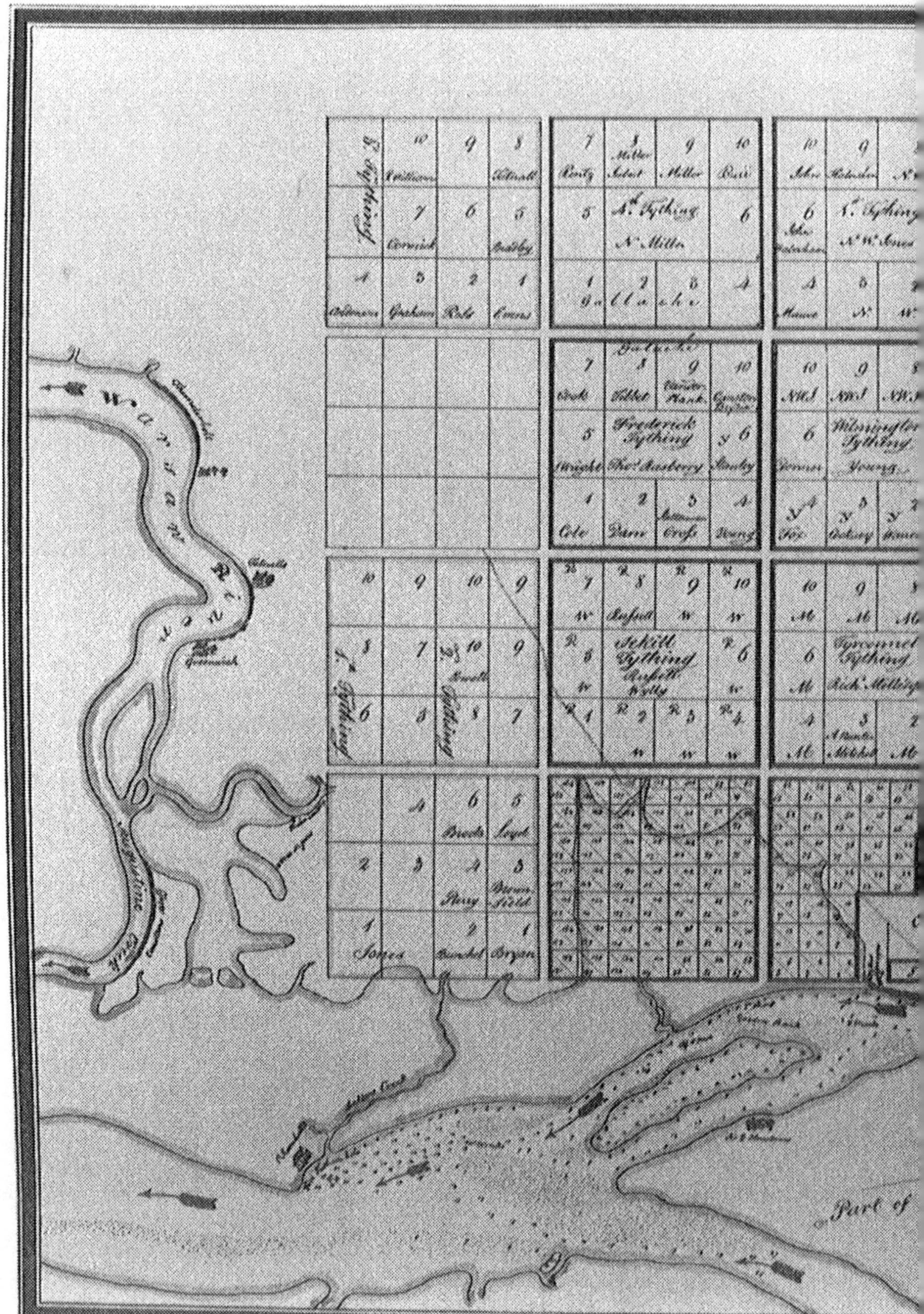

Garden and Farm Lots of Savannah. Lithograph by J. F. Minis, 1904. The original manuscript map by John McKinnon, ca. 1798, from which this lithograph was made, is filed in the Minis Family Papers 2725, Southern Historical Collection, Library of the University of North Carolina at Chapel Hill.

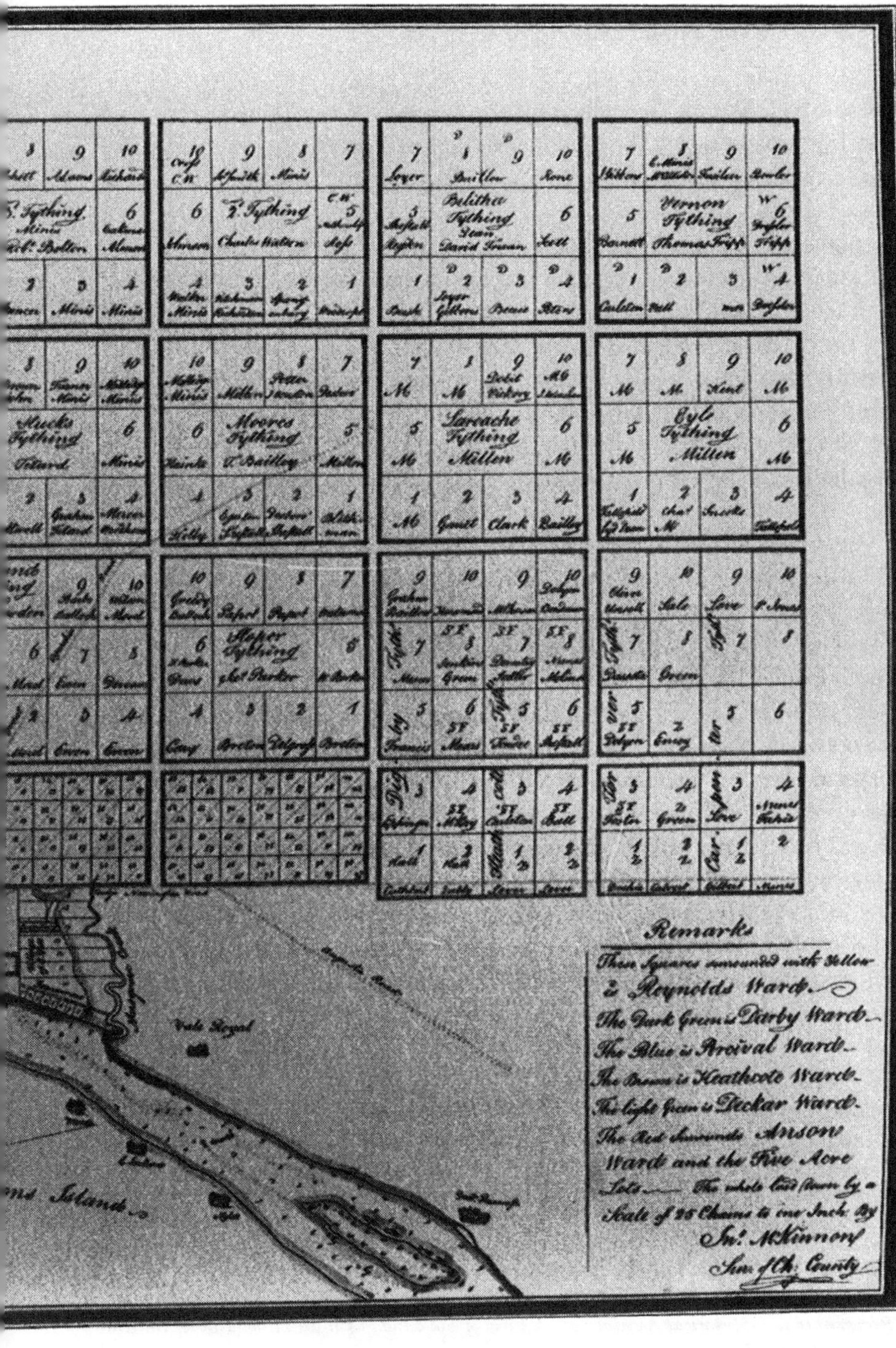
Remarks
Those Squares surrounded with Yellow
is Reynolds Ward.
The Dark Green is Darby Ward.
The Blue is Percival Ward.
The Brown is Heathcote Ward.
The light Green is Decker Ward.
The Red Surrounds Anson
Ward and the Five Acre
Lots. The whole laid down by a
Scale of 25 Chains to one Inch By
Jn. McKinnon
Sur. of Ch. County

and orderly made in Georgia until the advent of the land lottery system in the next century. The layout of the town of Savannah constitutes a marvel both of town planning and of survey precision for this early period. No other town in Georgia and few other towns in America have been planned with such elegant sophistication as Savannah. Few if any other lands, even towns, were laid out in Georgia with such precision for at least a century afterward.

Because Georgia was founded largely as a refuge for the destitute and impoverished, the Trustees set up a rigid and unrealistic system of land tenures based on grants in tail male. This factor probably more than anything else retarded the early growth of the colony. The Trustees were gradually forced to relax the restrictions until March 1750, when a resolution was adopted enlarging all grants to absolute inheritances.[29]

Noble Jones: First Public Surveyor

William Bull, with the aid of Oglethorpe, performed the first surveys in Georgia. The need for a full-time public surveyor for the colony was very apparent, and Oglethorpe appointed Noble Jones to the post.[30] The exact date of Jones's appointment is not known, but he is referred to as a "Surveyor" as early as March 1733/4.[31] Jones was born in London on June 20, 1702. He was apprenticed to his father as a carpenter in 1717. Master carpenters of the day were trained in many of the skills of architects and surveyors as well as building; hence his probable qualification for the job of public surveyor. Jones came to Georgia with the first group of colonists, his occupation entered as "carpenter" on the ship's list.[32]

Jones undoubtedly surveyed additions to the town proper of Savannah

29. Id. 226–49.

30. 29 Col. Rec. Ga. 71.

31. P. von Reck, An Extract of the Journals of Mr. Commissary Von Reck . . . 13 (London, 1734), *reprinted in* 4 Tracts and Other Papers, Relating Principally to the Origin, Settlement, and Progress of the Colonies in North America . . . (P. Force ed. 1844). For evidence that Jones had been public surveyor for at least four months previously, see Baine & De Vorsey, *The Provenance and Historical Accuracy of "A View of Savannah as It Stood the 29th of March, 1734,"* 73 Ga. Hist. Q. 784, 785–89 (1989).

32. S. Temple & K. Coleman, Georgia Journeys 12, 268, 297 (1961); Baine & De Vorsey, *The Provenance and Historical Accuracy of "A View of Savannah as It Stood the 29th of March, 1734,"* 73 Ga. Hist. Q. 784, 789–90 (1989).

as well as making a significant start on delineating its garden and farm lots, but scant record remains of these activities.

He was accorded the opportunity to survey a town from scratch when he undertook to lay out Ebenezer for the industrious German-speaking Salzburgers, who settled at Red Bluff on the Savannah River in what is now Effingham County. Jones arrived at the site on March 5, 1735/6. A plan closely resembling that of Savannah was adopted, "and the people were all set to work helping the surveyor by clearing the city area of underbrush, chopping out lines through the forest, carrying the surveying sticks, setting stakes, &c." The task of demarcating two wards was accomplished in a week. On March 13 "everyone was assigned his house lot through lottery, and a uniform, simple, yet comfortable style of architecture was prescribed."[33]

Barely three months after Jones had completed the work at Ebenezer, Oglethorpe ordered him to lay out the town of Augusta, some 150 miles upriver from Savannah. Jones was directed to lay off "40 House Lotts each of an Acre, the largest Streets not narrower than 25 yards; a Square in the Center, & Lotts for Public Buildings on each Side [of] the Square." All the public lots together were not to be less than four acres in area and the common was to cover six hundred acres. Beyond the common forty farm lots were surveyed out to accompany the house lots in the town. Thus Augusta's original design was much like that of Savannah and Ebenezer, but the plan underwent considerable mutation immediately after the Revolution.[34]

Unfortunately, Jones's attainments were largely overshadowed by frequent dissatisfaction with his performance. Besides having the full-time position of surveyor, he was also a conservator of the peace, constable, rum agent, Indian agent, and forest ranger. In addition to his regular surveying duties of running boundaries and making plats, the Trustees instructed Jones to keep them informed as to the number of acres cleared on each tract, the crops planted, and the manner of cultivation.[35]

Because he had so many jobs to fill, Jones was continually behind in

33. P. von Reck, Von Reck's Voyage 37, 66–67 (K. Hvidt et al. eds. 1980).

34. Journal of the Earl of Egmont 168 (R. McPherson ed. 1962); Orders and Instructions from Oglethorpe to Noble Jones and Others (June 14, 1736) (Egmont Mss. in Phillipps Collection, Ser. No. 14202, at 4, UGL).

35. 29 Col. Rec. Ga. 36, 54; E. Coulter, Wormsloe: Two Centuries of a Georgia Family 45, 49 (1955).

his surveying work, and soon a great many complaints were filed against him. Samuel Mercer wrote the Trustees in 1735 that Jones had promised to survey his land but had never done so. Mercer declared that because settlers could not get their land laid out, they were unable to plant crops and instead were working at odd jobs in Savannah and nearly starving.[36] Thomas Christie reported that Jones was "idle in surveying, so that he drove many from the Colony to Carolina to seek for land, being delayed by him in running out their land, till their money was spent."[37] Robert Parker, Jr., who had come over at the expense of the Trustees and therefore was entitled to have his land surveyed at Trust expense, complained that he had offered Jones five guineas of his own money in addition to what Jones would receive from the Trustees to survey his land but that it still had not been done.[38] In exasperation, colonial officers verbally instructed some colonists to select a piece of ground to settle on until a survey could be made, at which time the officers would see that the settlers secured good title.[39]

Jones's surveying competence also came under attack. According to one allegation, he was "so unskilful as to attribute one mans ground to another mans Lot, which being improved by the first Occupant, was afterwards claimed & detained by the second, to the ruin of the first, and to the discouragement of others from improving, who knew not but their lotts might have the same fate."[40] William Stephens, upon examining his forty-five-acre farm lot, found that Jones had marked the two corners that lay next to the road but had failed to run out or mark any of the other lines.[41] It was also charged that Jones was constantly behind in returning his plats and certificates to the register[42] and that he "follow'd more the business of hunting after run away Servants [indentured servants] than his proper Employ of Surveyor."[43]

The Trustees early confronted Jones with the charges being leveled

36. E. Coulter, Wormsloe: Two Centuries of a Georgia Family 46 (1955).

37. Id.; 5 Col. Rec. Ga. 464.

38. E. Coulter, Wormsloe: Two Centuries of a Georgia Family 47 (1955).

39. 4 Col. Rec. Ga. 106–7.

40. E. Coulter, Wormsloe: Two Centuries of a Georgia Family 46 (1955); 5 Col. Rec. Ga. 85, 87.

41. 4 Col. Rec. Ga. 70–71.

42. 21 Col. Rec. Ga. 321; 29 Col. Rec. Ga. 44.

43. 5 Col. Rec. Ga. 87.

against him.[44] Jones countered by saying that it would be a full-time job for any person just to draw plans and certify them, that the pay Oglethorpe allowed was not sufficient to hire an adequate number of assistants, and that he had originally taken on the job "(by way of trial) for a Year."[45] As proof of the insufficiency of his pay, Jones stated that the assistant whom he employed to delineate the farm lots belonging to the inhabitants of Derby Ward in Savannah had charged him more than Jones himself could receive for surveying the whole township.[46] Apparently, however, his excuses were insufficient. On October 21, 1738, Oglethorpe, acting on instructions from the Trustees, dismissed Jones from his job as public surveyor.[47]

In spite of Jones's failure as surveyor, he was subsequently to perform many valuable services for Georgia. In 1750 he was appointed register of the province. In this position he was entrusted with keeping records of grants, leases, and other conveyances. Previously the work had been done so miserably or not at all that confusion reigned supreme. The provincial authorities now felt that "no person in the Colony was better qualified to execute this weighty Trust" than Noble Jones, who, though it was not his duty, had kept private accounts of many land grants and had always generously assisted people in need of information.[48] In later years Jones also served as a member of the Governor's Council and frequently as president of that august body, as a judge of the Court of Oyer and Terminer, and as treasurer of the province. In fact, as one of the foremost men in the colony, he held—at one time or another—almost every office in the provincial government except the governorship. He died on November 2, 1775, at the age of seventy-three, apparently the last surviving member of the first group of colonists that came to Georgia.[49] Sadly not one of his maps or plats is known to exist today.

44. 2 Col. Rec. Ga. 226; 29 Col. Rec. Ga. 61, 70–71; 32 Col. Rec. Ga. 244; Letter from Benjamin Martyn to Noble Jones (May 15, 1735) (Egmont Mss. in Phillipps Collection, Ser. No. 14207, at 412, UGL); E. Coulter, Wormsloe: Two Centuries of a Georgia Family 48–49 (1955).

45. 21 Col. Rec. Ga. 321–22, 470; E. Coulter, Wormsloe: Two Centuries of a Georgia Family 49 (1955).

46. 21 Col. Rec. Ga. 322; E. Coulter, Wormsloe: Two Centuries of a Georgia Family 49 (1955).

47. E. Coulter, Wormsloe: Two Centuries of a Georgia Family 50 (1955).

48. Id. 51.

49. Id. 104. For more extensive biographies of Noble Jones see E. Coulter, Wormsloe:

Other surveyors were also employed by the Trustees during this period. Though some were designated as "assistants" or "deputies," others worked completely independent of Jones's supervision. Most worked for only very short periods of time. These men included Samuel Auspourger, Hugh Ross, Christopher Ford, Thomas Ellis, John Calwell, and others.[50] Auspourger, a native of Switzerland, had immigrated to Purrysburg, South Carolina, in 1734. Two years later he was appointed "Ingenier of the Fort, and Land Surveyor" at Frederica on St. Simons Island, Georgia. His pay was set at three shillings per day while that of his deputy, John Calwell, as well as laborers hired to assist them, was set at two shillings per day.[51] Although Fort Frederica ceased to be used for military purposes in the 1750s and the town itself died out in the early 1800s, the outlines of the fortifications, the compact town lots, and the orderly grid pattern of the streets are (after extensive archaeological excavations) clearly discernible today.

Only a few scattered, happenstance references about the services of most of the additional surveyors and deputies during Jones's tenure remain. The one exception is Hugh Ross. Ross, like Samuel Auspourger, was a resident at Purrysburg. He was employed to lay out the garden and farm lots for the town of Ebenezer.

John Martin Bolzius, Lutheran minister and a leader of the Salzburgers, kept a diary in which he recorded a rather detailed account of Ross's work.[52] Initially Bolzius manifested a contemptuous and irritated

---

Two Centuries of a Georgia Family 39–107 (1955) and S. Temple & K. Coleman, Georgia Journeys, 268–91 (1961). Noble Jones was progenitor of the Jones and DeRenne family of Wormsloe, Isle of Hope, Chatham County, Georgia. Two members of this family—George W. J. DeRenne and Wymberley Jones DeRenne, father and son—assembled what became the largest single collection of published works dealing with Georgia history. This collection, containing many rare and hard-to-find items about Georgia, is now housed at the University of Georgia Libraries. It proved extremely valuable to the author in his research for this book. The tabby ruins of Noble Jones's home are intact today at Wormsloe, which has been designated a state historic site.

50. 4 Col. Rec. Ga. 87; 5 Col. Rec. Ga. 70, 149, 252; 21 Col. Rec. Ga. 322; 22 Col. Rec. Ga. (pt. 2) 174; 29 Col. Rec. Ga. 206; Letter from Benjamin Martyn to Thomas Causton (Aug. 3, 1738) (Egmont Mss. in Phillipps Collection, Ser. No. 14209, at 21, UGL).

51. 2 Col. Rec. Ga. 236, 254, 271; 3 Col. Rec. Ga. 399; 5 Col. Rec. Ga. 54, 70, 149, 209, 243, 247, 250; 22 Col. Rec. Ga. (pt. 2) 280; 39 Col. Rec. Ga. 477–79 (unpub. ms., GDAH); E. Coulter & A. Saye, A List of the Early Settlers of Georgia 62 (1949); [B. Martyn], An Impartial Inquiry into the State and Utility of the Province of Georgia 75–78 (London 1741), *reprinted in* 1 Coll. Ga. Hist. Soc. 153, 191–92 (Savannah 1840).

52. Without itemizing sources, the following text concerning Ross is taken from 3 Detailed

attitude toward Ross, describing him as "long-winded," "vexatious," "annoying and obstinate." It was true that Ross was very dilatory in laying out the Salzburgers' lands, but, as Bolzius gradually began to realize, he was working under adverse conditions for substandard pay. In addition, the Salzburgers themselves, being Alpine mountainfolk unacclimated to the sweltering miasmic swamps of semitropical Georgia, were not inclined to assist with the survey.

The regular, close-knit arrangement of the farm lots of Savannah produced serious drawbacks that had apparently become manifest by the time Ross began his work at Ebenezer. Because of rigid adherence to uniformity in the case of Savannah, much time and effort had been wasted in laying out tracts on poor, unproductive ground. Many of the settlers who drew undesirable lots refused to occupy them and, instead, drifted to other colonies or back to England. Therefore, in the case of Ebenezer, a more flexible plan was adopted. Although the farm lots were still to be laid out in contiguous, rectangular tracts (in this instance, long, narrow strips measuring 6.60 by 75 chains), Ross was directed to cluster them closely about the rich creek bottomlands and to avoid the nearby pinelands as much as possible.

Ross began his work on a hot, sultry day in July 1737 along the banks of the Savannah River and continued it intermittently for the next two and a half years. He had been instructed by Oglethorpe "to survey all plantations in a square of sixteen English miles." From the beginning Ross was "accompanied by Salzburger men knowledgeable" in surveying matters, but they soon wearied of "his excessive demands and . . . unbearable way of dealing with people." For example, he carried prepared provisions "for several days or even a week" and took "no time out for cooking for himself or for those accompanying him." The work went on "from early dawn until well into the night, with only short breaks for breakfast and dinner." For such privations, Ross offered his help paltry pay. Most of the Salzburgers were already "quite weak from the fever and would soon suffer a relapse were they to work in swampy areas all day and in the still-

Reports on the Salzburger Emigrants . . . 134–36, 158, 167–68, 178–80, 193 (S. Urlsperger ed., G. Jones & M. Hahn trans. 1972); 5 Detailed Reports on the Salzburger Emigrants . . . 7, 27–28, 132, 175, 183, 185, 191–92, 197, 202, 210, 220–22, 251, 259–61, 266–67, 277–78, 285 (S. Urlsperger ed., G. Jones & R. Wilson trans. 1980); 6 Detailed Reports on the Salzburger Emigrants . . . 282, 289, 317, 318 (S. Urlsperger ed., G. Jones & R. Wilson trans. 1981); 8 Detailed Reports on the Salzburger Emigrants . . . 279 (S. Urlsperger ed.; M. Hoffman-Loerzer, R. Wilson, & G. Jones trans. 1985); and 22 Col. Rec. Ga. (pt. 2) 118.

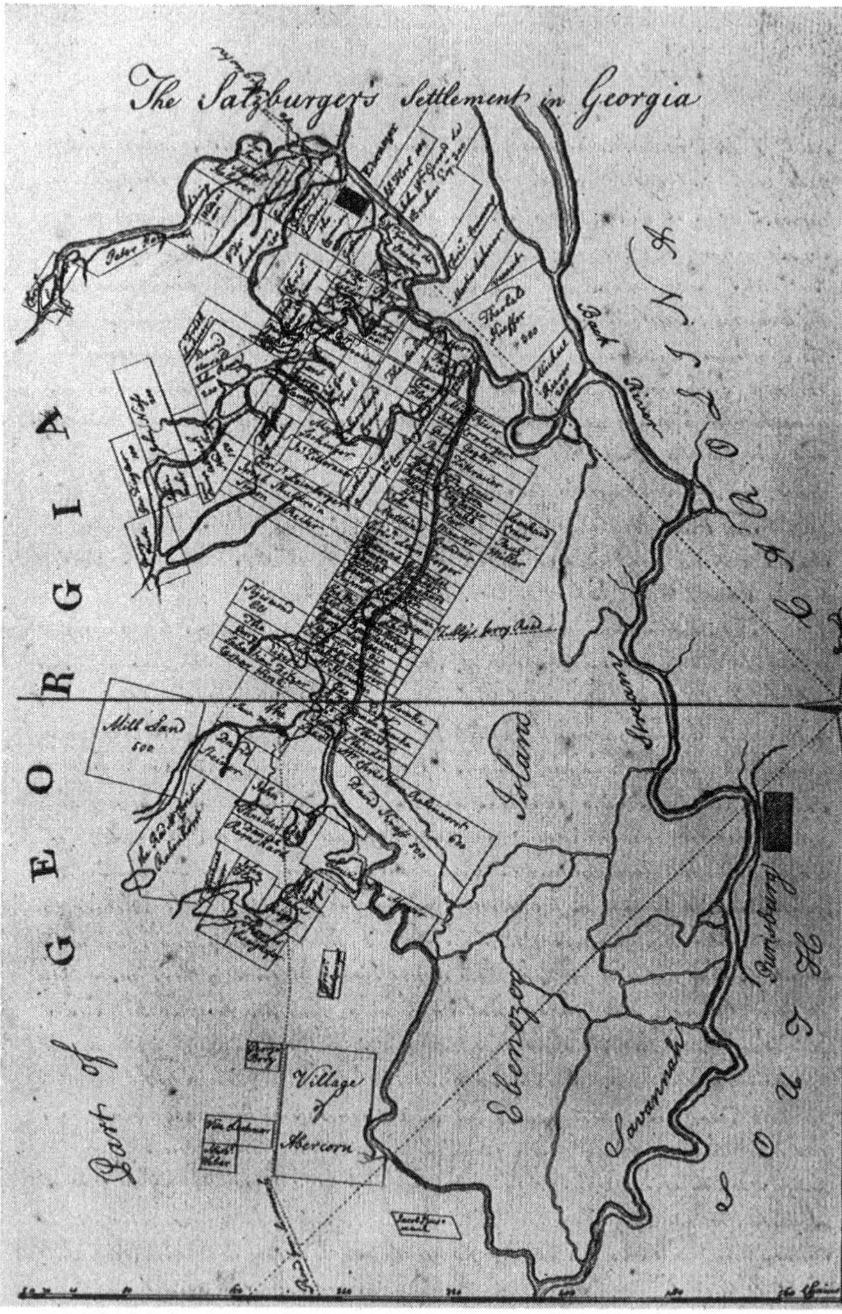

Plat by William Gerard DeBrahm of land grants in the vicinity of Ebenezer. Notice the rectilinear pattern of the layout of the farm lots, which cluster closely along the rich bottomlands of creeks. The villages of Purrysburg and Abercorn also appear. By permission of the British Library, London.

continuing heat and then camp out at night," being "sustained by nothing but the cold food" carried as provisions. In one instance Ross brought along two young indentured Swiss servants whose services he had purchased in Charleston. Though both were afflicted with fever, they were forced to assist with the survey. No wonder Reverend Bolzius considered it "a great blessing" that Oglethorpe had ordered Ross not to employ the Salzburgers unless he paid for their labor! Things got so bad that Bolzius could "find almost no one willing to work for him," and eventually the Salzburgers adopted the procedure of going with Ross in relays.

Ross often absented himself for weeks at a time, leaving no word as to his whereabouts and breaking promises as to when he would return. On one occasion in August 1738, however, Bolzius apparently considered his reason for leaving valid. Ross claimed he had "unavoidable business" to tend to on his own plantation near Purrysburg. Besides, continued Bolzius, it was "probably too hot" for surveying at that time of the year, and "he fears the snakes, which are now very dangerous." A few days later Bolzius became disconcerted when he found that Ross had extended the surveys into an area of pinelands that had been reserved by the Trustees for future granting. He suspected that Ross had done this "in his own interest" because it was "ten times easier" to survey "in such pine forests than on good soil that is covered with many trees, bushes, thorns, and reeds." These surveys were subsequently abandoned in favor of more productive lands across Abercorn Creek, but again the Salzburgers had difficulty persuading Ross to do the work. When asked to survey an island overgrown with reeds and tangled vines, he excused himself from this difficult task by insisting that he was not permitted to go across the creek as long as he could find enough land for the Salzburgers on their own side of the creek. Eventually, however, he was persuaded to perform this job. The Salzburgers who accompanied him reported that the soil on the island was productive but that it was "very difficult to survey" because of the dense undergrowth.

In the latter part of 1738 Oglethorpe became tardy in paying and provisioning Ross so that the Salzburgers themselves found it necessary to supply him, lest he leave and refuse to come back. With four Salzburgers, "his boy," and a supply of imported meat given him by Bolzius, Ross set out one Thursday and returned the following evening "full of anger." He had not been satisfied with his help. Furthermore, the meat, which Bolzius had purchased in Savannah as "New York beef," turned out to be horse meat.

Apparently, Ross had virtually finished the fieldwork for the record makes little mention of him for two and a half years. Then, in July 1741, Bolzius received a letter in which Ross reported that he was working on a map of Ebenezer and the surrounding plantations. As soon as the map was finished, he wished to come to Ebenezer and put the final touches on the surveys so he could receive full payment for his work.

Later Surveyors under the Trustees

After Jones's dismissal, Georgia seems to have done without a regularly appointed surveyor for four years. To provide the necessary surveys, Oglethorpe employed Jones's assistants or hired a surveyor from South Carolina and frequently had to employ Jones himself.[53] Matters apparently sank to a low ebb. One colonist during this period, without naming the individual he was alluding to, contemptuously referred to "the Person . . . who supplies the Place of" surveyor as hardly deserving that name for he "scarce knows a Point of the Compass."[54] There obviously was great need for a competent and full-time surveyor, but the salary offered (£50 a year) was too small to induce anyone to accept the job.[55] Thus Jones's earlier contentions that the office of public surveyor was no mere part-time position were fully substantiated. In addition, the War of Jenkins' Ear, which broke out in 1739 and heavily involved Georgia, probably diverted the colonists' attentions away from domestic concerns to the more pressing needs of defense against the Spaniards.

When relative peace returned to the southern frontier, the Trustees agreed to raise the surveyor's salary to £100 a year, and Joseph Avery became surveyor of the province in October 1742.[56] Avery, originally a resident of Scotland, had "always had a Mathematical and Mechanical Inclination." From his youth he had "been bred in his Majestys Dockyards in Britain to Ship Building, and in the knowledge of all other Maritime Affairs." In 1721 he charted the southern coast of England and published a "well approved" map of this work. He styled himself as "Surveyor

53. 6 Col. Rec. Ga. 22–23; 22 Col. Rec. Ga. (pt. 2) 174, 471–72; Acct. of Cash pd. by Col. Stephens & Hen. Parker on Acct. of Surveying from 1 Oct. 1741 to 31 Jany. 1741/2 (Egmont Mss. in Phillipps Collection, Ser. No. 14206, at 123B, UGL).

54. 23 Col. Rec. Ga. 258.

55. Id.; 5 Col. Rec. Ga. 420; 22 Col. Rec. Ga. (pt. 1) 80; 22 Col. Rec. Ga. (pt. 2) 471–72.

56. 23 Col. Rec. Ga. 415–16, 432–33; 30 Col. Rec. Ga. 253, 260, 274, 283; Letter from Joseph Avery to Harman Verelts (Oct. 27, 1742) (Egmont Mss. in Phillipps Collection, Ser. No. 14206, at 271, UGL).

of the Coast of Britain." He later emigrated to Charleston, South Carolina, where he occupied himself as a shipbuilder and timber merchant for several years. Avery first attracted the attention of the Trustees when he offered to construct shipyards in Georgia after the financial failure of his South Carolina ventures. This proposal was not acted upon, but soon thereafter the Trustees did hire him to compile a map of the northern portion of Georgia. He executed this task so adroitly that he was made public surveyor of the colony.[57]

Avery had his work cut out. As he wrote the Trustees in October 1742, "the Garden and Farm Lots of the Town of Savannah are not Run out as they ought, but remain in Confusion . . . and some of the people complain that they never had any Run out, and so they cannot know where they are." Stating that all these lands would have to be resurveyed, he noted that summer and fall was a less than desirable time for surveying "perticularly in the Woods, heads of Rivers &c. for fear of snakes." Therefore, he planned to do fieldwork primarily in the winter and spring, employing the summer season for drawing plats and writing journals of improvements made by the settlers.[58]

Shortly after Avery's appointment, a shipment of surveying instruments arrived from England for use in the colony. The following inventory of these instruments provides a glimpse of the types of equipment used by colonial Georgia surveyors:

> A large circumferentor for surveying in Woods, with a Staff. Ball and Socket Spare.
>
> Needle and Glasses &c. in a Wainscot Box.
>
> Two strong Chains for Woods with 10 Sticks shod with Steel.
>
> A Case of Instruments with a pair of Socket Compasses.
>
> A large round Protractor, to lay off Angles, and one dozen of black Lead Pencils.[59]

In addition, it was mentioned that "Mr. Avery has bought a Telescope for his Brass Quadrant, and a Spirit Level."[60]

57. Capt. Joseph Avery's Proposals (Feb. 1740/1) (Egmont Mss. in Phillipps Collection, Ser. No. 14205, at 266, UGL); Memorial of Joseph Avery to the Trustees (July 15, 1742) (Egmont Mss. in Phillipps Collection, Ser. No. 14213, at 19, UGL); 23 Col. Rec. Ga. 271–72; R. Tooley, Tooley's Dictionary of Mapmakers 28 (1979); South Carolina Gazette (Charleston), Dec. 15, 1737, at 3, col. 2.

58. 23 Col. Rec. Ga. 411–12.

59. Id. 454; 30 Col. Rec. Ga. 266; Letter from Harman Verelst to William Stephens (Aug. 21, 1742) (Egmont Mss. in Phillipps Collection, Ser. No. 14213, at 63, UGL).

60. 30 Col. Rec. Ga. 266.

One of Avery's more enduring memorials is the village of Vernonburg, which the English contemporaneously called "the Dutchtown." Located on a peninsula formed by the Vernon River just below Savannah, it was settled by Germans. They insisted, over Avery's objections, that the individual farms be measured out into long, narrow, contiguous strips. This innovative plan imitated a method of land layout already used by the Salzburgers who had settled around Ebenezer. Its purpose was to give equal shares of good and poor land to all the owners and to allow settlers to live near one another at the same end of the strips for the sake of mutual assistance. In addition, the plan facilitated the building of communal fences around adjoining farms. These long, narrow strips are apparent to this day in the layout of Vernonburg.[61]

Avery died prematurely in October 1744, cutting short a surveying career that had made a promising start in Georgia. Thomas Ellis, who for some years had acted as a deputy surveyor in the colony, was appointed to fill the vacancy. For the first year Ellis served merely at the pleasure of the Trustees, but in November 1745 his position was made permanent.[62] It was decided that Ellis should not receive a regular salary but should be paid only for the work he actually did—"Which Method will most probably keep him steady in his Duty, He not being entitled to any Wages till it appears he has done his Work to the Satisfaction of those from whom he is to receive his Pay." In case of an error, the survey was to be rerun at the expense of the surveyor.[63]

With the growth in population of the colony there was a corresponding increase in the duties of the public surveyor so that in 1749 the provincial authorities found it necessary to appoint another surveyor, Henry Yonge, to serve jointly with Ellis.[64] Henry Yonge (1712–78) was the son of Francis Yonge, who had held the post of surveyor general of Carolina and the Bahamas. Henry Yonge was born in London and received his education in England. He spent much of the early part of his life in South Carolina, where he was employed as a surveyor in the 1730s and 1740s. In 1747 he received from the Trustees a grant of land on Skidaway Island, Georgia.

61. 1 Col. Rec. Ga. 440; 5 Col. Rec. Ga. 680–81; 23 Col. Rec. Ga. 475–76; 24 Col. Rec. Ga. 20, 80–81, 116; G. Jones, The Salzburger Saga 83–84, 86–87 (1984).

62. 2 Col. Rec. Ga. 466, 468; 6 Col. Rec. Ga. 49, 117, 126–27, 129–30; 24 Col. Rec. Ga. 337; 31 Col. Rec. Ga. 14, 69; 2 W. Stephens, The Journal of William Stephens 157 (E. Coulter ed. 1959).

63. 6 Col. Rec. Ga. 130.

64. Id. 240–41, 357.

Henry Yonge. Original oil portrait in possession of Philip H. L. Yonge of London, England. Photograph by Steve Wakeham.

He named this quiet, secluded, sea-island retreat Cedar Groves and lived there for most of the remainder of his life. His prosperity and standing increased to the extent that in 1755 he was elected to the lower house of Georgia's first assembly. Sixteen years later he became a member of the Governor's Council.[65]

After twenty years of benevolent if somewhat idealistic administration, the Trustees now found themselves financially unable to continue the maintenance of the colony. In June 1752 they surrendered their charter to the crown.[66] They gave up not only their authority to govern the province, which had another year to run, but also their right to grant lands, which had been placed in the hands of the corporation forever.[67] One of the last official acts of the Trustees was the confirmation of Thomas Ellis and Henry Yonge as joint surveyors of the province.[68]

Upon the surrender of the charter, the administration of the colony passed under the control of the British Board of Trade and Plantations. The Privy Council thereupon issued a proclamation directing all officers in the colony to continue the functions of their respective offices "untill His Majesty's Pleasure shall be further known, or other Provision made for the due Government."[69]

Looking back over the twenty-year administration of the Trustees, one can see that the office of public surveyor had been exercised somewhat erratically. Perhaps such performance was adequate because Georgia's growth under the Trustees had been painfully slow. During the last eleven years of this period, an average of only thirty-four grants per year had been issued, and during the entire period only about 153,000 acres of land had been granted.[70] If all this land had been laid out into one square tract, it would have formed a lot less than sixteen miles to a side. The new system to be instituted would establish a firmer and more prac-

65. Id. 202, 399; 7 Col. Rec. Ga. 89, 285–86; 11 Col. Rec. Ga. 427; Eugene C. Hicks, Sir Ellis Hicks (1315), Captain John Ward (1598), John Wright (1500), Philip le Yonge (1215), and 7,812 Descendants 679–81 (1982); Philip Keyes Yonge, The Yonge Family in America (July 1960) (unpub. typescript in the P. K. Yonge Library of Florida History, University of Florida, Gainesville, Fla.).

66. A. Johnson, Georgia as Colony and State 99–100 (1938); 1 C. Jones, The History of Georgia 450–59 (1883).

67. A. Saye, A Constitutional History of Georgia 1732–1968, at 46 (rev. ed. 1970).

68. 2 Col. Rec. Ga. 516; 31 Col. Rec. Ga. 262.

69. 26 Col. Rec. Ga. 349, 368–73.

70. J. McCain, Georgia as a Proprietary Province 269–71 (1917).

tical foundation that would greatly accelerate the growth of the young colony.

Land System under the Royal Government

It was not until two years later, on August 6, 1754, that King George II, upon the recommendation of the Lords Commissioners of Trade and Plantations, established a civil government for Georgia, to be made up of a governor, a colonial assembly, and other necessary government offices.[71] Henceforth, until Georgia severed its ties with the mother country, all original land grants ran in the name of the king and were signed by the royal governor.[72] The title of "surveyor general" was bestowed upon the official charged with administering grants and public land surveys. In time, the duties of this office increased to such an extent that the surveyor general performed very little of the actual surveying, the office having become primarily administrative with the fieldwork left to deputies.

John Reynolds was appointed first royal governor of Georgia. Henry Yonge and William Gerard DeBrahm (Thomas Ellis had recently died) were commissioned joint surveyors general, at a salary to each of £50 per year.[73] Thus the somewhat novel approach of having two persons jointly fill the office of surveyor general was followed for the next decade.

Governor Reynolds found the entire system of ownership and disposition of lands in Georgia in a state of confusion when he took office in 1754.[74] The Trustees, because of their "many visionary ideas," had been neither consistent nor businesslike in their land policies. The result was an appalling ambiguity and irregularity of land titles. The Trustees had imagined that people in Georgia might subsist and even become affluent on little farms like the ones in England. This objective proved illusory, however, because of the difference in the lands, the climate, and the

71. 1 C. Jones, The History of Georgia 459–67 (1883).

72. The proceedings of the governor and council, which were taken up largely with matters involving land granting, can be found in Col. Rec. Ga., vols. 7–12. An addendum is in Coll. Ga. Hist. Soc., vol. 10. For forms of land grants under the royal government see 28 Col. Rec. Ga. (pt. 1) 63–67.

73. Commissions Book B-1, 32 (GDAH); 6 Col. Rec. Ga. 458; 7 Col. Rec. Ga. 317, 560, 784; W. DeBrahm, DeBrahm's Report of the General Survey in the Southern District of North America 15 (L. De Vorsey ed. 1971).

74. W. Abbot, The Royal Governors of Georgia 1754–1775, at 15 (1959).

William Gerard DeBrahm. Courtesy of the Library of the Religious Society of Friends, London.

crops that could be grown. "Rivetted to this chimaera" (to use the words of a contemporary), Oglethorpe had "established many little Townships" and "minced all the lands round them into whimsical figures of five, and forty five Acres which he put people in possession of by a written order directed to the Constable." In many cases persons held titles merely by the verbal permission of Oglethorpe or other officials.[75]

Soon after taking office, Governor Reynolds, in accordance with his royal instructions, issued a proclamation directing all persons holding lands of the Trustees to take out new grants in His Majesty's name for the lands they claimed within a specified period of time.[76]

Three months later the assembly passed an act "To Prevent Fraudulent Deeds and Conveyances." This legislation provided that "all Conveyances of Lands Tenements Negroes and other Chattels or Hereditaments whatsoever or Mortgages of the same . . . shall be registered in the Register

75. 28 Col. Rec. Ga. (pt. 1) 6, 24.

76. Proclamations Book H, 1754–1794, at 5, 7 (GDAH); 27 Col. Rec. Ga. 99–101, 112; P. Bryant, Entry of Claims for Georgia Landholders passim (1975); Saye, *Commission and Instructions of Governor John Reynolds, August 6, 1754*, 30 Ga. Hist. Q. 125, 149–50 (1946).

of the Records Office of this province."[77] This was the first formal enactment passed in Georgia that provided for the recording of deeds and other types of conveyances and established the priority of duly recorded deeds over earlier unrecorded ones.

The parsimonious attitude of the Trustees toward land granting was wholeheartedly abandoned by the royal government in favor of the more liberal headright system.[78] On the first Tuesday of each month—called "Land Day" or "Land Tuesday"—the governor and council members met in their chamber in Savannah to receive petitions for land. A person desiring land would appear before the council and present a written statement specifying the number of persons in his family (including indentured servants and slaves) for whom he had received no land from the colony. As the head of a family he was entitled to one hundred acres for himself and fifty additional acres for each dependent, white or black. If the council considered a person capable of cultivating more than this amount of land the governor was authorized to sell that person up to one thousand additional acres at the nominal price of one shilling for every ten acres. After 1763 veterans of the Seven Years' War (known as the French and Indian War in America) were entitled to grants of fifty to five thousand acres for their service, depending on the rank they had held during the war. In his petition the applicant included a general description of the location of the land he desired and a pledge that it was intended for his own use, not for speculation. Upon swearing to the truth of his statements, he was issued a warrant from the governor directing the surveyor general to have the tract surveyed.

The applicant took the warrant to the surveyor general, who issued a precept ordering a deputy to lay out the tract. The applicant usually accompanied the deputy to point out the exact lines for marking off and to provide chainmen for the survey. The chainmen generally consisted of the applicant himself and anyone he could procure to assist in the work. Chainmen were required to be sworn before entering upon their duties. It also behooved the applicant to furnish extra laborers as well as victuals

77. Cobb 159; 18 Col. Rec. Ga. 77.

78. The following text concerning the land-granting procedures under the royal government is taken from Am. St. Papers, 1 Public Lands 37; 7 Col. Rec. Ga. 127, 136–37, 295–97, 299–300, 337–38, 611–12; 9 Col. Rec. Ga. 238–39, 323; 37 Col. Rec. Ga. 210–11 (unpub. ms., GDAH); W. Abbot, The Royal Governors of Georgia 1754–1775, at 15–16 (1959); and Saye, *Commission and Instructions of Governor John Reynolds, August 6, 1754*, 30 Ga. Hist. Q. 125, 150–52 (1946).

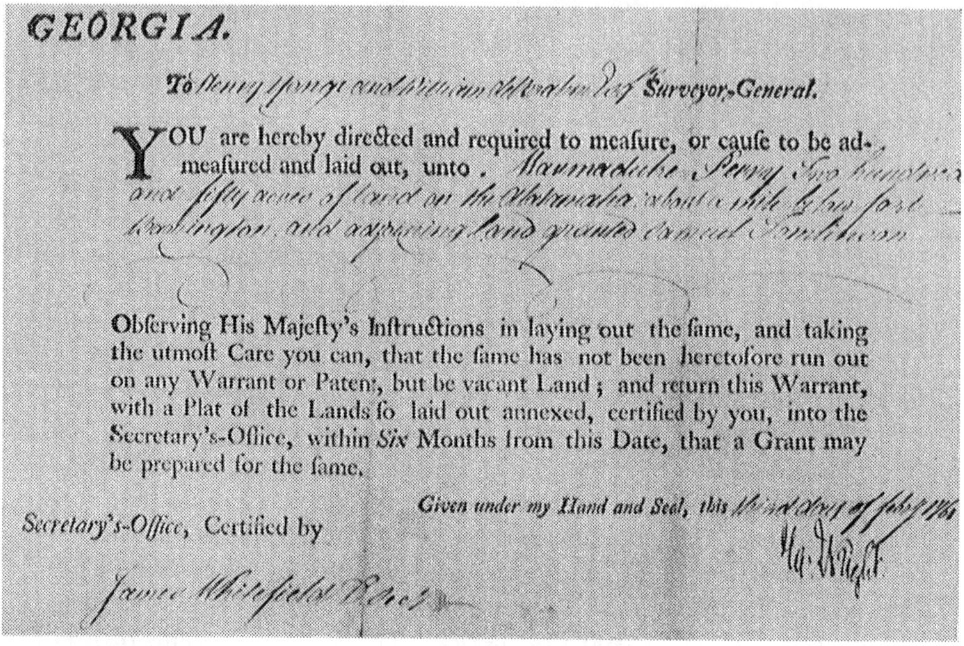
GEORGIA.

To Surveyor-General.

YOU are hereby directed and required to measure, or cause to be admeasured and laid out, unto .

Observing His Majesty's Instructions in laying out the same, and taking the utmost Care you can, that the same has not been heretofore run out on any Warrant or Patent, but be vacant Land; and return this Warrant, with a Plat of the Lands so laid out annexed, certified by you, into the Secretary's-Office, within *Six* Months from this Date, that a Grant may be prepared for the same.

Given under my Hand and Seal, this

Secretary's-Office, Certified by

Land warrant directed to Henry Yonge and William Gerard DeBrahm and signed by Governor James Wright. Courtesy of the Georgia Department of Archives and History.

for the entire crew. If he did not, the deputy was authorized to charge him an additional £0.2.6 for each one hundred acres surveyed. After the survey was completed, a plat certified by the surveyor general was returned to the attorney general's office within seven months. The attorney general prepared and sent a fiat, along with the plat, to the colonial secretary. The secretary then made out the grant, which was signed by the governor in council, countersigned by the clerk of the council, sealed by the secretary, and finally recorded by the register. Upon paying his fees, the petitioner held title to the land subject to the condition that he cultivate three out of every fifty plantable acres or that he otherwise improve the property in a manner acceptable to the council. As a person's family or slaveholdings increased, additional grants could be obtained.

Anyone who objected to the issuing of a grant (for example, if someone claimed to have received a prior grant to the land) could file a caveat with the governor and council, who would try the conflicting claims.

A land grant was not required to be contiguous to previously granted

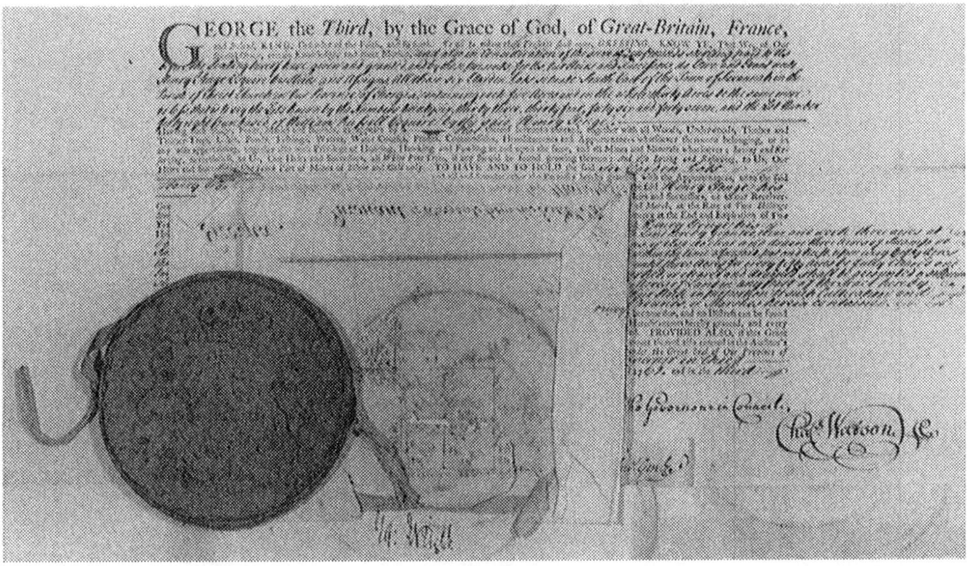
GEORGE the *Third*, by the Grace of God, of *Great-Britain*, *France*,

Land grant signed by Governor James Wright. Courtesy of the Georgia Historical Society (Tattnall-Jackson Papers, Collection 787, Box 1, Folder 4).

lands; nor did a grant necessarily have to all lie in one body; nor did a person have to take out, at one time, a grant for all the land to which he or she was entitled. As a matter of practice, each settler tended to isolate his holdings from those of others so that he could fence and use his granted lands for cultivation, while leaving ample adjoining public land (in an era before stock and fence laws) to graze livestock on. The surveyors were instructed to lay the lands off so that each grantee received some productive and some unproductive lands. In addition, the length of the tracts was not to exceed three times the width or to extend along the banks of any river. Instead, the length was to extend back from the river so "that . . . Grantees may have each a convenient Share of what Accomodation the . . . River may afford for Navigation or otherwise."[79]

This was the origin of the headright system in Georgia, so called because the land was granted on a per capita basis. That is, every free citizen or head of a family could receive a certain quantity of land, with an additional amount for each free white person in the family and for each

79. 34 Col. Rec. Ga. 62–63, 148, 158, 290–91 (unpub. ms., GDAH); Saye, *Commission and Instructions of Governor John Reynolds, August 6, 1754*, 30 Ga. Hist. Q. 125, 152 (1946).

Negro he brought with him.[80] The system was modeled after a similar one that had been used in the other southern colonies for some years.

It is of interest to delve briefly into some of the facets of the land-granting process—a concern that occupied the major portion of the time of the governor and council.

To discourage speculation, the law specified that settlers should improve their lands—they could not merely let them stand idle. By far the most common mode of improvement was clearing and cultivation, but other uses carried equal efficacy: range for cattle,[81] lumber yards,[82] and landing places;[83] cutting of timber;[84] erection of sawmills,[85] gristmills,[86] and wharves;[87] and charitable uses such as churches[88] and orphanages.[89] In 1761 Samuel Miller petitioned for two hundred acres of pine barren land on the north side of the Medway River near Sunbury "for the Conveniency of making Pitch and Tarr and sawing Timber."[90] Later the same year David Cunningham requested a grant of a portion of an island in the Savannah River "for the Purpose of . . . landing and Curing . . . Fish."[91] Town lots were granted on the condition that the grantee erect a suitable structure such as a house, a store, a warehouse, or a wharf. William Ewen, in 1755, received a grant of half of a trust lot in Savannah

80. King, *A Sketch of the History of Land Titles in Georgia,* in Report of the Second Annual Meeting of the Georgia Bar Association 126, 140 (1886). Between 1755 and 1769 the governor and council heard an average of 359 petitions for land per year. A. Gallay, The Formation of a Planter Elite 91, 226 n. 35 (1989). Women who were heads of households could receive land grants, though grants to women were exceedingly rare. Caldwell, *Women Landholders in Colonial Georgia,* in Forty Years of Diversity 183 (H. Jackson ed. 1984).

81. 7 Col. Rec. Ga. 371; 8 Col. Rec. Ga. 179–80, 503, 624; 11 Col. Rec. Ga. 5.

82. 7 Col. Rec. Ga. 815–16, 865; 8 Col. Rec. Ga. 15.

83. 8 Col. Rec. Ga. 143, 549.

84. Id. 48, 56, 179, 695–96; 9 Col. Rec. Ga. 10, 59, 94–95, 189, 320, 374; 11 Col. Rec. Ga. 102–3.

85. 8 Col. Rec. Ga. 114, 730, 762; 9 Col. Rec. Ga. 191, 199, 255, 294; 10 Col. Rec. Ga. 43–52.

86. 8 Col. Rec. Ga. 730; 9 Col. Rec. Ga. 255, 267; 10 Col. Rec. Ga. 828, 928; 12 Col. Rec. Ga. 125.

87. 7 Col. Rec. Ga. 736–37, 868–69; 8 Col. Rec. Ga. 24, 29, 30, 32, 102, 125, 130–31; 12 Col. Rec. Ga. 253.

88. 7 Col. Rec. Ga. 183; 9 Col. Rec. Ga. 29; 10 Col. Rec. Ga. 747–48, 788; 11 Col. Rec. Ga. 65, 146, 304; 18 Col. Rec. Ga. 263.

89. 9 Col. Rec. Ga. 259–61.

90. 8 Col. Rec. Ga. 504.

91. Id. 606.

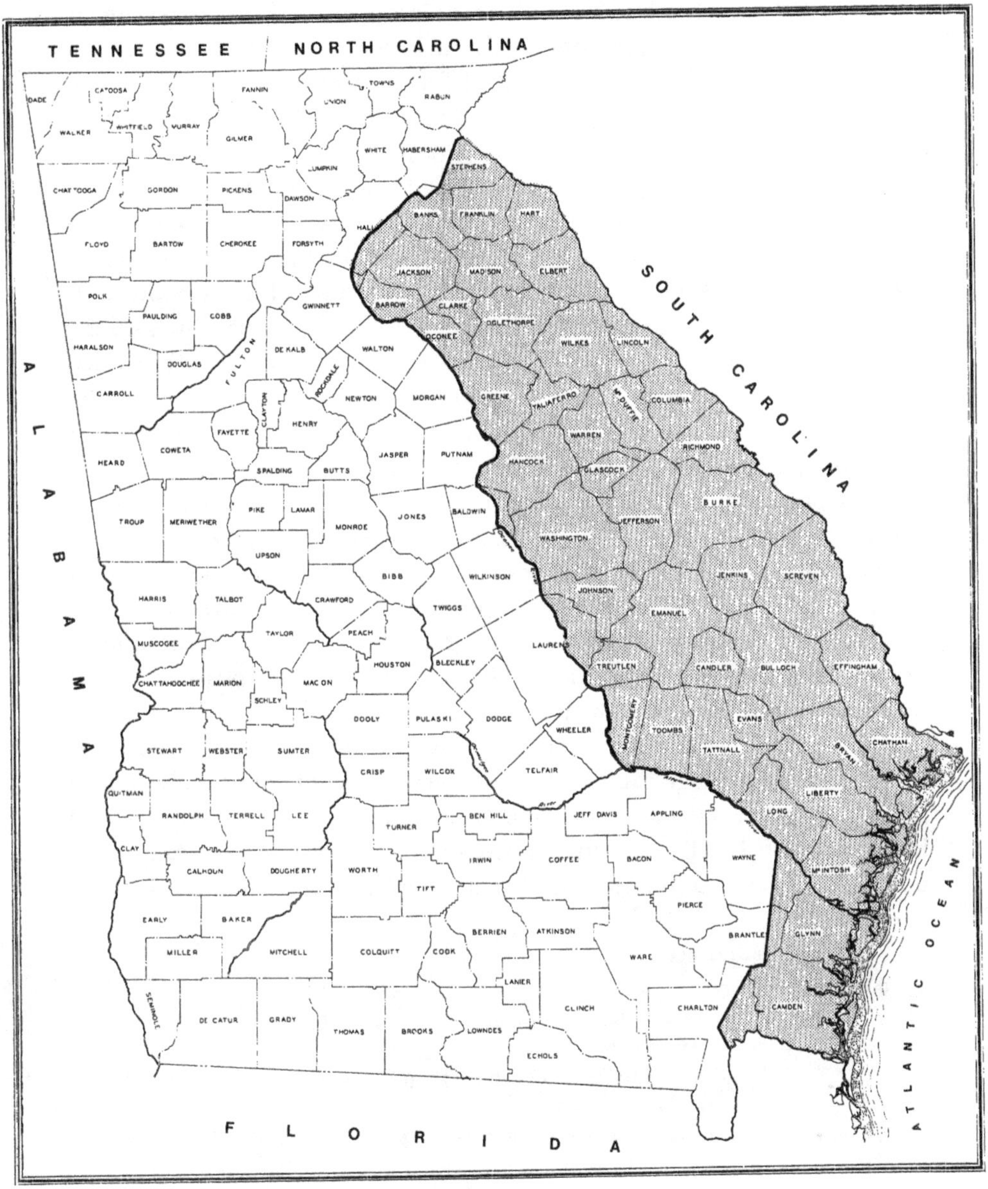

Lands distributed under the headright system.

for erecting a potter's kiln.[92] A few years later Michael Switzer received a grant of river frontage under the bluff in Savannah for building and repairing boats.[93] Despite the threat of forfeiture for failing to improve property, virtually no evidence exists to indicate that such a process was ever carried out.

Now and then a settler who had received a warrant discovered, upon going to have the land surveyed, that it was undesirable,[94] or that someone else had already laid claim thereto,[95] or that the land lay over the Indian boundary.[96] The settler then often surrendered his warrant and took out another one, in lieu of the original, for other lands. Christopher Hudson obtained a warrant for and surveyed a tract on the Ogeechee River near the Indian boundary. Upon settling the land, however, he found he "could not continue there on Account of the Trouble he met with from the Indians." He subsequently resigned this warrant for a warrant for other lands closer to white settlements.[97] In numerous cases the petitioner either failed to get the land surveyed within the prescribed time (often through fault of the surveyor) or failed to take out the grant within the prescribed time, whereupon he requested and received an extension of time for completing the granting process.[98] Sometimes a person died before taking out a grant on a warrant he possessed, and his heirs would petition for and receive the grant.[99] Often a petitioner who had a tract laid out between previously granted lands was unable to secure the full acreage called for in his warrant. At a later date, he would apply for and receive an additional warrant for the balance of the land he was due.[100] In all such diverse situations, the council was extremely indulgent in accommodating the peculiar circumstances of each settler.

92. 7 Col. Rec. Ga. 142–43.

93. 8 Col. Rec. Ga. 150.

94. 7 Col. Rec. Ga. 321; 8 Col. Rec. Ga. 20, 22, 79, 195, 318, 503; 9 Col. Rec. Ga. 308; 12 Col. Rec. Ga. 4, 64–65, 102–3, 164.

95. 8 Col. Rec. Ga. 30, 129, 142, 154, 202, 263, 573; 9 Col. Rec. Ga. 32, 196, 367–68; 11 Col. Rec. Ga. 88, 190–91, 215.

96. 9 Col. Rec. Ga. 333, 444, 534; 10 Col. Rec. Ga. 125, 764; 11 Col. Rec. Ga. 98; 12 Col. Rec. Ga. 38.

97. 9 Col. Rec. Ga. 81.

98. Id. 662–65; 8 Col. Rec. Ga. 22, 51, 94, 116, 219–20, 405, 590–92, 679; 10 Col. Rec. Ga. 98–114, 229–41.

99. 8 Col. Rec. Ga. 60, 528, 624–25, 667; 9 Col. Rec. Ga. 83, 89–90, 160–61, 490–91; 10 Col. Rec. Ga. 73–74, 218, 232, 971–72; 11 Col. Rec. Ga. 75–76.

100. 7 Col. Rec. Ga. 377; 8 Col. Rec. Ga. 274, 589, 632; 9 Col. Rec. Ga. 51, 195.

In spite of the legal implication that a grant had to be obtained before one could settle on the land, in practice many otherwise qualified settlers occupied tracts of land—sometimes for many years—on the basis of a warrant and survey or as mere squatters. Very little censure of these activities is found (except when they occurred beyond the Indian boundary) so they apparently were generally tolerated. Nathaniel Miller, in 1757, petitioned for land above the fork of Brier Creek. He had lived on and cultivated this land for three years but had never bothered to take out a warrant for it because of the considerable distance to Savannah.[101] A decade later Jesse Brasher was granted land where "he was settled . . . and had made great Improvements."[102]

The fact that taxes and quitrents could be avoided by delays in taking out grants undoubtedly encouraged squatting. Although the occupant ran some risk of losing the land, his occupation (especially if accompanied by a warrant and survey) generally gave him enough priority, in the eyes of the governor and council, to take out a grant if his possession seemed threatened.

In some cases, the practice of squatting became a nuisance—particularly on Indian lands. Acts were passed imposing fines, whippings, and prison sentences on those "Persons . . . of loose disorderly Lives . . . who Build Hutts and Intrude on the Kings Lands without any Right or Authority . . . and also . . . Trespass on the . . . Hunting Grounds of the Indians and Occasion Quarrels and Disturbances among them."[103] In 1759 the governor ordered that "sundry Persons illegally settled . . . on Ossabaw and Sapala Islands . . . forwith . . . remove themselves" therefrom.[104]

The policy of including slaves in the numbering for headrights led to certain abuses. A slave dealer often imported a cargo of slaves, obtained grants for them, and then sold them. The purchaser then acquired headrights for the same slaves. There are instances of settlers receiving one hundred or more headrights for slaves at one time.[105] To put a stop to such practices, the council resolved in 1760 not to grant any more lands for Negro headrights upon which grants had already been issued.[106]

101. 7 Col. Rec. Ga. 556.
102. 10 Col. Rec. Ga. 32.
103. 18 Col. Rec. Ga. 588.
104. 8 Col. Rec. Ga. 11.
105. Id. 67, 146, 367; 9 Col. Rec. Ga. 23, 606.
106. 8 Col. Rec. Ga. 450.

The deputy surveyors were instructed to watch for and report any sites suitable for worthwhile public uses such as ferries,[107] forts,[108] landing places,[109] schools,[110] and future towns.[111] When such sites were discovered, the council ordered that they be withheld from granting to individuals. The town of Brunswick was laid off on such a reserved site.

Rejections of petitions for warrants and grants were infrequent. The most common reasons for rejections were that a warrant or grant had already been issued to someone else for the specified land,[112] or that the land had been set aside for a public purpose,[113] or that the council did not consider the petitioner capable of settling and improving the land.[114] In 1760 John Gray's petition for 560 acres in six scattered parcels was denied because the council did not think "it prudent to grant so many small Tracts to one Person."[115] Five years later the governor ordered John Cuthbert's grant for land on Cumberland Island not to pass because Cuthbert "had left the Province . . . and was settled with his Family in Charles Town." It was further ordered that if Cuthbert did not move back to Georgia within three months, the land would be subject to grant to anyone who should apply for it.[116]

Henry Yonge and William Gerard DeBrahm: Joint Surveyors General

Henry Yonge and William Gerard DeBrahm were jointly to fulfill the office of surveyor general during the first decade of the royal regime. The contrast between the two men is striking. Yonge was a man of public affairs who took on a variety of political and civic duties. He was pas-

107. Id. 245.

108. Id. 511; 10 Col. Rec. Ga. 278.

109. 7 Col. Rec. Ga. 459, 534–35.

110. Id. 108; 9 Col. Rec. Ga. 323; 10 Col. Rec. Ga. 215.

111. 7 Col. Rec. Ga. 125; 10 Col. Rec. Ga. 215, 427–28, 690–94; 11 Col. Rec. Ga. 155, 385–87.

112. 8 Col. Rec. Ga. 184, 200, 278, 478, 506, 509, 616, 636; 9 Col. Rec. Ga. 143, 493; 10 Col. Rec. Ga. 115–16, 175, 208–9, 243, 327, 344, 561, 725.

113. 8 Col. Rec. Ga. 265; 9 Col. Rec. Ga. 493; 10 Col. Rec. Ga. 37, 174, 242.

114. 9 Col. Rec. Ga. 84–85.

115. 8 Col. Rec. Ga. 234–35.

116. 9 Col. Rec. Ga. 417, 430.

sionately devoted to his king and to maintaining the colonial status quo. DeBrahm, on the other hand, was a man of the scientific world, indifferent to political events except so far as they impinged upon his scientific endeavors. (He opposed the American Revolution, not on any grounds of policy but because it interrupted his research.) DeBrahm was an adept land surveyor but in addition showed great expertise as an engineer, botanist, astronomer, geographer, meteorologist, student of ocean currents, alchemist, and historian.[117] His writings (many of them published) were extensive and varied, and he was known for the exceptionally fine, accurate maps he produced of Georgia and the southern colonies.[118]

DeBrahm apparently was born in southern Germany in 1717 and, like Yonge, received a first-rate education.[119] In the summer of 1751 he led a party of German emigrants to Georgia and immediately made an impression on colonial officials, both in Georgia and in South Carolina.[120] He was employed to design and oversee the erection of fortifications for Charleston and Savannah. While serving as one of the joint surveyors general of Georgia, he served as interim surveyor general of South Carolina during 1755–56.[121]

At the close of the French and Indian War in 1763, Great Britain suddenly found itself possessed of vastly increased territories in North America about which it knew little. To acquire necessary geographic information regarding these territories, the British government in 1764 divided the North American continent into a Northern District and a Southern District.[122] It is a credit to DeBrahm's abilities as a surveyor and geographer that later the same year he was relieved of his job as one of the joint surveyors general in Georgia and appointed surveyor general of the Southern District of North America as well as of the newly acquired province of East Florida.[123] DeBrahm was to supervise the sur-

117. W. DeBrahm, DeBrahm's Report of the General Survey in the Southern District of North America 7 (L. De Vorsey ed. 1971).

118. Reproductions of some of DeBrahm's maps and plats can be found in W. DeBrahm, DeBrahm's Report of the General Survey in the Southern District of North America (L. De Vorsey ed. 1971). De Vorsey's introduction contains a listing of DeBrahm's published writings, but unfortunately none of these give any details about his land surveying experiences in Georgia.

119. Id. 7.

120. Id. 9–10.

121. Id. 11–18, 25–27.

122. Id. 5–6.

123. Id. 33.

veying and mapping of "all His Majesty's territories on the Continent of North America, which lye to the south of Potomac River, and of a line drawn due west from the Head of the main branch of that River as far as His Majesty's Dominions extend." [124]

In his later years DeBrahm lived successively in St. Augustine, Charleston, and London but made several excursions to France and Germany. In 1791 this restless genius moved to Philadelphia, where his long and versatile career ended with his death in 1799.[125]

Yonge's abilities as a land surveyor were hardly inferior to DeBrahm's. As attested by Georgia's most eminent twentieth-century historical geographer, Yonge's "extant plats and maps rank with the very best of mid-eighteenth-century American cartography." [126]

Considering the exceptional abilities of Yonge and DeBrahm, it is probably not inordinate to conclude that on the whole the quality of surveying in Georgia during the later colonial period equaled or exceeded that in the other American colonies. Unfortunately, the almost complete loss of the Georgia colonial survey records during the Revolution makes comparison on a broad spectrum difficult.

Georgia's first assembly in 1755 enacted a number of legislative reforms, several of which dealt directly with the office of surveyor general. Under the Trustees the records of surveys in the colony had become scattered because no place had been officially designated for their deposit.[127] In February 1755 Yonge and DeBrahm, having declared that the office of surveyor general could not be properly conducted until these records were brought together in one central place, petitioned the Governor's Council to order that all such records be deposited in the office of the surveyor general. They specifically requested the plans of Savannah, Frederica, Augusta, Ebenezer, Darien, and several of the outlying villages. The council accordingly ordered all surveyors and other persons having such records to turn them over to the surveyors general and the plans that had been kept in the council chamber also be delivered to them.[128]

Later the same year Yonge and DeBrahm learned that Noble Jones,

124. Id.

125. Id. 43–58.

126. De Vorsey, *Henry Yonge*, in 2 Dictionary of Georgia Biography 1102, 1103 (K. Coleman & C. Gurr ed. 1983).

127. 7 Col. Rec. Ga. 291.

128. Id. 128–29.

the first surveyor of the province, had kept private notes regarding his surveys, which were the only extant records of many of the earliest surveys in the colony. They petitioned the council either to have the notes turned over to the surveyors general or to have transcripts made of them. The council resolved to ask Jones to deliver any records and plans he had regarding surveys. For this he was to "be allowed Ten Guineas to buy a Piece of Plate as a small Acknowledgement for his Service."[129]

Two other significant actions were taken by the council with regard to the office of surveyor general. Apparently until this time the surveyors had kept their records at their private homes, but in March 1755 Trust Lot X in Anson Ward, Savannah, was set aside as a site for the surveyor general's office.[130] The council also set up a schedule of fees to be charged for running out lands, for copying plats, for obtaining warrants and precepts, and for performing other surveying functions.[131]

South Carolina Grants Lands South of the Altamaha

With Great Britain's victory in the French and Indian War France lost all its dominions in North America except two small islands south of Newfoundland and two islands in the Caribbean. Great Britain received Canada, the French possessions east of the Mississippi River, and the Spanish province of Florida. Spain received all the French possessions west of the Mississippi along with the city of New Orleans.

By royal proclamation dated October 7, 1763, the northern boundary of Florida was fixed at the St. Marys River and from the head of that river in a straight line to the junction of the Flint and Chattahoochee rivers, thence up the Chattahoochee River to the thirty-first parallel, thence due west to the Mississippi River.[132] By the same proclamation the lands lying between the Altamaha and St. Marys rivers were annexed to Georgia. Three months later Georgia's southern boundary was further defined as the north boundary of Florida, and its western boundary was fixed at the Mississippi River.[133]

129. Id. 291.

130. Id. 131–32, 313, 346; M. Hemperley, English Crown Grants in Christ Church Parish in Georgia 35, 207 (1973). The historic Owens-Thomas house stands on the site today.

131. 7 Col. Rec. Ga. 136–37.

132. Am. St. Papers, 1 Public Lands 36; Marbury & Crawford 667.

133. 9 Col. Rec. Ga. 215; Marbury & Crawford 672.

The preliminary articles of peace ending the French and Indian War had been announced in December 1762, and word that Spain was about to cede Florida to Great Britain reached the colonies early the following year.[134] While the British government was considering the best method of apportioning and disposing of this territory, a scheme was devised in Charleston for monopolizing the lands lying south of the Altamaha River.[135] The Carolina grant originally extended as far south as twenty-nine degrees north latitude, while the Georgia grant originally covered only the lands between the Savannah and Altamaha rivers. Thus South Carolina held legal title to the lands south of the Altamaha. Because of the contest with Spain, however, neither it nor Georgia had ever exercised any real authority in the area. In 1758 British prime minister William Pitt had prohibited settlement south of the Altamaha without crown approval.[136]

Realizing that the area would probably be annexed to Georgia, Governor Thomas Boone of South Carolina issued warrants in April 1763 for over three hundred thousand acres of land to fewer than two hundred South Carolinians. These warrants for grants of up to sixteen thousand acres to a single individual were all made on the same day to speculators who obviously had no intention of becoming permanent residents in the area.[137]

The Georgia Assembly complained about these acts, and Governor James Wright dispatched a member of his council to Charleston with a letter of protestation and caveat to Governor Boone against issuing warrants or making surveys of lands south of the Altamaha.[138] Governor Boone, however, refused to receive the protest and caveat and his secretary refused to allow the document to be recorded in the public records of the province.[139] Wright then appealed to the Board of Trade, which promptly disapproved of Governor Boone's actions.[140] This disapproval

134. 2 C. Jones, The History of Georgia 28 (1883).

135. Id.

136. Am. St. Papers, 1 Public Lands 51; 9 Col. Rec. Ga. 41; 28 Col. Rec. Ga. (pt. 1) 129, 409; 1 Correspondence of William Pitt 269–70 (G. Kimball ed. 1906); Flippin, *The Royal Government in Georgia, 1752–1776* (pt. 5), 10 Ga. Hist. Q. 1, 12–13 (1926).

137. 28 Col. Rec. Ga. (pt. 1) 409, 412.

138. Am. St. Papers, 1 Public Lands 52; 9 Col. Rec. Ga. 39–43; 17 Col. Rec. Ga. 49–50.

139. 9 Col. Rec. Ga. 43–44; Letter from James Habersham to William Knox (April 19, 1763), *printed in* 6 Coll. Ga. Hist. Soc. 11–12 (O. Ashmore et al. eds. 1904).

140. Am. St. Papers, 1 Public Lands 53–56; 28 Col. Rec. Ga. (pt. 1) 408–14, 427–31.

came too late, however, because many more warrants had been issued and grants had been taken out on a number of them.

James Habersham, president of the Governor's Council, reported seeing great numbers of people from South Carolina (some armed), with surveyors, passing by Thunderbolt on their way south to execute the warrants.[141] Governor Wright informed the board that "the Creek Indians are greatly alarmed at seeing a number of armed men Surveying these Lands & marking Trees."[142] A year later, after the area had been annexed to Georgia, Wright described to the board the surveying procedures used by the South Carolinians. He declared that the surveys had been made in the greatest haste "for fear of being interrupted by the Indians" and "to get the Grants Signed before they were forbidden." Wright asserted that the tracts along rivers had not been surveyed but that the method employed was to "Stop their Boat & Set up a Stake or Mark a Tree, and then Row a guessed distance, & there Stop again, & put in another Stake (which they Carried in their Boats ready Mark't & Notched) or Mark another Tree." "Few if any" inland tracts had been surveyed but "only a Corner & a few Trees Mark't." All this was done "at Random without ever Stretching a Chain upon the Land" so that "our Surveyors don't know how or where to Lay out any lands in that part of the Country with any Certainty, as they can find no Lines to Regulate their Surveys by." Wright expressed concern that "there may Probably be a much Larger Quantity Contained within the Lines than Expressed in the Plots returned."[143]

Although Governor Boone at length apologized for his actions, the Board of Trade—over the strenuous objections of Governor Wright—determined that the grantees had acquired vested rights in the premises and, after the area had been annexed to Georgia, directed Georgia to confirm the grants.[144] In November 1764 instructions were issued to Georgia surveyors working south of the Altamaha to lay out the lands so that they did not interfere with any Carolina grant.[145]

In March 1765 the Georgia Assembly passed an act to confirm the

141. Letters from James Habersham to William Knox (April 6, 1763, & April 19, 1763), *printed in* 6 Coll. Ga. Hist. Soc. 10–12 (O. Ashmore et al. eds. 1904).

142. 28 Col. Rec. Ga. (pt. 1) 428.

143. 28 Col. Rec. Ga. (pt. 2) 38, 79–80. See also 28 Col. Rec. Ga. (pt. 2) 71, and Letter from James Habersham to William Knox (April 4, 1765), *printed in* 6 Coll. Ga. Hist. Soc. 28–29 (O. Ashmore et al. eds. 1904).

144. 2 C. Jones, The History of Georgia 34–36 (1883).

145. 9 Col. Rec. Ga. 238, 272, 276.

South Carolina grants on the condition that within six months after the governor of Georgia had received the crown's approval of the act, the grantees appear before the governor and council and prove that they had settled the lands in accordance with regulations for land grants in Georgia. In addition, the act specified that the claimants had to register their grants in the register's office and their plats in the surveyor general's office. If the lands had not been surveyed in accordance with the regulations laid down for surveys in Georgia, the grantees had to have the lands resurveyed at their own expense. Failure to comply with these provisions within a specified time would render the grants void.[146]

Although approving of the broad stipulations of the act, the Privy Council in London disallowed it on technical grounds.[147] It instructed the governor of South Carolina to transmit to the governor of Georgia transcripts of all grants, warrants, and other proceedings concerning the lands south of the Altamaha and ordered that these transcripts be entered on the records in the proper offices of Georgia.[148] The Privy Council also recommended that the Georgia Assembly adopt a revised act to meet the objections to the act of March 1765 and that a court of commission be set up to settle the matter.[149] At the next session of the Georgia Assembly, both houses agreed to pass a law in line with the royal instructions. No such law was passed, however, and apparently the matter was dropped, for there is no record of any further proceedings with regard to it.[150]

A resolution commending Governor Wright for his prompt actions in the affair was adopted by both houses of the assembly; the committee of the lower house was headed by Henry Yonge.[151] Nonetheless, the

146. 18 Col. Rec. Ga. 627.

147. Acts of the Privy Council of England, 6 Colonial Series, 448–49 (J. Munro ed. 1912); 10 Col. Rec. Ga. 460.

148. Acts of the Privy Council of England, 5 Colonial Series 113–14 (J. Munro ed. 1912); Am. St. Papers, 1 Public Lands 58; 10 Col. Rec. Ga. 396.

149. Acts of the Privy Council of England, 5 Colonial Series 114 (J. Munro ed. 1912).

150. 17 Col. Rec. Ga. 426. Despite the royal disallowance, the act of March 1765 was carried in every major postrevolutionary statutory compilation of Georgia laws down to 1851. See, for example, Cobb 655. As late as 1804 Georgia was still passing special acts to confirm titles to individual tracts obtained under the South Carolina grants. See Clayton 196. In the surveyor general records of the Georgia Department of Archives and History are preserved three volumes of transcripts of grants issued by South Carolina to lands south of the Altamaha.

151. 14 Col. Rec. Ga. 98, 113–14, 134; 17 Col. Rec. Ga. 83, 106–7.

South Carolina grants continued to vex Georgia surveyors for some years to come.[152]

Treaty at Augusta in 1763

The English had lost no time taking advantage of their greatly enhanced position in North America at the close of the French and Indian War. By a treaty signed at Augusta on November 10, 1763, the Creeks ceded the lands between the Savannah and Ogeechee rivers as far northward as the Little River, and the western boundary of the lands between the Ogeechee and St. Marys was moved inland a few miles.[153] Although part of this territory along the Savannah River, especially around Augusta, had long been settled by whites, it had never been officially ceded by the Indians.[154]

Some difference of opinion erupted between colonial officials and the Indians over the exact location of the lines as defined in the 1763 treaty. It was not until 1768, after several adjustments and changes had been made, that the exact lines were agreed upon.[155] The boundary ran from Williams Creek, about fifty miles above Augusta, down to the St. Marys River, a total distance of 270 miles. Deputy surveyors Samuel Savery and Roderick McIntosh blazed it out in the latter part of the year 1768.[156] It took three months to perform this arduous task, during which, as Savery later described, he had been "exposed to all the inclemencies of the season to the great Injury of his Health" and two of his horses, worth £22.10.0, had been "killed with fatigue on that Occasion."[157] The

152. 9 Col. Rec. Ga. 341, 641; 10 Col. Rec. Ga. 396, 561, 628–29; 28 Col. Rec. Ga. (pt. 2) 79–80.

153. 28 Col. Rec. Ga. (pt. 1) 456; 39 Col. Rec. Ga. 379 (unpub. ms., GDAH); 11 St. Rec. N.C. 156–207; J. Alden, John Stuart and the Southern Colonial Frontier 176–91 (1944); L. De Vorsey, The Indian Boundary in the Southern Colonies 1763–1775, at 149–51 (1966).

154. 28 Col. Rec. Ga. (pt. 1) 454; L. De Vorsey, The Indian Boundary in the Southern Colonies 1763–1775, at 144 (1966); J. McCain, Georgia as a Proprietary Province 176–77 (1917).

155. L. De Vorsey, The Indian Boundary in the Southern Colonies 1763–1775, at 152–56 (1966).

156. Id. 156–57, 174. The concise field notes of the survey are published in W. DeBrahm, DeBrahm's Report of the General Survey in the Southern District of North America 151 (L. De Vorsey ed. 1971).

157. 15 Col. Rec. Ga. 162–63, 402–3.

marked lines were formally ratified by the Creeks at Augusta on November 12, 1768.[158]

In March 1770 Savery presented a memorial to the assembly stating that he had delivered his account the preceding year for running the boundary but had failed to receive full compensation. Savery conceded that the fee generally charged by surveyors was twenty-one shillings per day, but this usual fee was for surveying "about Plantations" where men could "be comfortably Accommodated." His work, by contrast, had been "of excessive Labour, risque and fatigue," and he had charged twenty-five shillings per day—an amount he considered moderate in view of the hardships and losses he had sustained.[159] It was more than three years, however, before the assembly provided the full compensation Savery requested.[160]

Frauds and Abuses, Elapsed Plats, and Other Matters

With the Spanish menace from Florida erased, Georgia entered an era of real and sustained economic growth. Settlers came flocking in from all directions, making land a valuable commodity. It is not surprising, therefore, that frauds and abuses should arise with regard to the survey and granting of the king's lands.

In January 1765 certain deputy surveyors were found to be laying out tracts on rivers with a much greater frontage on the river than was permitted by law. In addition, some settlers were having tracts of up to five hundred acres returned as three or four hundred acres by employing Negroes and white persons, not sworn, as chain carriers.[161] Henry Yonge promptly instituted a crackdown, issuing the following instructions to his deputies on January 31, 1765: (1) before performing any survey, timely notice to be given to the adjoining owners so that they might attend the survey; the adjoining owners to be requested to furnish their plats for use as reference; if marked trees referred to in the original plat vary

158. L. De Vorsey, The Indian Boundary in the Southern Colonies 1763–1775, at 157 (1966).

159. Id.; 15 Col. Rec. Ga. 162–63, 402–3.

160. 19 Col. Rec. Ga. (pt. 1) 124, 504.

161. 9 Col. Rec. Ga. 270–72; 14 Col. Rec. Ga. 187–88; 17 Col. Rec. Ga. 160; 28 Col. Rec. Ga. (pt. 2) 105.

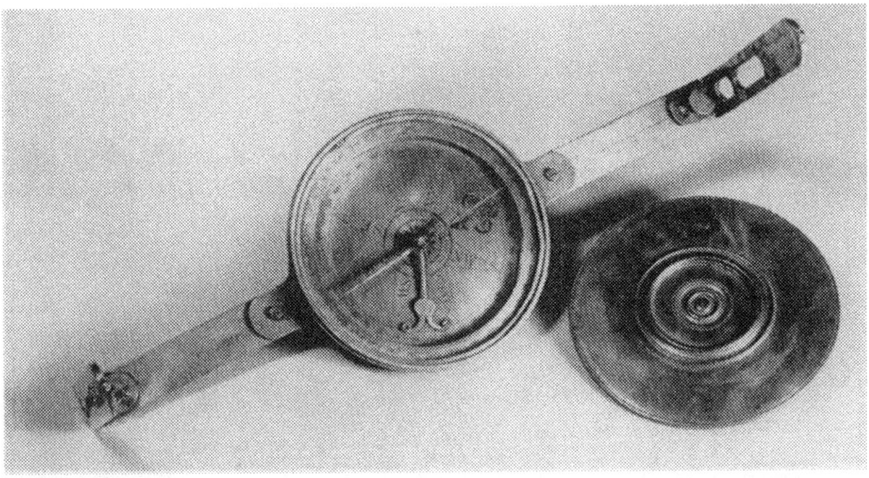

Surveyor's compass (or circumferentor) made ca. 1721–22 by Matthew Walker of Dublin, Ireland. According to the *Savannah Daily Journal and Courier,* April 10, 1855, at 3, col. 1, this compass was the first ever used to run a line in Georgia. Courtesy of the Georgia Historical Society (Artifacts Collection 1361AT).

Another early surveyor's compass with socket joint and mahogany carrying case. Notice that the compass is equipped with level bubbles. Courtesy of the Georgia Historical Society (Artifacts Collection 1361AT).

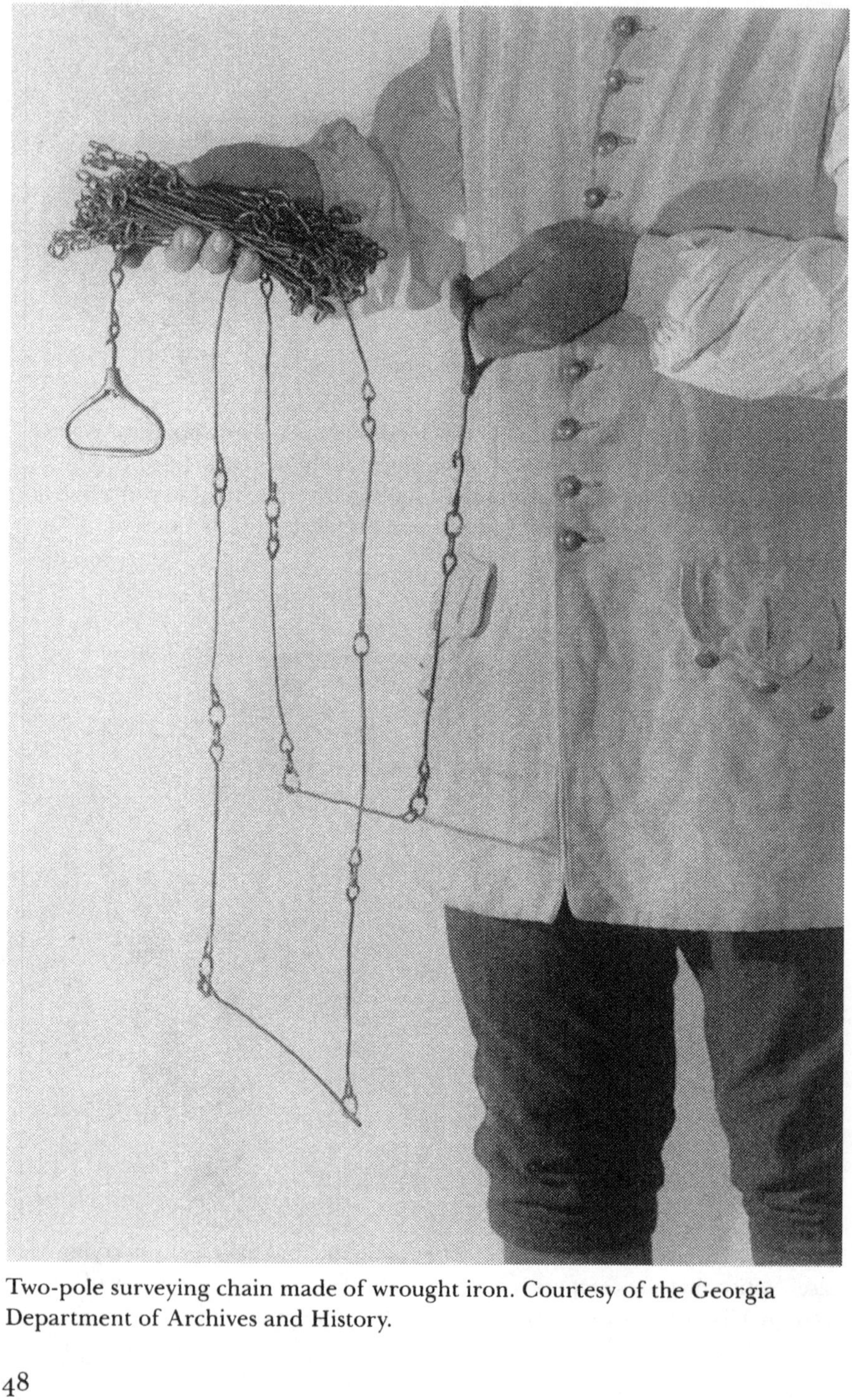

Two-pole surveying chain made of wrought iron. Courtesy of the Georgia Department of Archives and History.

from the courses given in the plat, the marked trees to govern, and where marked trees are wanting, the same course to be continued; (2) all surveys to be entered in a book or on sheets of paper along with the applicants' names, date of warrant, date of survey, and names of adjoining owners; these surveys to be returned to the surveyor general between January 1 and March 1 of each year "to enable him to form a General Correct Plan of the Province"; (3) persons acting as chain carriers to be sworn before entering upon their duties; (4) no survey to be performed without a precept signed and sealed by the surveyor general; no warrant to be executed after the time of its expiration; surveys and precepts to be returned to the surveyor general's office within one month of the time directed for the return of the warrant and plat to the secretary's office; (5) unless prevented by high rivers or "Swamps very deep in Water," all lines to be measured and marked except those of adjoining tracts known to be already measured and marked; rivers, creeks, and roads forming boundaries to be drawn in as accurately as possible on plats; no warrant to be executed in part unless prevented from surveying the whole by the lines of former surveys or by the survey being of an island containing less than the supposed number of acres; (6) a "considerable Number" of station trees to be marked on each line; trees or "Stakes 3X" to mark corners; stations and corners to be designated on the plat along with notations as to whether they are newly marked or are "double marked" (that is, containing only the marks of the present survey or containing the marks of both the present survey and an old survey); no corners or stations to be designated on plats unless made or seen by the deputy or shown on the original plat of an adjoining tract; figures of swamps, runs of water, foundations of tar kilns, and noted paths or roads to be designated on the plats; (7) the breadth of tracts laid out on rivers not to exceed one-third the length and the length of such tracts not to extend along the banks of the rivers but to run back from the river; inland tracts to be laid out into exact squares with sides and angles equal unless prevented by the bounding lines of former surveys; (8) plats to be laid down to a proper scale, to identify corners, stations, marks, and boundaries correctly, to state the total area of the tract, to contain an estimate of the quantity of swampland, corn land, and barren land, to give the course and distance of each line, to give the names of adjoining owners, to give the parish in which the land lies, and to contain a certification that the land was vacant at the time of the survey; plat and endorsement on warrant to bear the date the survey was completed; (9) in making surveys south of the Altamaha River, care to be taken not to run out or interfere with the lines of any South Carolina grant; reports to be made to the surveyor general about the South Carolina surveys describing the corners and lines found, the completeness of these lines, the manner of the surveys, and whether the surveys are in accordance with the usual instructions; (10) upon request, copies of fieldworks and plats to be loaned to other deputies for their use.[162]

A month later, at the urging of Yonge, the assembly passed "An act to prevent frauds and abuses in the admeasuring and laying out his majesty's lands." This act provided that (1) no deputy surveyor should make any survey of lands with chain carriers not sworn to measure justly and exactly according to the best of their knowledge, which oath every deputy surveyor in the province was empowered to administer; (2) all deputy surveyors were to take an oath before the governor faithfully to execute their offices and abide by the instructions issued by the surveyor general; (3) every deputy surveyor, before entering into the execution of his duties, was to post a bond of £200 sterling; and (4) where any undue or fraudulent survey was made by a deputy, such deputy should forfeit the sum of £100 sterling for each offense—half to be paid to the person reporting the fraud or abuse and half to the party who would suffer damages as a result of the fraud or abuse.[163]

An additional problem surfaced in 1766. In September of that year Yonge presented a memorial to the governor and council stating that in the surveyor general's office there were over three hundred plats that had been certified by surveyors but not returned to the colonial secretary; that the time directed for their return had lapsed (in many cases by several years); that lists of them had repeatedly been published in the public gazettes with notice that after a certain time they would be granted to any qualified petitioner who should request them but that very little heed had been paid to the notices; that many of the elapsed plats were of pinelands surveyed solely for the purpose of cutting the timber, the applicant having no intention of taking out a grant; that such plats might cause future inconvenience and hardships to persons applying for large warrants and including in the survey smaller tracts not settled within the lines; that under disguised representations elapsed plats had been obtained as lands never surveyed, which rendered haz-

162. 9 Col. Rec. Ga. 273–76.
163. 18 Col. Rec. Ga. 624.

ardous the regranting of such elapsed plats without a resurvey of the land; and that it was a common practice to sell lands upon the survey and return of the plat, sometimes through two or three different hands, to the manifest detriment of the perhaps ignorant purchaser who then found the lands granted to another. Yonge requested that the governor and council take measures to counter such practices.[164] It was thereupon ordered that in all cases and without any further notice the elapsed plats be granted as vacant lands to any qualified person who should apply for them, provided the applicant paid up all arrears in taxes and quitrents. It was also ordered that, in the future, no warrants for land were to be renewed or the time of survey extended unless a good reason was shown for not making the survey within the prescribed time.[165]

The appointment of William Gerard DeBrahm as surveyor general of the Southern District of North America in 1764 had left Henry Yonge (whom Governor Wright described as "very ingenious, able & Cleaver in his Office, a Well disposed Man, and very usefull in the Assembly") the sole surveyor general of Georgia.[166] Yonge must have been kept very busy supervising all the surveys for a rapidly growing colony that sprawled from above Augusta down to the St. Marys River. In 1766 the noted Dutch engineer, surveyor, and cartographer Bernard Romans was appointed Yonge's principal deputy.[167] Romans, however, stayed in Georgia less than two years and then left to join DeBrahm's exploits in East Florida. He was replaced by Thomas Shruder about 1769.[168]

In July 1771 a schedule of fees was adopted for services performed by the surveyor general and the deputies. It indicates the types of work Georgia surveyors were doing near the end of the colonial period:

SURVEYER GENERAL'S FEES.

For a Copy of every Warrant and Precept thereon	£0.1.4
Every hundred Acres of Land Surveyed	0.2.3
A new Precept or prolonging an Old one	0.1.4
Each plat, record of that plat, and Certificate	0.6.9

164. 9 Col. Rec. Ga. 634–36.

165. Id. 636.

166. 28 Col. Rec. Ga. (pt. 2) 31.

167. 10 Col. Rec. Ga. 171, 338, 352, 414, 489, 501; Siebert, *Bernard Romans,* in 8 Dictionary of American Biography (pt. 2) 126 (D. Malone ed. 1963).

168. 11 Col. Rec. Ga. 330, 384, 430; 19 Col. Rec. Ga. (pt. 1) 190, 490; Georgia Gazette (Savannah), Mar. 8, 1769, at 2, col. 2.

Survey of a Town Lot	0.3.0
Plat of a Town Lot record and Certificate of the same	0.6.9
For an Attested Copy of a plat	0.4.0
For a Search (that is, every different Name)	0.0.8
For a Certificate	0.1.0
For business done by Special order of the Council and not herein provided for	0.1.4
For a Copy of a Warrant without a Precept and Certifying the Same	0.1.4

N.B. If a Tract of Land or Town Lot is ordered and Surveyed for any person and afterwards ordered to be passed to another person, the latter is to pay all fees due thereon by the person or persons to whom it was before ordered.

DEPUTY SURVEYER'S FEES.

For every Survey under Twenty Miles distant from the place of abode	£0.6.9
For every Survey p^r 100 Acres	0.2.3
For a plat of each Survey certified and returned to the Surveyor general's Office	0.6.9
For every Survey above Twenty Miles from his place of abode	0.6.9
For every Twenty Miles exceeding that distance	0.6.9
For running out a Town lot if Cleared	0.3.0
For the Same if not Cleared	0.6.0
For a Certificate of the Same returned to the Surveyer general's Office	0.6.9[169]

In addition, colonial surveyors oversaw the laying out of roads; supervised the erection of wharves, lighthouses, fortifications, and other public structures; and made surveys of creeks, coves, and inlets to determine their suitability for navigation.

As a sidelight to the surveying scene in Georgia during this period, John Portrees announced in the June 16, 1763, issue of the *Georgia Gazette* that he was keeping a school on the plantation of a Mr. Gibbons and that among the subjects he taught were arithmetic, geometry, mensuration, plane and spherical trigonometry, navigation, and surveying.[170]

169. 11 Col. Rec. Ga. 430–31. See also 19 Col. Rec. Ga. (pt. 1) 402–3.

170. Georgia Gazette (Savannah), June 16, 1763, at 2, col. 2.

Treaty at Augusta in 1773

By the decade of the 1770s agriculture and commerce were flourishing in Georgia, enhanced to a large degree by the skillful administration of Governor Wright. Between 1761 and 1773 the colony's population—swelled by immigrants primarily from Virginia, the Carolinas, Scotland, and Ireland—more than tripled. Vacant lands fit for cultivation, though once plentiful, were now becoming scarcer.[171]

The difficulty finding suitable tracts of grantable lands and abundant rumors of an imminent Indian cession enticed a number of persons—described as "idle and disorderly"—to wander onto Indian lands and make settlements on them.[172] The resulting friction with the Indians induced the surveyor general in November 1772 to issue strict orders to the deputies not to survey any lands outside the present Indian boundary.[173]

Propitiously, a timely series of circumstances was in the making which Georgia was able to press to its own advantage. A sharp drop in prices resulting from the South Carolina public trading act of 1762 coupled with an acute shortage of deer, which provided hides used by the Indians in trade, had put the Creeks and Cherokees in dire financial straits with white traders, who had extended very liberal credit terms to them. Accordingly, the Creeks and Cherokees agreed to another land cession on the condition that Georgia assume the accumulated obligations (estimated to be between £40,000 and £50,000).[174]

The treaty of cession, signed at Augusta on June 1, 1773, opened over 2.1 million acres to white settlement.[175] This cession, which became known as the New Purchase, was in two tracts—one a fertile, well-watered domain along the Broad and Little rivers between the Ogeechee and Savannah rivers; and the other, a less productive expanse of towering pines and blackgum bottomlands between the Ogeechee and Altamaha rivers, just west and southwest of previously ceded lands.

171. 12 Col. Rec. Ga. 401–3; K. Coleman, The American Revolution in Georgia 1763–1789, at 10 (1958); L. De Vorsey, The Indian Boundary in the Southern Colonies 1763–1775, at 169 (1966).

172. 12 Col. Rec. Ga. 347–48; 38 Col. Rec. Ga. (pt. 1) 9 (unpub. ms., GDAH); L. De Vorsey, The Indian Boundary in the Southern Colonies 1763–1775, at 160–61, 168–69 (1966).

173. 12 Col. Rec. Ga. 336.

174. J. Alden, John Stuart and the Southern Colonial Frontier 301–5 (1944); L. De Vorsey, The Indian Boundary in the Southern Colonies 1763–1775, at 157–72 (1966).

175. Marbury & Crawford 601.

Immediately after the treaty was signed, Governor Wright dispatched surveyors to run out and mark the boundaries and to ascertain the quality of the lands.[176] William Bartram, the articulate Philadelphia naturalist, accompanied the surveying party that marked part of the boundary of the northern portion of the cession, and he left a short but captivating account of the expedition. The group Bartram accompanied left Augusta on June 7, 1773. A few days later they arrived at the Buffalo Lick, the starting point of the survey, located in what is now southeastern Oglethorpe County. The party then followed a course through the present-day counties of Oglethorpe, Clarke, Madison, Franklin, and Hart, probably returning to Augusta early in July. Bartram gave the following account:

> A company of surveyors were appointed, by the governor and council, to ascertain the boundaries of the new purchase; they were to be attended by chiefs of the Indians, selected and delegated by their countrymen, to assist, and be witnesses that the articles of the treaty were fulfilled, as agreed to by both parties in Congress.
>
> Col. Barnet, who was chosen to conduct this business on the part of the Georgians, a gentleman every way qualified for that important trust, in a very friendly and obliging manner, gave me an invitation to accompany him on this tour. . . .
>
> The preparatory business of the surveyors being now accomplished, Mr. J. M'Intosh, yet anxious for travelling, and desirous to accompany me on this tour, we joined the caravan, consisting of surveyors, astronomers, artisans, chain-carriers, markers, guides and hunters, besides a very respectable number of gentlemen, who joined us, in order to speculate in the lands, together with ten or twelve Indians, altogether to the number of eighty or ninety men, all or most of us well mounted on horseback, besides twenty or thirty pack-horses, loaded with provisions, tents, and camp equipage.
>
> The summer season now rapidly advancing, the air at mid-day, about this region, is insufferably hot and sultry. We set off from Augusta, early in the morning, for the Great Buffalo Lick, on the Great Ridge, which separates the waters of the Savanna and Alatamaha, about eighty miles distant from Augusta. At this Lick the surveyors were to separate themselves, and form three companies, to proceed on different routes. . . .
>
> After four days moderate and pleasant travelling, we arrived in the evening at the Buffalo Lick. . . .
>
> We were detained at this place one day, in adjusting and planning the sev-

176. 38 Col. Rec. Ga. (pt. 1) 60 (unpub. ms., GDAH).

eral branches of the survey. A circumstance occurred during this time, which was a remarkable instance of Indian sagacity, and had nearly disconcerted all our plans, and put an end to the business. The surveyor having fixed his compass on the staff, and about to ascertain the course from our place of departure, which was to strike Savanna river at the confluence of a certain river, about seventy miles distance from us; just as he had determined upon the point, the Indian Chief came up, and observing the course he had fixed upon, spoke, and said it was not right; but that the course to the place was so and so, holding up his hand, and pointing. The surveyor replied, that he himself was certainly right, adding, that that little instrument (pointing to the compass) told him so, which, he said, could not err. The Indian answered, he knew better, and that the little wicked instrument was a liar; and he would not acquiesce in its decisions, since it would wrong the Indians out of their land. This mistake (the surveyor proving to be in the wrong) displeased the Indians; the dispute arose to that height, that the Chief and his party had determined to break up the business, and return the shortest way home, and forbad the surveyors to proceed any farther: however, after some delay, the complaisance and prudent conduct of the Colonel made them change their resolution; the Chief became reconciled, upon condition that the compass should be discarded, and rendered incapable of serving on this business; that the Chief himself should lead the survey; and, moreover, receive an order for a very considerable quantity of goods.

Matters being now amicably settled, under this new regulation, the Colonel having detached two companies on separate routes, Mr. M'Intosh and myself attaching ourselves to the Colonel's party, whose excursion was likely to be the most extensive and varied, we set off from the Buffalo Lick, and the Indian Chief, heading the party, conducted us on a straight line, as appeared by collateral observation, to the desired place. We pursued nearly a north course up the Great Ridge, until we came near the branches of Broad River, when we turned off to the right hand, and encamped on a considerable branch of it. At this place we continued almost a whole day, constituting surveyors and astronomers, who were to take the course, distance, and observations on Broad River, and from thence down to its confluence with the Savanna. . . .

We at length happily accomplished our line, arriving at the little river [Bartram is mistaken here. This should read "Tugaloo River"], where our hunters bringing in plenty of venison and turkeys, we had a plentiful feast at supper. Next morning we marked the corner tree, at the confluence of Little River [Tugaloo River] and the Savanna; and, soon after, the Indians amicably took leave of us, returning home to their towns. . . .

The surveyors having completed their observations, we set off next day

> on our return to Augusta, taking our route generally through the low lands on the banks of the Savanna.[177]

After the surveys had been completed, Philip Yonge (son of Henry Yonge) prepared a map of the northern portion of the New Purchase, and Andrew Way (a deputy surveyor) made a map of the southerly portion, the survey of which had been completed in November 1773.[178]

New Procedures Attempted for Disposing Public Lands

Early in 1774, the British government attempted to implement a new method of land granting for Georgia and several other colonies. This move was brought about by a desire to improve the administration of royal lands in America as well as an interest in improving the efficiency of the empire as a whole. It was decided that, prior to granting, lands in the latest cessions should be surveyed into tracts of twenty thousand acres each and then subdivided into smaller tracts of from one hundred to one thousand acres. Instead of being granted free, the tracts were to be sold. Each head of a family could purchase one hundred acres for himself, plus fifty additional acres for each family member or slave he brought in. The total amount granted to one person was not to exceed a thousand acres. The particular tract each person or family received was to be determined by lot.[179]

Although this procedure for disposing of public lands anticipated by thirty years the land lottery system, which under the state government was to become the predominant form of land granting, it proved to be unworkable. In a letter to the Board of Trade dated May 18, 1774, Governor Wright conceded that the method was "the most proper, and Effectual" way of ascertaining the exact quantity and quality of ungranted lands. But he had been forced to abandon it after having had a few of the twenty-thousand-acre tracts surveyed because of the great expense

177. W. Bartram, Travels through North and South Carolina, Georgia, East and West Florida 33–35, 39–41, 45–46 (London 1792).

178. L. De Vorsey, The Indian Boundary in the Southern Colonies 1763–1775, at 165, 175 (1966); J. Goff, Placenames of Georgia 307–11 (L. Utley & M. Hemperley eds. 1975).

179. Proclamations Book H, 1754–1794, at 167 (GDAH); 34 Col. Rec. Ga. 632, 637–38 (unpub. ms., GDAH); 38 Col. Rec. Ga. (pt. 1) 19, 32–33, 80 (unpub. ms., GDAH); 2 L. Labaree, Royal Instructions to British Colonial Governors 1670–1776, at 533–37 (1935).

involved. This problem was compounded because settlers preferred to choose the land for themselves, and they had "raised . . . many difficultys and objections" against its being laid out into small tracts before they were able to make their choice.[180]

At about the same time the government in London decided that it would facilitate the settling and improving of vacant lands in the older sections of the colony if each parcel of vacant land was surveyed. Under this plan all ungranted lands in the colony were to be surveyed and laid off into districts. The districts were to be divided into tracts of between one hundred and one thousand acres each, and the surveyors were to make a map of each district showing the tracts marked and numbered thereon. The surveyor general was then to make a report describing the quality of each tract. Any lands found fit for public use were to be reserved from granting.[181]

Upon receiving instructions to put this ponderous plan into effect, Henry Yonge submitted to the governor and council that almost insurmountable difficulties would be encountered in attempting to carry it out. Vacant tracts were scattered over the entire province, and a general survey of all the lands in Georgia—granted as well as ungranted—would be necessary to distinguish granted lands from vacant lands. Yonge contended that such a survey would be prohibitively expensive and that the deputies were neither experienced in judging the quality of land nor competent in selecting sites to be set aside for public uses. Even if the funds could be raised, Yonge insisted that the project would be largely a waste of money for, in his estimation, five-sixths of the land in Georgia was unsuited for cultivation. Many, many years would therefore elapse before settlers could be found who would be willing to take up most of the tracts.[182]

The governor and council unanimously agreed with Yonge and sent his objections to the Privy Council in London. Wright pointed out in October 1774 that he had no funds to pay the expenses of making such surveys and no means of raising such funds. The onset of the Revolution cut short any serious attempts to implement the plan.[183]

In his memorial to the governor, made on May 17, 1774, concerning

180. 38 Col. Rec. Ga. (pt. 1) 275–77 (unpub. ms., GDAH).

181. 12 Col. Rec. Ga. 399–404.

182. Id.

183. 38 Col. Rec. Ga. (pt. 1) 328–30 (unpub. ms., GDAH).

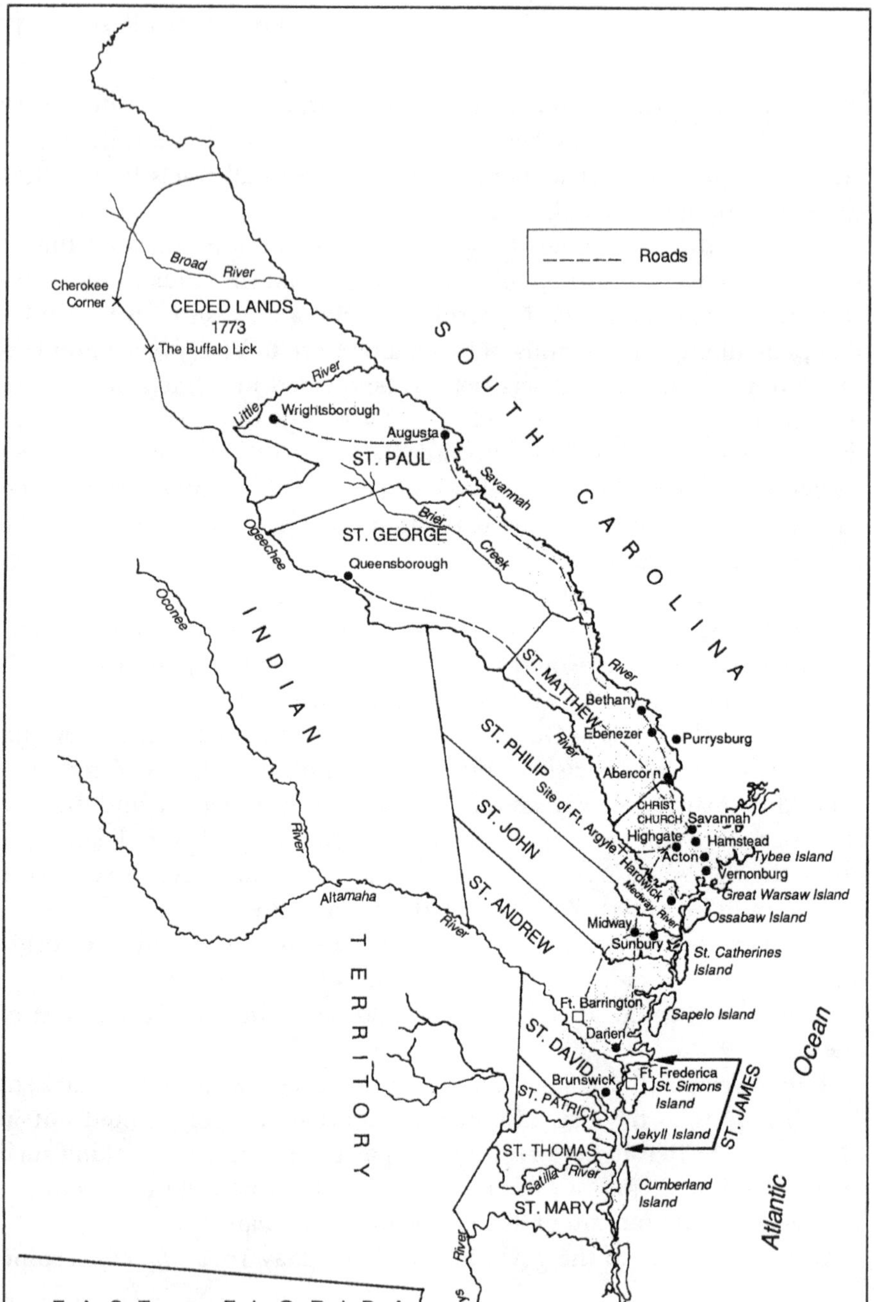

Georgia in 1775. The shading indicates the approximate area of white settlement. Adapted from Harold E. Davis, *The Fledgling Province* (1976).

the proposed survey of vacant lands, Yonge summarized the extent to which plantable lands had been taken up and settled. Exclusive of the late Indian cession, he declared, there were very few if any places within the present limits of the province that a vacant tract of five thousand acres, all lying in one body and fit for cultivation and improvement, could be found. Yonge asserted his firm opinion that no more than three hundred thousand acres of plantable lands remained vacant. These, he continued, were mainly in small tracts, ranging in size from one hundred to five hundred or at most a thousand acres and lay chiefly in the interior and distant parts of the province, either between former surveys or along the narrow branches and rivulets that extended into the pinelands.[184]

The council apparently recognized the growing shortage of vacant lands, for on January 9, 1775, it resolved that all future surveys of lands on rivers and navigable creeks should be run four chains back for one chain in front instead of three chains back for one chain in front as had been allowed previously.[185]

184. 12 Col. Rec. Ga. 401–2.

185. Proceedings and Minutes of the Governor and Council of Georgia . . . , in 10 Coll. Ga. Hist. Soc. 1, 10 (L. Hawes ed. 1952).

Chapter 2. Headright Grants under the State Government

Revolutionary War Grants

During the Revolution both the Loyalist and the rebellious factions attempted to maintain some semblance of government machinery, with all its various functionaries. Philip Yonge, the son of Henry Yonge, became England's surveyor general in Georgia after his father's death in May 1778 and held that office until his own death in January 1782.[1] George

1. Letter from James Wright to Lord George Germain (Jan. 22, 1782), *printed in* 3 Coll. Ga. Hist. Soc. (pt. 2) 363 (G. W. J. DeRenne et al. eds. 1873); Royal Georgia Gazette (Savannah), Jan. 24, 1782, at 3, col. 1. The plight of the Yonges, illustrative of the hardships suffered by all American Loyalists during the Revolution, is vividly portrayed in a memorial presented after the war to the British government by William John Yonge, son of Henry and brother of Philip. The memorial states: "Henry Yonge was Surveyor General upwards of thirty years [*sic*] in the . . . Province of Georgia, and upon the commencement of the Rebellion in America he used his utmost endeavour to suppress and discountenance it, by which conduct he incurred the Resentment of the usurpers of Government and was confined to his own Plantation upon Parole for many months and, early in the Year 1778, was banished with his Son Philip Yonge . . . , who were permitted to take only one Negro Servant each, and were at very great Expense obliged to purchase a Vessel in order to convey themselves to New Providence in the Bahama Islands, but they were taken on their passage by a British Privateer and carried into St. Augustine where they incurred heavy Charges in recovering their said Vessel in the Court of Vice-Admiralty there, and when the said Henry Yonge by being separated from his Family and Estate died a few months afterwards it was generally supposed that his Death was occasioned by his Affliction and Sufferings." Memorial of William John Yonge to the Commissioners Appointed . . . for enquiring into the Losses and Services of the American Loyalists (Jan. 28, 1788) (A.O. 13/37 at 754–55, Public Record Office, London, England).

Philip Yonge had been declared "dangerous to the liberties of America" by the Georgia Council of Safety on June 26, 1776, and after the death of his father at St. Augustine he set sail for England. On this voyage he was captured by a French fleet under Count d'Estaing, had his possessions confiscated, and was sent to Philadelphia as a prisoner of war. He eventually returned to Savannah under an exchange of prisoners and served as captain of a company of royal militia at the siege of Savannah in 1779. By the time of his death in 1782 the Americans had seized his plantations, slaves, and other property, leaving his widow and four children practically destitute. Today Yonge Island, South Carolina (near Charleston), and the P. K. Yonge Library of Florida History at the University of Florida in Gainesville

McKenzie was then appointed surveyor general and held the office until the British evacuated Savannah in July 1782. Thus he was the last surveyor general of Georgia under British rule. It is likely that neither Philip Yonge nor McKenzie was able to conduct the duties of the office to any significant degree. Yonge spent several months as a prisoner of war of the Americans and later served as an officer in the royal militia. McKenzie soon went into exile with other Loyalists at St. Augustine.[2] The available records indicate that Governor Wright signed his last Georgia land grant on June 6, 1775, bringing to a close a period of nearly twenty years of land granting under the royal government.[3]

The Georgia rebels set up a temporary government on April 15, 1776, to function until a more permanent government could be established.[4] Georgia's first constitution, adopted February 5, 1777, created eight counties. Seven of them were formed from the old parishes that had existed before the Revolution, and the eighth (Wilkes) from the northern portion of the lands the Creeks and Cherokees had ceded in June 1773. Table 1 lists the parishes and indicates the counties formed from them.

The revolutionary Executive Council appointed one of its members, Thomas Chisholm, the first surveyor general of the state of Georgia on May 22, 1777. He qualified for the post on February 9, 1778.[5]

The first land act of the newly formed government, entitled "An act for opening a land-office, and for the better settling and strengthening this state," was passed on June 7, 1777.[6] It along with an amendatory act passed three months later[7] provided for a headright system of land granting almost identical to that under the royal government. Each head of a family could receive two hundred acres of land plus fifty additional acres for each family member and fifty acres for each Negro up to ten.

---

pay homage to the Yonge family name. Henry Yonge and other members of his family once owned land on the island, and the P. K. Yonge Library was endowed by his great-great-great grandson. See Philip Keyes Yonge, The Yonge Family in America (July 1960) (unpub. typescript in the P. K. Yonge Library of Florida History, University of Florida, Gainesville, Fla.).

2. 39 Col. Rec. Ga. 421–22 (unpub. ms., GDAH).

3. M. Hemperley, The Georgia Surveyor General Department 14 (1982).

4. K. Coleman, The American Revolution in Georgia 1763–1789, at 76–79 (1958).

5. Minutes of the Executive Council, May 7 through October 14, 1777 (pt. 1), in 33 Ga. Hist. Q. 318, 325–26 (L. Hawes ed. 1949); 2 Rev. Rec. Ga. 28.

6. 19 Col. Rec. Ga. (pt. 2) 53; Marbury & Crawford 316.

7. 19 Col. Rec. Ga. (pt. 2) 70; Marbury & Crawford 318.

Table 1. Transition from Parishes to Counties

Parish	*Date of Creation of Parish*		*County*
			Wilkes[a]
St. Paul	March 15, 1758	became	Richmond
St. George	March 15, 1758	became	Burke
St. Matthew	March 15, 1758		
		became	Effingham
St. Philip[b]	March 15, 1758		
		became	Chatham
Christ Church	March 15, 1758		
St. John	March 15, 1758		
St. Andrew	March 15, 1758	became	Liberty
St. James	March 15, 1758		
St. David	March 25, 1765		
		became	Glynn
St. Patrick	March 25, 1765		
St. Thomas	March 25, 1765		
		became	Camden
St. Mary	March 25, 1765		

Source: 18 Col. Rec. Ga. 258, 689; 1 Rev. Rec. Ga. 284–85.

[a] Created from the northern portion of the New Purchase of 1773.

[b] Divided by the Constitution of 1777 so that part went to Effingham County and part went to Chatham County.

The applicant paid the costs of surveying and granting plus the sum of two shillings per hundred acres. Tracts of up to two thousand acres could be obtained by anyone willing to establish a mill, bloomery, or forge. The only requirement as to the shapes of the tracts was the broad stipulation that they "be surveyed and laid out . . . in either a square or oblong figure, the length not to be more than double the breadth" except such tracts as "lie between lands already granted" or which "lie between the forks of rivers or creeks" and then to be bounded by the previously granted lands or by the rivers or creeks. Another land act was passed in January 1780 with some minor additional provisions.[8]

8. 19 Col. Rec. Ga. (pt. 2) 130; Marbury & Crawford 319.

Because of the dearth of records, it is impossible to judge accurately the extent to which the state government attempted to carry on the process of land granting during the Revolution. Obviously such activities were very limited, to say the least. It is known that a committee room was "fitted up" for the use of the surveyor general; that a schedule of fees was adopted to be charged for surveying public lands; that George Seigar was appointed deputy surveyor general; that several deputy surveyors were appointed; that detailed instructions were issued by Chisholm to these deputies; that a few land warrants were issued by the Executive Council; that a few surveys resulted from these warrants; and that a few caveats resulted from the surveys.[9] No direct evidence indicates that land grants were actually issued by the state during this period, though the available records strongly suggest that some were. It is not known when or under what circumstances Chisholm left office. William Few, who was later to represent Georgia at the federal Constitutional Convention, subsequently served as surveyor general "for 6 or 8 Months"; the exact date is unknown.[10] During 1780 and 1781 most of Georgia was overrun by the British and Tories and the state government seems to have practically ceased to function.

In the wartime conflagration and chaos, most of the recorded plats of colonial grants issued up to that time were, in some unknown manner, lost. However, some eighteen hundred plats, about one-fourth the original number, were re-recorded after the war. The colonial grant books survived, and these along with the existing plats were absorbed into the state land records now housed in the surveyor general section of the Georgia Department of Archives and History.[11]

9. Instructions from Thomas Chisholm to William Greer (n.d.) (File II, GDAH); Minutes of the Executive Council, May 7 through October 14, 1777, in 34 Ga. Hist. Q. 19, 20, 118 (L. Hawes ed. 1950); 2 Rev. Rec. Ga. 9, 29, 62–63, 118, 225–26, 237–38, 240–43.

10. Memoir of William Few, 15 (William Few Collection in Mary Telfair Letters, GDAH); *printed in* 7 Magazine of American History 343, 347 (C. Jones ed. 1881).

11. M. Hemperley, The Georgia Surveyor General Department 14–18, 57 (1982); Hitz, *The Earliest Settlements in Wilkes County,* 40 Ga. Hist. Q. 260, 272–80 (1956). An 1815 act authorized persons holding lands obtained under colonial grants for which the plats had been lost to apply to have new plats made. The county surveyor was to resurvey the tract from the information contained in the grant and from "such other evidence" as he could procure. The resulting plat was to stand in the place of the original lost one and was to be recorded in the county records as well as in the surveyor general's office. Lamar 414. Numerous plats resulting from this act are on record today.

Early State Grants

The provisional articles for a treaty of peace ending the American Revolution were entered into by Great Britain and the United States on November 30, 1782, and the definitive treaty was signed in Paris on September 3, 1783. By this treaty the western boundary of the United States was recognized as the center of the Mississippi River. The southern boundary was fixed at a line beginning at the middle of the Mississippi River on the thirty-first parallel; thence along that parallel to the middle of the Chattahoochee River; thence southward along the middle of the Chattahoochee to its junction with the Flint River; thence in a straight line to the head of the St. Marys River; thence down the middle of that river to the Atlantic Ocean.[12] Thus Georgia, on paper at least, retained her prerevolutionary territory stretching from the Atlantic Ocean to the Mississippi River.

Georgia, the largest and least populated state in the confederation, seemed a land of limitless opportunities. Wave after wave of immigrants, primarily from Maryland, Virginia, and the Carolinas, poured in. These "restless, land-hungry, gambling, hard-drinking gentry" generally avoided the pine barren region, which lay immediately to the west of the coast.[13] That extensive primeval solitude consisted of mile after dreary mile of pines growing in coarse, sterile sand feebly held together by prickly palmettos and wiregrass. Despite its proximity to the older settled regions, the pine barrens were to be one of the last areas of the state to see extensive exploitation by humans.

It was the rich, fertile lands of the piedmont that beckoned the settlers of the 1780s. This area of rolling red hills, broad-leaved forests, and huge granite outcroppings had been barely breached on the eve of the Revolution, and the breach was now to become a torrent. Settlers spoke of this region in glowing terms. The attitude of Lydnor Corby typified this feeling. Corby, who had emigrated from Virginia to Wilkes County,

12. 10 G. Bancroft, History of the United States of America 591 (1874); 8 Stat. 80. Florida was returned to Spain at the end of the war, and for a time Spain disputed Georgia's title to the lands lying between the Mississippi and Chattahoochee rivers and 32° 30′ north latitude and 31° north latitude. By a treaty signed in 1795, Spain gave up the contest and recognized the boundaries as fixed in the Treaty of 1783. 8 Stat. 138. See Am. St. Papers, 1 Public Lands 57; Harcourt v. Gaillard, 25 U.S. (12 Wheat.) 523, 6 L. Ed. 716 (1827); Henderson v. Poindexter's Lessee, 25 U.S. (12 Wheat.) 530, 6 L. Ed. 718 (1827); and Burnett, *Papers Relating to Bourbon County, Georgia, 1785–1786*, in 15 Am. Hist. Rev. 66, 67–68 (1909–10).

13. E. Coulter, College Life in the Old South 2 (1979).

Georgia, shortly after the Revolution, wrote friends back home: "A great body of our Lands are good, & want only industry, to render us a flourishing people. you woud be amazed to see how low the first kind of Land sells [for] here. for cash a man with a few hundred pounds might . . . make a fortune. you may assure my old friends who are living on hillsides & old fields [and on] Stony ridges that they had better visit Georgia & get land that will produce whatever is committed to it in abundance."[14]

Just a few years later, the cotton gin invented by Eli Whitney would make the piedmont the wealthiest section of the state—the center of politics, culture, and refinement.

For Georgians land was to be the first order of business for half a century after the Revolution. The situation with regard to public lands in Georgia was unique among the states. The other original states had disposed of most of their public domains, except their western lands, during colonial times. Upon relinquishment of their western lands to the central government, they removed themselves largely from the public land business. Georgia, however, at the close of the Revolution was chiefly a vast, unsettled territory. Over 95 percent of its area, not counting its western lands, was still public domain. It was probably inevitable, therefore, that land policy should exert a major influence over the state's economy and people.[15]

Two crucial issues faced Georgia with regard to its public lands. The first was the question of ceding its western territory to the federal government. The signing of the Articles of Agreement and Cession in 1802 resolved this problem. The other issue, which became far more contentious and protracted, involved the disposition of the Indians who still occupied most of the lands Georgia claimed as its own. Their continued presence was considered an obstruction to economic progress and development. Reflective of this feeling is a letter Governor George M. Troup wrote to John C. Calhoun in 1824. Troup declared that of all the original states, Georgia was the only one whose political organization was incomplete, her civil polity was deranged, her military force could not be reduced to systematic order, the extent of her actual resources could not be counted, the great work of internal improvement was suspended—

14. Letter from Lydnor Corby to Garrett Minor (April 16, 1786) (Garrett Minor Papers, Manuscripts Division, Library of Congress, Washington, D.C.).

15. M. Heath, Constructive Liberalism: The Role of the State in Economic Development in Georgia to 1860, at 69 (1954). Georgia's "western lands" constituted generally the state's territory west of its present western boundary.

all because Georgia was not in possession of her vacant territory.[16] The problem was aggravated because Georgia Indians, especially the Cherokees, were fast acquiring the political and cultural skills of their white neighbors. Thus they promised to become permanent fixtures on the land unless decisive preventive measures were taken. As a result, a radical shift in policy occurred after the United States purchased the Louisiana Territory in 1803. Heretofore the general policy after each land cession had been to relegate the aboriginal inhabitants to their remaining lands until the next land cession. The demands now, however, were for their outright removal to lands beyond the Mississippi River. The United States, under the terms of the Articles of Agreement and Cession, obligated itself to expedite Indian removal, but no definite timetable had been set.

Georgia's resolute demands for immediate removal of the Indians, along with the question of the cession of its western lands, were to involve the state in many long, acrimonious controversies with the federal government and were to supply in one form or another a dominant issue in every political campaign for the next fifty years.[17]

Legislation dealing with the disposal of the public domain quickly became the primary concern in the assembly that gathered and elected Lyman Hall as governor on January 7, 1783. A number of provisions were introduced and debated at length in an effort to formulate an act to replace the wartime land acts, now regarded as largely obsolete.[18]

Wealthy landowners and leaders from the older coastal communities generally favored substantive fees for grants, rigorous conditions of settlement, and state control over issuing warrants. Georgia, though rich in land, was financially destitute. Since property in the older, developed regions sustained the greatest burden of taxation, leaders from these regions favored a land-granting procedure that would yield revenue. Similarly, they sought a policy that would minimize pressures on the Indians so as to avoid the heavy expense resulting from altercations on the frontier. Furthermore, fearing policies that might depress land values as well as encourage competition for their products, they opposed any policies that tended to open the rich interior to easy settlement.[19]

16. E. Harden, The Life of George M. Troup 203–7 (Savannah 1859).

17. M. Heath, Constructive Liberalism: The Role of the State in Economic Development in Georgia to 1860, at 70 (1954).

18. Id. 85, 88–89.

19. Id.

Representatives from the poorer counties and the backcountry generally took the opposite view on these issues. They pointed out that much of the ungranted land within the state consisted of "barren heaths and dreary wastes," which were not worth applying for at any cost. They emphasized the poverty of the state's propertyless class and of many of the immigrants and the need for rapid settlement of the frontiers for protection against Indians. Backcountry leaders, in particular, favored a process in which land warrants would be issued by local court justices rather than by a centralized state authority as advocated by leaders of the coastal regions.[20]

On the question of local versus state control over issuing warrants, the views of the coastal regions were expressed by Governor Hall and his council in an advisory statement on the proposed land act. They reasoned that land policy involved the disposal of "those great and invaluable sources of wealth" of the state and therefore should be administered by representatives of all the people and not according to the pleasures of the justices of individual counties. Administration of land granting by local justices would deny the inhabitants of the eastern and southern counties a voice in the most important concern of the state. Hall declared that the entire control of land administration machinery should therefore be centered in the governor and council (as it had been in colonial times) or in a special court consisting of one representative from each county, or in a combination of the latter and the governor and council. Such centralization, he declared, would reduce costs and assure a much desired uniformity of practice.[21]

Backcountry leaders, though admitting the greater likelihood for the commission of errors and even frauds by local land courts, expressed concern that centralized administration would lead to control of the land-granting process by wealthy landowners, most of whom were concentrated in the older counties.[22]

Because of a flood of immigration, the political balance in the assembly was rapidly tilting in favor of the newer counties. This and a pervasive desire to promote rapid settlement led to the adoption of a liberal land policy designed to make grants easily available to all who came.[23]

20. Id.
21. Id.
22. Id.
23. Id.

The first postwar land act passed the assembly on February 17, 1783.[24] It and an amendment passed on August 1, 1783,[25] became the fundamental law by which all headright grants of the state were issued. The procedure thus instituted was much like that followed in colonial times, though more dispersed in its administration.[26]

Under the headright law, each head of a family was allowed two hundred acres plus fifty additional acres for each family member or slave, the total amount not to exceed one thousand acres. The applicant received the first two hundred acres free but paid a fee, on a sliding scale, of one shilling to four shillings six pence per acre for each additional acre. Fees for lands were dropped entirely in 1785[27] but were reinstated in 1831. The 1831 act provided grants between zero and fifty acres for $2.50, between fifty and three hundred acres for $3.00, and between three hundred and one thousand acres for $4.00.[28] In all cases the applicant paid the cost of surveying and paperwork.

Any U.S. citizen who was a resident of Georgia was entitled to a grant. Before receiving the grant the applicant was required to settle on the tract for twelve months and improve three out of every one hundred acres. Strict application of the requirements for settling and improving was found impractical, and the requirements were modified in 1785. Beginning that year the applicant had to settle on the land and improve three out of every one hundred acres within eighteen months of receiving the grant, or the land would be subject to a treble tax.

To obtain a headright grant, the applicant had to appear before the land court in the county where the land he desired was located, describe the tract for which he wished to obtain a warrant, and take an oral[29] oath declaring that he was legally entitled to a grant.

24. 19 Col. Rec. Ga. (pt. 2) 201; Marbury & Crawford 323.

25. 19 Col. Rec. Ga. (pt. 2) 280; Marbury & Crawford 328.

26. Without fully itemizing sources, the following text concerning the procedure for headright granting is derived from 19 Col. Rec. Ga. (pt. 2) 201–15, 280–84, 292–304, 434–41; Marbury & Crawford 260–64, 323–39; Cobb 660–80; 1836 Ga. Laws 78; 1855–56 Ga. Laws 257; Ga. Code of 1863 §§ 2336–49; Ga. Code of 1882 §§ 2364–77; Holtzendorf v. Hilton & Dodge Lbr. Co., 141 Ga. 148 (1913); Roberts v. Hilton & Dodge Lbr. Co., 136 Ga. 473 (1911); Vickery v. Scott, 20 Ga. 795 (1856); Sanders v. Davison, 16 Ga. 537 (1855); Robert v. Palmer, 14 Ga. 349 (1853); M. Heath, Constructive Liberalism: The Role of the State in Economic Development in Georgia to 1860, at 84–97 (1954); and Hitz, *Georgia Bounty Land Grants,* 38 Ga. Hist. Q. 337 (1954).

27. 19 Col. Rec. Ga. (pt. 2) 434; Marbury & Crawford 334.

28. Cobb 678.

29. The court justices of Taliaferro County employed the unique practice of reducing each

At first, the land courts consisted of a majority of the justices of the peace in each county. They were required to sit at the county seat on the first Monday in each month and for as many days thereafter as were necessary to receive applications for land. When Franklin and Washington counties were opened up for settlement in 1784, a special land court composed of the governor or president of the Executive Council and three members of the Executive Council was set up at Augusta to receive petitions for land in these two counties. This measure was only temporary, however, and the next year the land courts in these counties were constituted in the same manner as those in other counties. A 1789 act provided for land courts to be made up of three or more justices of the peace, the one with the senior commission presiding. The duties of the land courts devolved exclusively upon the county inferior courts in 1856 and upon the county ordinaries in 1868, there being very little vacant land left in the state by that time.

If the land court approved, the applicant was issued a warrant for a survey that described, as far as practicable, the tract desired. The form of a warrant was as follows:

> By the Court of Justices of the County of _____ To A. B. County Surveyor for the Said County:
> You are hereby Authorized and Required to admeasure and lay out, or cause to be admeasured and laid out Unto C. D. a tract of land which shall contain _____ Acres in the said County of _____ [here was described the buttings and boundings of the land] taking especial care that the same has not heretofore been laid out to any other person or persons, And You are hereby also directed and required to record the Plat of the same in Your Office and transmit a Copy thereof together with this Warrant to the Surveyor General within the term of three Months from this date. Given under my hand as Senior Justice of the said Court this _____ day of _____ 178_____ [30]

Each county had an elected county surveyor who was empowered to employ deputies to assist him in his duties.[31] The applicant for land pre-

---

applicant's oath to writing, which the applicant signed before being issued a warrant. Two volumes of these signed oaths, along with transcripts of the warrants, are filed in the office of clerk of the superior court of Taliaferro County. In other counties the land court minutes bear merely a brief notation that the applicant took the prescribed oath.

30. 19 Col. Rec. Ga. (pt. 2) 207; Marbury & Crawford 325.

31. The office of county surveyor has become, in modern times, incompatible with the needs of society and, in fact, is perfectly antagonistic to the professional development of land surveying. This state of affairs came about, however, only upon the dissipation of the

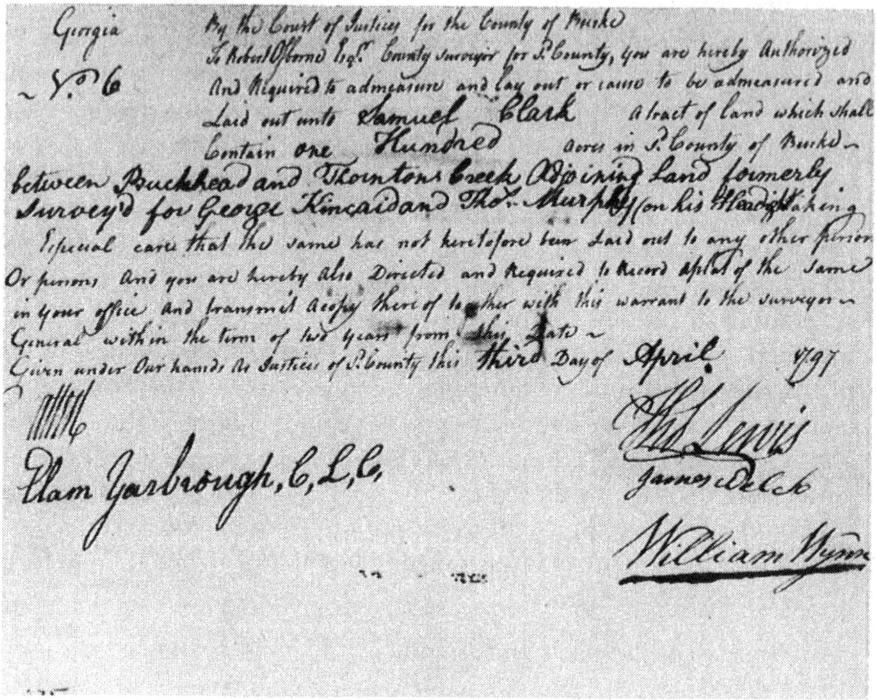
Georgia By the Court of Justices for the County of Burke
To Robert Osborne Esqr. County Surveyor for Sd. County, you are hereby Authorized
~ No. 6 And Required to admeasure and lay out or cause to be admeasured and
Laid out unto Samuel Clark A tract of land which shall
Contain one Hundred acres in Sd. County of Burke ~
between Buckhead and Thorntons Creek Adjoining land formerly
Surveyd for George Kincaid and Thos. Murphy (on his Headright) taking
Especial care that the same has not heretofore been laid out to any other person
Or persons And you are hereby also Directed and Required to Record a plat of the Same
in your office And transmit a Copy thereof together with this warrant to the surveyor ~
General within the term of two years from this date ~
Given under Our hands As Justices of Sd. County this third Day of April 1797
Elam Yarbrough, C.L.C.
Thos. Lewis
James [illegible]
William Wynn

Land warrant issued by the Burke County land court in 1797. Courtesy of the Georgia Department of Archives and History.

sented his warrant to the county surveyor or to a deputy who was responsible for making sure the land was vacant, notifying the adjoining landowners that a warrant had been taken out on the land, and surveying and making plats thereof. If two or more warrants were simultaneously presented for the same land, the oldest was given preference; other-

public domain, which rendered the office obsolete, coupled with the failure to legislate it out of existence. A very different perception of the office existed in the eighteenth and nineteenth centuries. Public lands could come under private ownership only after they had been demarcated into individual farms and grants issued. The pivotal role of the county surveyor in this process elevated his office to one of considerable prestige. Its importance is demonstrated by the fact that, at one point or another, practically every family living in the headright area of Georgia during the decades immediately following the Revolution found the services of the county surveyor indispensable. Despite isolated instances of incompetence and fraud on the part of some of the officeholders, generally only the most qualified men were selected for the position. It is unjust to judge the office of county surveyor of old in the light of the disrepute it has assumed in the modern era.

wise the first warrant to be presented to the county surveyor was given priority, even if it was a junior warrant.

The applicant usually accompanied the surveyor to show him the exact bounds to be laid out to supplement the vague description given in the warrant. The applicant was allowed great latitude in fixing his bounds, for the headright regulations passed after the revolutionary war (in contrast to those in force before and during the war) contained no stipulations as to the shapes of the tracts. In addition, the applicant was obliged to furnish chainmen, who usually consisted of the applicant himself and one other person—perhaps a close relative or neighbor. The high correlation on headright plats between the names of chainmen and the names of the applicants or owners of adjacent tracts, together with the fact that county surveyors and their deputies seldom used the same chainmen on consecutive jobs, confirms the almost universal practice of employing inexperienced laymen rather than a permanent crew for chaining purposes.

The county surveyor kept an office at the county seat and was required to record all plats in the office within two months of the date of the warrant and to advertise the survey for at least three months thereafter. In addition, he posted a bond of £2,000 (later set at $1,000) to assure honest and faithful work. The county surveyor was at first elected by the legislature, later by the justices of the peace in each county, and still later by the voters at large in each county.

Within three months after recording the plat in his office, the county surveyor transmitted a copy of it, along with the executed warrant, to the state surveyor general. When the grant fee and office costs were paid, the surveyor general made a copy of the plat in a record book kept in his office. He retained the original plat in his files and transmitted a copy of it to the secretary of state's office. This copy was attached to the grant, which was then signed by the governor. This done, the grant was returned to the secretary of state's office for registering and affixing the state seal. It was then transmitted back to the county surveyor to be recorded in the county records and delivered to the grantee. If the grant fee and office fees were not paid into the state treasury within one year of the date of the warrant, the grant would be deemed lapsed and could be issued to any qualified person who should apply for it.

Anyone objecting to the issuing of a grant could file a caveat with the county surveyor, who would give thirty days' public notice of the time and place for hearing the caveat. By the 1783 headright act, caveats were tried

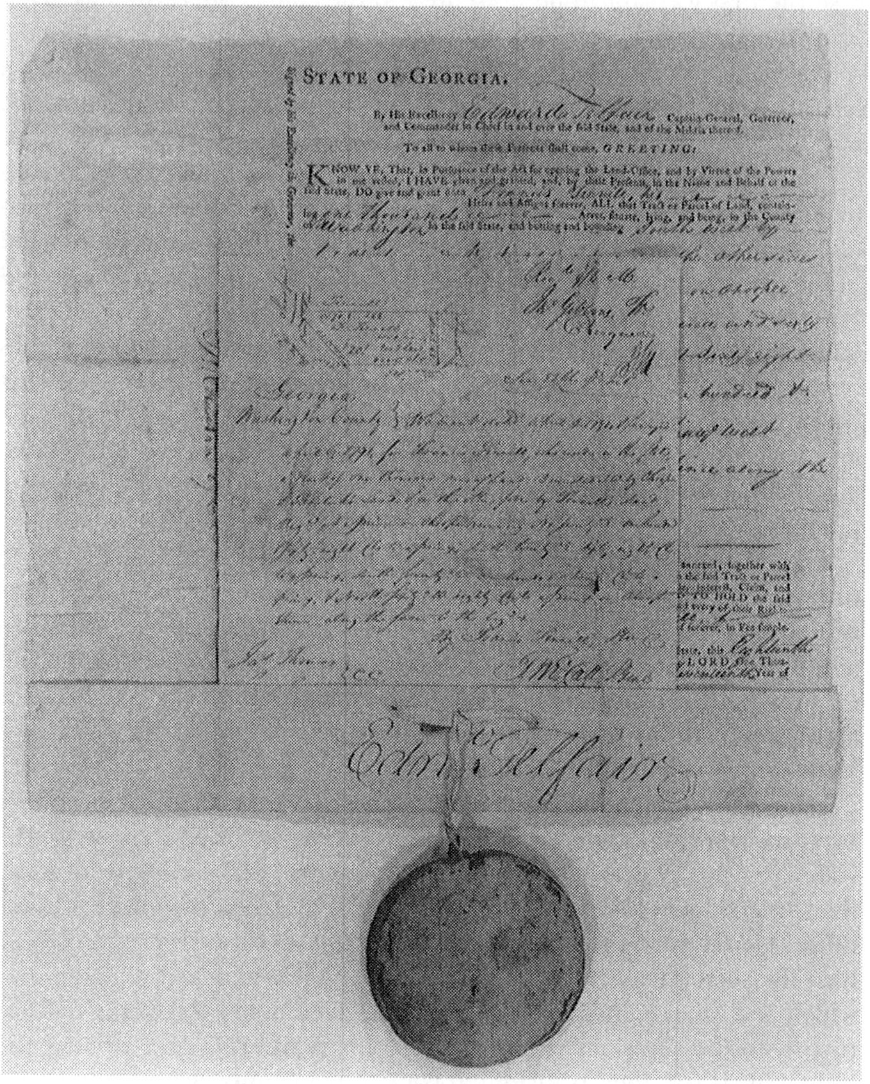

Headright grant signed by Governor Edward Telfair. Courtesy of Special Collections, University of Georgia Libraries.

by twelve freeholders, duly sworn, who were chosen by the land court. At first, the verdict was final, but soon an appeal to the governor and Executive Council was allowed (and to the governor alone when the Executive Council was abolished in 1789). In 1836 jurisdiction over appeals of caveats was transferred from the governor to the superior courts. This general procedure was invalidated by the supreme court in 1853, which held that, under the Constitution of 1798, the superior courts possessed *original* jurisdiction in all cases respecting title to land, including caveats.[32] Thereafter caveats were tried in the superior courts and appeals went to the supreme court itself.

When a caveat was filed, the signing and sealing of the grant were stayed until the determination of the caveat. If a caveat was overruled, the granting process proceeded.

All surveys that had been run out on lands not formally ceded by the Indians, all surveys that had been run across the lines of previous grants, and all surveys of land that had been previously granted were deemed void. In such cases, if the surveyor was negligent or dishonest, he was obliged to pay a fine, half of which went to the informer and half to the state.

In spite of the legal limitation of one thousand acres per person, a number of legitimate grants were made out in excess of this amount because the limitation as to quantity pertained only to warrants and not to grants. Some warrants were sold or transferred before the grant was taken out. A transferee might consolidate several warrants and obtain a single grant for all of them.[33]

In addition to headright grants, the state had at various times during the war offered bounty grants to those who had served honorably in the Georgia Line, the Minute Battalions, the militia, those inhabitants who enlisted in military units of other states, and those who could not "be convicted of plundering or distressing the country." The original acts and resolutions pertaining to bounties allotted to each rank of each class a specified number of acres ranging from two hundred acres for a private in the Georgia Line to seventeen hundred acres for a brigadier general in the militia. An act of February 25, 1784,[34] however, increased the amount by 15 percent for each class. The first warrants for bounties were dated

32. Robert v. Palmer, 14 Ga. 349 (1853).

33. Doe v. Roe, 31 Ga. 593 (1860); Patterson v. Winn, 24 U.S. (11 Wheat.) 380, 6 L. Ed. 500 (1826).

34. 19 Col. Rec. Ga. (pt. 2) 292; Marbury & Crawford 330.

May 15, 1784, and they all conformed to that later act, thus giving a private in the Georgia Line 230 acres and a brigadier general in the militia 1,955 acres. For several years after the war, bounties were also offered to those who enlisted in the militia to fight Indians.[35]

A person applying for a bounty grant would first obtain a voucher from his commanding officer or (if he were merely a qualified citizen who had not served in the military) from the commanding officer of the district in which he lived. The voucher attested to the person's services or status during the war. The person then presented the voucher to a land court, and the granting process proceeded in the same manner as that for obtaining headright grants. A person receiving a bounty grant was entitled to a headright grant as well.[36]

Several grants were made to individuals by special acts of the legislature for rendering exceptional assistance during the war. For example, the French Vice Admiral Count d'Estaing was granted twenty thousand acres along the North Fork of the Oconee River in Franklin County in 1784 for his services at the siege of Savannah in 1779.[37] In succeeding years land grants were also made to counties, schools, churches, religious bodies, orphanages, and to the University of Georgia by special acts of the legislature.

35. K. Coleman, The American Revolution in Georgia 1763–1789, at 187, 248–49 (1958).

36. 19 Col. Rec. Ga. (pt. 2) 303–4; 2 Rev. Rec. Ga. 587. Many misconceptions exist among patriotic societies and novice historians about bounty grants. It is widely assumed that all those who received such grants served honorably in the military. In actuality, two-thirds of the persons who obtained bounty grants fell in the class of those who could not "be convicted of plundering or distressing the country." The law was loosely interpreted as to what constituted "military service" and some persons who did receive grants based on so-called military service merely enlisted and never saw actual duty. Furthermore, many persons who secured vouchers and land warrants for military service sold the vouchers and warrants before taking out the grant—a perfectly legal practice. The transferees thus obtained the grants, which, nonetheless, were designated as "bounty" grants in the official records. Finally, a number of unqualified persons procured fraudulent vouchers from military officers for the purpose of obtaining illegal grants. See Hitz, *Georgia Bounty Land Grants,* 38 Ga. Hist. Q. 337 (1954).

37. 19 Col. Rec. Ga. (pt. 2) 448; 2 Rev. Rec. Ga. 639–40, 717, 795–96; CARTE d'une partie de l'Etat de GEORGIE Pour montrer la position géographique des Terres de M^{r} le Cte D'ESTAING (n.d.) (Papers relating to the sale of lands in Franklin County, Georgia, from the estate of Le Comte d'Estaing to Francis Louis Taney of Maryland, Rare Books and Manuscripts Division, New York Public Library, Astor, Lenox, and Tilden Foundations, New York, N.Y.).

Treaties at Augusta in 1783 and at New York in 1790

During the Revolution the Indians had generally sided with the British. Soon after the cessation of hostilities Colonel Elijah Clarke led his regiment and a motley assortment of former Tory soldiers who had deserted the British cause on a military campaign into the Cherokee Nation. As a result of this expedition, the chiefs, headmen, and warriors of the Cherokee Nation were forced to sign a treaty at Augusta on May 31, 1783, ceding a tract of land between the Tugaloo and the upper Oconee rivers, just to the northwest of the lands that had been organized as Wilkes County.[38]

Six months later, a treaty was signed between Georgia and a minority faction of the Lower Creeks. This treaty purported to cede a large strip of land between the Ogeechee and Oconee rivers as far southward as the Altamaha River and nearly as far northward as present-day Athens.[39] The majority faction of the Creeks, led by the half-breed Alexander McGillivray, disputed this treaty and also claimed title to the lands the Cherokees had ceded in the treaty of May 31, 1783. Complicating matters still further was the question of whether the state or the central government now had authority to make treaties with the Indians, since Georgia had entered into the Articles of Confederation and Perpetual Union along with the other states in 1781.[40] Despite Georgia's dubious claims to these lands, the northern cession was organized as Franklin County and the cession between the Ogeechee and the Oconee as Washington County by an act approved on February 25, 1784.[41]

When the United States Constitution was adopted in 1788, the federal government was given exclusive authority in Indian matters, including treaty making. At the cost of much money and many presents, McGillivray and his band of Creeks were persuaded to sign a treaty with the United States at New York (then the national capital) in 1790. Under this treaty they recognized the cession of the lands between the Ogeechee

38. Marbury & Crawford 603; H. McCall, The History of Georgia 545–48 (1909 reprint); 2 W. Stevens, A History of Georgia 411–15 (Philadelphia 1859).

39. Marbury & Crawford 605.

40. M. Heath, Constructive Liberalism: The Role of the State in Economic Development in Georgia to 1860, at 71–74 (1959); Wright, *Creek-American Treaty of 1790: Alexander McGillivray and the Diplomacy of the Old Southwest,* 51 Ga. Hist. Q. 379, 383–84 (1967).

41. 19 Col. Rec. Ga. (pt. 2) 292; Marbury & Crawford 330.

and the Oconee and relinquished their claims to the lands that had been ceded by the Cherokees in 1783.[42] Much to the (groundless) dismay of Georgians, this treaty implied that the United States would guarantee to the Creeks their remaining lands forever.

Georgia had authorized the survey of the boundary line of these latest cessions as early as 1783,[43] but it was not until 1798 that the lines were actually run and marked. The survey was conducted by a group of Creek Indians and white commissioners headed by Benjamin Hawkins, U.S. agent for Indian affairs south of the Ohio River.[44] The Treaty of 1790 specified that the artificial lines were to be marked by a line of felled trees twenty feet wide and the trees on each side chopped so that the boundary would be rendered distinct and well known. The lines were completed on February 1, 1798. Hawkins wrote the secretary of war: "I am happy in being able to assure you that there was no diversity of opinion among us, and that the line was closed in perfect harmony."[45]

42. Marbury & Crawford 622; C. Kappler, Indian Affairs (Treaties), S. Doc. No. 319, 58th Cong., 2d Sess. 25 (1904).

43. 19 Col. Rec. Ga. (pt. 2) 440; 3 Rev. Rec. Ga. 404, 565; Marbury & Crawford 258; K. Coleman, The American Revolution in Georgia 1763–1789, at 242 (1958).

44. The report of the commissioners appointed . . . to accompany Col. Benjamin Hawkins . . . for the purpose of running the Temporary Boundary Line Between the white Inhabitants of this State and the Creek Indians (Feb. 22, 1798) (Telamon Cuyler Collection, UGL); The Letters of Benjamin Hawkins, in 9 Coll. Ga. Hist. Soc. 1, 242–44, 286–87, 343–44 (O. Ashmore et al. eds. 1916); Patterson v. Jenks, 27 U.S. (2 Pet.) 216, 7 L. Ed. 402 (1829). The commissioners involved in surveys of boundaries between sovereign entities (such as surveys of Indian boundaries and state boundaries) were individuals or a group of men separate and distinct from the surveyors. Their purpose was to organize the expedition, to procure supplies, to supervise the surveyors, and to settle political issues and disputes arising during the course of the survey. The primary requirement for an appointment as a commissioner was political standing, and the appointees often had little technical knowledge of surveying. On some of the smaller surveys, such as those of Indian reserves, the same individual acted as both commissioner and surveyor.

45. The Letters of Benjamin Hawkins, in 9 Coll. Ga. Hist. Soc. 1, 287 (O. Ashmore et al. eds. 1916). Despite Hawkins's optimistic assessment of the outcome of the survey, portions of the line were subsequently found to have been run in a faulty manner. In May 1800 Levin Wailes filed a report declaring it "well known to all the Inhabitants on the Frontiers of Franklin & Jackson Counties" that the commissioners erred in their determination of the source of the Appalachee River, one of the terminal points of the survey. From this erroneous point, they ran their experimental line for several miles in a northeasterly direction until the next terminal point—Currahee Mountain—came into view. Finding that they were on a course that would take them far to the southeast of the mountain, the commissioners packed up their equipment and headed to the mountain's summit. From there they pursued a southwesterly course, on a bearing roughly computed from their partial traverse. This line, intended to intersect the starting point, veered so far south that it intersected

Opening of the Land Office for Franklin and Washington Counties

In spite of the chaos, immigration to Georgia had continued during the Revolution, and the pace greatly quickened upon the coming of peace. Granting lands and hearing caveats were to be the biggest items of executive business during the immediately succeeding years. Richard Call, a Virginian who had come to Georgia as a land speculator during the war, was appointed surveyor general by the Executive Council on October 1, 1782.[46]

Because no organized governments existed in Franklin and Washington counties when they were opened up for settlement in 1784 and there was a tremendous backlog in demand for land grants (held up by the war), it was necessary to set up a special land court to receive applications for land in those two counties. John Habersham, president of the Executive Council, and four of its members met in Augusta on April 6, 1784, to receive the petitions and to issue warrants. Public notification was given that petitions could be made from April 13 to April 27, on every day of the week except Thursday, Friday, and Sunday.[47] By late April the small river town on what was then Georgia's northwestern frontier had become overrun with recently discharged soldiers, expectant settlers, hopeful immigrants from other states, speculators, rough backwoodsmen, and an

---

the experimental line at a sharply oblique angle, about sixteen miles from the beginning point of the experimental line. The commissioners thereupon discontinued their work, determining that the lines thus marked should stand as the boundary. The result was that a considerable portion of territory ceded by the treaties of 1783 and 1790 was left on the Indian side of the line. See Report of Levin Wailes of the Boundary Line (May 26, 1800) (Franklin County Boundary Papers, Telamon Cuyler Collection, UGL).

46. 2 Rev. Rec. Ga. 381, 437–38, 459; 3 Rev. Rec. Ga. 245, 496. From 1777 to 1861 the surveyor general was elected by the Executive Council annually, then by the Senate annually, and finally by the General Assembly biennially. The Georgia Constitution of 1861 provided for a secretary of state, a surveyor general, a treasurer, and a comptroller general, but it contained the provision that the General Assembly could consolidate any two of these offices. The General Assembly thereupon, on November 22, 1861, passed an act consolidating the office of surveyor general with the office of secretary of state, and it has remained so ever since. 1861 Ga. Laws 72. Today the existing field notes, district maps, plats, copies of land grants, and other materials concerning the original surveys of the state's public domain are filed in the surveyor general records of the Georgia Department of Archives and History. Land surveyors should find it helpful to make a careful study of the records in this office that pertain to the localities wherein they work.

47. Gazette of the State of Georgia (Savannah), April 15, 1784, at 2, col. 1; 2 Rev. Rec. Ga. 787–99.

intermingling of Indians, all bent on grabbing up warrants for as much land as possible or at least sharing in the spectacle that was taking place.

To establish priority in issuing warrants, the names of those who had filed their petitions by April 27 were to be deposited in a box and an equal proportion of tags with numbers written on them in another box. A child was then to be assigned to each box—one to draw the names and the other to draw the numbers—and the warrants would be given out accordingly.[48] In addition, the two thousand or more warrants in five different classes had to be prepared and registered before they could be issued.

David Rees, the clerk of the land court, wrote an engrossing eyewitness account of the wild confusion and disorder that occurred at the meeting of the land court in Augusta.[49] In a communication to Governor John Houstoun, Rees noted that both the drawing of names and the delivering of warrants were scheduled to commence on May 17. Foreseeing the turmoil that would result from such a procedure, he advised that the delivering of warrants be postponed several days until the drawing had been completed and the warrants filled in. The land court failed to take action, however, and the "populace came down upon us . . . on the 17th . . . with eager expectations both of being spectators at the drawing of the Lottery, and immediately receiving their warrants." It was, of course, impossible to do all this at once so the mob was put off for a few days while the land court employed several extra clerks and worked daily from early morning until past midnight. By the twenty-second the patience of the crowd began to wear thin. A "few evil disposed persons" hinted to the people, "whose minds, by the Juice of the Cane were already inflamed," that there was "a more concise way" of obtaining warrants. The crowd "immediately herded" and "look'd formidable." Imagine, declared Rees, "the President & members of Council, four or five in number, convened on the business of their Countrymen—their Secretary or Clerk, with his assistants, in an adjoining apartment ready to faint with excessive heat & fatigue, not suffered to open the Door or Window, whereby a little fresh air might be admitted, and 1500 or perhaps 2000 men, set on by designing villains, approaching the doors with threats & menaces, crying

48. 2 Rev. Rec. Ga. 790.

49. Letter from David Rees to John Houstoun (June 2, 1784) (John Houstoun Papers, Georgia Historical Society, Savannah, Ga.). Portions of the letter are reprinted in K. Coleman, The American Revolution in Georgia 1763–1789, at 218–19 (1958) and in 2 W. Stevens, A History of Georgia 355–58 (Philadelphia 1859).

out aloud, that their warrants they would have at any rate." The council posted an urgent proclamation promising to have the warrants ready by May 24. This pledge seemed to appease the crowd and many went home, but a Creek chief called the Fat King was displeased. He, "being some what intoxicated . . . and making others so by lavish treats . . . kicked up another dust," pulled down the proclamation, and insisted that the warrants be given out the next day. The crowd, "heated with liquor and mad with disappointment," was now "noisy as ever, and ready for any mischief." In desperation, the president verbally promised to use every exertion to have the warrants ready by the twenty-fourth. This added assurance brought about a second calm.

By May 24 the claims for warrants had been laid up in numerical order and affairs were proceeding smoothly when someone in the crowd proposed that individuals seeking single warrants should go ahead of those seeking several warrants. Rees "dreaded the confusion" that would result from such a procedure and vigorously asserted his misgivings, to no avail. "This having taken place, the general vociferation was, why cant I get my warrant, as well as another" and "such a one has got his and by G——d I'll have mine." "My office was no longer one at my command," bewailed Rees. The "breach being made—the Torrent soon widened it—and he thought himself the happiest person" who could grab the most warrants. The alphabetical order was quickly obliterated and "no trace of regularity or decorum left." "They soon had the Warrants—from my Table—on the floor—at the door—and on the highway." About four or five hundred warrants disappeared in this outrage, and the land court had to issue duplicates. Orders were sent to the surveyors in Washington and Franklin counties not to survey any lands on warrants for which the duplicates had been issued. Nevertheless, a number of rioters managed to use both their original and duplicate warrants to obtain two surveys and grants. Many of the missing warrants were later returned "under the pretense of" having been "taken away by mistake." In several instances applicants who returned warrants swore that instead of, for example, 287½ acres of bounty land for which their warrants had been made, they were entitled to 575 acres. Others "on second thought" deposed that they had originally "petitioned or meant to petition" for land in the other county.

In concluding his report, Rees bitterly lamented to the governor: "Your love of integrity and justice, and your sublime Ideas of Honor . . . would be sadly chagrined to observe how familiar, and cheap Sacred Oaths

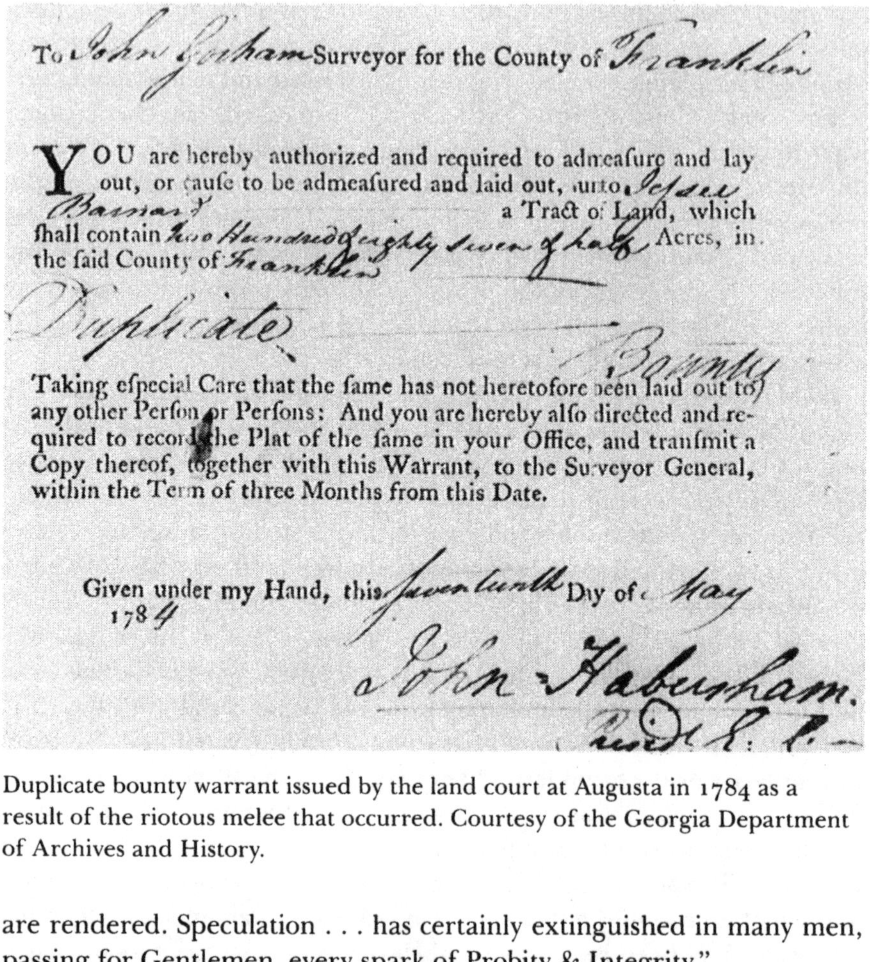

To John Gorham Surveyor for the County of Franklin

YOU are hereby authorized and required to admeaſure and lay out, or cauſe to be admeaſured and laid out, unto Jesse Barnard a Tract of Land, which ſhall contain two Hundred Eighty seven & half Acres, in the ſaid County of Franklin

Duplicate

Bounty

Taking eſpecial Care that the ſame has not heretofore been laid out to any other Perſon or Perſons: And you are hereby alſo directed and required to record the Plat of the ſame in your Office, and tranſmit a Copy thereof, together with this Warrant, to the Surveyor General, within the Term of three Months from this Date.

Given under my Hand, this seventeenth Day of May 1784

John Habersham.

Duplicate bounty warrant issued by the land court at Augusta in 1784 as a result of the riotous melee that occurred. Courtesy of the Georgia Department of Archives and History.

are rendered. Speculation . . . has certainly extinguished in many men, passing for Gentlemen, every spark of Probity & Integrity."

So great was the deluge of petitions for land that the governor and Executive Council were compelled to meet alternately at Augusta and Savannah throughout the summer and fall of 1784, and the surveyor general remained in Augusta until November.[50] Because of the great number of land applications, the act creating Franklin and Washington counties had provided that the county surveyor of each might divide his county into as many as six districts and for each district appoint a surveyor, who would make his returns to the county surveyor.[51] Nevertheless,

50. 2 Rev. Rec. Ga. 667–68, 679, 684, 716.

51. 19 Col. Rec. Ga. (pt. 2) 297.

Joseph Pannell, surveyor of Washington County, insisted that the six district surveyors he had appointed were inadequate for the task and that two additional ones were needed for each district; otherwise many of the warrants would lapse.[52] The Executive Council, although sympathetic to Pannell's plea, concluded that it was powerless to give him permission to override the express terms of the act. A month later, however, it relented and allowed him to appoint the additional surveyors.[53]

A valuable experiment in rectangular land subdivision, which may have set some precedent for the future, came about as a result of the act creating Franklin and Washington counties. Georgia was about to embark upon the ambitious project of building a state university.[54] Being money poor but land rich, the state ordered forty thousand acres of land of the first quality, in separate parcels of five thousand acres each, to be laid off for the endowment of "a College or Seminary of Learning"—the University of Georgia. Four of the tracts were to be located in Franklin County and four in Washington County.[55] Two weeks after the passage of the act, and a full month before the meeting of the land court in Augusta, the Executive Council determined that the legislature intended these surveys to take precedence over all others. Accordingly, eight warrants, each for five thousand acres, were immediately issued to the trustees of the university.[56] The tracts, in the shape of squares measuring 223.61 chains to a side, were laid off by district surveyors under the supervision of the county surveyors during the spring of 1784.[57]

Two years later, the trustees of the university directed surveyors to subdivide four of the parcels into square lots by "running six paralel lines through the whole tract, intersecting the same with six other paralel lines at right angles, that the whole may be divided as nearly as may be into forty nine tracts of" 102.04 acres each. An accurate plat of each 5,000-acre tract so subdivided was to be returned to the trustees within three months. Thus the subdivisions of these college lands bore strong super-

52. Memorial of Joseph Pannell to John Houstoun (n.d., but ca. April 1784) (Telamon Cuyler Collection, UGL).

53. 2 Rev. Rec. Ga. 637, 796–97.

54. Without completely itemizing sources, the following text concerning the lands with which the university was endowed is derived from Minutes of the Board of Trustees of the University of Georgia (Feb. 13, 1786, Nov. 29, 1799, Nov. 26, 1800, Feb. 10, 1816, & May 21, 1816) (Records Management & University Archives, UGL).

55. 19 Col. Rec. Ga. (pt. 2) 299–301, 363–71.

56. 2 Rev. Rec. Ga. 609–10; 3 Rev. Rec. Ga. 563–64.

57. Plat Book A, pages 151, 241–45 (GSGD).

ficial resemblance to the lottery surveys that were to be made by the state in the next century.

One of the quickest ways to make money in the eighteenth century was to boom a town. Ten of the 102.04-acre lots on one of the richest of the college surveys were set aside as a reserve for the town of Greensboro. To give this site added prominence and thus increase the sales potential of the lots, the legislature made Greensboro the county seat of the newly created county of Greene.[58]

The selling of lands—even for the benevolent purpose of financing a college—was next to impossible in a commonwealth where one could obtain up to a thousand acres free. The university trustees leased out many of the lots and, half a century later, were still struggling to sell some of them. In time they had two additional five-thousand-acre tracts laid out into grids.[59]

Because there was almost no public support for higher education during this period, the university board of trustees was barely able to maintain its existence. As a result, the two remaining five-thousand-acre tracts were frittered away through neglect. One lay between the north and south forks of the Oconee River. In the same act that had created the university, this especially fertile territory had been temporarily set aside as an exclusive reserve for bounty-seeking officers, soldiers, and seamen of the Revolution. The courts determined that all other grants in this reserve, including the university grant, were void.[60]

The remaining five-thousand-acre tract lay in the fork of the Keowee and Chattooga-Tugaloo rivers at the head of the Savannah River—an area claimed by Georgia as part of Franklin County but claimed by South Carolina as well. By the Beaufort Convention, signed in 1787, this territory was allotted to South Carolina. The convention, however, allowed any Georgia grants that had been made in the region to be confirmed by South Carolina if the claimants registered their grants within twelve months.

In due time the grant for the university tract was sent to the proper office in South Carolina, but for some unknown reason it was not recorded. A committee of the trustees of the university reported in 1799

58. 19 Col. Rec. Ga. (pt. 2) 484; Plat Book 1784–1861, p. 5, Probate Court, Greene County courthouse, Greensboro, Ga.; Deed Book JJ, 1824–1828, pp. 328–29, Office of the Clerk of Superior Court, Greene County courthouse, Greensboro, Ga.

59. Clayton 142; Lamar 1058, 1060, 1063.

60. 19 Col. Rec. Ga. (pt. 2) 299; Athenian (Athens), Dec. 15, 1829, at 4, cols. 1–2.

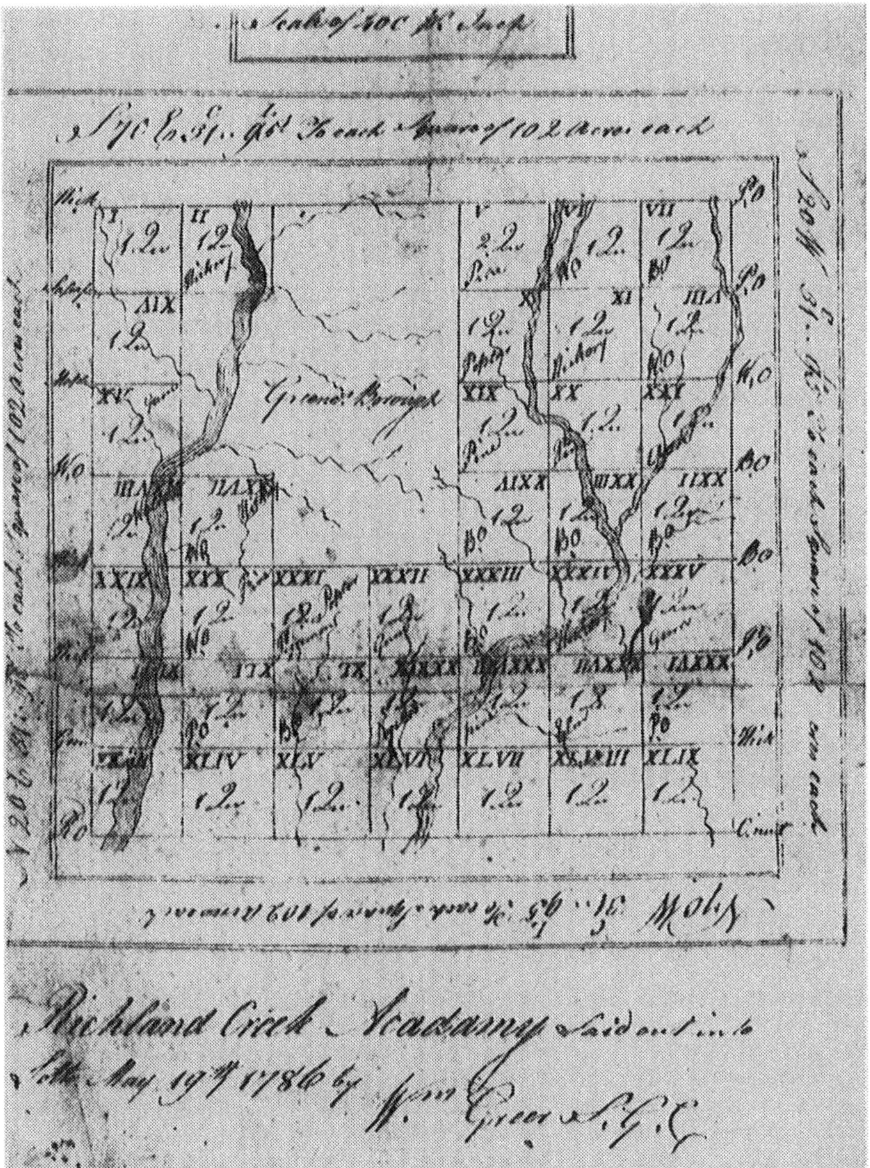

Plat of the Richland Creek Tract, one of the five-thousand-acre tracts the University of Georgia was endowed with, as subdivided into forty-nine lots. Ten of the lots were set aside as a reserve for the town of Greensboro. The tract was located in original Washington, later Greene County. From Plat Book, 1784–1861, p. 5, Probate Court, Greene County courthouse, Greensboro, Ga.

that this tract had been settled by South Carolinians and that it was of considerable value. Appeals were made to the South Carolina legislature and suits commenced in the United States circuit court for its recovery, but all were in vain. After eight or ten years the matter was dropped.[61]

Four-Mile Purchase

When Benjamin Hawkins belatedly ran the line of the Indian cession in 1798, several grants for lands supposedly in Franklin County actually proved to comprehend lands beyond the Indian boundary in Cherokee territory. Ordinarily this would have rendered the grants void, but the settlers in this case had obviously acted in good faith. Forcibly to evict them would result in the loss of the costly improvements they had made.[62]

Georgia's entreaties to the United States to endeavor to purchase the lands (called Wofford's Settlement after one of the principal inhabitants) resulted in a treaty signed at Tellico in Tennessee in October 1804. By its terms the United States agreed to pay the Cherokees $5,000 in goods and merchandise and a $1,000 annuity in return for a strip of land four miles wide, running along Hawkins's line and of sufficient length so as to enclose the lands of the encroaching settlers.[63] The boundaries of this cession, run out by James Blair of Franklin County and the Cherokee chiefs James Vann and Katahee, originally extended twenty-three miles sixty-four chains. Some years later, however, with the gratuitous consent of the Cherokees, the boundaries were enlarged by ten miles twelve chains to take in additional white encroachments. The Four-Mile Purchase (as it came to be known) thus enclosed a rectangular-shaped strip of land four

61. E. Coulter, College Life in the Old South 3–4, 21–23 (1979); A. Hull, A Historical Sketch of the University of Georgia 12–13 (1894). Oddly, no attempt was made to use any of the five-thousand-acre tracts for the site of the university itself. Instead, a location on the North Fork of the Oconee River, just twelve miles from Indian country in what was then Jackson County, was purchased from a private individual. The University of Georgia and the city of Athens have occupied the locale ever since.

62. U.S. Bureau of Ethnology, Smithsonian Institution, Fifth Annual Report of the Bureau of Ethnology 183–88 (1883–84); Flowers, *The Wofford Settlement on the Georgia Frontier*, 61 Ga. Hist. Q. 258 (1977); Georgia Department of Archives and History, *Wofford's Station*, Georgia Magazine, Apr.–May 1968, at 24.

63. Am. St. Papers, 2 Indian Affairs 506; C. Kappler, Indian Affairs (Treaties), S. Doc. No. 319, 58th Cong., 2d Sess. 73 (1904).

miles wide and thirty-three miles seventy-six chains long, located largely in what is now Banks and Hall counties.[64]

Strangely, the Tellico treaty was misplaced or lost while in transit to Washington. Not until early in 1824, nearly twenty years after its signing, was attention called to it by a Cherokee delegation visiting Washington. The United States Senate immediately provided for ratification.[65] The Four-Mile Purchase was attached to Franklin and Jackson counties and surveyed under the headright system.[66]

In 1812 Georgia unilaterally annexed the lands between the southwestern portion of the Four-Mile Purchase and the ridge dividing the waters of the Oconee and Chattahoochee rivers, alleging that this area had been ceded to it by the terms of the 1783 treaty with the Cherokees.[67] This accession completed the state's acquisitions of its territory to be surveyed under the headright system.

Pine Barrens Fraud

Land speculation and land fraud are about as American as baseball and apple pie. Under the administration of a weak, debt-ridden state government, Georgia's loosely supervised public domain became the target of one of the largest blue-sky swindle operations in American history—an enigmatic scheme known as the Pine Barrens Fraud.[68]

64. 1 The Papers of Chief John Ross 68, 151–52 (G. Moulton ed. 1985); Flowers, *The Wofford Settlement on the Georgia Frontier,* 61 Ga. Hist. Q. 258, 262–65 (1977).

65. Foreman, *The Lost Cherokee Treaty,* 33 Chronicles of Oklahoma 238 (1955).

66. Clayton 330, 398, 476, 585.

67. Lamar 194.

68. The Pine Barrens Fraud constitutes one of the great uninvestigated areas of Georgia history. Only two researchers have made serious attempts to explain it: former Secretary of State Samuel G. McLendon in his *History of the Public Domain of Georgia* and Absalom H. Chappell in his *Miscellanies of Georgia.* Other narratives are derived almost exclusively from these two sources. The explanations by both McLendon and Chappell leave much to be desired. McLendon used the novel and painstaking approach of carefully compiling and analyzing thousands of land grants from the period filed in his office but overlooked other germane sources. As a result, he drew several erroneous conclusions, including his continual assertion that all grants in excess of a thousand acres were illegal, that there were no protests against the fraud while it was being committed, and that no surveys were made by the perpetrators. Chappell, an eminent attorney who lived during the first three-quarters of the nineteenth century, composed his brief narrative primarily from the recollections

Speculation has always been more or less endemic to public lands. Georgia presented an especially inviting field for speculation and fraud following the Revolution because the central authority for surveying and granting that had existed in colonial times under the governor and council and the surveyor general had been swept away; the requirements for improving property as a requisite for granting had been dropped; there had been a massive influx of immigrants into a vast, sparsely settled wilderness; and constraints for obtaining grants had been loosened.

In the 1780s the practice of buying and selling land warrants became widespread. Nothing in the law prohibited such trafficking; it had gone on to some extent even in colonial times. Although warrants were limited to a thousand acres, a person might buy up several warrants, present them to the county surveyor, and, by following the procedures set down in the law, acquire a single grant in his own name for thousands of acres of land. Thus warrants came to be traded like commodities. The brisk demand for them was partially filled by their procurement through the use of bribes and manipulative schemes in the land courts. The way was considerably smoothed for such activities by the growing practice of acquiring land for speculative purposes. A distinct land boom was under way, and practically everyone was indulging in it to some degree.[69]

Left to interpret the law on their own, the land courts adopted the loose practice of renewing transferred and lapsed warrants and did so even for warrants issued by land courts in other counties. Hence a person might buy up several warrants made out to various persons in various counties, present the mass of them to the land court of a particular county, and obtain a single new warrant for the total acreage represented by the

---

of persons who had been witnesses to the fraud and perhaps also from personal experiences gained through litigation and other legal problems resulting from it. His explanation, though generally accurate (and delightfully written), is lacking in depth.

The subject admittedly is difficult to piece together. In contrast to the Yazoo Fraud, for example, no comprehensive contemporaneous investigation of the Pine Barrens Fraud took place. Most of the transactions of the land courts (where the frauds originated) were oral. Thus other than the huge profusion of grants, few contemporary records exist to explain the phenomenon. Some historians have given up on the Pine Barrens Fraud as hopeless to unravel. Such an attitude is not justified. It can be satisfactorily explained by a careful study of the available records.

The present narrative, derived largely from previously untapped sources, attempts to shed new light and perhaps will inspire further investigations.

69. M. Heath, Constructive Liberalism: The Role of the State in Economic Development in Georgia to 1860, at 99 (1954).

several old warrants (some or all of which may have expired in the meantime). This became such an accepted practice that land courts routinely issued new warrants merely on the face of old ones, raising few questions about the legitimacy of the old warrants. An individual desiring land in a county whose land court refused him a warrant might therefore obtain—by whatever means possible—a warrant or warrants in other counties, have them renewed in the county where he desired land, and thereby secure a grant in that county. By such devices, grants to individuals totaling 13,700, 14,300, 15,600, 15,800, 20,000, 21,000, 32,000, and 50,000 acres were made in 1787 during the first administration of Governor George Mathews.[70]

These practices continued under the succeeding administration of Governor George Handley. In 1788, in an attempt to stem the rising tide of malfeasance, Handley issued an executive order directing the surveyor general not to pass grants on warrants in excess of one thousand acres.[71] The land courts were notified of this order, but a simple expedient was quickly devised to circumvent it. One merely had a series of warrants made out, each for a thousand acres or less; had the county surveyor lay the tracts off into separate (even contiguous) blocks of a thousand acres or less; and then acquired a separate grant for each tract.

For five years after Governor Handley's order, no significant action was taken to check the various artifices and more or less corrupt methods that were often employed to secure warrants. As a prologue for what was to come, during these years grants ranging from 10,000 to 267,000 acres—all in allotments of 1,000 acres each—were made to Francis Tennille, Robert Middleton, John Gardiner, Joseph Ryan, and about twenty-two other individuals.[72] Statehouse officers received fees for signing grants in direct proportion to the number and size of grants, which probably abetted an official attitude toward sanctioning such illegal grants.[73]

The stage was now set for organized corruption on a grand scale. In the early 1790s a company of unprincipled land speculators, who had previously been operating in South Carolina, came to Georgia and began collaborating with some of the state's citizens. Many of the Georgia collaborators sat on justices courts or were county surveyors and were there-

70. Id.

71. Executive Council Minutes (June 4, 1788) (GDAH); S. McLendon, History of the Public Domain of Georgia 172–73 (1924).

72. Plat Books O and P (GSGD).

73. Marbury & Crawford 226.

fore familiar with the various ways to manipulate the granting process. Included in this nefarious combination were Francis Tennille, Thomas Davis, Abner Davis, James Shorter, Richmond Dawson, Thomas Dawson, James Dawson, Joseph Cooper, Thomas Cooper, William McKissock, Mack McKissock, Robert Commens, Eleazer Commens, Benjamin Kitchen, Micajah Vasser, Patrick Crookshank, and Joseph Ryan. The plot this consortium concocted was to fabricate warrants in their own names for all the vacant lands in the state that had been ceded by the Indians. They proposed then to have these lands run out and granted to themselves.[74] To do so they conspired to gain control of the land courts and then to elect their own members as county surveyors, who at that time were chosen by the land courts.

Although few counties were left untouched by this scheme, its perpetrators found the remote, sparsely populated areas most amenable to their designs. There, among ignorant, unsuspecting settlers scattered thinly over an immense wilderness, they could rig justices' elections with comparative ease. In addition, relatively few grants existed in such areas to interfere with the large surveys that were made.[75]

Montgomery County, which was cut out of the lower part of Washington County in December 1793, became the arena for the conspirators' most prolific activity. Montgomery County was primarily a vast pine barren solitude, scarcely settled at the time.[76] The fact that most of the frauds were perpetrated in such regions accounts for their appellation. Extensive machinations were also conducted in Franklin, Washington, Greene, Wilkes, Richmond, Effingham, Liberty, McIntosh, Glynn, and Camden counties.[77]

The conspirators commenced operations on a large scale about May 1793 and continued until October or November 1794. Warrants and plats, representing in aggregate acreages far exceeding the actual exist-

74. W. Primrose, Report of William Primrose Explanatory of Washington County, Georgia 5–6 (n.p. 1838) (pamphlet filed in the Papers of the North American Land Company, University of Virginia, Charlottesville, Va.).

75. A. Chappell, Miscellanies of Georgia (pt. 2) 46 (1874).

76. Though no evidence exists to support the theory, it is possible that the act to create Montgomery County was introduced and ramrodded through the legislature by the conspirators, their intent being to create a sinecure for their ambitions. The legislative journals for 1793, except for some excerpts printed in contemporary issues of the *Augusta Chronicle*, have been lost.

77. S. McLendon, History of the Public Domain of Georgia 46–54, 58 (1924).

ing area of the regions, were churned out at an astounding pace and sent to the surveyor general for grants. They were all made in the amount of one thousand acres each to comply with the policy of the executive department not to pass grants in excess of that amount. Huge chunks of territory were sliced up into square tracts—the easiest method of layout. In many instances the conspirators laid down these large surveys into one connected plan, indifferent to the smaller grants that had been made to others within the lines. Having no thought of settling the lands themselves but desiring merely to secure grants to sell to others, they performed the surveys in great haste and with considerable inaccuracy. An investigator in the 1830s reported that the marks made by these surveyors that still existed were distinguishable from others in that they were located considerably higher up on the trees—indicating that they had been made by a person on horseback.[78] When resurveys were attempted in 1799 of four large tracts in Montgomery County originally granted to Thomas Dawson in 1794, it was found that only the outer boundaries had been properly marked on the ground. The interior lines had been marked very poorly or not at all. According to the original surveys, three of the four tracts consisted of forty-five contiguous grants of 1,000 acres each and the fourth consisted of sixty-four contiguous grants of 1,000 acres each. The resurvey revealed, however, that the three "45,000"-acre tracts contained, respectively, only 17,578 acres, 17,578 acres, and 22,281¼ acres; while the "64,000"-acre tract contained only 27,224 acres.[79]

It soon became evident that the actual surveying of the lands on the ground was a useless formality. Indeed, it became impossible for the surveyors and deputies to keep pace with the tremendous deluge of warrants being turned out. Surveying was eventually dispensed with altogether.

An examination of the records reveals that Richmond Dawson and James Shorter "received" respectively 1,412,000 and 1,219,000 acres. Both were county surveyors. They made the surveys for each other or directly for themselves. As proof that most of the surveys were fictitious, the records indicate that Shorter "surveyed," for himself, 52,000 acres in Montgomery County on August 29, 1793. Thomas Cooper "surveyed"

78. W. Primrose, Report of William Primrose Explanatory of Washington County, Georgia 6–9 (n.p. 1838) (pamphlet filed in the Papers of the North American Land Company, University of Virginia, Charlottesville, Va.).

79. Montgomery County Deed Book F, 1799–1801, pp. 14–16 (GDAH).

100,000 acres on August 6, 1793, and 118,000 acres on August 18, 1793. He surveyed both allotments for James Shorter and "received" a total of 277,000 acres himself.[80]

In addition to the grants made to Dawson, Shorter, and Cooper, nineteen other persons each received grants ranging from 250,000 to 500,000 acres, thirty-three persons each received 100,000 to 250,000 acres, twenty-eight persons each received 50,000 to 100,000 acres, and forty persons each received 10,000 to 50,000 acres.[81]

An extraordinary feature of these grants (thousands of which still exist) is that the names of the grantees, surveyors, and chain carriers, the date of warrant, date of survey, date advertised, and other information are *printed* on their faces. Furthermore, the exact same information is repeated through hundreds of consecutive grants. Thus the remarkable process of running the grants and plats off on printing presses was employed to obviate the tedious job of filling them in by hand. To enhance their sales potential, corner and station trees on many of the plats were denominated as oak, hickory, walnut, dogwood, and buckeye—trees indicative of rich soil. In many cases the lands were designated as "good rice & indigo land"—in regions of sterile sand where only pine trees and wiregrass had ever grown.

The activity of the surveyor general's office increased so much that, in the two years of Governor Mathews's second term, over twenty volumes containing more than nine hundred grants per volume were registered by the secretary of state. In some of these volumes upwards of two hundred consecutive pages contain grants to one individual. The plat books are also full of curious matter. In one (Plat Book AA), for example, the first 1,196 plats are made out to James Shorter in the amount of a thousand acres per plat. All of Shorter's plats are dated between August 4 and September 6, 1794. In some of the other plat books, two hundred to four hundred consecutive plats, each of a thousand acres, are made out to the same person. Most of these plats are certified by only eighteen or so individuals, all of whom were grantees of the huge tracts. In numerous instances the grantee certified his own plats.

The Executive Minutes show that the governor daily signed away grants —both fraudulent and legitimate—to hundreds of thousands of acres. For example, on January 3, 1795, the same day the Yazoo Act passed in

80. S. McLendon, History of the Public Domain of Georgia 46–52, 58 (1924).
81. Id. 46–52.

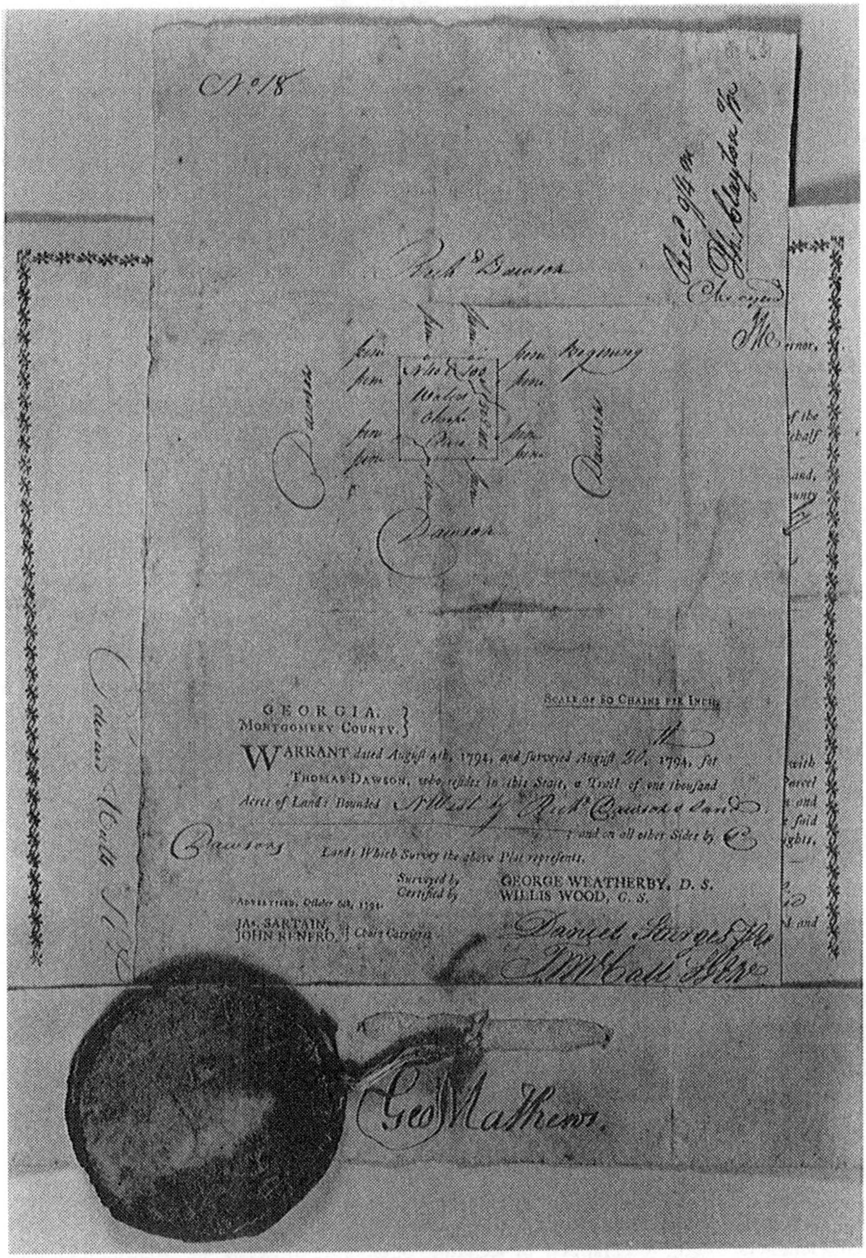

Pine Barrens Fraud grant for one thousand acres in Montgomery County, issued to Thomas Dawson in 1794 and signed by Governor George Mathews. Much of the information that normally would have been filled in by hand is, instead, printed. Courtesy of the Georgia Historical Society (Thomas Dawson Papers, Collection 203, Box 3, Folder 9, Item 49).

the legislature, Governor Mathews signed grants to over 450,000 acres in Montgomery County alone.[82]

In a memorandum to the governor dated March 23, 1795, the surveyor general made the belated and somewhat simple-minded observation that he was "induced to believe that the quantity of Land" that had passed his office for Montgomery, Glynn, and Camden counties exceeded "the quantity contained in . . . those Counties; yet, plats for Extensive Bodies of land" which were "on land probably already Granted" were being brought forth almost daily to be signed into grants.[83]

A report issued by the surveyor general in 1839 shows that the total area of the headright counties of Georgia was approximately 11 million acres. Nonetheless, according to the records in his office, a total of 29,912,793 acres had been granted by the state in those counties.[84]

It must not be supposed, as has been generally asserted, that the inhabitants of the affected counties stood silently by while the malfeasors carried out their scheme. Petitions and affidavits began pouring into the executive department almost from the moment the operation commenced. These quaint, handwritten documents, with scores of signatures attached, constitute, after the ravages and suppressions of two centuries, practically all that will ever be known about the inner workings of this plot. They set forth an intriguing tale.

In one of the first of many petitions to be filed, the inhabitants of Effingham County angrily charged that at the regular meeting of their land court in May 1793 Joseph Ryan, "who was concerned with many others," received warrants for three hundred thousand acres "though opposed by some of" the honest justices present. And in spite of the policy of the Effingham County land court to meet only once every three months, John McCall, the sheriff, who was acting as clerk of the court, set up a meeting for the following month at Indian Bluff.[85] He advertised

82. Id. 52–54.

83. Letter from Thomas McCall to George Mathews (Mar. 23, 1795) (GSGD).

84. Georgia Commissioners on State Finances, Report of the Commissioners Appointed by Authority of the Legislature, on the Subject of the State Finances (pt. 2) 54–55, 63 (Milledgeville, Ga. 1839); A. Chappell, Miscellanies of Georgia (pt. 2) 52–53 (1874). McLendon in *History of the Public Domain of Georgia,* p. 58, made a somewhat faulty analysis from the report by failing to include in his total existing area of the headright counties the areas of counties that had been cut out of older counties after 1796.

85. In the early 1790s Effingham County consisted of all of what is now Effingham and Bulloch counties and parts of Screven, Emanuel, Candler, and Bryan counties. Indian Bluff

this meeting very discreetly. Few people heard of it, and few attended. This gave "Ryan, Flournoy, and others, all non Inhabitants of" the county "and allmost strangers to every man in" the county the opportunity to obtain nearly a million acres of land "without any questions asked, or warrants proved." The few inhabitants who were at the meeting received warrants. When they applied to have the warrants run out, however, the county surveyor informed them that for him to do so they would have to tag along while he was running the "great surveys" so as to run theirs when they happened to reach the locations they desired. This, the petition protested, imposed a severe hardship on the citizens, for they "were busy in their crops, and on publick duty after the Indians." The petition revealed that one of the justices had "confess'd they were hired to sit" at the land court and had "boasted of cash received" for his services. The petitioners had learned that the surveyors were running the lines of these large surveys through some of their "Pattant Lands." In "order to prevent such unwarrantable proceedings," some of the petitioners had assembled and followed the surveyors, intercepting and breaking "the chains & compasses of two of them."[86]

Accompanying the petition were several affidavits. In one, a witness who had been present at the land court at Indian Bluff alleged that "Mr. Ryan Petitioned said Court for a quantity of Land . . . founded . . . upon old warrants from Washington County, the validity of which were not examined by" the court. The informant further testified that he saw the four court justices "sign Blank Warrants by quiers for several hours."[87]

In another affidavit a witness alleged that Thomas Wylly and Jesse Bell showed up together at the Indian Bluff meeting and had a secretive conversation with John McCall. Wylly then told McCall to "pick his Court." McCall replied that he had already done so. The court thus assembled consisted of Wylly, Bell, McCall, and one other individual. John Moore, the county surveyor, who was acting as court clerk, produced a petition of Robert Flournoy, praying for the renewal of two hundred warrants

---

was located on the west bank of the Ogeechee River about where the present Effingham-Screven boundary strikes that river. At the time of the Pine Barrens Fraud, Indian Bluff was near the center of Effingham County.

86. Petition of the Inhabitants of Effingham County to Edward Telfair (n.d., but ca. July 1793) (Telamon Cuyler Collection, UGL).

87. Affidavit of Gross Scruggs sworn before Paul Bevill (June 27, 1793) (Telamon Cuyler Collection, UGL).

of one thousand acres each.[88] Moore later "put his hand into his pocket from whence he drew a hand full of Dollars, and put them into Wyllys pocket."[89]

In still another affidavit William Sheppard testified that several days after the land court at Indian Bluff, he casually mentioned to John McCall that he planned to renew a warrant at the next land court. McCall, "without any hesitation," informed Sheppard that he could have it done right away. Immediately McCall took out a ready-made warrant, filled in the blanks, and handed it to Sheppard. Upon examining the warrant, Sheppard was surprised to find that it bore the signatures of three justices (even though no one else was present) and was dated the same day as the land court meeting at Indian Bluff. Sheppard surmised that McCall was merely attempting to relieve him of the necessity of attending the next land court to prevent him from observing the surreptitious dealings that would take place.[90]

In yet another affidavit a deponent reported taking lodging one night with several strangers when, "in the rounds of Conversation, the subject of those large surveys" came up. One of the parties present declared "that a certain Jackson—a deputy surveyor . . . of Washington County, had counterfeited Ben Tennells, Clerk of the land courts hand and made warrants, to the amount of several hundred thousand acres which he sold for half dollar per Thousand." Further, it was believed that "large Quantities of those warrants were purchased for the purpose of being laid in" Effingham County.[91]

An emotional petition from an aroused citizenry of Liberty County charged that on July 1, 1793, "at the instance of a few Individuals," a land court was convened at Sunbury attended by eleven justices. Then, "to the concern of all present, but those interested in the Speculation," three of the justices renewed land warrants from other counties "to the enormous amount of Eight Hundred thousand acres of Land." The other eight justices viewed "the Speculation as destructive & not warranted by

88. Affidavit of John London sworn before John King (June 29, 1793) (Telamon Cuyler Collection, UGL).

89. Affidavit of H. Gindrat sworn before Charles Treutlen (July 10, 1793) (Telamon Cuyler Collection, UGL).

90. Affidavit of William Sheppard sworn before John London (June 29, 1793) (Telamon Cuyler Collection, UGL).

91. Affidavit of Robert Scott sworn before Nathaniel Lunday (July 3, 1793) (Telamon Cuyler Collection, UGL).

Law." They, being a majority, "declared against such renewals, which by Law was final." Notwithstanding, the three justices continued to hold the court "and did renew & sign the aforementioned warrants together with an additional number to the amount of nearly one million . . . acres," which would "include every acre of unlocated land in the County." The petition stated that almost the entire western portion of Liberty County consisted of ungranted lands, which, before this alarming proceeding, the subscribers had confidently anticipated would be "settled by poor industrious people who would not only have cultivated the Ground, but have protected the interior settlements . . . from the future ravages of the Indians." Yet the recipients of these large warrants were "residents of other Counties & come into this for no other purpose then private interest, and leave it as soon as that interest is accomplished." The petition noted that the chief justification proffered by the speculators for these large grants was the fees and taxes the state would collect. Declaring that people were the real wealth and strength of a country, it queried whether such fees and taxes were worth "the risque of breaking up one of" the "most valuable counties" in the state by giving preference to those few individuals who would monopolize the land and thwart settlement, leaving the frontiers "exposed to . . . a savage enemy" for many years to come.[92]

In its presentments for the July term in 1794, the Liberty County grand jury took the highly unusual step of citing public officials of a neighboring county. James Spalding, Samuel Wright, Raymond Demere, and William Limbert, justices of the peace for Glynn County, were charged with "suffering themselves at a land court" the preceding June "to be corrupted and bribed, contrary to their Oaths and duty as Magistrates, in shamefully renewing land warrants many of which" were "believed to be counterfeited to the amazing amount of Sixteen hundred thousand acres." Roswell King, county surveyor of Glynn, was charged with "receiving a bribe to grant . . . deputations to certain persons who had warrants renewed at the last land Court" in his county. The citizens of Glynn County were implored to "take immediate steps to obtain a suspension of their Corrupt Justices and county Surveyor" and to "bring to exemplary punishment the perpetrators of such illegal practices."[93]

92. Petition of the Inhabitants of Liberty County to Edward Telfair (July 6, 1793) (Telamon Cuyler Collection, UGL).

93. Minutes of the Superior Court 1794–1800, p. 76, Office of the Clerk of Superior Court,

On September 1, 1794, a delegation of Montgomery County citizens addressed a passionate petition to the governor complaining that the "hardships Injuries & insults" they had undergone were so great that they could endure them no longer. While still a part of Washington County, some of their lands had been run by speculators "no less than Seven Times." As soon as their county had been formed, these same speculators took "a Delluge of Land Warrants from other Counties amounting to the enormous Some of nearly three Millions of acres of Land," which, in the opinion of the petitioners, were "mostly counterfeit." "Part" of their land court then passed and renewed these warrants. Afterward many of the "Surreptitious Speculators" were appointed deputy surveyors and ran through their "plantations & Lands & Choped & hacked . . . Trees in so many places" that it was "now difficult, & in a few years" it would "be impossible" for them to find their "ancient boundries." Because of "the Land being thus fraudulently taken" from them, others were discouraged from settling in their region to "assist . . . in curbing the ambitious Ferocity, Cruelty & inhumanity of the Savagees." They were thus "left almost alone, being but few in number to defend" their "little Quantity of Land" and could "have no assistance" from those outsiders who pretended "to own 9 tenths" of their county. To prevent "troublesome, perplexing & vexatious Law suits" and "every Species of misery & misfortune," the petition begged the governor not to sign the grants for "these large Surveys" which were dated since May 1, 1794, until the next meeting of the General Assembly, so that they could have an opportunity to lay their grievances before that body.[94]

In July 1793 the governor ordered the surveyor general to suspend the passing of grants based on suspected illegal warrants and surveys for Effingham and Liberty counties until further orders or until legislative action should be taken.[95] When the legislature met the following November, steps were taken to impeach several of the corrupt county officials.[96] Incomplete journals make it impossible to determine the exact outcome

---

Liberty County courthouse, Hinesville, Ga.; Georgia Gazette (Savannah), Aug. 7, 1794, at 2, cols. 1–2.

94. Petition of the Inhabitants of Montgomery County to George Mathews (Sept. 1, 1794) (Keith Read Collection, UGL).

95. Executive Minutes (July 13, 15, & Sept. 23, 1793) (GDAH).

96. Augusta Chronicle & Gazette of the State, Nov. 23, 1793, at 3, col. 2; Id., Dec. 7, 1793, at 2, cols. 1–3; Id., Dec. 21, 1793, at 1, cols. 1–2.

of these actions, but apparently they were futile. The legislature also directed Governor Mathews to investigate the matter. In an opinion solicited by Mathews regarding how to conduct this investigation, however, the attorney general declared that land courts determined the legality of warrants and surveys and that, under the land acts, the governor was powerless to make inquiries except when proceedings came before him on an appeal of a caveat. Furthermore, it was the opinion of the attorney general that, except under such circumstances, the governor was obligated to sign all grants presented to him without questioning their legal status.[97] Numerous caveats were being filed against the grantees of the huge tracts, but the land courts either declined to hear them or found in favor of the defendants. Appeals were thwarted by the county surveyors, who refused to send the proceedings to the governor.[98]

At its session beginning in November 1794, the legislature (the same one that was to pass the Yazoo Act) passed an act empowering the governor to stay all transactions involving suspected illegal surveys and to require the parties thereto to appear before him in the executive chamber for questioning. The act also prohibited county surveyors from laying out lands under warrants that did not appear to be founded on the laws of the state and prohibited land courts from renewing transferred warrants.[99] Three days later, another act was passed declaring void all surveys and grants made contrary to the laws and welfare of the state.[100] Thus was presented the anomalous spectacle of the same legislature suppressing one type of land fraud and ushering in another (see next section).

By this time, however, the initial phase of the Pine Barrens Fraud had run its course. Since the nature of the plot had become general knowledge within Georgia, it was no longer feasible to sell the grants there. The perpetrators therefore departed for distant points to the north, to Europe, and elsewhere to unload their ill-gotten gains on innocent, unsuspecting victims.

A second phase of the Pine Barrens Fraud was carried out in 1795 and 1796 by a group composed partly of some of the instigators of the first phase but primarily of persons who had acted as chain carriers in the first phase. This second phase, conducted on a much smaller scale than

97. Executive Minutes (May 20 & 22, 1794) (GDAH).

98. Petition of Samuel Fulton to George Mathews (Sept. 1, 1794) (Telamon Cuyler Collection, UGL).

99. Marbury & Crawford 339.

100. Id. 266.

the first, was apparently purely a paper transaction, for it elicited little contemporary comment.[101]

In the meantime, the local populace of the counties had hit upon an effective method for combating the fraud. During this era the taxes on small farms were nominal, but the taxes on hundreds of thousands of acres—especially non-income-producing acres—could be devastating. A speculator was reluctant to pay taxes on property to which his title was doubtful and even more reluctant to pay taxes on property he knew did not exist. Legal sections of newspapers of the 1790s are full of notices of the huge tracts being put up for sale for nonpayment of taxes.[102] Thus the grantees were forced into the position of either paying the yearly taxes or seeing their property sold at auction, and most chose the latter. Since few or no bids were proffered at these auctions (everyone was well aware of the nature of the grants), the counties gradually wound up the purchasers of most of the tracts. The effect, therefore, was to accomplish what an impotent state government was unable to do: nullify the grants at the cost of canceling the accrued taxes owed on them. Grants for non-existent lands were, of course, nullified without the loss of a single penny in taxes.

An act of February 1796 provided for popular elections of county surveyors, thus freeing them from the control of the land court justices.[103]

All this was now of little consequence to the speculators. They had their paper grants to sell to others. The tax sales did, however, relieve the lands, to a marked degree, of numerous title complications that would otherwise have resulted. The large grants finally ceased to be issued when James Jackson (of Yazoo Fraud fame) took office as governor in January 1798.

In the years that followed, surveyors generally were able to recognize easily the lines marked as a result of the Pine Barrens Fraud and for the

101. W. Primrose, Report of William Primrose, Explanatory of Washington County, Georgia 6–7 (n.p. 1838) (pamphlet filed in the Papers of the North American Land Company, University of Virginia, Charlottesville, Va.).

102. See, for example, Augusta Chronicle & Gazette of the State, Nov. 2, 1793, at 2, col. 2; Id. Nov. 16, 1793, at 4, col. 1; Id. July 26, 1794, at 5, cols. 1–3; Id. July 11, 1795, at 3, col. 2; Id. Oct. 17, 1795, at 3, col. 3; Id. April 15, 1797, at 3, col. 3; Id. Sept. 2, 1797, at 3, col. 3; Id. April 7, 1798, at 3, col. 4; Id. Dec. 15, 1798, at 3, col. 3; Id. May 18, 1799, at 4, cols. 2–4; Columbian Museum & Savannah Advertiser, Mar. 13, 1798, at 2, col. 1; Id. May 31, 1799, at 2, col. 2; Georgia Gazette (Savannah) June 18, 1795, at 2, col. 2; and Id. Oct. 11, 1798, at 2, col. 1.

103. Watkins 611.

most part ignored them—overlapping the areas with legitimate surveys and grants.[104] In time adverse possession, if nothing else, would confirm these titles.

A demonstration of the perplexities that sometimes resulted, however, is found in a resurvey of lands lying astride Buffalo Swamp in Glynn County. Isidore Stouf, surveyor of Chatham County, performed this resurvey in 1807 for use as evidence in a suit before the federal court in Savannah—a suit in which the adjoining landowners claimed an overlap between their properties. Stouf noted that "it is evident and generally known by Planters in Glynn County, that the lands on Buffaloe Swamp have been run at different times by Speculators," who "have cut the land by so many false lines as to throw a general confusion in the surveys which leaves the greatest uncertainty upon the rights of" such claims.[105]

The subsequent history of the Pine Barrens Fraud involves the ped-

104. W. Primrose, Report of William Primrose, Explanatory of Washington County, Georgia 5 (n.p. 1838) (pamphlet filed in the Papers of the North American Land Company, University of Virginia, Charlottesville, Va.); Daily Georgian (Savannah), May 24, 1839, at 2, col. 5.

105. Plat representing a tract of land in the possession of John Carnochan (Dec. 4, 1807) (Mein v. Carnochan, U.S. Circuit Court for the District of Georgia, Record Group 21, Atlanta Regional Branch of the National Archives, East Point, Ga.). Isidore Stouf was born in Paris, France, about 1766. By 1790 he was living in Savannah, Georgia, after reportedly having served as king's engineer in Senegal and Santo Domingo. He was employed by the United States government to conduct engineering studies in connection with the erection of fortifications to guard Savannah and St. Marys. During 1795–97 he conducted studies for the commissioners of pilotage for the port of Savannah on methods to improve the navigability of the lower Savannah River. In 1797 he became a United States citizen. He served several terms as city surveyor of Savannah and county surveyor of Chatham. In addition, he operated an academy and a mercantile business. In 1819 Stouf prepared a report for the state of Georgia on the feasibility of cutting a canal from the Altamaha to the Turtle and Sapelo rivers, which was read before the Georgia House of Representatives. In the summer of 1821 Stouf laid out the 14th District of Monroe County for the state. He died of dropsy in Savannah on June 22, 1822. Stouf was a surveyor and engineer of the first caliber. He published several maps of Savannah and the immediate vicinity, besides executing many splendid plats of property boundaries.

See Instructions from Henry Knox to Paul Hyacinte Perrault (April 11, 1794), and Letter from Paul Hyacinte Perrault to George Mathews (April 30, 1794) (both in George Mathews Papers, Georgia Historical Society, Savannah, Ga.); Chatham County Civil Superior Court Minutes 1796–1799, p. 183, Office of the Clerk of Superior Court, Chatham County courthouse, Savannah, Ga.; Death Register for Chatham County (June 1822) (Georgia Historical Society, Savannah, Ga.); Daily Georgian (Savannah), Dec. 16, 1819, at 2, cols. 3–5; Georgian (Savannah), June 27, 1822, at 3, col. 1; M. De Lamar & E. Rothstein, The Reconstructed 1790 Census of Georgia 53 (1985); and L. De Vorsey, The Georgia–South Carolina Boundary 86–87 (1982). The author is indebted to Michael V. Jacobs of Savannah for preparing the foregoing sketch.

dling of this stupendous mass of phony grants throughout the United States and Europe and the countless futile attempts that those duped into buying the grants made to locate their supposed lands and establish tangible claims thereto. This fallout was to reverberate throughout the nineteenth century and into the twentieth on such a massive scale that it constitutes a separate study.

The fabricators had little difficulty retailing their "grants" for practically the entire nation was engrossed in a wild orgy of land speculation at the time. Most of the grants were sold in large batches of perhaps several hundred to a single individual or land company that in turn sold them in smaller batches to others. By this method the grants gradually permeated throughout the speculative web of society. Few doubted their genuineness for they masqueraded under the autographic signatures of the governors and other statehouse officials, were appended with the great seal of the state, and had been prepared with all due formalities.

Significantly, throughout the nation in the mid-1790s, pamphlets and newspaper ads began appearing that advertised the lands for sale and described their quality in glowing terms. The authors of the spurious grants even executed affidavits for their vendees, attesting to the high quality, utility, and healthfulness of the lands. These affidavits were inserted in the promotional literature.[106]

In February 1794 Thomas Smyth of Georgia informed William Patterson of Baltimore, Maryland, that he had 300,000 acres of land in Franklin County "surveyed & ready to be Pattented." He did not, however, have "money sufficient to carry it through the public offices so as to obtain . . .

106. R. Harper, Observations on the North American Land Company 63–73 (London 1796). A description of the method of selling the grants is found in a letter from Robert Fielding to William Constable, both of New York, dated September 20, 1796. The letter states: "When I had the Honor some Time since of soliciting you to purchase 100,000 acres of Georgia Lands situated in Liberty County, I did not know the proprietor held any more. those being sold, he desires me to offer 600,000 acres more . . . and engages them to be as good Lands as any in the State, not subject to Litigation like the Yazoo Lands. his price 10 cents per acre, for payment of one third in cash or merchandize, one third by Bond at 1 year, and, the remaining third by Bond at 2 years. if Sir, this is worth your notice I shall be very happy in having made you the tender. perhaps, if it should not, you will have the Goodness to refer me to such Gentlemen as you think may become Purchasers. The proprietor is now in Town and will depart for Georgia in 8 days but before he goes he would be glad fully to satisfy any Gentleman inclined to buy, of the Legality both of Sale and Tenure." Letter from Robert Fielding to William Constable (Sept. 20, 1796) (Constable-Pierrepont Papers, Rare Books and Manuscripts Division, New York Public Library, Astor, Lenox, and Tilden Foundations, New York, N.Y.).

Grants." Smyth therefore asked Patterson to advance him $1,500 to assist in the execution of this object, for which he would give him 100,000 acres of the land. Patterson eventually purchased several large tracts in Franklin County from Smyth, one tract consisting of 200,000 acres. Thus, Patterson, initially an unwitting accessory to the crime, became one of its earliest victims.[107]

Patterson attempted to have agents living in Georgia sell the lands, but when the agents got wind of the character of the grants, they curtly brushed him aside.[108] After other land agents hired by Patterson reported that they could not find the lands, he hired a professional surveyor—Robert McRea—to journey to Georgia to locate them. McRea spent several days in January 1797 in Louisville, the state capital, searching the official land records. He wrote that, in doing so, he encountered "Considerable difficulty in tracing out the surveys . . . owing to the Returns of the surveyors being verry imperfect in their description of their Situation, boundaries & the particular waters where executed." In addition, "many of them were part of some large Surveys," the outer boundaries only of which had been run, the interiors having been "laid out in the Chamber" (that is, the plats of the interiors had been drawn from imagination).[109]

Four months later McRea wrote from Washington, Georgia, that he had just "returned from the Frontiers of Franklin & Jackson Counties after a verry fatiguing Rout, and fruitless enquiry for the Lands." During this trek he had been unable to find anyone who could lead him to the lines. He had made two futile attempts himself to locate the 200,000-acre tract and had "repeatedly in vain" approached "J. Dabbs," the deputy who surveyed the tract. All "reasoning and address" he could muster "together with the offer of any reasonable & generous reward proved ineffectual." Nor could he find the plat books or field works of John Gorham, former surveyor of Franklin County, who had certified Patterson's plats. Neither could he discover any "Chainners or persons who attended the Surveys." Thomas Smyth had promised to show the lands to McRea but he would "not appear." McRea was informed that Smyth

107. Letter from William Patterson to Thomas Cumming (Mar. 2, 1794) (Cumming Letters, Special Collections, Reese Library, Augusta College, Augusta, Ga.).

108. Letter from Eleazer Early to William Patterson (Oct. 18, 1796) (Cumming Letters, Special Collections, Reese Library, Augusta College, Augusta, Ga.).

109. Letter from Robert McRea to Thomas Cumming (Jan. 18, 1797) (Cumming Letters, Special Collections, Reese Library, Augusta College, Augusta, Ga.).

had "gone to Europe perhaps to get out of reach, knowing he could not comply," for it had been "hinted" by several persons that "there never was any Legal survey made of the" 200,000-acre tract. J. Dabbs, however, insisted that the outer boundaries of the tract had been surveyed. McRea had asked "the best woods men on the frontiers" about how to proceed, but they were unwilling to offer any advice. McRea added that all the lands Patterson bought from Smyth were in a similar predicament.[110]

Gradually Patterson came to the realization that he would be unable to effect any claim to the lands. For more than twenty years afterward he prodded the legislature to make a substitution of other lands, charging that it was the "wretched and dishonest conduct of the state government" and "the loose manner of taking up & conveying Lands" in Georgia that had induced him to make the purchases. His requests went unheeded.[111]

Robert Morris of Philadelphia, a signer of the Declaration of Independence and a notorious land speculator, unwittingly bought 2.3 million acres worth of the feigned Georgia grants.[112] In August 1794 the French minister to the United States, acting on intelligence received from Georgia, published a letter in Paris newspapers warning Frenchmen of the "crowd of speculators" in this country "who survey poor or barren land, which are either in possession of the rightful owners, or out of the limits of the United States, in their last treaties with the Indians." The "pretended proprietors," nevertheless, obtained "plots and grants, or titles attested by the government, and their officers of the state appointed for the purpose; but the laws of" Georgia had "provided that these grants obtained by fraud . . . shall be of no account." The letter went on to state that the "imposters sell their pretended lands to merchants, who resell them . . . to Mr. Robert Morris, and he again to French families, who already have come, or who design to Emigrate, to this country."[113]

Morris vehemently denied the allegations that the Georgia grants were fake. He and several associates had been gluttonously buying up millions of acres of land throughout the United States which they expected

110. Letter from Robert McRea to Thomas Cumming (May 22, 1797) (Cumming Letters, Special Collections, Reese Library, Augusta College, Augusta, Ga.).

111. Letters from William Patterson to Thomas Cumming (Oct. 16, 1802 & Dec. 30, 1818) (both in Cumming Letters, Special Collections, Reese Library, Augusta College, Augusta, Ga.).

112. R. Harper, Observations on the North American Land Company 39 (London 1796).

113. Id. 92–93; Dunlap & Claypool's American Daily Advertiser (Philadelphia), June 15, 1795, at 1, cols. 1–4.

to sell for exorbitant profits. They borrowed huge sums of money and sold shares to underwrite the costs of these exploits. In 1795 these land jobbers combined their assets, including Morris's Georgia grants, and formed the North American Land Company—the "largest land trust ever known in America."[114] When agents were sent south to explore the lands, the true nature of the Georgia grants was quickly discovered. Having vastly overextended their speculative enterprises, Morris and his associates were encountering great difficulties in selling their lands fast enough to pay off the enormous debts that had piled up. Rapidly approaching bankruptcy, they were soon to find their way to debtors' prison. Morris evidently told his adherents that they would keep the Georgia grants and sell them "in the same way as they had bought them."[115]

Morris died a bankrupt and broken man in 1806. Not until the end of the nineteenth century was the colossal nightmare of legal entanglements that resulted from the excesses of his frenzied speculation resolved.[116] As late as the 1840s, the Georgia grants were listed as part of the assets of the North American Land Company—then still frozen in bankruptcy proceedings.[117]

In 1801 George Sibbald, a Maryland businessman residing in Georgia, announced the formation of the Georgia Asylum Company. Sibbald had "acquired" over 500,000 acres of pine barren land in Georgia—140,000 acres of which he had purchased at a tax sale in Montgomery County for thirty dollars. Sibbald proposed to bring poor immigrants in from

114. R. Arbuckle, Pennsylvania Speculator and Patriot 165–84 (1975); B. Chernow, Robert Morris: Land Speculator 170–98 (1978); Deed Book 3-B, p. 108, Office of the Clerk of Superior Court, Chatham County courthouse, Savannah, Ga. This deed contains an elaborate table itemizing Morris's Georgia grants and setting forth the chain of title for each.

115. R. Arbuckle, Pennsylvania Speculator and Patriot 175 (1975).

116. North American Land Company's Estate, 83 Pa. St. Rep. (2 Norris) 493 (1877); Estate of the North American Land Co., 60 Pa. St. Rep. (10 P. F. Smith) 247 (1869); Moss's Appeal—Halsey's Appeal, 43 Pa. St. Rep. (7 Wright) 23 (1862); B. Chernow, Robert Morris: Land Speculator 222–25 (1978).

117. Vaughan v. Barclay, 6 Wharton 392 (Pa. 1841). The saga of Morris's Georgia grants didn't end there. Deeds between Morris and various creditors and trustees of the North American Land Company can be found recorded in courthouses throughout eastern Georgia. The last one observed by the author is dated October 10, 1882. In it, James Dundas, trustee in bankruptcy, quitclaimed to Thomas Finn of Long Island, New York, all remaining "lands" of the North American Land Company in Washington, Johnson, Jefferson, Effingham, Bulloch, and Franklin counties. Deed Book 3, p. 86, Office of the Clerk of Superior Court, Effingham County courthouse, Springfield, Ga. The deed is also recorded in the courthouses of Washington and Emanuel counties.

northern states and from Europe to settle these lands. He likened the climate of the region to that of southern France and Italy and claimed that cotton, wheat, sweet potatoes, and peaches, as well as exotic crops such as figs, almonds, olives, oranges, and lemons, could be profitably grown there. His company was to have lands cleared, houses built, and sawmills, gristmills, cotton gins, and blacksmith shops in operable condition by the time the settlers arrived to purchase the lands. Sibbald even received a temporary suspension of taxes on the lands from the legislature, provided he brought in a requisite number of settlers within a given time.[118] Needless to say, this harebrained scheme never got off the ground. A few years later the legislature ordered Sibbald's "lands" sold for the accumulated taxes owed, implicitly recognizing that it would be difficult to find buyers.[119]

One of the more determined efforts at establishing actual claims to grants growing out of the Pine Barrens Fraud came in the late 1830s. Certain Philadelphia magnates, who held large blocks of the grants and owned stock in companies that held such grants, sent William Primrose south to "recover" the lands from the "squatters" who occupied them. Primrose spent a year and a half in Georgia compiling a substantial array of title abstract information and retaining John M. Berrien, former attorney general of the United States, to prosecute the claims.[120] These

118. Clayton 673–74; J. Dorsey, Footprints along the Hoopee 10–11 (1978); G. Sibbald, Notes and Observations on the Pine Lands of Georgia passim (Augusta, Ga. 1801).

119. Clayton 689, 693. It is of interest that in 1801 Sibbald donated a two-hundred-acre tract of land on which to establish the seat of Bulloch County. Bulloch had been cut largely from territory that had been a part of Effingham County, and the deed of gift states that the tract was to be "admeasured and laid off out of . . . lands belonging to the said . . . Sibbald . . . which are known . . . by the name of Ryan Survey." Apparently this was a reference to Joseph Ryan, mentioned above as one of the ringleaders of the Pine Barrens Fraud in Effingham County. Sibbald had obtained, through sheriff's sales, thousands of acres of land in Bulloch County that had belonged to Joseph Ryan. See Columbian Museum & Savannah Advertiser, May 31, 1799, at 2, col. 2; Id. Sept 27, 1806, at 2, col. 1; Georgia Republican (Savannah), Sept. 30, 1806, at 2, cols. 1–2. The city of Statesboro was built on the site of Sibbald's donation. The deed is recorded in Deed Book A, 1796–1806, p. 85a, Office of the Clerk of Superior Court, Bulloch County courthouse, Statesboro, Ga. The deed has been reprinted, with minor errors of transcription, in Bulloch County Historical Society, The Story of Bulloch County 41–42 (1973); and in A. Register, Bulloch County, Georgia Genealogical Source Material 15–16 (1985).

120. W. Primrose, Report of William Primrose Explanatory of Washington County, Georgia passim (n.p. 1838) (pamphlet filed in the Papers of the North American Land Company, University of Virginia, Charlottesville, Va.).

actions aroused considerable excitement among the local settlers. A battle of letters raged in the newspapers for several weeks.[121] Public meetings were held in which the inhabitants denounced Primrose and threatened to wage "war" on his sponsors if necessary to protect their titles. Berrien withdrew and Primrose was run out of the state. Nearly twenty years later, Primrose's sponsors attempted to have the sheriff of Bulloch County serve a writ of mandamus on the clerk of court to compel him to record certain of the grants. Riots broke out and, rather than comply with the order, both the sheriff and the clerk resigned. The courthouse door was nailed shut to prevent anyone from getting inside to record the documents.[122] Ultimately the legislature passed an act prohibiting clerks of the superior courts from recording deeds derived from the " 'Primrose Grants' or any" other "such grant."[123]

An example of later literature that cropped up to promote sales of Pine Barrens Fraud grants is a pamphlet published in 1865 by Robert L. Pell of New York. Pell claimed to own 4 million acres of "valuable land for timbering" in Georgia, part of which he had "purchased" from creditors of the North American Land Company. He left no doubt as to the origin of the titles, however, when he listed the original grantees as James Shorter, Richmond Dawson, Joseph Cooper, Thomas Cooper, Robert Middleton, and Thomas Flournoy. Offering to sell the lands, Pell alleged that the soil was fertile, the climate healthy, and the region possessed "opportunities and facilities for the creation of wealth, unsurpassed by any section of equal extent in the entire South." Attached were copies of letters from attorneys, solicitors, and clerks of court in the counties of Montgomery, Washington, Emanuel, Laurens, and Tattnall. Apparently oblivious to the pernicious consequences that might result (and perhaps not willing to pass up any opportunity to promote a region that progress had generally bypassed), the letters corroborated Pell's claims and attested to the validity of the titles.[124] Pell managed to sell a major portion of his lands,

121. Daily Georgian (Savannah), May 3, 1839–June 14, 1839.

122. Savannah Morning News, April 2, 1860, at 2, col. 2; D. Brannen, Life in Old Bulloch 11, 18–19 (1987). Significantly, the superior court records for Bulloch County for 1857, the year the attempt was made to record the documents, are almost a total blank, indicating that the office was idle.

123. 1857 Ga. Laws 58.

124. R. Pell, An Account of the Lands Belonging to Robert L. Pell in the State of Georgia passim (Kingston, N.Y. 1865).

as evidenced by deeds recorded in the Washington and Emanuel county courthouses.[125]

These are only a few of the more noteworthy episodes that resulted from the Pine Barrens Fraud. Throughout the 1800s and into the 1900s, the bogus grants were regularly vended in the North and West. In Georgia the successive governors, surveyors general, and secretaries of state frequently received letters inquiring about the grants.[126] During the early 1900s the secretary of state received so many queries that he published a letter explaining that subsequent adverse possession and nonpayment of taxes rendered all of the grants void.[127] Periodically proposals were made to enact legislation to put an end to the confusion, but no one could come up with an exact formula that would lay the matter to rest and at the same time not upset numerous legitimate titles. The citizens who traded in the grants regularly and dutifully mailed copies of their deeds to the clerks of court in the counties where the lands supposedly existed, to be duly entered into the deed books. Eventually long, unbroken chains of title could be established from these records, extending from the immediate grantee far back in time to those names that had become synonymous with fraud and corruption—Francis Tennille, James Shorter, Richmond Dawson, Thomas Dawson, Joseph Cooper, Joseph Ryan, and others. A prodigious amount of money and effort was expended in attempts to establish these claims, and it all came to naught.

One witness to these later doings eloquently summed them all up by stating,

> For many a bootless pilgrimage from distant States and sections was made years afterwards by the sufferers and their agents in search of those fabled lands. In vain, however, did they thread the woods and interrogate the trees. No land marks, no corners or stations could they ever find responsive to

125. Deed Book A, pp. 188–98; Deed Book B, p. 597; Deed Book D, pp. 373, 426 (all in Office of the Clerk of Superior Court, Washington County courthouse, Sandersville, Ga.); Deed Book E, pp. 199–202, 155–56, 189–91; Deed Book AA, pp. 17–18 (all in Office of the Clerk of Superior Court, Emanuel County courthouse, Swainsboro, Ga.).

126. P. Cook, Report of the Secretary of State of Georgia 4–6 (1890); King, *A Sketch of the History of Land Titles in Georgia,* in Report of the Second Annual Meeting of the Georgia Bar Association 126, 146–47 (1886); S. McLendon, History of the Public Domain of Georgia, 175–77 (1924); Atlanta Constitution, Dec. 20, 1897, at 7, col. 5; Savannah Evening Press, Jan. 9, 1913, at 5, col. 4; Savannah Morning News, Oct. 24, 1894, at 8, col. 5; Id. Feb. 1, 1895, at 6, col. 3; Id. Dec. 1, 1909, at 9, cols. 3–5; Id. Dec. 5, 1909, at 4, cols. 2–6.

127. P. Cook, Annual Report of the Secretary of State of the State of Georgia for Year Ending December 31, 1911, at 14–18 (1912).

> the well drawn, false speaking charts they brought along with them. No oaks, and hickories, no walnut, dogwood or buckeye, nor any kindly soil did they ever encounter to cheer their wearisome explorings or raise their sunken spirits. But barren wastes spread out sad and interminable before their eyes, and the tall sighing pines sounded a lugubrious sympathy in their ears. The golden dreams they had been made to cherish were dispelled for ever. Reluctantly they awoke to the bitter reality of being the victims of a great concocted turpitude, and with heavy hearts wended their way back to their far off homes, full of indignation, and cursing and hating more than ever before the villains and villainies of the world.[128]

Yazoo Land Fraud

The most celebrated land fraud scheme in this period—indeed, in American history—was the great Yazoo Land Fraud.[129] Few events in history, except the Civil War, have shaken the foundations of the state as much as this fiasco did. Perpetrated by a corrupt legislature in 1795, it was to have nationwide repercussions for the next twenty years. In addition, one of the most important constitutional decisions in American history resulted from it.

The story of the Yazoo Fraud has been told many times. The subject is not entirely relevant to this history because the lands involved lay completely outside the present limits of Georgia. Therefore, only the principal events will be described.

Actually two Yazoo sales took place. Land speculation was rampant in America in the years immediately following the Revolution, and individuals combined to form land companies to indulge in this speculation. In 1789 the South Carolina Yazoo Company, the Virginia Yazoo Company (which, incidentally, was headed by Patrick Henry), and the Tennessee Company presented petitions to the Georgia legislature requesting permission to purchase large tracts of land in what are now the states of Alabama and Mississippi. The legislature thereupon obligingly passed a bill granting 25,400,000 acres of territory for $207,580 to these three companies. The lands were to be reserved as a preemption for two years.

128. A. Chappell, Miscellanies of Georgia (pt. 2) 51 (1874).

129. So called because the land lay in close proximity to the Yazoo River, which is located in present-day Mississippi.

At the end of that period, if the stipulated amounts had been paid, grants were to be issued giving the companies title in fee simple.[130]

A surge of protests from public-minded citizens erupted over this deal, but the protests died down when this first Yazoo sale soon fell apart. Much of the territory was still claimed by Spain, and Spanish intrigue coupled with threats from the Indians discouraged white settlement. President George Washington issued a proclamation ordering that the laws and treaties protecting Indian rights be observed. Under these circumstances it was difficult to entice settlers to purchase the lands. In 1791 the companies defaulted, whereupon Georgia reasserted title.[131]

The failure of this venture did not diminish the avariciousness of its promoters. In November 1794 the Georgia Company, the Georgia-Mississippi Company (formerly the South Carolina Yazoo Company and later called the New England–Mississippi Land Company), the Upper Mississippi Company (formerly the Virginia Yazoo Company), and the Tennessee Company—being better organized and financed than their previous counterparts—asked the legislature for a grant of an area that covered about three-fifths of present-day Alabama and Mississippi. It is noteworthy that among the leading stockholders in these companies were two United States senators, two congressmen, three judges, a territorial governor, a United States attorney, a future state governor, and a future congressman. Most of these men were from states other than Georgia.[132]

The requested bill passed the legislature, but Governor Mathews vetoed it on the basis that it was not the proper time to dispose of the western territory, that the sum offered was too low, that the quantity of land reserved for citizens of the state was too small, and that if so large a territory were sold to individuals they could monopolize it and retard settlement.[133] A new bill, slightly altered, was immediately introduced, rushed through the legislature, and signed by the governor on January 7, 1795.[134] This act conveyed the territory, estimated to contain 35 million acres, to the four companies for a total of $500,000, or less than one and a half cents per acre.

130. Am. St. Papers, 1 Public Lands 133; Haskins, *The Yazoo Land Companies,* 5 Papers Am. Hist. Assoc. 395, 399–400 (1891); 2 W. Stevens, A History of Georgia 463–66 (Philadelphia 1859); Watkins 387.

131. C. Magrath, Yazoo Law and Politics in the New Republic 4–5 (1966); Haskins, *The Yazoo Land Companies,* 5 Papers Am. Hist. Assoc. 395, 400–414 (1891).

132. C. Magrath, Yazoo Law and Politics in the New Republic 5–6 (1966).

133. Am. St. Papers, 1 Public Lands 156; Executive Minutes (Dec. 28, 1794) (GDAH).

134. Am. St. Papers, 1 Public Lands 152; Watkins 557.

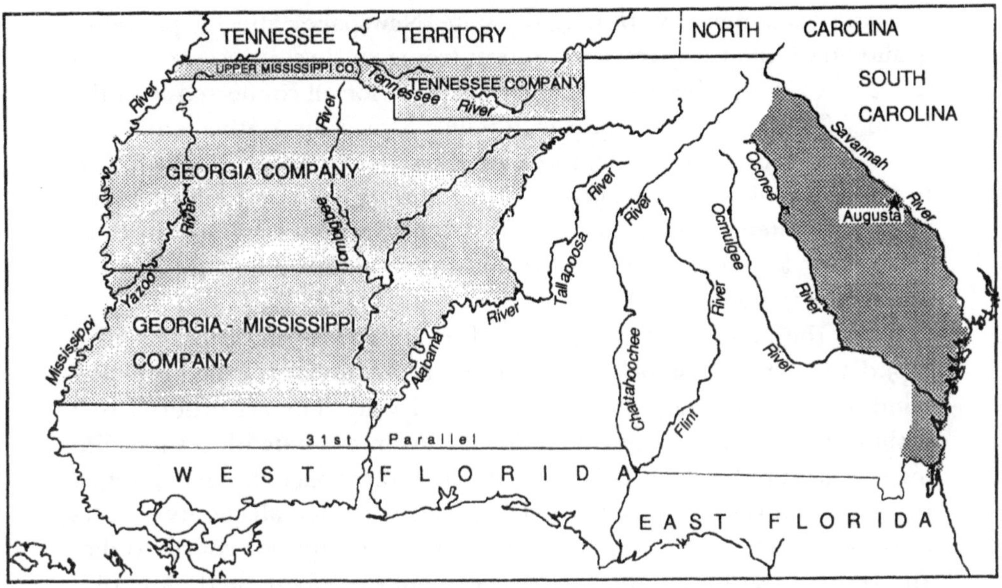

Land sales made under the Yazoo Act in 1795.

With but one exception, every member of the legislature who voted for the bill was a shareholder in one or more of the companies.[135] In addition, many had been offered money, Negroes, barrels of rice, and other bribes.[136] Weakness rather than corruption, however, has usually been cited as the reason for Governor Mathews's signing the bill.[137] The sale had been negotiated in the greatest secrecy and was passed under the altruistic-sounding title "An act . . . for appropriating a part of the unlocated territory of this State for the payment of the late State troops, . . . declaring the right of this State to the unappropriated territory thereof, for the protection and support of the frontiers . . . , and for other purposes." Thus its real meaning was camouflaged from the public.[138]

When the true nature of the sale became known, a frenzy of indigna-

135. Haskins, *The Yazoo Land Companies,* 5 Papers Am. Hist. Assoc. 395, 418 (1891).

136. Am. St. Papers, 1 Public Lands 144–49.

137. Haskins, *The Yazoo Land Companies,* 4 Papers Am. Hist. Assoc. 395, 416 n. 1 (1891). For Governor Mathews's explanation of his motives in signing the Yazoo Act see his message to the legislature recorded in Executive Minutes (Jan. 12, 1796) (GDAH).

138. Mayor etc. of Savannah v. State of Georgia, 4 Ga. 26, 38 (1848). As a result, a provision was inserted in the Georgia Constitution of 1798 prohibiting statutes from containing matter different from what is expressed in the title thereof. Cobb 1114. The provision has been carried down to the present constitution. Ga. Const. of 1983, art. 3, § 5, para. 3.

tion and excitement swept over the state. Newspaper attacks, petitions, grand jury remonstrances, and protests from private citizens were almost universal. Georgia became a dangerous place for all connected with the fraud. Most did not dare appear in public. Governor Mathews and several others found it expedient to leave the state. One legislator who fled to South Carolina was tracked down by irate citizens and killed. James Jackson, United States senator from Georgia, resigned his seat in Congress and came home to run for the state legislature and to lead the opposition to the Yazoo Act.[139]

When the new legislature convened in January 1796, it lost no time repudiating the Yazoo deal. In a verbose act, commonly called the "Rescinding Act," the Yazoo sale was declared void. The act ordered that within three days all records and documents in the state offices pertaining to the affair be publicly burned and all references to them expunged from the indexes of the books.[140] The act also directed all county officers to present their books at the next session of the superior courts so that any records therein relating to the sale could be obliterated. Any officer failing to do so was to be suspended and rendered ineligible to hold office, and any officer who in the future should attempt to record transactions relative to the sale was to be suspended and fined $1,000. The act went on to prohibit the entering of records pertaining to the sale as evidence in the courts to establish a right to the territory and empowered the governor to refund all money collected from the sale. Two days later the enrolled copy of the Yazoo Act and other papers pertaining to it were burned in front of the statehouse in Louisville.[141]

In 1798 a new state constitution was adopted, and the Rescinding Act was written into it.[142] But this did not end the matter. The land companies had sold a number of the tracts and several court suits followed. One of these, *Fletcher v. Peck,*[143] eventually reached the Supreme Court of the United States. John Peck had conveyed to Robert Fletcher for the sum of $3,000 a fifteen-thousand-acre tract that had been included in the Yazoo purchase and passed down through a chain of conveyances

139. Jackson, *Sketch of the Yazoo Speculation,* in 11 Coll. Ga. Hist. Soc. 33, 35 (L. Hawes ed. 1955); A. Johnson, Georgia as Colony and State 179 (1938).

140. Am. St. Papers, 1 Public Lands 156; Marbury & Crawford 573.

141. W. Foster, James Jackson: Duelist and Militant Statesman 118–22 (1960); 2 W. Stevens, A History of Georgia 491–92 (Philadelphia 1859).

142. Marbury & Crawford 25; Cobb 1116.

143. 10 U.S. (6 Cranch) 87, 3 L. Ed. 162 (1810).

from the Georgia Company. After the passage of the Rescinding Act, Peck found he could not make good title, whereupon Fletcher sued. The case was brought up from the Circuit Court for the District of Massachusetts and argued twice before the United States Supreme Court. In the historic opinion delivered in March 1810, Chief Justice John Marshall held that Georgia's attempt to nullify the Yazoo sale was unconstitutional as an impairment of contract and a violation of natural rights and principles. The decision, the first in which the Court held void a state statute, set a precedent for the federal review of state statutes.

Georgia, however, had washed its hands of the whole Yazoo affair in 1802 by ceding its western lands to the United States.[144] The United States gradually bought up the claims, ultimately paying a total of $4,282,151.12 in 1818 as final settlement of the matter.[145]

Rounding Out the State Boundaries

In the twenty years following the end of the American Revolution, Georgia established—at least on paper—its boundaries, which, with only a few trifling changes, have remained in effect to the present day. The negotiations and maneuverings that led to the final definition of the state's boundaries are intricate and often confusing. Space permits only a general outline of these events, but the reader may refer to other sources for a thorough account.[146]

It will help first to understand that during colonial times, and indeed well into the nineteenth century, geographic knowledge of the frontier regions of the state was very scanty and inaccurate. Hence a number of gross blunders—some bordering on absurdity—were made in drawing up the treaties, conventions, laws, and resolutions defining the state's boundaries.

Before the Revolution there had been little apparent need to define colonial boundaries any further than the vague descriptions given in colonial charters and commissions issued to royal governors. The situation had changed radically by the 1780s, however. Settlers were flocking into the interior. The fertile upper reaches of the Savannah River, on both

144. See next section.

145. Am. St. Papers, 3 Finance 281–83; 3 Stat. 116, 192.

146. The notes that follow cite sources which discuss the state's boundaries in greater detail.

the Georgia and the South Carolina sides, became an area of especially prolific settlement and thus of primary concern to both states.

In June 1785 South Carolina petitioned the Continental Congress, stating that boundary disputes had arisen between it and Georgia and requesting that a court be appointed to resolve the differences.[147] The Continental Congress proceeded with this business, but several postponements ensued which finally induced the two states to settle the matter on their own.[148] Accordingly, each appointed three commissioners, who met at Beaufort, South Carolina, and on April 28, 1787, signed a document known as the Beaufort Convention.[149]

One of the major points of contention was whether the triangular-shaped area between the Keowee and the Chattooga-Tugaloo rivers at the head of the Savannah River belonged to Georgia or to South Carolina. Georgia's charter fixed its boundary as the most northerly branch of the Savannah River. The Keowee River (the eastern branch) rises farther north than the Chattooga-Tugaloo River (the western branch), but the latter has a stronger flow. Perhaps for this reason the commissioners assumed that the Chattooga-Tugaloo River was the most northerly branch of the Savannah, or perhaps they merely wanted to settle the matter quickly and avoid court expense. At any rate, the convention fixed the Chattooga-Tugaloo River as the boundary, thereby conceding the disputed area to South Carolina. Because Georgia had already granted land in this area to settlers, South Carolina agreed to confirm these grants, provided the claimants registered them within one year. South Carolina also relinquished to Georgia all claim to the lands lying south and southwest of the Savannah River, as well as to all islands in that river. This agreement was thereafter ratified by Georgia,[150] by South Carolina,[151] and by the United States.[152]

The boundary as fixed in the Beaufort Convention has remained in effect to the present day, although in the 1850s it occasioned a mild disputation between Governor Howell Cobb of Georgia and Governor John H.

147. 19 Col. Rec. Ga. (pt. 2) 528–32; 28 Journals of the Continental Congress 408–10 (J. Fitzpatrick ed. 1933); 2 W. Stevens, A History of Georgia 384 (Philadelphia 1859).

148. 31 Journals of the Continental Congress 622–29, 642, 650–54 (J. Fitzpatrick ed. 1934); 32 Journals of the Continental Congress 13, 440, 463 (R. Hill ed. 1936).

149. Am. St. Papers, 1 Public Lands 60; Marbury & Crawford 662.

150. Marbury & Crawford 337; Cobb 150.

151. 1 T. Cooper, The Statutes at Large of South Carolina 411 (Columbia, S.C. 1836).

152. 33 Journals of the Continental Congress 467–76 (R. Hill ed. 1936).

Means of South Carolina;[153] and in 1922 the U.S. Supreme Court was called on to clarify certain points.[154] In 1990 a suit was settled in the U.S. Supreme Court regarding the precise location of the Georgia–South Carolina boundary in the lower reaches of the Savannah River—an area where erosion and alluvion had greatly changed the river's course over the years.[155]

After the division, in 1729, of the colony of Carolina into North Carolina and South Carolina, the former's chartered limits encompassed generally what is now the states of North Carolina and Tennessee, while the latter's included what is now South Carolina, Georgia, Alabama, and Mississippi. After Georgia was cut from South Carolina in 1732, it was believed that South Carolina possessed a narrow strip of land about twelve miles wide that extended across the northern portion of present-day Georgia, Alabama, and Mississippi. By 1776 South and North Carolina had fixed the thirty-fifth parallel as the boundary between their western lands.[156]

During the Revolution the general agreement was that all the states should hand over their western lands to the central government. In 1787, after the Beaufort Convention had been signed, South Carolina ceded to the United States its twelve-mile-wide strip (which in reality did not exist). North Carolina ceded its western lands in 1789, thus leaving only Georgia (of the original thirteen states) retaining western lands.[157]

Georgia had offered the southern half of its western lands to the United States in 1788,[158] but the offer was declined because of the conditions imposed.[159] Notwithstanding, the United States proceeded in 1798

153. H. Cobb, Governor Cobb to Governor Means, on the Boundary between Georgia and South Carolina passim (Athens, Ga. 1852); I. Hayne, Report to His Excellency Governor Means on the Subject of the Augusta Bridge, and the Navigation of the Savannah passim (1853?).

154. State of Georgia v. State of South Carolina, 257 U.S. 516, 66 L. Ed. 347 (1922).

155. Georgia v. South Carolina, 497 U.S. —, 111 L. Ed. 2d 302 (1990); L. De Vorsey, The Georgia–South Carolina Boundary passim (1982).

156. Coulter, *The Georgia-Tennessee Boundary Line,* 35 Ga. Hist. Q. 269, 272–73 (1951); M. Skaggs, North Carolina Boundary Disputes Involving Her Southern Line passim (1941).

157. Howard v. Ingersol, 54 U.S. (13 How.) 381, 14 L. Ed. 189 (1851).

158. 31 Journals of the Continental Congress 507 (J. Fitzpatrick ed. 1934); 33 Journals of the Continental Congress 690, 692–93 (R. Hill ed. 1936); 34 Journals of the Continental Congress 14, 188 (R. Hill ed. 1937); K. Coleman, The American Revolution in Georgia 1763–1789, at 264–65 (1958); Watkins 370.

159. 34 Journals of the Continental Congress 320–26 (R. Hill ed. 1937); Watkins 757.

to organize Georgia's western lands as the Mississippi Territory.[160] Georgia protested but was caught up in the quagmire of the Yazoo Fraud at the time.[161] In addition, there was a growing realization on the part of Georgians that the western territories were too vast and distant for successful state administration. It would be far wiser to concentrate efforts toward the settlement and development of the large public domain within the state's more immediate borders.

These considerations, coupled with continuous prodding from the federal government, finally induced action. By the Articles of Agreement and Cession entered into on April 24, 1802, Georgia ceded to the United States all right and title it had to the lands lying west of a line "beginning on the western bank of the Chattahoochee river" where it crossed the boundary between the United States and the Spanish province of Florida; thence up the Chattahoochee "along the western bank thereof, to the great bend thereof, next above the place where a certain creek or river, called 'Uchee,' (being the first considerable stream on the western side, above the Cussetas and Coweta towns) empties into the said Chattahoochee river; thence, in a direct line, to Nickajack, on the Tennessee river; then crossing the said last mentioned river; and thence, running up the said Tennessee river, and along the western bank thereof, to the southern boundary line of the State of Tennessee."[162]

In return, the United States ceded to Georgia the nonexistent strip South Carolina had ceded to the United States in 1787, paid Georgia the sum of $1,250,000, and agreed to extinguish the Indian title to Georgia's remaining lands "as early as the same can be peaceably obtained." The execution of this latter provision was to be a major source of contention between Georgia and the federal government until final removal of the Indians from Georgia soil in the 1830s.

In 1803 Georgia laid out a new county called Walton, the boundaries of which were located entirely within territory later conceded to be part of North Carolina.[163] At the time, however, Walton County was believed to be in Georgia. It sent representatives to the Georgia legislature every

160. 1 Stat. 549; 2 Stat. 69.

161. Am. St. Papers, 1 Public Lands 111–12; Georgia General Assembly, An Address and Remonstrance of the Legislature of the State of Georgia passim (n.p. 1800?) (DeRenne Collection, University of Georgia Libraries, Athens, Ga.).

162. Am. St. Papers, 1 Public Lands 125; Clayton 48.

163. 1803 Ga. Laws 16; Clayton 128. This Walton County should not be confused with the Walton County that was laid out some years later in north central Georgia.

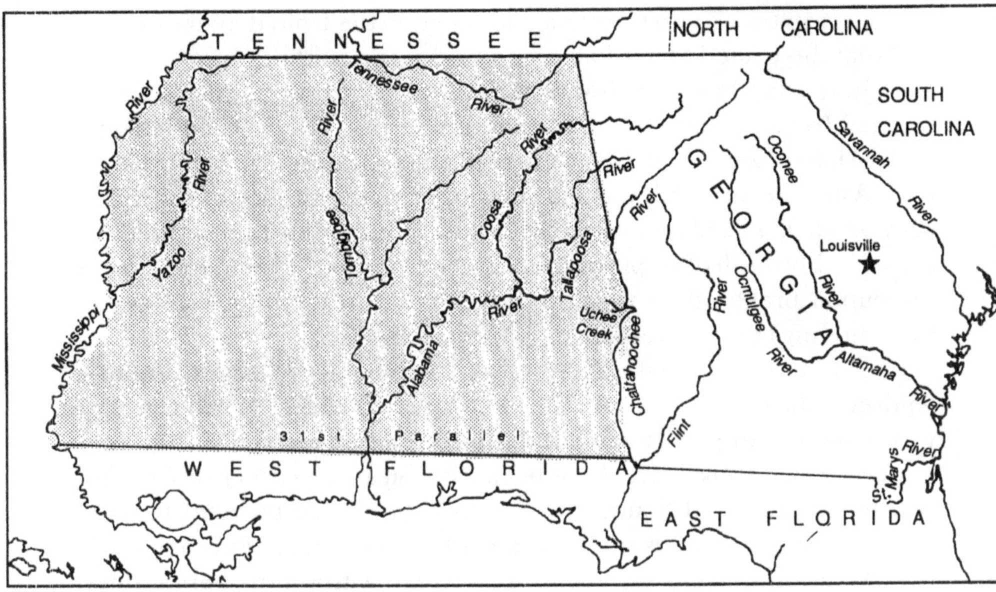

Lands ceded by Georgia to the United States in 1802.

year from 1804 to 1811.[164] Violence and controversy soon erupted between Georgia settlers who moved into the area and the North Carolina residents, underscoring the urgent need to ascertain the exact boundary between the two states.[165] Consequently, a correspondence was initiated between Governors John Milledge of Georgia and James Turner of North Carolina. The thirty-fifth parallel, the line North Carolina recognized as its southern boundary on the border with Georgia, had been readily accepted by Georgia as its true northern boundary by the Articles of Agreement and Cession. Georgia thus abandoned the head of the Chattooga-Tugaloo River which lay north of the thirty-fifth parallel and had been fixed as the state's northern boundary by the charter of 1732, as amended by the Beaufort Convention. Georgia (and North Carolina) apparently believed that the thirty-fifth parallel was north of the head of the Chattooga-Tugaloo River and that the area between these two points

164. M. Skaggs, North Carolina Boundary Disputes Involving Her Southern Line 173–75, 200 (1941); see also the Georgia House and Senate journals for the years 1804–11.

165. Am. St. Papers, 1 Miscellaneous 439–49; Am. St. Papers, 2 Miscellaneous 72–79; 15 Annals of Cong. 340, 456–57 (1806); 16 Annals of Cong. 968–91 (1806); 21 Annals of Cong. 2273–88 (1810); M. Skaggs, North Carolina Boundary Disputes Involving Her Southern Line 164–75, 180, 182, 194–95, 198–202 (1941).

was the strip that South Carolina had ceded to the United States in 1787 and that the United States had ceded to Georgia in 1802.[166]

In June 1807 Professor Josiah Meigs[167] and Surveyor General Daniel Sturges (representing Georgia) and Professors Joseph Caldwell and James Hall (representing North Carolina) met at Buncombe Court House (now Asheville) in North Carolina to determine the location of the thirty-fifth parallel. The observations they made indicated that the parallel lay considerably south of its previously assumed location. The commissioners thereupon broke off further proceedings and headed home to report these findings to their respective governments.[168]

In 1811 Georgia hired a surveyor and astronomer of national note—Andrew Ellicott—to ascertain the thirty-fifth parallel.[169] North Carolina was requested to participate but declined. Ellicott marked a large rock on the eastern bank of the Chattooga River near the corner of Georgia, North Carolina, and South Carolina as defining the thirty-fifth parallel.[170] The rock, which is still visible, has been known as "Ellicott's Rock" ever since. Modern measurements show it to be within a few hundred feet of the true thirty-fifth parallel. Shortly after Ellicott marked the rock, a dispute over pay erupted between him and Governor David B. Mitchell, and Ellicott left the state.[171] Upon receiving the results of Ellicott's work, Governor Mitchell announced to the Georgia legislature in November 1812: "By this report it appears that no part of the territory heretofore

166. Coulter, *The Georgia-Tennessee Boundary Line,* 35 Ga. Hist. Q. 269, 273–74 (1951); M. Skaggs, North Carolina Boundary Disputes Involving Her Southern Line 178 (1941).

167. Josiah Meigs (1757–1822), an attorney, newspaper editor, and educator, was a native of Connecticut and a graduate of Yale College (now Yale University) where he served as professor of mathematics. In 1800 he was appointed president of the University of Georgia and held the position for a decade. It was during this tenure that he was hired to assist in locating the Georgia–North Carolina boundary. In 1812 President Madison appointed him surveyor general of the United States, and later he became commissioner of the General Land Office of the United States. See Starr, *Josiah Meigs,* in 12 Dictionary of American Biography 506 (1937).

168. M. Skaggs, North Carolina Boundary Disputes Involving Her Southern Line 183–90 (1941).

169. Id. 198–201; 1810 Ga. Laws 156; C. Mathews, Andrew Ellicott: His Life and Letters 218–26 (1908).

170. Wright, *Ellicott's Rock: Where Georgia Begins,* Georgia Magazine, Mar. 1971, at 12. The true corner between the three states is the center of the river, a few feet to the west of the rock.

171. M. Skaggs, North Carolina Boundary Disputes Involving Her Southern Line 200–201 (1941).

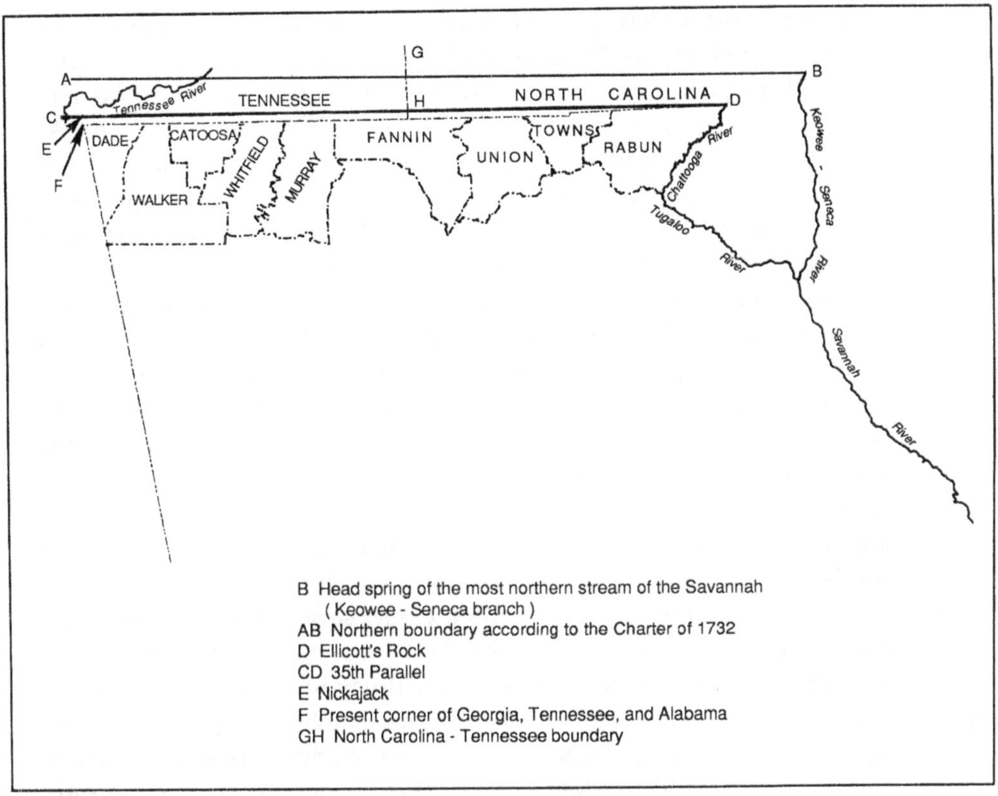

Northern boundary of Georgia. Adapted from Coulter, *The Georgia-Tennessee Boundary Line,* 35 *Ga. Hist. Q.* 269 (1951).

claimed by this state as Walton county remains to Georgia."[172] Thereafter Walton County passed out of existence.[173]

At this juncture Tennessee entered the picture. The state of Tennessee had been formed from North Carolina's western cession to the federal government and was admitted to the Union in 1796. In 1817 the Tennessee legislature requested Georgia's cooperation in locating the boundary between the two states, whereupon Georgia promptly obliged. A few months later commissioners of the two states met at Nickajack, an abandoned Indian town near the present northwest corner of Georgia. Rep-

172. 1812 Ga. Senate Journal 11.

173. M. Skaggs, North Carolina Boundary Disputes Involving Her Southern Line 200 (1941).

resenting Georgia were James Camak (professor of mathematics at the University of Georgia), Thomas Stocks, and Hugh Montgomery. Representing Tennessee were John Cocke and James G. Gaines. Agreement had already been reached that the thirty-fifth parallel was the correct boundary between Georgia and Tennessee. The next problem was to determine the point at which the western boundary of Georgia intersected the thirty-fifth parallel. The Articles of Agreement and Cession of 1802 defined Georgia's western boundary as up the Chattahoochee to the first big bend above Uchee Creek, "thence, in a direct line, to Nickajack, on the Tennessee river; then crossing the said last mentioned river; and thence, running up the said Tennessee river, and along the western bank thereof, to the southern boundary line of the State of Tennessee." Thus the commissioners who drew up the Articles of Agreement and Cession clearly had assumed that the western boundary of Georgia would strike the Tennessee River south of the Georgia-Tennessee boundary and then run up that river to the Tennessee line. In actuality, however, the river in this area is completely north of the thirty-fifth parallel.[174]

The commissioners erected a rock eighty-seven chains south of the Tennessee River and due south of the center of Nickajack at a point their observations indicated to be on the thirty-fifth parallel. They then proceeded to blaze a line eastward and, completely misapprehending the location of the Tennessee–North Carolina boundary, continued on until they were within thirty-seven miles of the northeast corner of Georgia. Tennessee immediately ratified the line as run but Georgia never has done so officially, although it did adopt a resolution authorizing the governor to have the maps and papers pertaining to the survey recorded in the surveyor general's office.[175]

As it subsequently turned out, this line was about half a mile south of the thirty-fifth parallel. The error was discovered in 1826 when Georgia ran the Georgia-Alabama boundary. James Camak, who had participated in the survey with Tennessee, acted as mathematician in this survey and noted the discrepancy.[176]

In preparation for the surveys for the 1820 land lottery, Georgia in

174. Battle, *The Georgia-Tennessee Boundary Dispute,* in Report of the Nineteenth Annual Session of the Georgia Bar Association 87, 103–5 (1902); Coulter, *The Georgia-Tennessee Boundary Line,* 35 Ga. Hist. Q. 269, 275–76, 300 (1951).

175. Coulter, *The Georgia-Tennessee Boundary Line,* 35 Ga. Hist. Q. 269, 276–80, 299–306 (1951).

176. Id. 280–81.

1818 secured North Carolina's cooperation in running its boundary with that state. The following year Benjamin Cleveland, Allen Daniel, James Camak, and Timothy Tyrrel (representing Georgia) and Jesse Franklin, James Mebane, and Robert Love (representing North Carolina) marked a line from Ellicott's Rock in a westerly direction for thirty miles along what they believed to be the thirty-fifth parallel but which—as discovered later—angled slightly southward from that parallel. North Carolina ratified this line in 1821 and passed an act defining the rest of the Georgia–North Carolina boundary as running due south from the termination of this line for a distance of 661 yards and thence westward (apparently along the line that had been run by the Georgia-Tennessee commissioners in 1818) to the common corner of Georgia, North Carolina, and Tennessee. Georgia, however, never passed any reciprocal legislation.[177]

In 1820 the northeastern corner of Georgia was surveyed into land lots and districts; and in 1831 and 1832 the Cherokee lands, which made up the northwestern portion of the state, were laid off into land lots, districts, and sections. In each case the surveys were run to the erroneously marked lines along Georgia's northern border, the state making no issue as to the correctness of the lines.

Such in a nutshell is how Georgia's eastern and northern boundaries evolved. The intervening decades have seen numerous scattered attempts to have the northern boundary redefined, but nothing of substance has been accomplished.[178] Although Georgia has never expressly ratified any of the surveys of its northern boundary, the state would appear to be bound by the lines as marked, be they in error or not. The U.S. Supreme Court has ruled repeatedly that, as in the case of adjoining private owners, if a boundary between states has been run out, located, and marked on the ground and is afterward recognized and acquiesced in by the states for many years, that boundary is conclusive even though later found to be erroneously located.[179]

177. M. Skaggs, North Carolina Boundary Disputes Involving Her Southern Line 203–4 (1941). The Georgia Code of 1863 and all subsequent codes have defined the north boundary of the state as the thirty-fifth parallel. O.C.G.A. §§ 50-2-1 to 50-2-3.

178. Battle, *The Georgia-Tennessee Boundary Dispute,* in Report of the Nineteenth Annual Session of the Georgia Bar Association 87 (1902); Coulter, *The Georgia-Tennessee Boundary Line,* 35 Ga. Hist. Q. 269, 281–303 (1951); Hood, *Georgia's Northern Boundary,* 8 Ga. St. B.J. 197 (1971).

179. Georgia v. South Carolina, 497 U.S. —, 111 L. Ed. 2d 309 (1990); State of Arkansas v. State of Tennessee, 310 U.S. 563, 84 L. Ed. 1362 (1940); Commonwealth of Massachusetts v. State of New York, 271 U.S. 65, 70 L. Ed. 838 (1926); State of Michigan v. State

The southern boundary of Georgia had been defined on paper in colonial times and the western boundary was defined by the Articles of Agreement and Cession in 1802. An account of the surveys of these lines is deferred until later.

---

of Wisconsin, 270 U.S. 295, 70 L. Ed. 595 (1926). For discussion see Battle, *The Georgia-Tennessee Boundary Dispute,* in Report of the Nineteenth Annual Session of the Georgia Bar Association 87, 88–89, 111–17 (1902).

Chapter 3. Fieldwork and Office Work under the Headright System

The Discipline of Land Surveying

Annals recounting the day-to-day activities of the original surveyors of the headright domain of Georgia are, regrettably, very meager. Most of the colonial survey records were lost during the Revolution. The surveying and major part of the business of granting headrights under the state government were left to the administration of local officials. These officials, regarding their records as private property, made little or no provision for their preservation for future public reference. Although early county surveyors usually kept an office at the courthouse in the county where they worked, their records were scattered and lost after the repeal of the headright laws which rendered the office of county surveyor inactive.

Consequently, relatively little is known about the ideas, methods, procedures, adventures, and experiences of these early artisans who bordered out the eastern third of Georgia with notched trees and lightwood stakes. It is indeed unfortunate that the high reputation of men such as Noble Jones, William Gerard DeBrahm, Henry Yonge, and others must necessarily be based more on their small-scale cadastral surveys and public services outside the field of surveying than on their achievements in performing large-scale land surveys.

Scattered newspaper ads provide some details about the training early surveyors received. For example, in 1764 Henry Yonge sought to hire "a lad" who understood arithmetic, wrote "a tolerable hand," and had "a taste for drawing." The ad went on to state that, if the applicant desired to become an apprentice, he would "be properly instructed so as to qualify . . . for the employment of a surveyor."[1] In 1786 Claud Thomson solicited surveying work in and around Savannah, stating that he had

1. Georgia Gazette (Savannah), Nov. 22, 1764, at 1, col. 2.

"been regularly bred to that business" and had had experience in both Europe and America.[2]

A general knowledge of surveying must have been considered an important part of nearly every schoolboy's training, for the subject is mentioned with surprising frequency in school curricula of the late eighteenth and early nineteenth centuries.[3] The preparatory subjects of algebra, geometry, and trigonometry, as well as the complementary courses of astronomy, navigation, and drawing were also taught. Textbooks widely used in Georgia schools included Robert Gibson's *A Treatise of Practical Surveying* (various British editions, 1752–1839; various American editions, 1785–1822), Charles Davies's *Elements of Surveying* (various American editions beginning in 1830), Adrien M. Legendre's *Elements of Geometry* (numerous European and American editions, 1794–1890), John Bonnycastle's *An Introduction to Mensuration and Practical Geometry* (London, various editions, 1782–1840; Philadelphia, various editions, 1812–60), and James Atkinson's *Epitome of the Whole Art of Navigation* (various British editions, 1711–90).[4] In 1785 the trustees of Richmond Academy at Augusta ordered a "Set of surveyors' instruments, with azimuth compass" for instructional use at the academy.[5]

Antebellum college curricula dwelt heavily upon the liberal arts (languages, philosophy, theology, oratory, and belles lettres); but when the University of Georgia opened its doors for the first time for classes in the fall of 1801, prominently displayed among the traditional courses to be taught were "Algebra, Geometry, Mensuration of Superficies and Solids, Conic Sections, Plane and Spherical Trigonometry, with their application to Navigation and Surveying."[6] The university attempted to set up a school of civil engineering in the 1830s, but the necessary funds were not forthcoming and the program languished until after the Civil War.[7] Both

2. Gazette of the State of Georgia (Savannah), Feb. 2, 1786, at 2, col. 1.

3. E. Boogher, Secondary Education in Georgia 136, 138, 141, 143, 145–46, 162–69, 281, 285–86, 291–92, 311 (1933).

4. Id. 58, 156, 158, 410, 412, 416–17. Another textbook worthy of note, though it came along much later, is *Caldwell's Practical Land Surveying* by M. P. Caldwell (published at Macon, Georgia, in 1880 and again in 1905). The author was the county surveyor of Hall County, Georgia.

5. Minutes of the Board of Trustees of Richmond Academy (April 4, 1785) (Special Collections, Reese Library, Augusta College, Augusta, Ga.).

6. Minutes of the Board of Trustees of the University of Georgia (June 18, 1801) (Records Management & University Archives, UGL).

7. E. Coulter, College Life in the Old South 39 (1979).

Emory and Mercer colleges taught surveying by the 1840s.[8] It is obvious, however, that few surveyors of the public domain benefited from such highly technical training. Most continued to learn whatever basic skills were taught in the local academies and old field schools and then went on to master their profession by the apprenticeship method. Still, as a rule, they were better educated than the average citizen of their time.

Some general observations about the shapes of the tracts laid out under the headright laws can be made from a perusal of the original land-grant plat books filed in various courthouses and in the surveyor general records of the Georgia Department of Archives and History. These indicate a definite tendency on the part of the surveyors to lay the land off into square- or rectangular-shaped tracts, at least in the earlier years when vacant land was plentiful. Such shapes vastly simplified computations in an era in which mathematical know-how was less than it is today and automated computational devices were nonexistent. In a significant number of cases the tracts are oriented in the cardinal directions, although along rivers and large creeks the general tendency was to run the side lines perpendicular to the banks.

The applicant for the grant usually accompanied the surveyor to point out the exact bounds he desired. Because grants were limited in quantity but not quality, or strictly to location, there was naturally a strong proclivity to enclose only the most desirable lands and to pass over poorer lands—a practice that accounts for the almost endless variety of multifaceted polygons inscribed upon the landscape of eastern Georgia. Headright grants after the Revolution were not required to be in any particular shape, and even in colonial times, to avoid undesirable lands, settlers often received exemptions from the regulation that the length of tracts was not to exceed three times the width.[9] A typical example is found in a petition for eight hundred acres in St. Philip Parish that James Cochran presented to the governor and council in 1767. Cochran requested that his land be laid out "upon a Branch of Black Creek a little above Land granted George Ducker to include a Savannah about a Mile and [a] half long and about sixteen or eighteen Chains wide . . . further praying that the Land might be run out in a long Square that he might not take in more of the Pine Land than would serve for Plantation Use."[10]

8. H. Bullock, A History of Emory University 104–5 (1936); Catalogue of the Officers and Students of Mercer University 13 (Penfield, Ga. 1847).

9. 10 Col. Rec. Ga. 886–87; 11 Col. Rec. Ga. 152.

10. 10 Col. Rec. Ga. 336.

The small, odd-shaped, relatively worthless parcels that were left interspersed among granted lands often remained ungranted for many years, and squatters might settle on them unchallenged.

One of the leading problems arising from this disorderly array was that of determining which lands had been granted and which had not. The confused patchwork of vacant lands lying between appropriated lands often made it difficult to delineate public from private land. Areas where the land was desirable filled quickly with settlers and became hotbeds of contention and litigation because of numerous overlapping warrants and grants in favor of diverse persons. Often tracts were so poorly surveyed and marked as to be difficult to locate.[11] Lines running into river swamps routinely were surveyed on the ground only to the edge of the swamp but were projected on the plat as running to the river itself. Natural boundaries such as rivers and creeks, although protracted on the plats, were almost never traversed on the ground.[12]

In one of the few censurable acts of his surveying career, William Gerard DeBrahm in 1764 ran out a 764-acre tract for Thomas Red in St. George Parish (later Burke County) but left one of the lines unfinished. Perhaps DeBrahm was in a hurry to get on with his duties as surveyor general of the Southern District of North America, which commenced that year. Whatever the reason, Red obtained a grant and made considerable improvements within what he believed to be the lines of the grant. When the adjoining lands (which had been vacant) were surveyed some two and a half years later, however, it was discovered that, because of

11. Primary sources concerning such problems are numerous and scattered. Since the governor was empowered to hear appeals of caveats, the Executive Minutes, especially for the 1790s and early 1800s, are replete with examples. Another source is the records of trials of caveats in courthouses and in the surveyor general records of the Georgia Department of Archives and History. Undoubtedly some small parcels of land remain which were never granted, but poor records and the lapse of time make it difficult or impossible to determine the exact title status of such tracts. The author knows of a situation in which two landowners in Emanuel County discovered a 9.80-acre parcel of apparent "no-man's-land" between their lands. The parties simply split the tract between themselves and took possession of it.

12. These factors are evident not only from the original surveys but also from other contemporary documentation. In an 1854 letter to the surveyor general, the county surveyor of Liberty wrote that the lines for "many of the surveys made on rivers as you are aware of, were never originally run through, on account of the swamps in many places being impassable a large portion of the year, and sometimes the whole year; consequently the rivers and the distance to the same had to be represented in an imaginary manner: subsequently, when the lines are carried through, some overrun, and some fall short." Letter from William Hughes to P. M. Compton (Oct. 10, 1854) (Caveat File for Liberty County, GSGD).

Composite map of headright grants in the fork of the Broad River, showing original grantees and dates of grants. A number of overlaps and gores are noted. The area was initially part of Wilkes County and today is in southeastern Madison County. From David H. Robertson and Susan T. Jenkins, *Original Land Grants, Deeds and Early History of Fork District; Madison County, Georgia* (1983).

the unfinished line, Red had mistakenly built his improvements outside the grant.[13]

An extreme example of an overlap is found in St. Matthew Parish (later Effingham County). In the same year that Thomas Red was having DeBrahm run out his land, John Hangleiter obtained a grant for two hundred acres on Turkey Branch. Four years later, John Martin Dasher obtained a grant for one hundred acres, all but ten of which were subsequently found to run afoul of Hangleiter's grant. Of course, Dasher's grant had to be disallowed.[14]

In 1790 Robert Pogue had the county surveyor run out three hundred acres for him on the North Fork of the Oconee River in Franklin (later Clarke) County. More than once his plat was forwarded to the surveyor general for a grant but just as often it was returned because the survey did not indicate the title status of some adjoining land. Pogue died in 1791 without having received a grant, and his heirs removed to Virginia. Some fourteen years later Archilaus Bryant applied for a grant of ninety-six acres in the same general vicinity. Pogue's heirs filed a caveat, claiming that Bryant's survey overlapped Pogue's. Upon appeal to the executive, Bryant's title was sustained. Governor Milledge grounded his decision partly on the failure of Pogue's heirs to obtain a grant but principally because he could not ascertain whether Bryant's survey actually lay within the lines of Pogue's survey because of the imperfections exhibited by Pogue's plat.[15]

Sometimes caveat proceedings became so contentious that the surveyor general himself was sent to resurvey the premises. In one such instance the surveyor general determined that a 69¾-acre tract a party had recently attempted to obtain a grant for lay entirely within the lines of a much older grant. In the process the surveyor general discovered the reason why the party had assumed the 69¾-acre tract was vacant. One of the lines of the older grant was only vaguely marked and the original survey showed the older grant to contain 400 acres, whereas it actually contained 450 acres. The surveyor general also discovered a small overlap between the older grant and another adjoining grant.[16]

The long, drawn-out litigation resulting from such confusion con-

13. 9 Col. Rec. Ga. 223, 436; 10 Col. Rec. Ga. 206–7.

14. Plat Book D, p. 49, Office of the Clerk of Superior Court, Effingham County courthouse, Springfield, Ga.; 9 Col. Rec. Ga. 154, 608; 10 Col. Rec. Ga. 342, 516.

15. Executive Minutes (Nov. 2, 1805) (GDAH).

16. Plat Book B, p. 36 (GSGD).

sumed so much of the time of the governor and council in colonial times, and later the county land courts, that it often interfered with their primary function of issuing warrants. The frontier lawyer filled his purse for years from the proceeds of boundary suits.

With all its inherent problems and imperfections, the headright system did possess some rectifying mechanisms of sorts in addition to the sometimes expensive and contentious caveat process. Where grants were found to run afoul of each other, it was often convenient for one of the grantees simply to concede the overlapped area to the other and then obtain a grant of adjoining vacant land, equal in area to and in lieu of the overlapped area.[17] Thus when Hugh Kennedy and his brother received adjoining grants on the Savannah River in 1761, it was discovered that a sixty-acre overlap existed between their two tracts. Hugh, with the obliging approbation of the council, relegated the overlapped area to his brother and received, as compensation, a sixty-acre grant of adjoining vacant land.[18]

As long as it lasted, the abundance of vacant land constituted a redeeming quality. It was often cheaper for the defendant in a caveat case simply to resign his claim and seek out another tract rather than bear the cost of litigation. For example, in January 1818 Hardy Crawford had surveyed, pursuant to a warrant, a 33.4-acre tract on the waters of Greenbrier Creek in Greene County. The following March John D. Buckhannan entered a caveat against the passing of the grant. A hearing was set for the first Tuesday in May. When that date arrived, however, there were not enough justices present to form a court, and the hearing was postponed until the first Tuesday in June. Before the date of the second scheduled hearing, Crawford "paid cost and Relinquished all Clame to the . . . Land in dispute."[19]

Doubtless a great many boundary problems were worked out in private by the adjoining landowners affected. Although this generally was the cheapest and quickest method for settling such differences, it usually left no permanent record as to the ultimate decision—to the confoundment of later surveyors.

Seemingly the authors of the 1783 headright act intended for the county surveyors to act under the close supervision of the surveyor gen-

17. 8 Col. Rec. Ga. 562; 12 Col. Rec. Ga. 124–25.

18. 8 Col. Rec. Ga. 560–61.

19. Plat Book 1784–1861, pp. 240–41, Probate Court, Greene County courthouse, Greensboro, Ga.

eral. In actual practice this was never the case because the surveyor general was relegated to the role primarily of keeping records and supervising grants. In the earlier years, however, he did exercise some nominal supervision, for there exist fragmentary remnants of instructions issued by him to various headright surveyors. In March 1783 Surveyor General Richard Call instructed the county surveyors to lay out the tracts into square or oblong figures, the length not to exceed twice the breadth unless the land lay between lands already granted or unless it was bounded by rivers or creeks; to give notice to the adjoining landowners, before undertaking any survey, so that they might attend the survey; to swear in all chain carriers; to measure every line except those known to have been previously run and marked; to mark a "considerable" number of station trees along each line; and, if requested, to allow other surveyors to make copies of the field notes and plats.[20]

In 1784 the surveyors of Franklin and Washington counties were instructed to lay the lands out into square or oblong figures, the length not to exceed twice the breadth, and on rivers or large watercourses, two chains back from the river to one on the river. They were also instructed to mark distinctly the lines around each tract, to mark at least two stations on each line except lines designated by natural boundaries, and to record the magnetic variation on each plat. In addition, because duplicate warrants had been issued as a result of the riotous free-for-all at the land court in Augusta, the surveyors were instructed not to run out any lands on warrants that had the smallest erasure upon them.[21]

Headright plats were fairly uniform in the types of information they depicted and, to some degree, manifest the level of sophistication of the surveys they represent. Bearings were usually stated to the nearest degree, or occasionally to the nearest half degree, and sometimes the magnetic variation was noted. Distances were usually to the nearest chain or link. Areas generally were shown to the nearest acre and occasionally to greater refinements. No attempt was made to balance the field measurements mathematically as is commonly done in modern surveys. Instead, only the raw field data were denoted on each plat. Chopped trees and wooden stakes or posts were almost invariably used to mark corners and lines—even in the piedmont, where rock was plentiful. In

20. Instructions to the County Surveyor (Mar. 25, 1783) (GSGD).

21. Instructions from Richard Call to John Gorham, county surveyor of Franklin County (May 25, 1784) (GSGD); 19 Col. Rec. Ga. (pt. 2) 297; 2 Rev. Rec. Ga. 794.

addition, the plats set forth the parish or county in which the property was located, the beginning point, the scale, the names of adjoining landowners (or a designation that the adjoining land was "Vacant" if it was still public domain), and the quality of the land (for example, "Pine Land" or "Good oak & hickory land"). They also gave the names of the chain carriers, date of warrant, date of survey, and date the survey was advertised. Physical features such as bodies of water, marshes, and swamps were often depicted. Very rarely human improvements such as roads, paths, bridges, ferries, houses, and cleared fields were shown. A brief certification, followed by the county surveyor's signature, was inserted at the bottom. Because they were purely functional in nature, the plats displayed rather crude draftsmanship even for their times.

Some generalizations about the pace of surveying activity can be drawn from a careful analysis of the headright plat books. For example, between May 1813 and January 1818 Travis Thigpen, the county surveyor of Emanuel, certified an average of five plats per month. The average area per tract was 324 acres. At the same time Thigpen also served as clerk of the superior and inferior courts and must have given these jobs priority. Notations on his plats indicate that the fieldwork for the great majority of them was supervised by deputies. During his nearly five-year tenure as Emanuel County surveyor, Thigpen performed the fieldwork for only about twenty-five surveys. His three deputies—J. B. McFarland, Charles Thigpen, and John Chason (the latter later became Emanuel County surveyor)—performed the fieldwork for the rest. The deputies generally followed each other in succession so that during most of the period Thigpen had only one deputy at a time.[22]

Thigpen had previously served as surveyor of Montgomery and had helped to run the boundary between that county and Tattnall in 1810.[23] When Emanuel County was cut from Montgomery and Bulloch two years later, Thigpen found himself situated in the new county and had no difficulty securing the position as its county surveyor.

Perhaps Thigpen's most significant undertaking as county surveyor of Emanuel came in 1814 when, pursuant to certain legislative acts, he directed his deputy Charles Thigpen to circumscribe the lines of a one-hundred-acre tract near the center of the county for the location of a

22. Emanuel County Land Plat Book A, passim, Probate Court, Emanuel County courthouse, Swainsboro, Ga.

23. Clayton 616.

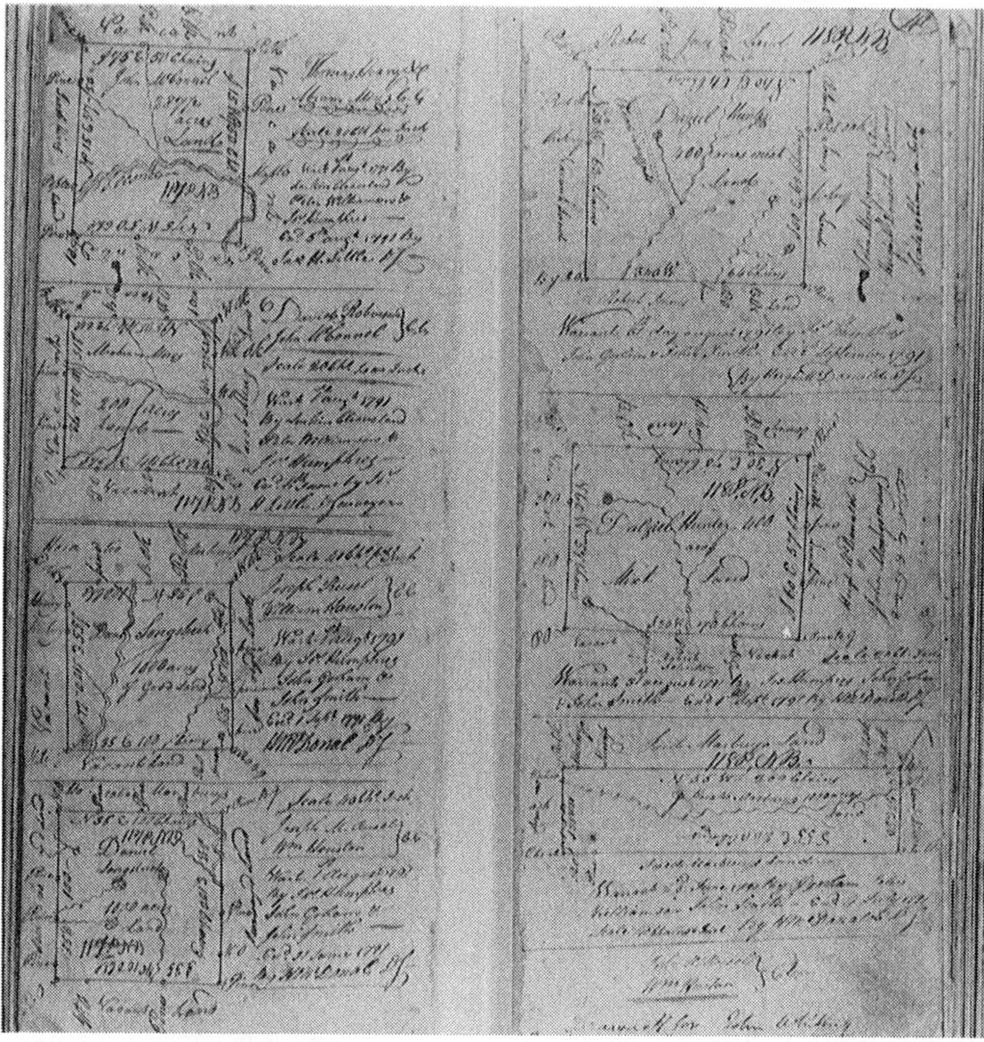

Franklin County headright plat book, 1785–1793, showing some rectangular-shaped grants. Courtesy of the Georgia Department of Archives and History.

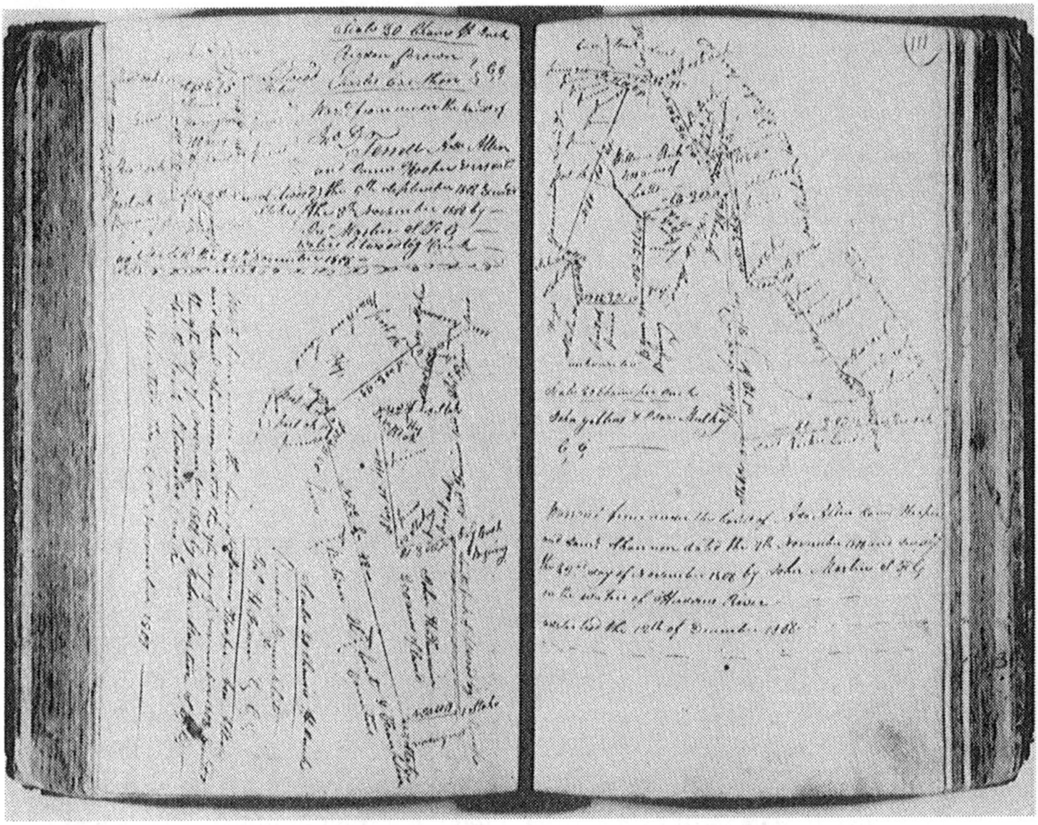

Franklin County headright plat book, 1787–1828, showing some irregularly shaped grants. Courtesy of the Georgia Department of Archives and History.

permanent county seat. Despite the special import of this survey, there is no indication that any extraordinary techniques or procedures were used in carrying it out. Certainly the plat of the survey is exemplary of other headright plats. The tract was laid out in the shape of a square and measured 31.63 chains to a side. It was oriented thirty-seven degrees from the meridian, reportedly to conform to the drainage pattern of the two principal rivers that traverse the region—the Ogeechee and the Ohoopee. In keeping with the homely informality of the times, two members of the courthouse and jail commission assisted as chainmen. Upon completion of the survey the tract was granted by the state to the commissioners, who had the town of Swainsboro laid out on it.[24] No doubt Thigpen or

24. Emanuel County Land Plat Book A, p. 84, Probate Court, Emanuel County courthouse, Swainsboro, Ga.; Lamar 197, 204, 210, 224.

his immediate successor performed this latter task, but records bearing on the matter have been lost.

Between March 1818 and December 1821 Thigpen's successor, Reuben Neel, each month surveyed an average of three tracts. These tracts averaged about the same size as those Thigpen surveyed. Neel apparently supervised all the fieldwork himself, for his plats make no mention of deputies and bear the notation "Surveyed by Reuben Neel C.S." rather than the mere "Certified" followed by the signature that appears on most of Thigpen's plats.[25] The largest tract surveyed by either Thigpen or Neel was a thousand acres, and the smallest was twenty-five acres.

In the latter half of the year 1819 Neel took a leave of absence from his surveying duties in Emanuel County to subdivide the 6th District of original Appling County into land lots as part of the state surveys for the 1820 land lottery.[26] During this six-month interval, headright surveying was suspended in Emanuel County.

An examination of the extant records in other headright counties indicates that the volume of work performed by Thigpen and Neel was not atypical of that performed by most surveyors who worked in areas of moderate population growth during the half-century following the Revolution. Of course, these figures do not include purely private surveys.

Following the almost wholesale loss of the colonial plat records during the Revolution, a concerted effort was made to gather up and record loose plats in private hands. In addition, numerous resurveys were performed to replace those plats that could not be found. Impetus was added in 1815 when the legislature specifically authorized such resurveys. The act directed that the resulting plats be recorded in the county records and in the surveyor general's office.[27] These resurveys provide invaluable quantitative data as to the precision of area determinations of the original surveys. The examples shown in Table 2, taken from the wonderfully detailed records of Effingham County, are typical.

Errors abnormally large by modern standards were the rule of the day. The most substantial errors occurred in tracts bordering swamps and marshes, where discrepancies of 50 to 100 percent—and in some cases even more—were frequent. In a decided majority of the cases, the areas

25. Emanuel County Land Plat Book B, passim, Probate Court, Emanuel County courthouse, Swainsboro, Ga.

26. Field Notes of Reuben Neel, surveyor of the 6th District of Appling County (GSGD).

27. 1815 Ga. Laws 36; Lamar 414.

Table 2. Comparison of Acreages as Shown by Original Surveys and Resurveys in Effingham County

Original grantee	*Date of original grant*	*Area shown on original survey* (acres)	*Date of resurvey*	*Area shown on resurvey* (acres)
Christian Rabenhurst	1759	500	1816	540
Michael Boreman	1761	200	1816	234
David Kraft	1756	500	1818	542
John Graham	1771	300	1819	317
Ludiwig Ernst	1764	100	1819	147
Joseph Clay	1772	500	1819	592
John Hall, et al.	1771	150	1819	153
Christian Leimberger	1759	100	1819	77
John Mulryne	1766	500	1820	485
Clement Martin, Jr.	1773	300	1820	462½
Ludiwig Weitman	1768	250	1820	244½
Frederic Helvenstine	1770	300	1821	310
John Rentz	1772	300	1828	336½
Alexander Cunningham	1774	350	1828	385

SOURCE: Plat Book C, pp. 389, 423; Plat Book D, pp. 48, 53, 55, 60, 92, 101, 112, 130, 148, 149, 214, 217 (both in Office of the Clerk of Superior Court, Effingham County courthouse, Springfield, Ga.).

NOTE: This tabulation is not intended to single out the work of colonial surveyors for deprecation. Resurveys of grants originally laid out by state surveyors reveal area inaccuracies of about the same consistency and with similar magnitudes.

shown on the resurveys exceed those in the original surveys—strong evidence that the original surveyors tended (intentionally) to lay off more land than was called for in the applicants' warrants.

If not kept under strict supervision, the deputies employed by county surveyors could do great mischief. In one instance, Benjamin Starrett, surveyor of Franklin County, went so far as to "Revoke Disallow and Forbid Timothy Terrell from making any surveys or . . . Returns to me or my . . . successors in office under any authority or Direction heretofore given him by me" and did "Revoke and set aside all Deputations commissions & Instructions heretofore given the said Timothy to act under me as my Deputy Surveyor." To apprise his successors and the public, Starrett recorded a certified notice to this effect in the Franklin County plat records.[28] A few years before, the surveyor general had noted the possible havoc deputies could wreak and had admonished the county surveyors to "Take special care" that the deputies "have the necessary qualification; and that they are men of reputation, as . . . you are answerable for the misdemeanors of a deputy of yours."[29]

Sometimes, to while away time or to add a little merriment to their work, surveyors doodled pictures, poetry, or other frivolities into their plat books. Thus today one finds fish swimming in the rivers depicted in the Wilkes County plat records; an Indian, a deer hunt, and a survey crew at work on a plat in Franklin County; and a flustered-looking devil glaring out of Devils Bay in Effingham County. (Perhaps this latter was to indicate the difficulty of surveying the grant and the surveyor's satisfaction at having triumphed over the adversity.) When George W. Clifton, county surveyor of Emanuel, purchased Plat Book C for the county, he inscribed in it, for the benefit of posterity, the price of the book ($4), the date of the purchase (November 6, 1845), his own birthday (May 3, 1814), and then—as the first order of business—proceeded to compose a poem of a religious nature.

Delineating the Metes and Bounds of Towns and Counties

Besides the everyday routine of measuring the metes and bounds of headright tracts, surveyors occasionally assumed kindred tasks such as staking out towns and county seats, surveying county boundaries, and producing small-scale geographical maps.

Brunswick, the only town of lasting endurance established in Georgia under crown administration, was located on a 2,034-acre site that had been set aside some years before by the royal council for such purposes. It was laid out by deputy surveyor George Mackintosh in 1771. The town proper, covering 383.5 acres, consisted of 556 building lots arranged, in the style of Savannah, around six squares. In addition, 50 wharf lots fronted on the Brunswick River.[30]

28. Franklin County Land Grant Plat Book R, 1787–1829, p. 160 (GDAH).

29. Copy of Letter from Thomas McCall to the County Surveyors of Greene, Franklin, Burke, Wilkes, Elbert, and Washington Counties (May 25, 1791) (Georgia Surveyors' Records—Miscellaneous, Telamon Cuyler Collection, UGL).

30. 11 Col. Rec. Ga. 155, 385–87; Dart v. Orme, 41 Ga. 376 (1870); An Accurate Survey of

Augusta afforded the earliest example of urban planning through state authorization. The town had been founded under the direction of the Trustees, but by the late colonial period the lot lines had become confused and public roads were thought to encroach onto some of the lots. In 1773 the colonial Executive Council directed the surveyor general "to cause an Accurate resurvey to be made of the Town . . . agreeable to the Original plan . . ., the Expenses to be defrayed by the Several Persons Interested therein."[31] Either this directive was not carried out or it failed to produce the desired results. Consequently, a state law of 1780, after enumerating several problems that had arisen, among them that "the Streets . . . of Augusta . . . are not regular or Strait," created a five-man commission "to lay out, measure, and post the same in the best and most regular way." The commissioners were also to subdivide part of the common above and below the town into one-acre lots, to lay out additional streets in alignment with existing ones, and to sell the lots at public auction. Lots were to be reserved for a courthouse and jail, schools, churches, and cemeteries. So that the town would be "regularly built," minimum dimensions were specified for houses, and they were to be "framed . . . in a workman like manner" and "placed in such part of the lot as . . . pointed out" by the commissioners.[32]

The British capture of Augusta shortly after the passage of this act prevented its effective operation, but it was reenacted, with modifications, when the war ended.[33] Under this latter act, the commissioners held their first meeting on September 16, 1783.[34] Noting that no plan of Augusta was known to exist, they ordered the freeholders be summoned to help ascertain the lines of the town.

Augusta had grown but little since it was first laid out by Noble Jones nearly half a century before. It still consisted of only forty lots on which had been built "few buildings and none of them very valuable." Geor-

---

the Town and Commons of Brunswick . . . Agreeable to the original Plan, by Geo. Purvis (1796) (GSGD).

31. 12 Col. Rec. Ga. 394–95.

32. 19 Col. Rec. Ga. (pt. 2) 131–34.

33. Id. 248, 560.

34. Without itemizing sources, the following text concerning the surveying of Augusta after the Revolution is derived from Minutes of the Board of Trustees of Richmond Academy (Sept. 16, 1783, through July 21, 1789) (Special Collections, Reese Library, Augusta College, Augusta, Ga.); Plat Book A, p. 149 (GSGD); and Robertson & Robertson, *The Town and Fort of Augusta,* in Colonial Augusta: Key of the Indian Countrey 59, 62–64 (E. Cashin ed. 1986).

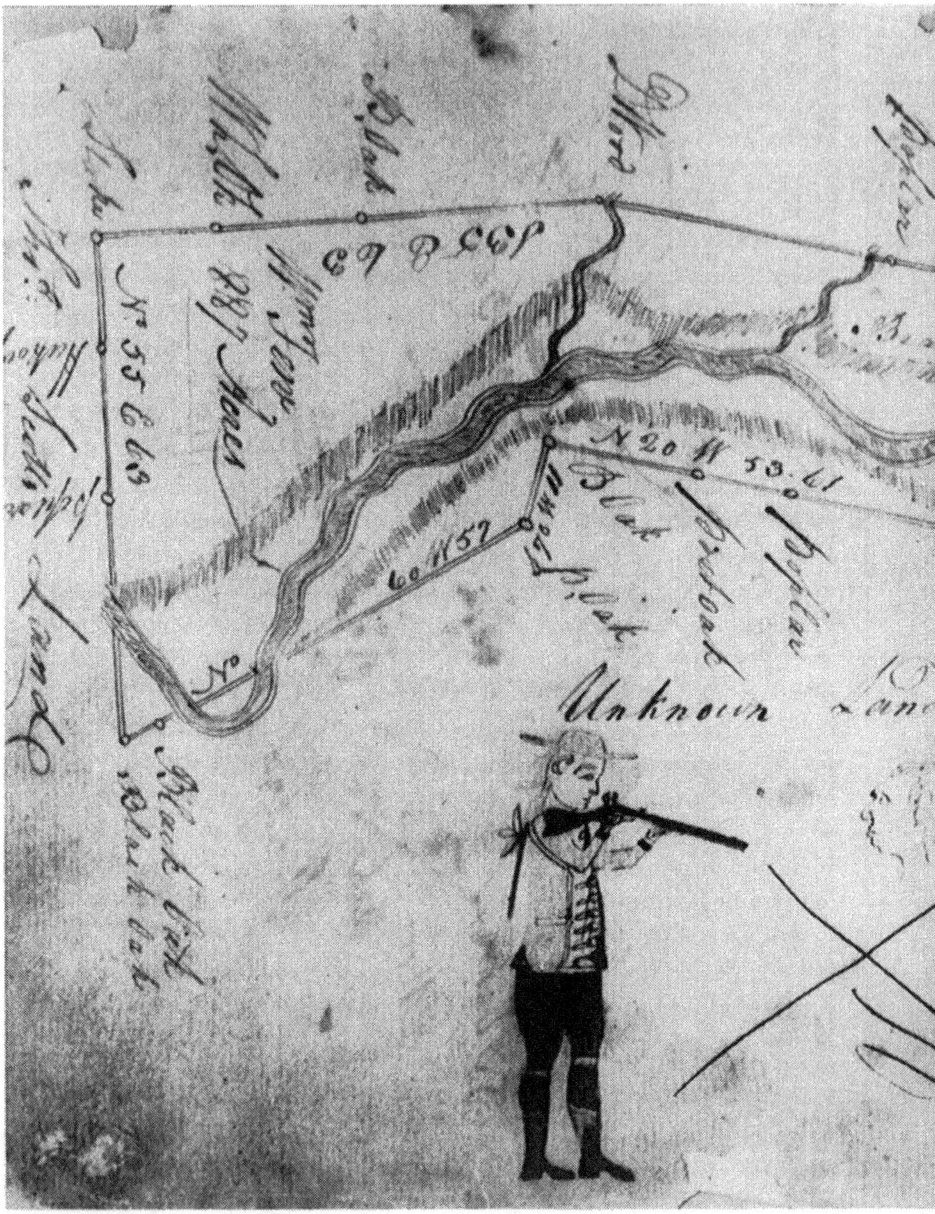

Plat of 887 acres in Franklin County surveyed for William Few in 1785. Notice the use of a staff compass and Gunter's chain. Notice also that the surveyors are using waist-high sticks to mark chaining points. Courtesy of the Georgia Department of Archives and History.

N 30 W 50
W 13
Hickory
Red Oak
Maple
No 65
Low Grounds
Unknown Land

gia's up-country was now growing rapidly, however, and the town needed to expand. John Walton, county surveyor of Richmond, was directed to make a resurvey of the town and commons and to lay off additional lots and streets. He was assisted by Robert Bonner, who was given the task of procuring "a sufficient number of stakes, for cornering the lots as they" were surveyed. Several encroachments onto the common and conflicts between property lines were discovered which had to be rectified as the work progressed.

One result of the adjustment of the lines of Augusta was the narrowing of Broad and Reynolds streets. Augusta's plan originally included a central square flanked by lots for public buildings on either side. By the 1780s, however, this entire area, along with the abutting streets immediately to the north and south, had been combined to form one large street known as Broad Street. This street had a breadth of 300 feet—too wide for convenience. Therefore, the lots to the north were extended into it, constricting it to 165 feet. A similar process was carried out for Reynolds Street. To augment travel to and from Augusta, the road running from Sand Bar Ferry, below the town, to Rae's Creek, above the town, was straightened, given a uniform width of 32 feet, and made to align with Broad Street. The commissioners specified that this road "be laid out by a compass and measured with a chain." In addition, "numbered oblong pieces of rock" were to be placed each quarter of a mile along its length. One of the advantages Augusta derived by being made the temporary state capital in 1786 was that the surveyor general took over direct responsibility for laying out its subsequent additions.

Between the Revolution and the Civil War, most of the urban establishments founded in Georgia were county seats, many of which, in spite of their often lofty pretenses, remained little more than minuscule hamlets for a great number of years. Eighteenth- and nineteenth-century Georgia was emphatically rural in character, and there was little demand for building lots inside urban enclaves.

Unfortunately, most of the details regarding the planning of Georgia's early towns will never be known. In most instances the only remaining evidence is a simple plat certified and dated by the surveyor in charge, and in some cases not even this much remains. It is safe to conclude that the surveyors were somewhat more scrupulous in their measurements but otherwise the techniques and equipment employed were essentially the same as those used for rural land surveys. In fact, given the infrequency with which towns were established and the almost total absence of resurveys of town lots in early county records, one may infer that most surveyors (except those around Savannah and Augusta) rarely if ever set foot within the confines of a town for the purpose of making surveys. Thus there was little opportunity for them to learn or develop complex techniques of layout.

The usual procedure for founding a county seat was for the legislature to pass an act empowering the county inferior court justices or certain named individuals (generally five in number and styled "courthouse and jail commissioners") to locate and plan the town. The county surveyor was generally assigned the task of determining the geographical center of the county. The town commissioners then selected the most eligible site within a mile or two of this location. The body was authorized to acquire a tract for a town reserve, usually fifty to one hundred acres in area, at the selected location. If the site was already in private ownership, title might be obtained by gift or purchase. Otherwise a grant would be forthcoming from the state, vesting title to the premises in the county. The county surveyor was then directed to survey and measure out the town.

There is nothing greatly sophisticated or unusual about the design of these early towns; in fact, in every case they were even less grandiose than the plans of colonial towns. Invariably they were of a simple, practical grid pattern, usually consisting of less than one hundred lots and frequently less than fifty lots. An embellishment sometimes incorporated was to have the main streets run up to and dead end at a large central square, reminiscent of the plan of Savannah and other colonial towns. Sparta and Danielsville are examples of this style of layout. Lot sizes generally ranged from one-half to one acre, while street widths commonly were 33, 49½, 66, 82½, 99, or 132 feet—corresponding respectively to two perches, three perches, four perches or one chain, five perches, six perches, and eight perches or two chains, the customary units of measure of that day.[35] In some towns the more modern units of street widths—40 feet, 50 feet, 60 feet, and so on—were employed. Alleys were rarely provided.

Proceeds derived from the sale of lots were used for the initial construction of county buildings, and the land within the reserve not immediately used for town layout provided space for future expansion. Even with the

35. This factor is of profound significance when attempting resurveys of town lots. For example, without access to the original plan, a surveyor might unwittingly assume that a street laid out three perches (49½ feet) in width is 50 feet wide, or a street laid out six perches (99 feet) in width is 100 feet wide—widths commonly used in modern layouts.

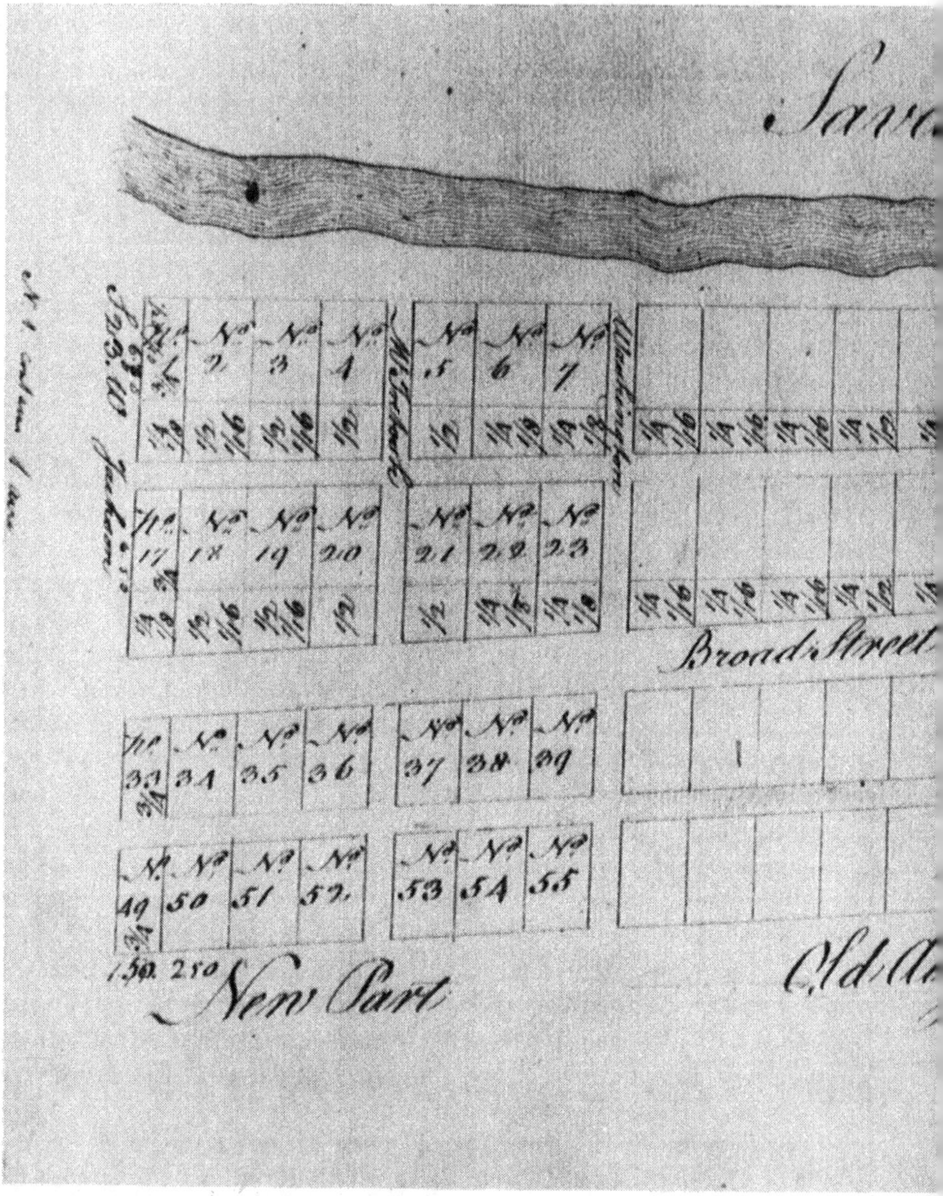

Plan of Augusta, ca. 1784. "Old Augusta Town" designates the original forty lots. "New Part" designates additions made after the Revolution. Broad and Reynolds streets have been narrowed. Courtesy of the Georgia Department of Archives and History.

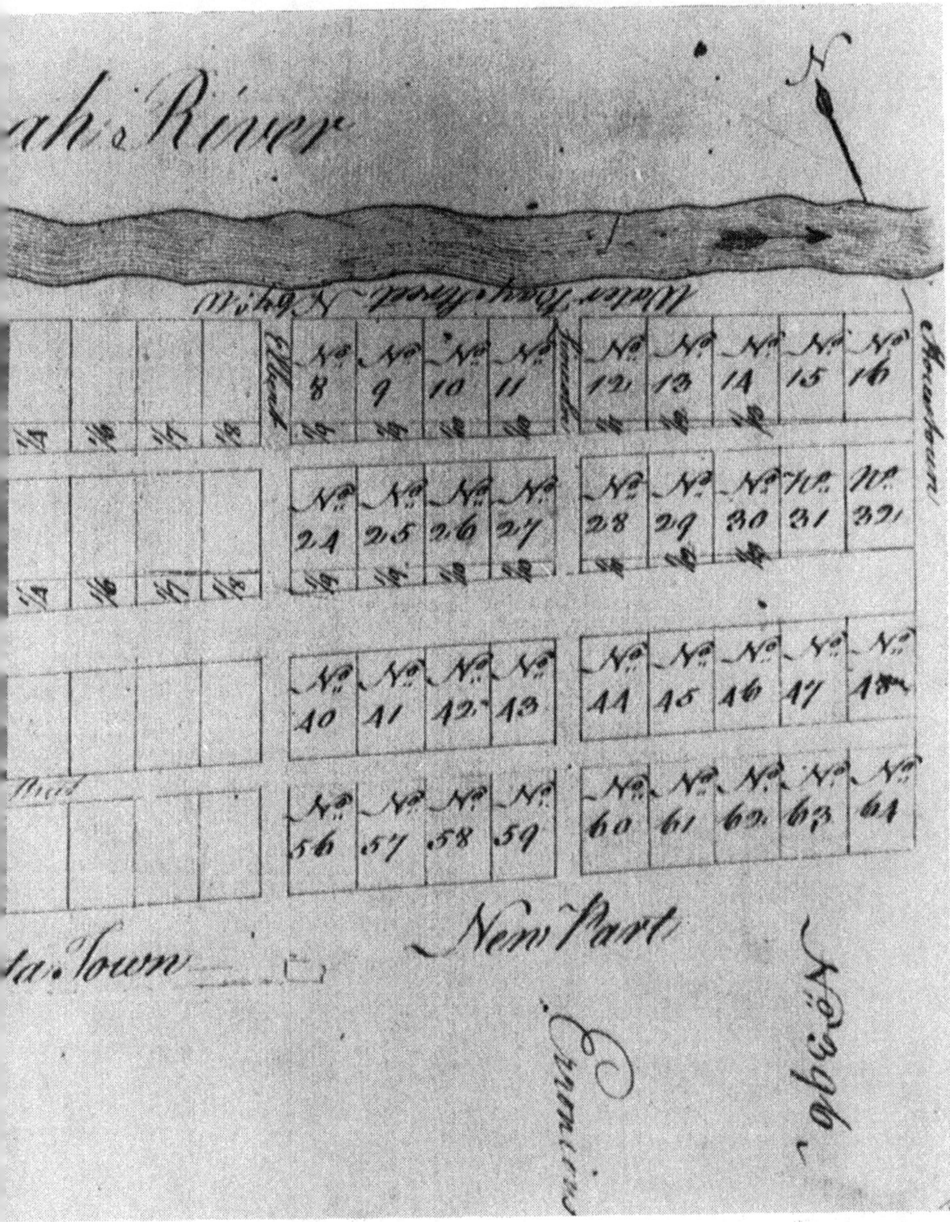
ah River
N
No. 8
No. 9
No. 10
No. 11
No. 12
No. 13
No. 14
No. 15
No. 16
No. 24
No. 25
No. 26
No. 27
No. 28
No. 29
No. 30
No. 31
No. 32
No. 40
No. 41
No. 42
No. 43
No. 44
No. 45
No. 46
No. 47
No. 48
No. 56
No. 57
No. 58
No. 59
No. 60
No. 61
No. 62
No. 63
No. 64
New Part
ta Town

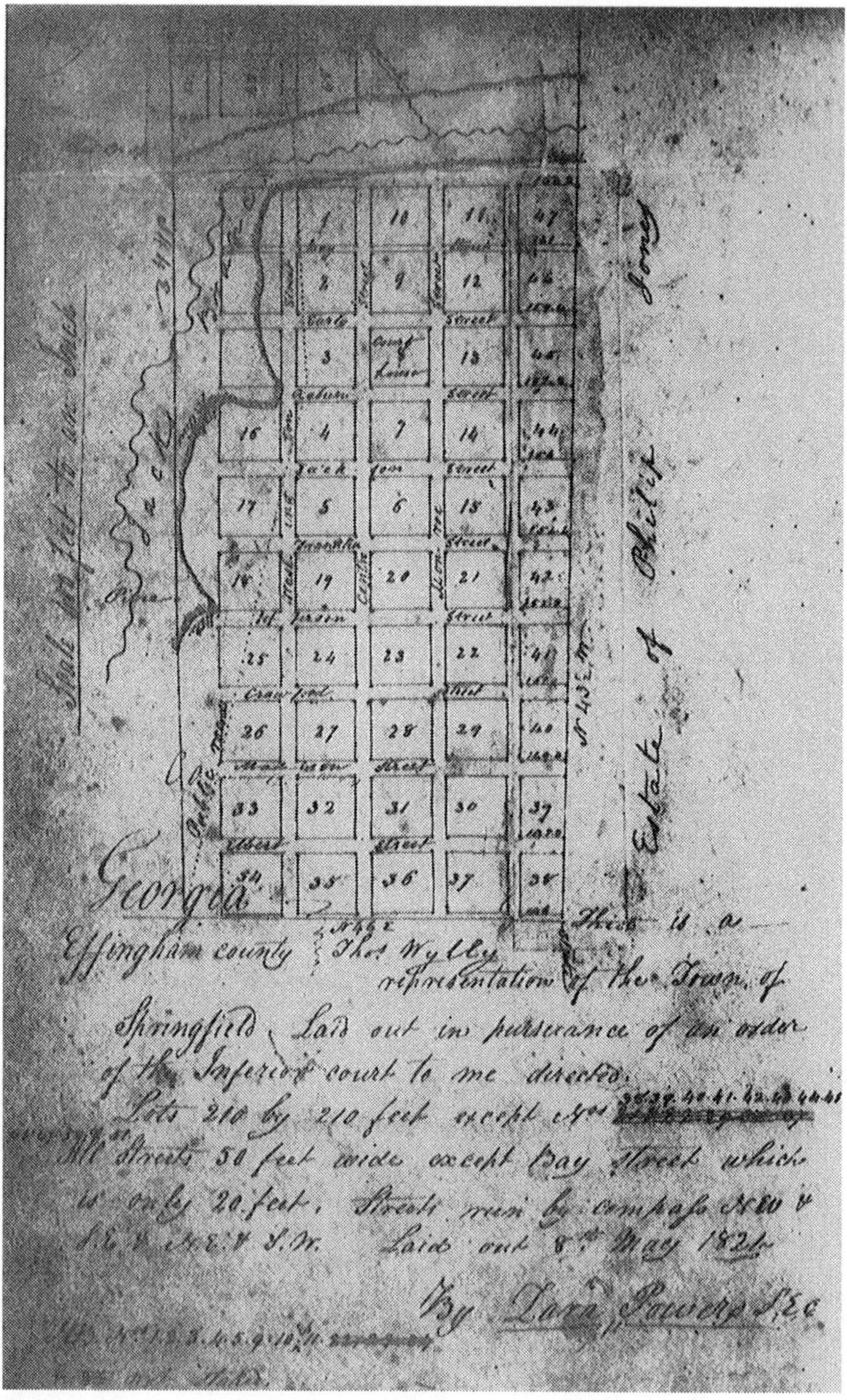

Plan of Springfield, laid out in 1821. From Plat Book D, p. 114, Office of the Clerk of Superior Court, Effingham County courthouse, Springfield, Ga.

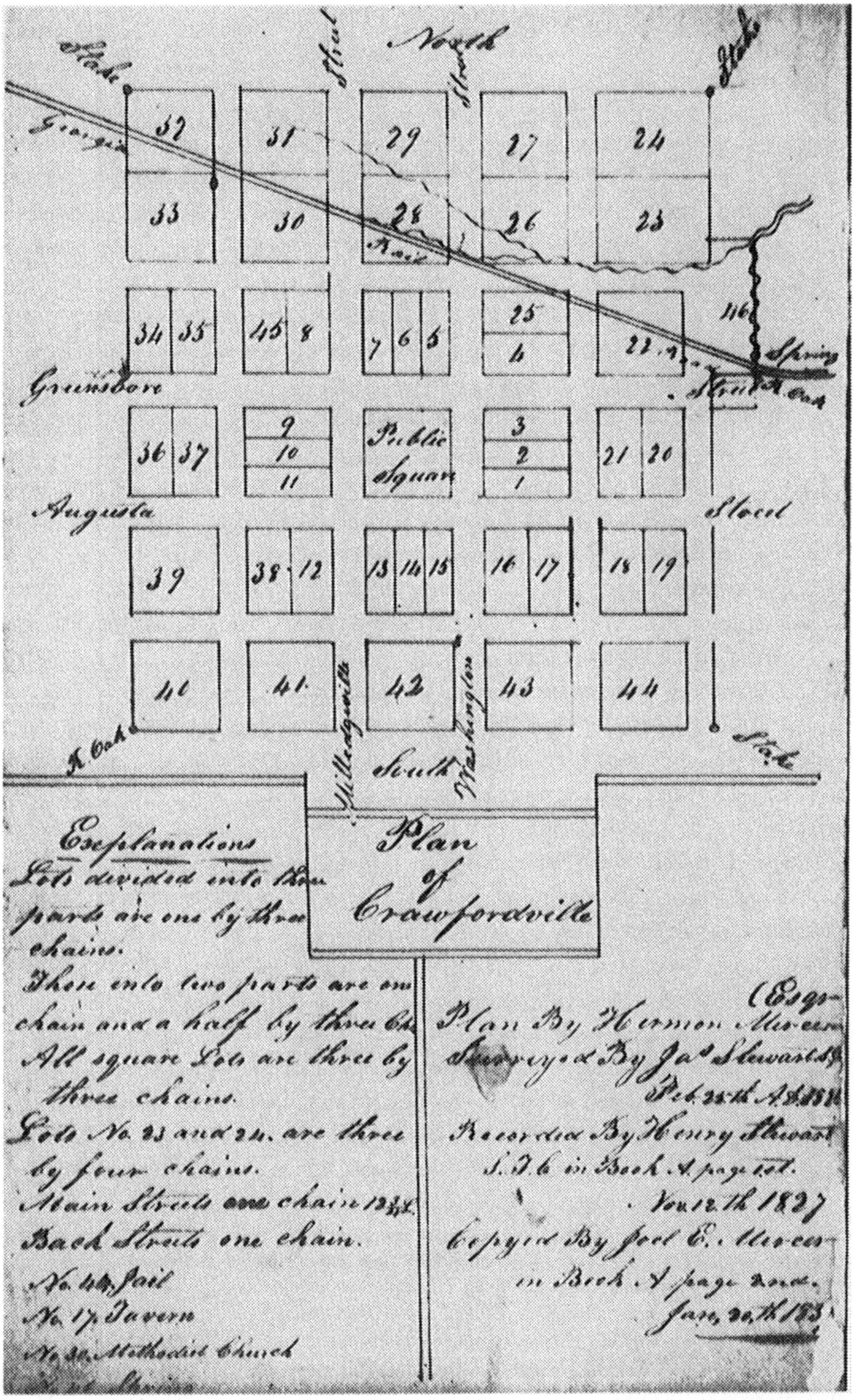

Plan of Crawfordville, laid out in 1826. From Surveyor Book 1836, p. 2, Office of the Clerk of Superior Court, Taliaferro County courthouse, Crawfordville, Ga.

limited scope of these early town layouts, a half-century or more sometimes elapsed before need was felt to enlarge them—and some of the towns eventually died out altogether.[36]

Although most surveyors in the eighteenth and nineteenth centuries rarely departed from the familiar and stereotyped techniques used in both rural and urban surveys, some did attempt a more scientific approach to their work. In 1816 John McKinnon, city surveyor of Savannah, filed a report with the mayor and aldermen suggesting methods for improving the uniformity and precision of town surveys. Because of the disappearance of old stakes and the lack of consistency between the original surveys and later resurveys, disputes had arisen regarding the location of some of the lot lines in Savannah. McKinnon observed that the only remaining original landmark in the town was a "plug in the bottom of the well at the central intersection of Bull and Broughton Streets." As a result, surveyors generally had to resort to the lines of old buildings and other "ancient fixtures" in performing resurveys. McKinnon noted that "no Standard measure" existed in the United States and that even the best measuring scales often varied as much as one and a quarter inches in twenty feet. This caused "a different result" to be "given by Surveyors on the same piece of Ground." To remedy this shortcoming McKinnon declared that in "all towns and cities there should be a Standard of measure in the first laying out, which measure, true or false, should never afterwards be altered."[37]

Noting that the compass and chain had been employed almost exclusively in surveys in Savannah, McKinnon urged that they be discarded. Instead, for laying off angles, he recommended the use of a "Theodolite, or a Sextant with a staff & counterpoint." For measuring distances he recommended the use of a wooden pole "graduated by a fixed standard whereby the measure will be the same at all times." In addition, McKinnon proposed that stones "with proper marks" be placed along the edge of the rights-of-way of each street, one at each extremity and one near the middle, to serve as tie points for future surveys; that the entire town

36. Early town plats are found in courthouse records and in the surveyor general records of the Georgia Department of Archives and History. Some have been published in J. Sears, The First One Hundred Years of Town Planning in Georgia (1979). A listing of all acts passed before 1851 concerning the founding and early administration of Georgia's towns can be found in Cobb 1144–1210.

37. Report of John McKinnon and Thomas McCall to the Mayor and Aldermen of Savannah (July 23, 1816) (Minutes of the City Council, Sept. 1812–Sept. 1817, pp. 330–43, Office of the Clerk of the City Council, City Hall, Savannah, Ga.).

be resurveyed and referenced to these points; and that the instruments used by the city surveyor "be made the standard measure and carefully preserved by which to graduate all future measures."[38] The mayor and aldermen failed to implement McKinnon's farsighted proposals.[39]

Georgia's twelve parishes, which existed before the Revolution, were arrayed in a single upright tier stretching from the St. Marys River on the south to the Broad River on the north. The Atlantic Ocean and the Savannah River formed their eastern limits, while their northern and southern boundaries were delineated by creeks flowing into these bodies of water or by noted Indian paths. The westerly limits were left somewhat ill-defined, but that was not of great consequence because most of the settlement clustered closely along the Savannah River and eastern seaboard. In all probability many of the artificial parish boundaries were never run out on the ground.

Following the Revolution, extensive migration into the hinterland made it necessary to define with precision the buttings and boundings of the growing proliferation of counties. Between 1777 and 1800 Georgia's original eight counties increased in number to twenty-four, and thirty years thereafter there were seventy-six. The acts providing for the creation of these new counties relied heavily on well-known roads and natural features such as rivers, streams, and ridges to demarcate boundaries but now routinely specified that the artificial lines were to be run out and marked by the county surveyors of the counties affected. The acts generally described the artificial lines as running from one physical landmark to another. Bearings or distances were specified only in extremely rare cases. The following examples are typical:

> beginning at Rae's mill on Little River, running a direct line from thence to Zimmerman's, on the road leading from the town of Washington to Barksdale's ferry on Savannah, from said Zimmerman's a direct line to Drury Cade's mill, on Broad River, from thence down Broad River to its mouth, thence down Savannah River to the mouth of Little River, thence up Little River to the beginning (Lincoln County, 1796);[40]
>
> beginning at Carter's bridge on the Ogeechee, in the county of Hancock, thence along the Washington road by McFarland's store, to Wilkes county line; thence to the mouth of Beaver dam [Creek], where it empties into William's Creek, in the county of Warren; thence a direct line to the twelve

38. Id.
39. Id.
40. Marbury & Crawford 170.

> mile post, near John Gibson's, on the road leading from Wingfield's Bridge to Powelton; thence to the residence of Francis B. Billingslea; thence to the mouth of Powder Creek, where it empties into Little River; thence to the twelve mile-post leading from Washington to Greenesborough; thence to the point where the road leading from Washington to the Scull Shoals intersects the Oglethorpe county line, near Green's mills . . . (Taliaferro County, 1825).[41]

Because the frontier had not as yet been tied together with accurate surveys, the legislature often had little to go on regarding the relative positions of the various features it denominated as county boundaries. Only when attempts were made to survey the lines out on the ground was the ambiguity or impossibility of some of the descriptions discovered. Remedial acts were then necessary to correct previous descriptions.

Two dozen or so plats of county boundaries made before the mid-nineteenth century are known to exist today. These quaint renderings depict essentially the same types of survey information as that shown on contemporaneous headright plats. Although evidence is vague on the point, it appears that only in extremely rare instances was the procedure employed of running a random line between terminal marks and then running the corrected or true line. In other words, the usual practice seems to have been for the surveyor to set up his compass at one of the terminal marks, fix the sights in as nearly a correct direction as he could determine (from whatever information was at his disposal and from his knowledge of the country), and then proceed to chop a line along the bearing so selected until a point was struck at or near the next terminal mark. The line so marked would thereafter be considered the correct boundary. Somewhat greater than usual care seems to have been taken in marking the lines (using the usual tree chops and lightwood stakes) and in noting names of property owners and physical features encountered along the way. This care is evident not only from the county boundary plats themselves but also from subsequent headright plats made of tracts along the boundaries. These records demonstrate conclusively that such lines of demarcation were distinct and well known for many years after they were surveyed.

When deputy surveyors Isaac Perry and James Oliver delineated the division between Burke and Screven counties in 1794, they did leave a

41. 1825 Ga. Laws 58; Dawson 129.

specific description of the marks they made. The boundary began at a sycamore tree, inscribed with an "X" and three chops, at Summerall's Old Ferry on the Savannah River (its present point of beginning). From there it ran South 48½° West a distance of twenty-seven miles three chains to the mouth of Buckhead Creek on the Ogeechee River (now in Jenkins County). Trees directly on course were "marked with blase and five notches fore and aft." Where branches, creeks, roads, or "noted places" were crossed, trees on each side of the line were notched with three chops. At the end of every mile a tree on each side of the line was notched with three chops and inscribed with the total number of miles measured.[42]

The crucial importance of having county boundaries plainly defined was underscored by problems resulting from confusion over the correct location of the artificial portion of the southwestern boundary of the 1773 Cherokee cession. The treaty specified that the line was to run from the source of the southernmost branch of the Ogeechee River (near Union Point in present Greene County), "thence along the ridge between the waters of Broad River and Oconee River . . . to the Buffaloe Lick; and . . . thence in a straight line to the tree marked by the Cherokees [the Cherokee Corner],[43] near the head of a branch falling into the Oconee River." The royal government had this line run in 1773, but bickering between the Indians and surveyors over its exact location caused a very zigzagged line to be marked.[44] The state formed Wilkes County from the cession in 1777. In 1783 the Cherokees and Creeks ceded the lands southwest of Wilkes County. This territory became Washington County. In 1784 the state ordered Henry Graybill, a deputy surveyor of Wilkes, to rerun the line. Graybill's Line, as it became known, ran straight from the source

42. Land Plat Book A, 1793–1812, p. 33, Office of the Clerk of Superior Court, Screven County courthouse, Sylvania, Ga.

43. The Cherokee Corner was a well-known landmark for many years. It was located six miles southeast of Athens in present Oglethorpe County on property currently owned by a church called (appropriately) Cherokee Corner Methodist Church. In addition to forming the terminal point of one of the cession lines, it became the corner of several adjacent land grants. Early records refer to it as a red oak, but later records state that it was a white oak "mistakenly" called a red oak. A granite marker stands on or near the site today.

44. See pages 54–56. The line ran through what is today Greene and Oglethorpe counties. The original map of the survey is in the Public Record Office, London, England. A description of this map can be found in L. De Vorsey, The Indian Boundary in the Southern Colonies 1763–1775, at 175–77, 258 (1966).

of the South Branch of the Ogeechee River along a bearing of North 30° West to the Cherokee Corner.[45] It lay a small distance southwest of the line that had been marked in 1773. Soon thereafter clashes erupted between Wilkes and Washington counties concerning jurisdiction over the strip between the two lines. Later counties cut from these original counties carried on the fracas. Warrants for land in the disputed area were issued by the land court at Augusta in 1784 (which had authority to issue warrants for Franklin and Washington counties only) and later by the land courts of counties along the two lines. County surveyors, lacking access to the warrant and plat records in other counties and unfamiliar with what lands were being granted by other counties, unwittingly ran tracts that overlapped those previously laid out by surveyors from the adjoining counties. Numerous caveats and lawsuits resulted. The parties in these suits, possessing warrants and surveys from opposing counties, each contended that the controversial strip was a part of the county from which he had received his warrant and that the other party's warrant was therefore void.

In 1793 the lower branch of the legislature appointed a committee to prepare a bill making Graybill's Line the official county line and validating all warrants for land between Graybill's Line and the 1773 line, regardless of which county had issued the warrants. In case of conflict, the oldest warrant or survey was to take precedence.[46] As far as can be determined, no legislation resulted from this action and litigation continued, finally petering out in the nineteenth century.[47]

Curiously, one of the few existing narratives of a county boundary survey is in the form of a poem written by John Goodman, surveyor of Bulloch County. Goodman composed this nontechnical, fifty-eight-line

45. Deposition of Stephen Heard sworn before M. Williamson and John King (May 28, 1784) (Georgia Indian Affairs—Wilkes Boundary File, Telamon Cuyler Collection, UGL); Petition of the Inhabitants of Greene County to Edward Telfair (Oct. 19, 1792) (Telamon Cuyler Collection, UGL); Petition of the Inhabitants of Wilkes County to Edward Telfair (Feb. 4, 1793) (Telamon Cuyler Collection, UGL); Executive Minutes (April 1, 1799) (GDAH); Augusta Chronicle & Gazette of the State, Sept. 15, 1792, at 1–2; Gazette of the State of Georgia (Savannah), Nov. 18, 1784, at 2, col. 2; 2 Rev. Rec. Ga. 796. Portions of Graybill's Line are still intact as property boundaries and can clearly be discerned on aerial photographs.

46. Augusta Chronicle & Gazette of the State, Dec. 7, 1793, at 1, col. 2.

47. The author is indebted to Frank Parker Hudson of Atlanta for sharing his extraordinary research on Graybill's Line and litigation respecting lands along the line. Much of Parker's information came from deed records and superior court minutes of the counties involved.

"epic" in 1843 following his resurvey, in concert with his Bryan County counterpart, of the boundary between Bulloch and Bryan counties. This boundary, which had originally been established in 1796, extended some twenty-three miles from the Canoochee to the Ogeechee River. The line still serves as an effective demarcation between the two counties.

Judging from Goodman's verses, the survey was uneventful to the point of being monotonous. The country was almost a complete wilderness, and he dwells upon gripes echoed by the modern surveyor: creeks "With briers and bushes withall very rough," swamps that were "almost too thick for a cooter or frog," and flats that would "do for a possom or cat." The survey party had to make at least three wide offsets to avoid such obstacles, but when they encountered the "Catail Pond . . . Of that portion" they were "quite fond" for "It was open and level and withall dry." A simple plat of the survey accompanies Goodman's rather appealing poem, both now filed in the Bulloch County courthouse. Even the plat does not escape his poetic finesse, for its certification reads "The foregoing is a true representation of the line/dividing the Counties of Bulloch and Bryan."[48]

A glimpse of the palliatives some surveyors seem to have found indispensable for relief from the vexations of their work is revealed in a bill that Claud Thomson, county surveyor of Chatham, presented to the state treasury in 1791. Thomson was seeking compensation for his half of the expenses in running the boundary between Chatham and Effingham counties. In addition to the £8.8.0 Thomson requested for six days' work, he had the temerity to subjoin a request for £1.2.0 for "Provisions, liquor, entertainment, &c. for selves & attendants, at Spencer's & Wiggons's Taverns." Since the survey had been ordered to begin near Spencer's Tavern, one wonders what condition these esteemed artisans were in when they got under way. At any rate, history does not record whether these remarkable amenities were compensated for.[49]

48. Plat Book B, pp. 268–70, Office of the Clerk of Superior Court, Bulloch County courthouse, Statesboro, Ga.; 1842 Ga. Laws 65.

49. Extract from the Minutes of the Superior Court of Effingham County for February Term, 1791, transmitted from John G. Neidlinger to Edward Telfair (Telamon Cuyler Collection, UGL); Bill from Claud Thomson to Edward Telfair for assisting in running the boundary between Chatham and Effingham counties (April 20, 1791) (Telamon Cuyler Collection, UGL); Letter from Claud Thomson to Edward Telfair (July 1791) (Telamon Cuyler Collection, UGL). The plat of the survey, certified by John Moore, surveyor of Effingham County, is in the surveyor general records of the Georgia Department of Archives and History. Less than a month after the survey, President George Washington, while

Small-Scale Mapmaking

In the eighteenth and nineteenth centuries commercial atlas publishers produced a number of small-scale maps and navigational charts of Georgia. In the beginning these were largely unreliable, based as they were on the tales of explorers and cursory observations of itinerant travelers and on the imaginations of the mapmakers themselves.[50] As time went on, however, the perserverance of the surveyors, who doubled as the colony's chief cartographers, made it possible to produce highly accurate maps of Georgia. Some of the small-scale maps made by Georgia's surveyor-cartographers were published, but a large number remained in manuscript form. Only the most significant published ones will be described here.

The first serious attempts at compiling truly authentic small-scale maps of Georgia were made by Joseph Avery, public surveyor of the province from 1742 to 1744. In 1741, while still a resident of South Carolina, Avery proposed to make "a new Map of the North Part of Georgia."[51] The Trustees, in great need of precise information about the colony, readily acquiesced in this proposal and executed an agreement with Avery to pay him ten shillings per day for his work and to furnish him four assistants.[52]

In October 1741 Avery was in Savannah to pick up additional supplies and showed his preliminary work to William Stephens, secretary of the colony. Stephens wrote that he was "extremely well pleas'd to see it" and had no doubt that Avery "would compleat it so as to shew that no Man in these Parts (if any where) was capable of excelling it." Stephens added that his knowledge of Avery's abilities "for many Years past" gave him "full Assurance of his Skill."[53]

---

on his southern tour of the United States, took lodging for a night at Spencer's Tavern. 6 G. Washington, The Diaries of George Washington 139–40 (D. Jackson & D. Twohig eds. 1979). As is well known, Washington had pursued a land surveying practice in his earlier years. For a detailed county-by-county list of acts and other information concerning their boundaries see P. Bryant, Georgia Counties: Their Changing Boundaries (2d ed. 1983).

50. On the earliest maps to show Georgia see De Vorsey, *Maps in Colonial Promotion: James Edward Oglethorpe's Use of Maps in 'Selling' the Georgia Scheme,* 38 Imago Mundi 35 (1986).

51. 4 Col. Rec. Ga. (supplement) 249.

52. Id. 249, 269; 23 Col. Rec. Ga. 269–70; Acct. of Cash pd. by Col. Stephens & Hen. Parker on Acct. of Surveying from 1 Oct. 1741 to 31 Jany. 1741/2 (Egmont Mss. in Phillipps Collection, Ser. No. 14206, at 123B, UGL).

53. 4 Col. Rec. Ga. (supplement) 269.

The discharge of the work was delayed three months, but, as Stephens explained, "the Coast was so water'd with great Variety of Creeks, which form'd many Islands; that 'twas not possible" for Avery "to bring it to Perfection within the Time" allowed by the agreement. In fact, Avery and his men had labored "with Diligence, not coming home oftner than once a Week or 10 Days for Provisions" and frequently working on Sundays.[54]

Avery finished his map of the northern part of Georgia (depicting primarily the coastal area from Port Royal Sound in South Carolina to St. Catherines Island) and then sailed to England to present it personally to the Trustees. The Trustees were clearly impressed, for six months later he was back in Savannah with a commission as public surveyor of the province and directions to begin a general survey of the southern part of Georgia. For the preceding four years, however, Georgia had been without a full-time surveyor, and the Trustees instructed Avery not to let the general survey of the lands to the south take precedence over surveying individual plots of land. This, unfortunately, delayed his start of the general survey, and he died before being able to carry it out.[55]

In addition to his map of the northern portion of Georgia, Avery left a well-executed cadastal map (showing individual property boundaries) of the same region, as well as several other small-scale surveys.[56] Because of the great value of these maps and the war with Spain then going on, the Trustees ordered them shipped to London not on a regular commercial packet but on a man-of-war. So the maps would not be spoiled in the passage over they directed them to be wrapped in an oilskin and placed in a box filled with sawdust.[57]

Avery's maps were not published immediately, apparently for fear that the Spaniards might use them in their prosecution of the war against Georgia. The maps seem to have been more or less forgotten until 1780, when they provided the basis for a published map entitled "The Coast, Rivers and Inlets of the Province of Georgia, Surveyed by Joseph Avery and others." This map constituted part of a much larger project known as *The Atlantic Neptune*—a collection of charts published by J. F. W. Des Barres for use by British forces engaged in the American Revolution.

54. 23 Col. Rec. Ga. 269–70.

55. 5 Col. Rec. Ga. 638; 23 Col. Rec. Ga. 285, 316, 407–14; 24 Col. Rec. Ga. 81–82; 30 Col. Rec. Ga. 274; 1 W. Stephens, The Journal of William Stephens 1741–1743, 63–64, 67–68, 130–31 (E. Coulter ed. 1958).

56. 2 Col. Rec. Ga. 466; 6 Col. Rec. Ga. 122–23; 24 Col. Rec. Ga. 338.

57. 1 Col. Rec. Ga. 494; 31 Col. Rec. Ga. 56.

The Atlantic Neptune has ever since been acclaimed by authorities as one of the great accomplishments in the history of cartography. The Avery–Des Barres map delineates rivers, coastal islands, marshes, and shoals, as well as water depths, with considerable accuracy.[58] The original copperplate from which the map was published was presented to the Georgia Department of Archives and History by the British consul in 1949.

The most prolific mapmaker in the southern colonies in the later 1700s was William Gerard DeBrahm. Scholars have roundly praised DeBrahm's mapmaking abilities, describing his work as "far superior to any cartographical work for the southern district that had gone before" and submitting that "with DeBrahm we turn from the amateur to the professional, from the general outlines of the region to topographical accuracy."[59]

In February 1753, less than two years after his arrival in Georgia from Germany, DeBrahm disclosed in the *South Carolina Gazette* his intention to publish by subscription a map of South Carolina and Georgia.[60] This project bore fruit four years later when Thomas Jefferys, royal geographer in London, published DeBrahm's map with the flamboyant, baroque title:

> A map of South Carolina and a Part of Georgia. containing the Whole Sea-Coast; all the Islands, Inlets, Rivers, Creeks, Parishes, Townships, Boroughs, Roads, and Bridges: As Also, Several Plantations, with their Proper Boundary-Lines, their Names and the Names of their Proprietors. Composed From Surveys taken by The Hon. William Bull, Esq. Lieutenant Governor, Captain Gascoign, Hugh Bryan, Esq; And the Author William De Brahm, Surveyor General to the Province of South Carolina, one of the Surveyors of Georgia, and late Captain Engineer under his Imperial Majesty Charles VII. Engrav'd, by Thomas Jefferys, Geographer to his Royal Highness the Prince of Wales.

This large, artistic map, consisting of four sheets, each measuring twenty-four inches by twenty-six and a half inches, was drawn to a scale of five miles to the inch. It portrays the coast from the northern boundary of South Carolina, southward to the St. Marys River, and extends west-

58. L. De Vorsey, Map Notes [accompanying] The Coast, Rivers, and Inlets of the Province of Georgia passim (Institute of Government, University of Georgia, Athens, Ga. 1981).

59. W. DeBrahm, DeBrahm's Report of the General Survey in the Southern District of North America 32 (L. De Vorsey ed. 1971).

60. Id.; South Carolina Gazette (Charleston), Feb. 5, 1753, at 5, col. 1.

ward to the Indian country. Considerable detail is shown of areas along the coast and large rivers, though information on the interior regions is, predictably, meager. Although DeBrahm incorporated data gathered from his own surveys, he was careful in the title to acknowledge his debt to others whose work he used in creating the map. DeBrahm's map went through several editions in both English and French and was the basis for several other published maps of the same region. It also appeared in widely circulated atlases published by Jefferys and his successor, William Faden.[61]

Without substantial financial backing, either from the government or from large private subscriptions, the work of compiling and publishing maps was a prohibitively expensive undertaking. A case in point was the attempt by Claud Thomson, county surveyor of Chatham, to publish a map of the backcountry of Georgia. In 1787 he announced that persons wishing to proffer their financial support for such an enterprise could apply at Zimmerman and Lang's store in Savannah, where they would be shown the preliminary draft along with the terms of subscription.[62] A few months later Thomson regretfully informed the public that he had decided to discontinue the project because the number of subscriptions was insufficient to cover the costs of the venture. Expressing his "warmest thanks" to those who had patronized the work, Thomson promised that all monies advanced toward the endeavor would be refunded.[63]

As the eighteenth century drew to a close, the detailed maps that had been produced under the auspices of the British government were becoming obsolete, and the more recent ones put out by geographers and atlas makers were too general to serve most practical needs. In 1791 the Richmond County grand jury lamented "the want of an actual survey of the state" and recommended that the legislature pass an act "making it the duty of the County Surveyors under the direction of the Surveyor General to take an accurate Survey of their respective Counties, laying down the several roads & water courses."[64] The state was too impover-

61. W. Cumming, The Southeast in Early Maps 227–28 (1958); W. DeBrahm, DeBrahm's Report of the General Survey in the Southern District of North America 32 (L. De Vorsey ed. 1971).

62. The Gazette of the State of Georgia (Savannah), June 7, 1787, at 2, col. 2.

63. Id., Jan. 31, 1788, at 2, col. 1.

64. Superior Court Minute Book 1790–1792, p. 206, Office of the Clerk of Superior Court, Richmond County courthouse, Augusta, Ga.; Augusta Chronicle, Aug. 10, 1791, at 1, cols. 1–2.

ished to give this proposal serious consideration, but events soon in the making were eventually to bring it about.

Daniel Sturges, surveyor general of the state from 1797 to 1809 and again from 1817 to 1823, set upon the task himself with a zeal and perseverance almost beyond comprehension. In 1801 he notified the public that, having "been frequently solicited by" friends to publish a map of the state, he had "devoted much time and labour" for the previous six years to that object. His projected map, to measure three and a half feet by four and a half feet, was to be sent to the engraver "within . . . four months" and the plates "prepared with all possible expedition."[65]

For six years nothing more was heard, and then the silence was broken by announcements in the Savannah and Augusta newspapers: "We are happy to learn, that Daniel Sturges, Esq. has nearly finished a Map of this State, and a Chart of the Sea Coast, for the use of Mariners." Exhorting "the public spirited citizens" to "encourage him by a liberal subscription," the newspapers confidently predicted that because of Sturges's "known talents, the public may justly anticipate an accurate map."[66]

For nearly a year and a half there was again silence, and then Sturges, through the newspapers, offered some explanation for the delays. He acknowledged that several years before he had notified the public that he had collected materials for drafting a large and correct map of Georgia, which he had intended to publish "in a short time." He explained that "some domestic occurrences, together with the pressure" of official duties resulting from the recent territorial acquisitions from the Indians, had caused the delay. Hoping that these words would "be received as a sufficient apology," especially since greater accuracy would be the result, he begged the indulgence of his subscribers and promised to use his utmost exertions to have the map finished "before the end of the succeeding year." Sturges went on to give notice that, since the cession of Georgia's western territory to the United States, he had decided to confine the map to the present limits of Georgia, although he had originally intended to embrace not only the western territory but the two Floridas as well. Reducing the scope of the map in this way, he warranted, would enhance its usefulness because it could be drawn to a larger scale.[67]

Poor Daniel Sturges, meticulous in the discharge of his public duties

65. Columbian Museum & Savannah Advertiser, Dec. 18, 1801, at 2, col. 3.

66. Republican & Savannah Evening Ledger, Oct. 29, 1807, at 3, col. 3.

67. Augusta Chronicle, Feb. 11, 1809, at 2, col. 4; Id. at 3, col. 1.

and scrupulous to a fault, was—alas—careless about his business affairs. His solo attempts to publish the map of Georgia probably contributed to his financial undoing. In 1810, the year Sturges had so confidently predicted he would finish the work, he not only had been ousted from his position as surveyor general but also was confined in the Baldwin County jail on a writ of execution in favor of Arthur Cheetham for the benefit of William Schley. Upon the motion of his attorney, it was ordered that Sturges's creditors be notified, agreeable to law, so as to allow him "the benefit of the insolvent acts."[68] The next year legal sections of the Milledgeville newspapers carried the disquieting notice: "WILL BE SOLD, at the Market-House in the town of Milledgeville, An Original Map of the State of Georgia, about 5 feet by 3, painted, and correctly and elegantly executed by Daniel Sturges, Esq. late Surveyor General of Georgia, and nearly ready for the Engraver, levied on as the property of the said Sturges, to satisfy Benj. Chairs', James Meriwether's and other's executions against him."[69]

Eleazer Early, brother-in-law of Sturges and sometime comptroller general of the state, purchased the map and had it copyrighted. Two years later Early informed the public that the "work is now in the hands of an eminent Engraver in Philadelphia, and will *certainly* be published with the least possible delay." Pointing out that Georgia was the "only state in the union of which no authentic Map" had yet been produced, Early felt "too well assured of general patronage to render the publication dependent upon any contingency."[70]

Sturges must have possessed the patience of Job, for after all these trials and tribulations he was still willing to devote another four years, under the sponsorship of Early, to completing the map. In October 1817, having just resumed his position as surveyor general, he had the pleasure of reviewing and correcting proof sheets of the map before they were returned to the engraver.[71] And so in 1818, Sturges's sorely beset map—the first published map of the entire state to be compiled by a Georgian—appeared from the plates of Samuel Harrison in Philadelphia.

The Sturges-Early map, measuring 56.5 inches by 44.5 inches was

68. Georgia Journal (Milledgeville), May 8, 1810, at 3, col. 5; Foreign Correspondent & Georgia Express (Athens), May 12, 1810, at 3, col. 3.

69. Georgia Journal (Milledgeville), Nov. 6, 1811, at 1, col. 4.

70. Republican & Savannah Evening Ledger, Jan. 7, 1813, at 4, col. 4; Id. April 28, 1814, at 3, col. 3.

71. Georgia Journal (Milledgeville), Oct. 14, 1817, at 3, col. 1.

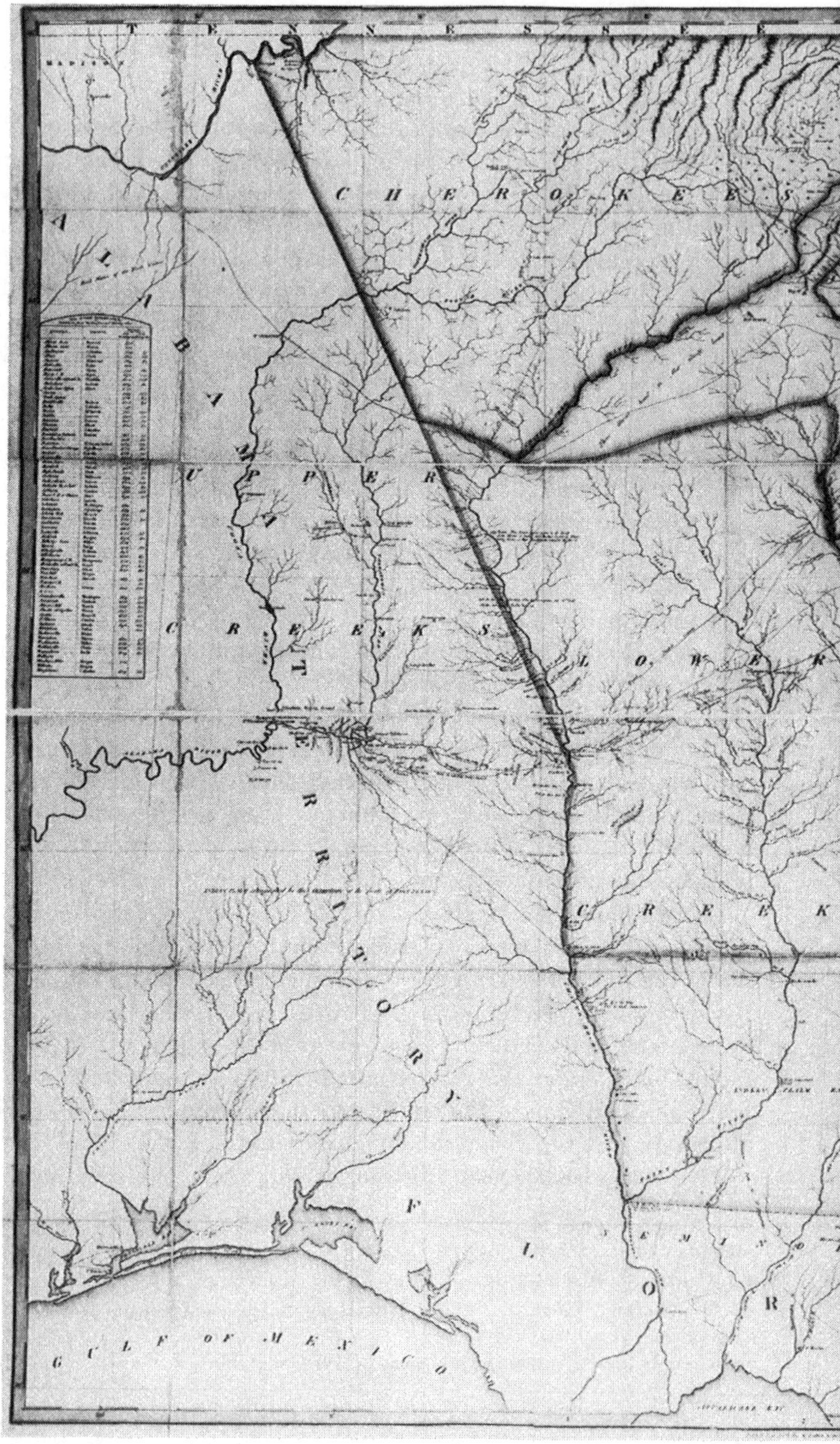

Sturges-Early Map of Georgia, 1818. Courtesy of the Library of Congress.

SOUTH
CAROLINA
Map
OF THE
State of Georgia
ELEAZER EARLY
ATLANTIC
OCEAN

drawn to a scale of eight miles to the inch. It delineates the boundaries of the thirty-nine counties then existing, as well as towns, roads, rivers, mountain ridges, and swamps. Westward of the Ocmulgee River (the Indian boundary at that time) are shown forts, Indian towns, and trails. Also included is the eastern portion of the Alabama Territory. Two editions of the map are known. The titles, imprints, and dates are the same on both, but in the later version changes and additions have been made in the southern Alabama portion.[72]

Georgia's expanding frontier of settlement, illustrated by successive maps of the state depicting settled regions extending farther and farther to the west, was nearly complete by the time Carlton Wellborn, former surveyor general, and Orange Green published their map in 1830. Only the Cherokee lands in the northwestern corner of the state are left devoid of features. In addition to the usual details, the Wellborn-Green map was the first to show land districts, and it includes the numbers of the land lots located in the corners of each district.[73]

In 1847 William G. Bonner published a surprisingly modern-looking map of Georgia. It was engraved by Sherman and Smith of New York and laid down to a scale of six miles to the inch. Crammed with a wide variety of minute details, it shows towns, counties, land districts, roads, railroads, bridges, forts, mills, rivers, mountain ridges, and swamps. There are also insets showing the street layouts of Augusta, Columbus, Macon, and Savannah. In addition, the map is adorned with well-executed illustrations of the statehouse, executive mansion, the Female College at Macon, Augusta Medical College, the University of Georgia, and the state seal. The map went through a number of later revised editions by Bonner, James R. Butts, and A. G. Butts.

Significance of Indians

There is, conspicuously, an almost total absence of comment in early records about Indians harassing headright surveyors. One may safely infer, therefore, that such molestations were extremely rare despite the reluctance of some Indians to leave their homes after the chiefs had

72. W. Ristow, American Maps and Mapmakers: Commercial Cartography in the Nineteenth Century 128–30 (1985).

73. Georgian (Savannah), July 27, 1830, at 3, col. 2; Id., Oct. 26, 1830, at 2, col. 1.

ceded their lands away and despite the attempts by some white settlers, either through greed or mistake, to lay out tracts on Indian lands. One of the earliest settlers relates that when walking in the woods he was always more fearful of rattlesnakes than of Indians.[74] Perhaps it is not surprising to find little hostility displayed by Indians toward surveyors. Indian removal beyond the Mississippi was not contemplated until 1803, and until the early 1800s the vast, sparsely settled wilderness provided ample space for them despite the white expansion. In addition, their lack of sophistication prevented the Indians from grasping the full meaning and permanency of white settlement until it was too late.

In May 1784 the special land court meeting in Augusta received a message from the surveyor general stating that a Mr. Gregg had been appointed to remark the Indian boundary and that it was dangerous for him to go without a guard. The governor and council thereupon requested Colonel Elijah Clarke to furnish a party, consisting of an officer and twelve men properly armed and accoutred, to accompany Gregg.[75]

In 1789 the Creeks were on the rampage in Glynn and Camden counties, and many of the settlers had to flee the area. That year Owen Owens of Savannah hired a surveyor to subdivide a two-thousand-acre tract in Glynn County—a job the surveyor performed but with reluctance because of the "great risque from the Indians."[76]

In October 1791 John Fitzpatrick appeared before a justice in Columbia County and made a deposition concerning Indian depredations. The previous August, he related, he was at the house of the Creek chief Alexander McGillivray in the fork of the Tallapoosa and Coosa rivers (now in Alabama but then in Georgia). In a room of the house he saw "a Surveying Compass with Instruments Suitable (Except the Chain)." Shortly afterward the half-breed brother-in-law of McGillivray, Monac, came into the room, took the instruments out of their case, and examined them. Fitzpatrick inquired about the instruments and learned that they were "the property" of Monac, who had obtained them from a party of six men he had found surveying. Monac and the warriors with him had been "Obliged to kill the Surveyor—he cried, and beged so much for his life." In fact "the whole of the Company with the Surveyor was Kill'd and . . .

74. T. Reese, Colonial Georgia 108 (1963).

75. 2 Rev. Rec. Ga. 798.

76. Petition of Owen Owens to George Walton (July 25, 1789) (Telamon Cuyler Collection, UGL).

there hair was brought in." Two days later, Fitzpatrick visited another of McGillivray's plantations, where he saw an Indian riding a bay gelding that appeared to be of great value. The Indian offered it for sale. Fitzpatrick questioned whether horses of such quality were raised in that region and was told that the horse had been taken at the same time as the surveying instruments. A few minutes later "the news . . . arrived and was proclaimed . . . that two American Scalps had been brought into the upper Towns, a few days before."[77]

Frontier altercations between the two competing races were, of course, to be avoided as much as possible. In 1778 the revolutionary Executive Council observed that it was questionable whether the lands in the fork between the Ogeechee River and Beaver Dam Creek (now in Taliaferro and Greene counties) had been included in the Indian cessions of 1773. Before the Revolution the royal government had prohibited surveys in this area "for fear of bringing on a dispute with the Indians." Notwithstanding, some persons had run and marked tracts within the area. The council ordered that no more lands be granted within the fork until its exact status had been determined.[78]

Successive land laws repeatedly imposed fines and severe punishments on those who knowingly attempted to survey lands beyond the Indian boundary and declared such surveys and any grants obtained under them void. A 1787 act inflicted a penalty of one hundred to five hundred lashes for the first such offense and decreed that the second be adjudged a felony.[79]

Despite all the pains, penalties, and forfeitures prescribed by law, it was next to impossible to stem the restless tide of frontiersmen who surged across the Indian boundary with eager expectations of getting first choice in lands and then having their claims quietly legitimized once a cession was made. In the 1790s Judge George Walton came under intense public criticism for ruling against the holder of such an illegal grant. As a result, he was compelled to offer an elaborate and somewhat apologetic vindication of his decision before the Wilkes County grand jury. Walton acknowledged that sanctioning grants over the Indian boundary

77. Deposition of John Fitzpatrick sworn before Thomas Hamilton (Oct. 7, 1791) (GDAH). Alexander McGillivray was the half-breed chief of the Creek Nation who signed the treaty at New York in 1790. His principal home was located at the Hickory Grounds in the fork of the Tallapoosa and Coosa rivers near the present-day site of Montgomery, Alabama.

78. 2 Rev. Rec. Ga. 66.

79. Marbury & Crawford 258.

by popular consent had become so commonplace that it was "the delusion of the times" that such grants would ultimately "be sustained."[80]

Although public sentiment, especially on the frontiers, often swelled in favor of the trespassers, officials did attempt to punish the offenders, or at least to thwart their actions. Judge Walton's decree was just one example. In 1790 the Greene County grand jury condemned "the practice of marking of Land Beyond the temporary line of this state" and accused settlers along the northern fringes of the Oconee River of actively engaging in and attempting to cover up such practices.[81] Two years later Governor Telfair ordered the surveyor general not to pass grants on 116 plats that had been sent in by the county surveyor of Greene because the lands represented by the plats were thought to extend over the Indian boundary.[82] In the summer of 1787, Greensboro, the county seat of Greene, had been burned to the ground and many of its settlers scalped and killed in a Creek Indian raid.[83] This catastrophe may have spurred Georgia's chief executive and the grand jury to take stronger than usual measures to enforce the law in that area.

Ultimate measures became necessary in 1803, when several families settled on Indian lands opposite Jackson County. Rumors abounded that other families would soon follow. A force of Jackson County militia, consisting of a commissioned officer and twelve dragoons commanded by a Colonel Hampton, was quickly assembled and dispatched to the area. They burned two cabins and two camps that had been erected by the squatters and ordered four other families to remove themselves from the lands immediately.[84]

Fraudulent Surveys and Grants

Besides the great land frauds that have been described, many others on a smaller scale must have been perpetrated, given the lax supervision of the public domain and the ample opportunity for complicity between the

80. Augusta Chronicle & Gazette of the State, Sept. 15, 1792, at 1–2.

81. Superior Court Minutes 1790–1792, p. 12, Office of the Clerk of Superior Court, Greene County courthouse, Greensboro, Ga.

82. Executive Minutes (Oct. 3, 1792) (GDAH).

83. Am. St. Papers, 1 Indian Affairs 23.

84. Letter from David Dickson to John Milledge (April 6, 1803) (Telamon Cuyler Collection, UGL).

land courts, the county surveyors, and prospective grantees. In fact in the 1790s, Governor Jackson denounced the "collusions" that had "been practiced pretty generally in obtaining lands." Later, while presiding over an appeal of a caveat brought on by chicanery, he angrily lashed out at "the practices used in too many instances . . . on such abominable land swindling" and charged that the case under consideration was probably no more reprehensible "than hundreds of others."[85]

In the early 1780s a dozen or so former officers of the Revolution made out hundreds of fake vouchers to twenty or more fictitious men, whom they alleged had served honorably in the war and thus were entitled to bounty grants. Accompanying the vouchers were petitions—all in the same handwriting—alleging that these twenty or more men had sold their vouchers to the officers and praying that their bounties be granted to the officers. Among the worst culprits who thus obtained thousands of acres illegally were General Elijah Clarke; future secretary of state Horatio Marbury; his brother Leonard Marbury; future governor Edward Telfair; John Gorham, surveyor of Franklin County; and Surveyor General Richard Call. Fraudulent military grants became so rife that in 1786 all provisions for bounties to veterans were repealed. Not until the 1820 land lottery was any preference again shown in the land acts for veterans.[86]

In July 1784 allegations were made to the special land court meeting in Augusta that many frauds were being committed in the surveys in Franklin and Washington counties. It was recommended that all grants for those two counties be stayed until the next meeting of the assembly. The Executive Council, however, felt that it had no authority to make such a move—a decision affected no doubt by the mad rush for land grants at that time.[87] The following year a bounty grant was issued to General Elijah Clarke for 1,150 acres of land in Washington County. On the back of the plat, in an unknown hand, someone scribbled, "Neither platt nor Warrant will bear too near a scrutiny."[88]

In 1789 David Harris petitioned the governor for relief, declaring that in June 1784 he had had the county surveyor run out a tract on Big Creek

85. Executive Minutes (May 2 & June 26, 1798) (GDAH).

86. Hitz, *Georgia Bounty Land Grants*, 38 Ga. Hist. Q. 337, 338 (1954).

87. 2 Rev. Rec. Ga. 667–68; The Papers of Lachlan McIntosh, in 12 Coll. Ga. Hist. Soc. 1, 137–38 (L. Hawes ed. 1957).

88. Plat of 1,150 acres surveyed for Elijah Clarke by Solomon Marshall (Nov. 14, 1785) (Loose Headright and Bounty Documents File, GSGD).

in Washington (later Greene) County and had obtained a grant for it; that a short time thereafter some designing persons had cut down and removed several of the corners and station trees and had made six or seven surveys, some wholly and others partially within his survey; that one of the surveys was said to have been run by William Greer, the district surveyor, and dated the same day as his survey; and that he was reliably informed that Greer was not within many miles of the place on that day or for several days thereafter.[89]

In 1818 Solomon Zitrour obtained a warrant for an eighty-five-acre parcel of bottomland lying astride Three Runs in Effingham County. He then procured Clem Powers, the county surveyor, to traverse out its bounds and to record a plat of it. Early the following year a "court organized according to law" declared that the warrant for the survey had "been illegally obtained & the survey null & void." The record is unclear as to the cause for the ruling, but less than three months later the tenacious Zitrour obtained a warrant for exactly the same parcel, less twenty acres now shown to belong to David Lovett, the adjoining landowner. The second survey depicts a bridge on the tract, crossing Three Runs, and bears the highly unusual stipulation, "The bridge is by order of the court granting the warrant to be held and concidered a public bridge & not included in the warrant."[90]

One can only conjecture what story lies behind a two-hundred-acre survey purportedly made for John W. Gibson in March 1822 and recorded in an old headright plat book filed in the Bulloch County courthouse. The plat bears a spurious-looking signature of Peter Cone, the county surveyor. Next to the plat appears the terse note: "This is a Forgery: I never set my name or authorized any Person to do so for me—Peter Cone." Perhaps Gibson, being ineligible to receive a grant, surveyed the two hundred acres himself, obtained access to the plat book, and forged the plat of the tract into it, thus giving the illusion of having received a legitimate grant.[91]

In 1858 one Hampton C. Parker filed suit in McIntosh Superior

89. Petition of David Harris to George Walton (April 29, 1789) (Telamon Cuyler Collection, UGL); Executive Council Minutes (April 28, 1789) (GDAH); Executive Minutes (July 24, 1790) (GDAH).

90. Plat Book C, p. 422 and Plat Book D, p. 59 (both in Office of the Clerk of Superior Court, Effingham County courthouse, Springfield, Ga.).

91. Plat Book B, p. 57, Office of the Clerk of Superior Court, Bulloch County courthouse, Statesboro, Ga.

Court against a William Hughes. The suit alleged that some years before Hughes had obtained a warrant for eight hundred acres of land and had surveyed the land himself, locating the eight hundred acres on the northern side of the Altamaha River swamp. It further alleged that Hughes, well knowing that the tract comprehended all the acreage called for in the warrant without taking in any part of the swamp, and without running, marking, or measuring any lines to the river, fraudulently platted the lines on paper as though they had been run all the way to the main channel of the river. Parker accused Hughes of having represented the lines running through the swamp to the river as much shorter than their true length and of having shown the river channel curving inward when it actually curved outward. As a result, Parker charged, Hughes was able to make the plat represent only a little over eight hundred acres when, in fact, the tract contained nearly fifteen hundred acres. Parker alleged that Hughes fraudulently procured the county surveyor to certify the survey, which he then transmitted to the surveyor general, thereby obtaining a grant for the tract.

Parker prayed that the entire grant be canceled or that the portion containing approximately seven hundred acres that was platted but not surveyed be declared void. Upon appeal to the Georgia Supreme Court, however, the suit was dismissed on the grounds that a grant from the state cannot be set aside in any proceeding to which the state was not a party.[92]

What was branded as one of the more outrageous schemes of land fraud was perpetrated in Columbia County in the 1790s—a decade in which land frauds reached almost epidemic proportions. William Fen, without possessing a warrant or any other right, made a partial survey (running only two lines), treacherously procured a certified plat to the land thus "surveyed," and, without further ado, took possession of this piece of turf. Some two years later Richard Randolph obtained a warrant for three hundred acres, which he had the county surveyor lay off within the locale that Fen occupied. Randolph then proceeded through the proper channels to secure a legitimate grant. Fen, perceiving that his possession was about to be thwarted, "borrowed" an unexecuted warrant that had been made out to Benjamin King two years before. In collaboration with King, Fen finished his partial survey, furtively altered his plat so as to make it conform to King's warrant, and then hurriedly sent the

92. Parker v. Hughes, 25 Ga. 374 (1858).

plat along with the warrant to the county surveyor's office so he could obtain a grant before Randolph obtained his. Upon appeal of Randolph's caveat against King, Randolph's title was sustained. Governor Jackson declared this an "extraordinary" case because King had no interest in the disputed land even though his name had "been dragged from Court to Court" in the numerous and protracted legal maneuverings, whereas Fen—the real culprit—was not made a party to the proceedings.[93]

Depletion of the Headright Domain

The surging tide of headright settlement over the eastern third of Georgia in the eighteenth and nineteenth centuries produced a hodgepodge of private freeholds interspersed among unappropriated public lands. In regions where the soil was uniformly rich, grants were usually taken up in rapid succession. The result was a closely compacted mass of private holdings between which few and relatively small chunks of vacant land remained. In poorer regions, granting went more slowly so that often there was a greater proportion of vacant land than appropriated land for many years after the initial grants had been made. Thus vacant lands became scarce as early as the 1790s in Wilkes County, in many of the counties cut from original Franklin County, and in the extreme upper portion of original Washington County, where wide expanses of fertile hills and valleys predominated. Several decades later vacant tracts were still fairly plentiful in the counties that had been cut from the lower part of Washington County and in Liberty, Glynn, and Camden counties—regions abounding in marshes, swamps, and pine barren wastes.

Two recurring difficulties became especially acute during the latter half of the nineteenth century. Numerous granted tracts were left by their owners in a wild, uncleared state, which rendered them almost indistinguishable from vacant lands. Sharp controversies erupted when attempts were later made to take out grants comprehending part or all of such tracts. The applicants (and the county surveyors) were often unaware that the tracts were already under private ownership. The hapless owner all too often did not discover the adverse claim until his title had been divested by long, continued possesion, or, at best, he was able to reassert his title only after an expensive and protracted legal proceeding.

93. Executive Minutes (June 26, Sept. 1, & Oct. 23, 1798) (GDAH).

The converse of this phenomenon was that a number of vacant tracts had originally been appropriated and improved by squatters whose heirs and descendants were now succeeding to the tracts. These second- and third-generation squatters were often oblivious to the origin of their ancestors' possession. The indiscriminate destruction of title records through courthouse fires and other mishaps rendered suits by the state to recover these tracts highly problematic. Even county surveyors, who maintained custody of the grant plat books and who were as familiar as anyone with the status of lands within their respective localities, often expressed bewilderment concerning these matters.

In the 1850s the General Assembly attempted to deal with these difficulties by requiring that ten days' notice be served upon any person in possession of land for which someone else intended to take out a warrant and by giving the person in possession, if qualified, preference over the person not in possession as to warrants for the land. Applications for land were required to be accompanied by a statement from the county surveyor certifying either that such notice had been given or that no one was in possession of the land. If the statement was not included, the secretary of state was prohibited from passing grants on the land.[94]

Careful study and comparison of current property lines of a given area in the headright system with the original grant lines of the same area indicates that a significantly large portion of the original lines are still intact as present-day lines, though they are not generally discerned as such. By contrast, in the areas of the state laid out under the rectangular system, most of the original land lot, district, and section lines are still recognized.

In 1858 the General Assembly repealed the headright laws so far as they pertained to Washington County,[95] and the following year it did the same for Franklin County.[96] In setting forth the reasons for repeal,

94. 1853–54 Ga. Laws 75; 1857 Ga. Laws 120; Jackson v. Moye, 33 Ga. 296 (1862). By the latter part of the nineteenth century vacant lands had become so scarce that few individuals had the opportunity to acquire headright grants even though they otherwise were entitled to them. Since many persons in possession of tracts under dubious or nonexistent titles were qualified for a grant anyway, it was preferable to allow them to take out grants to the lands they claimed rather than stir up litigation by allowing others to attempt to take out grants thereto.

95. 1858 Ga. Laws 94.

96. 1859 Ga. Laws 343. In 1829 all remaining vacant lands in Chatham County had been appropriated for educational and charitable uses in the county. 1829 Ga. Laws 103; Dawson 272; 1855–56 Ga. Laws 255. The headright laws were repealed for Camden County

the acts declared that "little if any ungranted land" remained in those counties; that most of the land had been granted before the year 1800; that in many cases the original grants had "been lost and the names of the grantees forgotten"; that the lands had been subdivided and passed "from generation to generation . . . without reference to the boundaries of the original grants"; that many of the recorded deeds in Washington and Franklin counties had been burned, making it difficult and frequently impossible "to trace up and locate the grants to much of the land therein"; and that applications for headright grants were "productive of much trouble and litigation."

The number of applications made each year for headright grants declined steadily for the state as a whole after the early 1800s and fell off sharply after the 1850s. Most of these later applications were for small parcels of inferior quality lands that had been passed by in earlier years. The headright laws had, for all practical purposes, lapsed into obsolescence by the turn of the twentieth century. On August 13, 1909, they were repealed in their entirety, thus closing this notable chapter in the state's history.[97]

---

in 1877 (1877 Ga. Laws 304) but were reinstated (perhaps for the benefit of individual parties) in Franklin and Camden counties in 1885. 1884–85 Ga. Laws 638, 652; 1897 Ga. Laws 574.

97. 1909 Ga. Laws 115. The last headright grant by the state was a four-and-one-quarter-acre tract in Camden County granted to John M. Holzendorf, Jr., on May 9, 1912. Although this grant came after the repeal of the headright laws, Holzendorf had made application before repeal. See Grant Book AD, p. 253 (GSGD); and Survey Deed Record Book G, p. 198, Office of the Clerk of Superior Court, Camden County courthouse, Woodbine, Ga.

Chapter 4. The 1805 and 1807 Land Lotteries

The Land Lottery System

At the beginning of the nineteenth century Georgia made two fundamental changes in its land policy. It adopted a rectangular system of land subdivision for surveying out the remaining two-thirds of the state, and it adopted the lottery as a means for distributing these lands.

The rectangular system was not without precedent. It was, with modifications, merely a return to the system that had prevailed at the founding of the colony. And after this original system had been abandoned, sporadic flirtations had been made with a rectangular system in the headright areas of the state. In addition, the rather obvious concept of laying out more or less uniformly shaped tracts over a broad expanse of territory prior to granting had existed in widely dispersed localities both long before and after the founding of Georgia. The system had been used by the Egyptians in their Nile Valley surveys, by the Romans, and by the Dutch as well as by New Englanders in the 1600s. It had also been adopted by the United States as part of the Northwest Land Ordinance of 1785.[1]

Precedents also existed for the use of the lottery in distributing public lands. Lotteries had been employed in allotting the original town, garden, and farm lots of Savannah and Ebenezer; on a very limited scale for allotting public lands late in the colonial period; and for distributing warrants at the land court at Augusta in 1784.

The primary obstacle to implementing a rectangular system, as Georgia now contemplated, was the extremely high initial cost of the surveys—a cost that could be recouped only after a considerable period of time. Then, too, the headright system admittedly had some advantages over a rectangular system: the tendency to equate land with labor

1. 1 R. Patton et al., Patton on Land Titles § 114 (2d ed. 1957); N. Thrower, Original Survey and Land Subdivision 8–10 (1966).

supply, simplicity of operation, and freedom of site selection on the part of the grantee. The chief drawback of the headright system was that it exposed land titles to all sorts of difficulties and lawsuits, for frequently the same surface was overlapped by conflicting warrants and grants in favor of diverse persons. In addition, it afforded a poor basis for resurveys, further subdivision, and legal descriptions. Finally, the mode of obtaining grants under the headright system tended to favor the wealthy and strong. Before a headright grant could be taken out, one first had to make a personal exploration of the lands sought—a burden onerous to widows, orphans, and the poor, who were the most proper objects of public bounty.[2]

Georgia adopted the rectangular system and the mechanism of the lottery, however, not so much as an escape from these shortcomings of the headright system as an indignant reaction against the two great land scandals of the 1790s—the Pine Barrens Fraud and the Yazoo Fraud. After having suffered these experiences, the people of Georgia resolved in all future dispositions of public lands to guard against every opportunity for fraud and corruption. Under the system that was devised, when lands were ceded by the Indians, they were surveyed at public expense into small lots of uniform size and were marked, numbered, and mapped. The field notes and maps were then returned to the surveyor general's office. A lottery, presided over by commissioners selected by the legislature, was then held so that the lots could be distributed by chance. This method eliminated most of the temptation and means for fraud and corruption because it was next to impossible to gain anything by bribing members of the legislature or other public officials. Nor could anything be gained by bribing surveyors to measure or mark lots falsely or to make forged or fictitious returns of surveys because no one could know who would draw any particular lot. In addition, the system gave everyone an equal and fair chance and instantly gave each grantee an unquestionable title. By operating on the basis of fairness, by assuring indubitable titles, and by preventing the accumulation of large bodies of land in the hands of speculative individuals and companies the lottery system greatly promoted the settlement and improvement of new regions.[3]

Georgia's district and land lot system is often compared to the federal township and section system. Strong superficial similarities are evident,

2. A. Chappell, Miscellanies of Georgia (pt. 2) 25–27 (1874).

3. Id.

but there are basic underlying differences between the two systems. Georgia's lottery lands lack the coordinate system of meridians and baselines. The federal system started where a base line and principal meridian intersected and worked from that point. Georgia's system started with "boxes" (counties), created smaller "boxes" (districts) within them, and then created still smaller "boxes" (land lots) within these. Though the lots are numbered under both systems, no standard system of numbering was specified by the Georgia land lottery acts until 1830. Thus Georgia's numbering technique is less uniform than that of the federal system.[4] Also, in many areas of Georgia's rectangular system, the lots are oriented either 45° or 60° from the meridian or, as in the case of original Wayne County, 13° from the meridian.

Georgia's land lotteries of 1805–33 were the largest ever held in the United States.[5]

Land Lottery Act of 1803

The United States moved quickly in making the first installment on its pledge in the Articles of Agreement and Cession to extinguish the Indians' title to Georgia lands. In a treaty signed on June 16, 1802, at Fort Wilkinson (located on the west side of the Oconee River just south of present-day Milledgeville), the Creeks ceded a narrow sliver of land on the western side of the Oconee-Appalachee rivers and another area known as the "Tallassee strip," about twelve miles wide, running from the Altamaha River to the St. Marys River.[6] In return, the United States agreed to pay the Creek Nation $25,000 in goods, plus an additional sum annually for ten years, and to furnish the nation with blacksmith tools and men to work them for three years.

At the regular session of the legislature in November 1802, a resolution was adopted calling for the governor to convene an extra session to dispose of the lands as soon as the treaty was ratified by the U.S. Senate.[7]

4. Hilliard, *An Introduction to Land Survey Systems in the Southeast,* Studies in the Social Sciences, June 1973, at 4. The 1830 land lottery act (see Chapter 8) was the first and only lottery act to specify a uniform numbering system. The provision applied only to square districts.

5. Hilliard, *An Introduction to Land Survey Systems in the Southeast,* Studies in the Social Sciences, June 1973, at 4.

6. Clayton 701; C. Kappler, Indian Affairs (Treaties), S. Doc. No. 319, 58th Cong., 2d Sess. 58 (1904).

7. Proclamations Book AAA, p. 226 (Mar. 27, 1803) (GDAH).

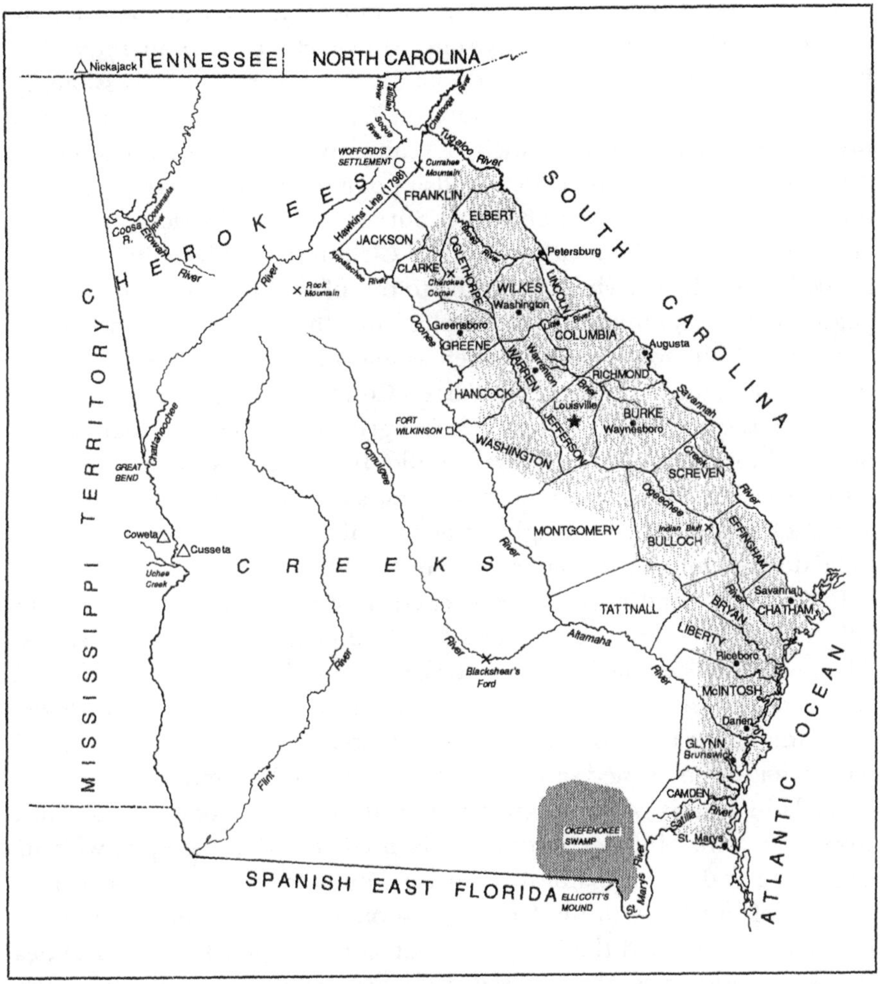

Georgia in 1802. The shading indicates the approximate area of white settlement.

The Warren County grand jury presentments a few years before may have inspired the land lottery system. At its October term in 1797, that body had recommended that lands obtained in future Indian cessions be granted in such a way as "will insure the Citizens of this State, an equal proportion" of the lands, "which . . . can only be done, by dividing the Territory obtained, into districts, and appoint the Surveyor Genl (or some other fit person) to make an actual Survey of each district, and make afair plat thereof, then each individual Claiming Shall have his Chance, by lot." After the drawing, according to this plan, "the first number shall begin at the Beginning of the district" and the district "laid off [into] so many P[l]otts southwardly & then as many Northwardly, as will make the Compliment."[8] Perhaps the Warren County grand jury had taken its cue from the act passed by the U.S. Congress the preceding year, which provided for rectangular surveys in Ohio. It was the first such act passed under the U.S. Constitution and the first since the Land Ordinance of 1785.[9] Obviously, the headright areas were too cut up by former surveys to be divided into a rectangular system.

In April 1803 the legislature convened at Louisville, a coarse, unkempt village populated by fewer than six hundred people, half of them slaves. The town had been hacked out of the wilderness the previous decade and crowned with the proud designation of the capital of Georgia. The legislators met in the two-story brick edifice in front of which the Yazoo Fraud papers had been burned seven years before. With the memory of this and other land frauds no doubt pressing hard upon everyone's conscience, Governor John Milledge optimistically predicted that Georgia, with its large area and varied natural resources, could "be the most flourishing and important State in the Union" as soon as its population increased. He therefore warned the legislative members to "guard all . . . avenues to private interest" in the disposal of the new cession so as to prevent the lands from concentrating "in the hands of a few companies"—a situation that would leave the state "rifled of its brightest prospects" for future development. Calling attention to the Pine Barrens Fraud, he observed that "millions of acres" had "been owned and sold by a few without the . . . addition" of a single settler to the state. Milledge deemed it "against the genious" of a republican form of government "to have large land holders;

8. Superior Court Minute Book A, pp. 73–74, Office of the Clerk of Superior Court, Warren County courthouse, Warrenton, Ga.; Augusta Chronicle & Gazette, Dec. 16, 1797, at 4, col. 3.

9. 1 Stat. 464.

and if events and carelessness" had previously brought about such a state of affairs, it was "time they should be checked." The governor offered few specifics for disposing of the land, but he did think it "prudent to reserve . . . some of the best Scites . . . for Towns or public institutions" since it would be difficult to acquire such spots after they fell into the hands of private individuals.[10]

The legislative journals are too brief to furnish substantial details about the formulation of this first land lottery act, but doubtless many—if not most—of the members of the legislature had a part in it. The basic provisions were hammered out in committees and then debated, often in lively animation, by the members as a whole. Most of the debates centered around such matters as the qualifications of the grantees, the method of granting, and the price of the grants.[11] Few contentions arose regarding the sizes of the lots or the methods for surveying them, although an amendment offered in the Senate to set aside a portion of the cession for military bounties surveyed into 640-acre lots was defeated.[12]

The final bill, signed on May 11, 1803, is too lengthy to be reproduced in full, but a summary follows:

Section 1. The cession to be divided into three counties. That part lying south of the Altamaha River to form one county called Wayne. Wayne County to be divided into three districts "in such manner" as the surveyor general should direct. That part of the cession lying west of the Oconee River to form two counties, the division line to be run "according to the true meridian from the Oconee [River] at Fort Wilkinson, south forty-five degrees west, to the Indian boundary." The territory to the north of this line to be called Baldwin County and the territory to the south to be called Wilkinson County. Baldwin and Wilkinson counties each to be divided into five districts, the district lines to be run "parallel" to the dividing line between the two counties. All districts in a given county to be "as near equal" in area "as circumstances will admit." District lines to be laid off by the surveyor general or by such surveyors as he should appoint.

Section 2. Districts to be divided into lots by lines running "parallel" to the district lines and by other lines crossing "at right angles." Lots in

10. Executive Minutes (April 18, 1803) (GDAH).

11. 1803 ex. sess. Ga. Senate Journal 17, 60, 65, 67–69, 71–86, 90–96, 110–11, 117–30 (ms., GDAH); 1803 ex. sess. Ga. House Journal 43–44, 47, 49, 52, 54–55, 57, 59, 61–81, 108–36 (ms., GDAH). The first thirty-nine pages of the House journal are missing.

12. 1803 ex. sess. Ga. Senate Journal 78–79 (ms., GDAH).

Baldwin and Wilkinson counties to measure "forty-five chains square" and contain 202½ acres each, except for fractional lots. Lots in Wayne County to measure "seventy chains square" and contain 490 acres each, except for fractional lots. Fractional lots to be reserved from drawing and sold in such manner as a future legislature should direct.

Section 3. One surveyor to be appointed by the legislature to each district.

Section 4. Each surveyor to post a $10,000 bond.

Section 5. Surveyors to subdivide the territory, marking the lines "plainly and distinctly, upon trees, if practicable, otherwise stakes may suffice." All lines to be run "with the utmost possible exactness, with a half chain, containing two perches of sixteen feet and one half each, consisting of fifty equal links, which shall be adjusted by a standard" kept in the surveyor general's office. The meanders of all watercourses that form natural boundaries of any of the lots and the meanders of all navigable rivers to be measured "as accurately as possible." The measurements; names of corner and station trees or stakes; and rivers, creeks, and other watercourses touched upon or crossed to be noted in field books. The lots to be numbered "in such order as the Surveyor General shall prescribe." Transcripts of field notes and a map of each district to be deposited in the surveyor general's office. Surveys to be completed within 90 days [later extended to 150 days] after the running of the Indian boundary.

Sections 6, 7. Surveyors to receive $2.75 for each mile of line run, out of which they must defray all expenses including compensation to chainmen and axmen. Governor to advance $300 to each surveyor before the survey begins, and the balance to be paid upon completion.

Sections 8, 9, 10. Upon completion of the survey, tickets with the numbers of each full-size lot written on them to be placed in a drum and drawn at random as prizes in a lottery supervised by five commissioners appointed by the legislature. Each unmarried white male citizen twenty-one years of age and older and each minor orphan or family of minor orphans entitled to one draw. Each white male citizen with a family and each widow with minor children entitled to two draws. Inferior courts to receive applications for draws at the county seats of the respective counties and to forward lists of the names of applicants to the governor.

Sections 11, 12. Prices of grants issued to successful drawers to range from $0.50 per hundred acres for pine lands to $9.00 per hundred acres for first-quality river lands. Any grants not taken out within five years of the drawing to revert to the state [fees were changed to $4.00 per hundred acres, regardless of land classification, and the time for taking out grants was reduced to twelve months by acts passed the following December. These acts, preceding the actual lottery, ensured that no one paid the fees originally stipulated by these sections].

Section 13. Governor to fill any vacancies caused by the death or resignation of surveyors or by the failure of any surveyor to perform his appointed task.

Section 14. Surveyors and chainmen to take an oath before beginning their work.

Section 15. Each surveyor to appoint a person to judge the quality of the lands of his district [this provision was subsequently repealed].

Section 16. Five commissioners to be appointed by the legislature to locate a suitable site at or near the head of navigation of the Oconee River on which to lay out a reserved tract for a town to be known as Milledgeville. The tract to be equal in size to sixteen of the 202½-acre lots.

Section 17. Headright surveys that may have been made on lands embraced within the latest cession to be void.[13]

The above provisions are particularly important because the 1803 land lottery act served as the prototype for each subsequent land act passed by the state of Georgia. All provisions regarding the way in which lands were to be measured and marked have been directly quoted in the above summary. Although this act and the later acts provided for systematic rectangular surveys of the new areas, it is significant that the surveyor general and the surveyors in the field were still left with wide discretion in laying out the lands. A comparison of these acts with the map on page 4 will help the reader understand this layout.

The district surveyors were chosen almost exclusively from the pool of county surveyors in the older counties. Although the compensation they received seems ridiculously low by modern standards, it apparently was adequate for the times; correspondence between the surveyors and government officials confirms the former's ardent willingness to carry out their prescribed tasks. Sometimes a surveyor defaulted in his duties. In such cases numerous requests were usually made by other surveyors for the appointment to fulfill the defaulter's duties. Because of unforeseen problems and delays, the deadlines for completing surveys were rarely met in the first or any of the later land lotteries.

The land lottery acts made no changes in the office of county sur-

13. 1803 ex. sess. Ga. Laws 3; Clayton 100.

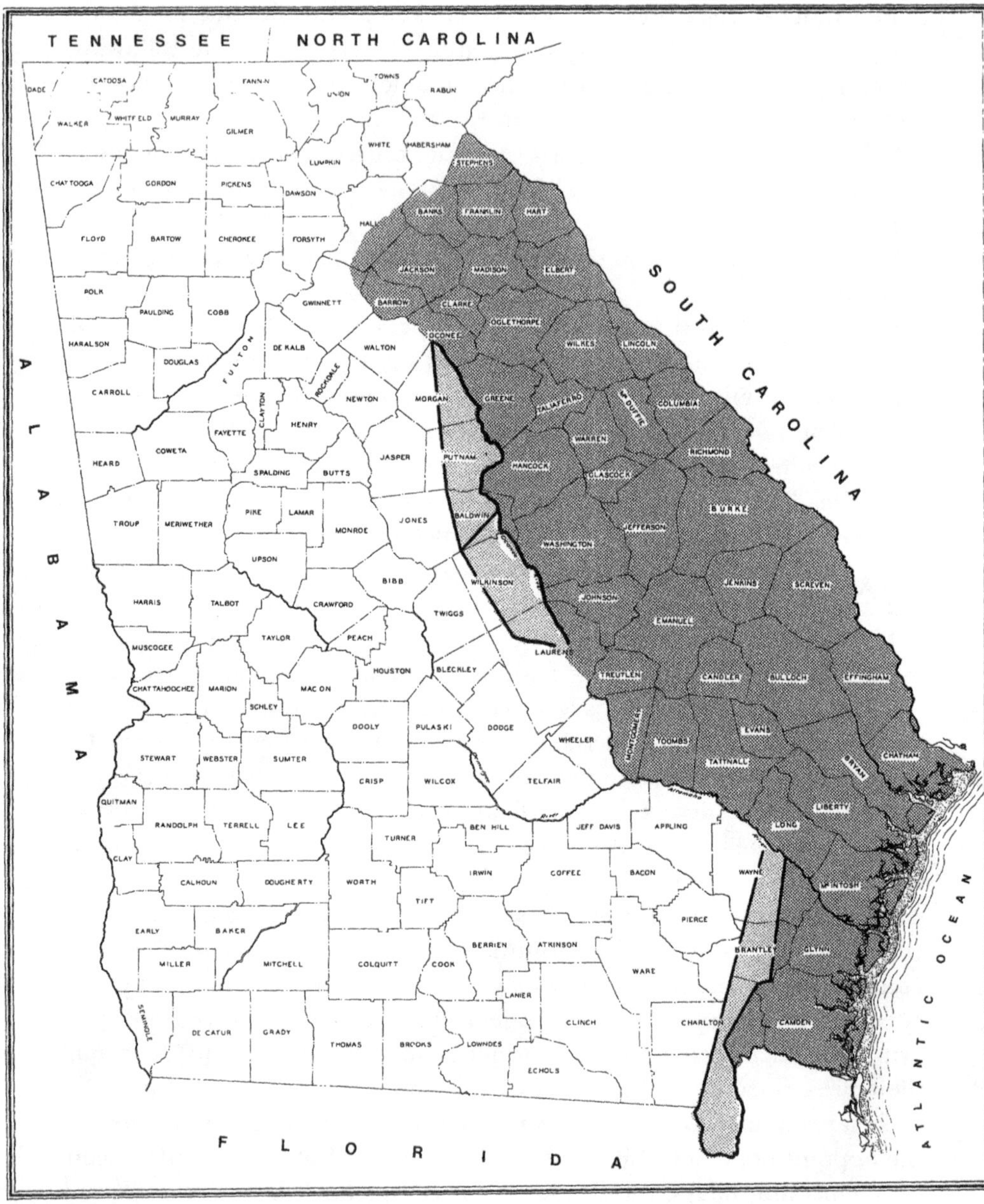

Lightly shaded area indicates lands distributed in the 1805 land lottery.

veyor because that office was created and regulated by a general statute rather than by the land acts. Thus each new county in the lottery areas acquired its own surveyor. He had little to do, however, compared to his counterpart in the headright counties.

Oddly, there appear to be no detailed contemporary accounts of the lottery proceedings. Nonetheless, from the legislative acts and a few meager gleanings gathered elsewhere, it is possible to construct a composite description of the lotteries. When the legislature passed a lottery act, newspapers around the state carried copies of it. Qualified persons who wished their names entered would apply to the justices of the inferior court in their county of residence and take an oath attesting that they were entitled to a draw or draws. The inferior court justices forwarded lists of qualified applicants to the governor, who, upon completion of the surveys, assembled the lottery commissioners at the state capital. Pieces of paper called "tickets," each bearing the designation of a particular lot (or "prize"), were placed along with blank pieces of paper in a rotary wooden drum (or "wheel"). The total number of pieces of paper in the wheel equaled the total number of applicants. In the 1805 lottery the names of all applicants were listed in books, roughly in alphabetical order. As the tickets were drawn from the wheel, the nature of each draw—whether a "successful" draw or a "blank"—was listed beside the applicants' names in the order in which the names were entered in the books. In all of the subsequent lotteries, the names of applicants were written on pieces of paper and placed in a wheel. One person then drew from the wheel containing the names of applicants and another from the wheel containing the designations of lots. The two corresponding tickets were then pasted together upon each draw. Also, to save time in some of the later lotteries, blank tickets were not placed in the prize wheels. Instead, when all prizes had been drawn, those applicants whose names had not been drawn were considered to have drawn blanks.

While the lottery surveys were being conducted, land speculators questioned and often even accompanied the surveyors to discover particularly valuable lots. Afterward these speculators hung around the lottery drawings with wolfish intent, waiting for one of the valuable prizes to be drawn so that they could immediately rush off and buy it from the lucky applicant at a bargain price before he realized the true value of his prize.

Probably the use of the word *lottery* in these proceedings is somewhat inaccurate because the applicants risked nothing, save for a small fee paid the inferior court justices for taking their applications.

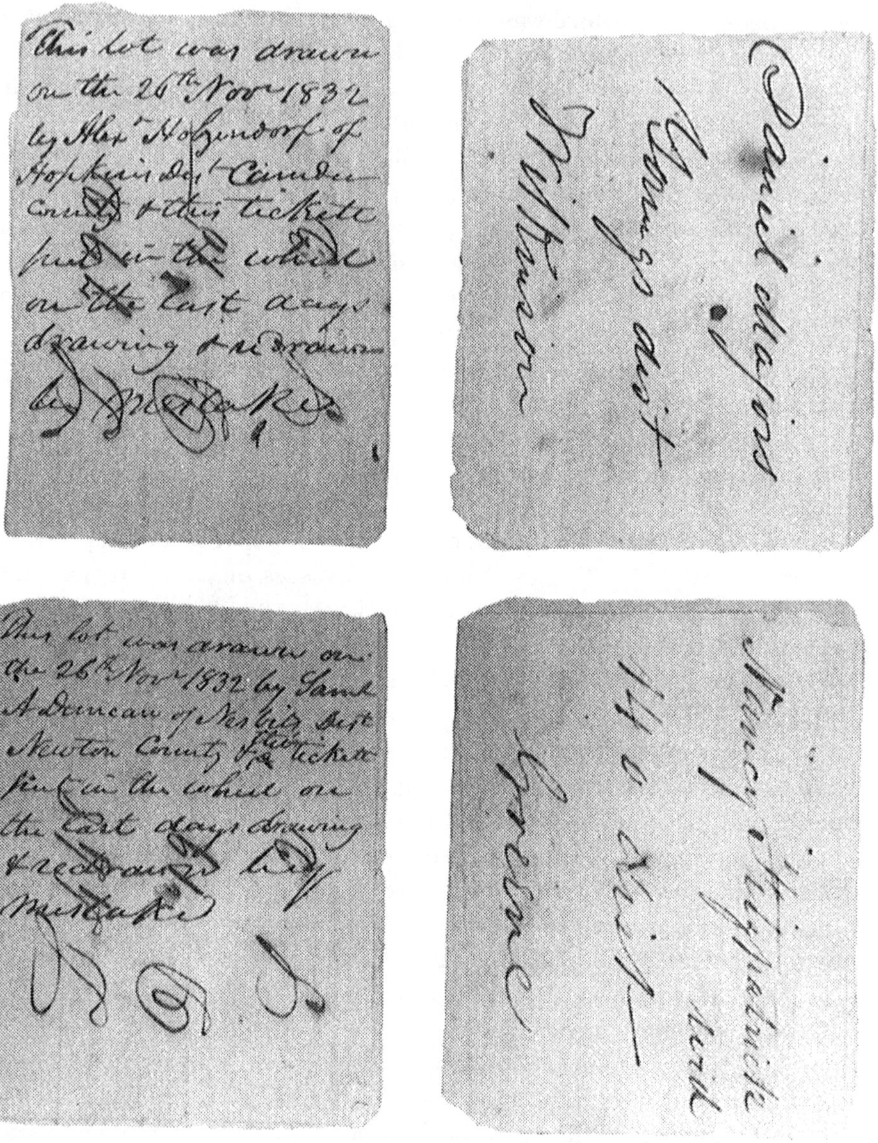

Land lottery tickets. Courtesy of the Georgia Department of Archives and History.

Surveys for Original Baldwin, Wilkinson, and Wayne Counties

Promptly after the passage of the land lottery act, both branches of the legislature repaired to the House chamber to elect surveyors, commissioners to superintend the lottery, and commissioners to locate a site near the head of navigation of the Oconee on which to lay out Milledgeville, the future state capital.[14]

During the early weeks of 1804 Benjamin Hawkins, John Clark, David Adams, and Jesse McCall, as commissioners; Thomas Freeman, as surveyor; and the Creek chiefs Tuskenehau Thucco, Efau Tustunnuggee, and Cussetuh Tustunnuggee blazed out the lines of the cession on the western side of the Oconee. The running of the boundary of the Tallassee strip was delayed until the spring and summer of 1804, however, because many Indians living in the region were opposed to the treaty.[15]

Levin Wailes, the most accomplished practitioner of the art of surveying in Georgia during this period, was appointed by the surveyor general, Daniel Sturges, to lay off the county and district lines of original Baldwin and Wilkinson counties and to subdivide the 2d District of Baldwin. Born in Maryland on March 9, 1768, he was descended from a family that included several professional surveyors. Though Wailes's early education was interrupted by the revolutionary war, he had learned much of navigation and mathematics on two voyages he made from Philadelphia to England. He later studied law. It was probably while in Philadelphia that he met Robert Morris, who in 1792 sent him to Georgia to represent his interests there. Wailes brought with him his sister Elizabeth and her husband, Philip Thomas, along with "an excellent surveyor's compass purchased in Philadelphia." Philip and Elizabeth Thomas were the parents of Edward Lloyd Thomas, who was later to make an indelible mark on the surveying scene in Georgia. Wailes had learned surveying from

14. 1803 ex. sess. Ga. House Journal 131–36 (ms., GDAH); 1803 ex. sess. Ga. Senate Journal 127–30 (ms., GDAH).

15. Journal of the Commissioners who attended the running of the line between the State of Georgia and the Creek Nation of Indians (Jan. 20, 1804, to Feb. 3, 1804) (Indian Affairs File, Telamon Cuyler Collection, UGL); Letters from David Adams to John Milledge (April 18, May 17, & Sept. 30, 1804), and Letter from Thomas Freeman to John Milledge (Sept. 30, 1804) (all in Telamon Cuyler Collection, UGL); Letters from John Milledge to Benjamin Hawkins (Oct. 27, 1803, & Feb. 27, 1804), to John Clark (Dec. 16, 1803), and to David Adams (April 26, 1804) (all in GLB, GDAH); 1804 Ga. Senate Journal 5.

John Thomas, Philip's father, and had established an excellent reputation. A colleague in Augusta wrote in 1806 that Wailes was "the most correct and best surveyor in the State of Georgia, or perhaps, elsewhere." He won the respect of Governor Milledge, who referred to him as "a good mathematician, whose skill in his profession, and probity in life" he could "readily vouch for." Wailes was elected surveyor of Richmond County but spent much of his time with his compass and chain in the backcountry wilds of Georgia. He was present at the state capitol in Louisville in 1803 when the first land lottery act was drawn up and likely played a key role in its evolution. Upon the conclusion of the second land lottery in 1807, the public land surveying business entered a twelve-year lull in Georgia, whereupon Wailes removed to the Mississippi Territory. In this new domain he put his valuable experience gained in Georgia to work by performing several large and important public lands surveys for the federal government. He spent the remainder of his life there and died on May 22, 1847.[16]

Detailed instructions were drawn up by Daniel Sturges[17] and issued

16. Letter from John Milledge to Henry Dearborn (July 5, 1806) (GLB, GDAH); Augusta Chronicle & Gazette of the State, Mar. 18, 1797, at 3, col. 2; Id., Sept. 9, 1797, at 3, col. 4; Id., Mar. 6, 1802, at 3, col. 3; Id., Sept. 24, 1803, at 3, col. 4; Id., Sept. 20, 1806, at 3, col. 3; Clayton 28, 160; C. Sydnor, A Gentleman of the Old Natchez Region 30, 32–40, 85 (1938).

17. Daniel Sturges, an important but obscure and relatively unknown figure in Georgia history, was born about 1765. He was the son of the Reverend Daniel and Eleanor Sturges of Norborne Parish, Virginia. The family apparently migrated to South Carolina about 1789 and then into Georgia the following year. Sturges, the subject, served as deputy surveyor general of Georgia during the 1790s and was elected surveyor general of the state on January 26, 1797. Two years later a contest and prize was announced for the best design for altering Georgia's great seal. Sturges, a Mason, submitted the winning sketch in which Masonic symbols were incorporated and won the thirty-dollar prize. His rendering, with slight modifications, remains the great seal of the state today. Sturges also compiled the first detailed published map of the entire state of Georgia—an undertaking that required over twenty years to complete. In 1807 he assisted commissioners and surveyors from Georgia and North Carolina in attempting to determine the boundary between the two states. The land records Sturges kept during his tenure indicate that he was a meticulous and exacting administrator who possessed great insight into the problems of surveying a vast wilderness. Nevertheless, the state Senate failed to reelect him when his sixth term expired on November 24, 1809. Possibly an overabsorption with his public duties had led to his neglecting his personal affairs. At any rate, he soon encountered severe financial difficulties, which sent him to debtors' prison. He made a comeback, however, and, upon the death of Surveyor General Edmund B. Jenkins in 1817, solicited and received from Governor David B. Mitchell the appointment to fill the vacancy. Sturges was reelected by the Senate and served until his death at Columbia, South Carolina, on September 17, 1823. His son Benjamin Hicks Sturges succeeded to the unexpired term.

Daniel Sturges was surveyor general during the first four of Georgia's six land lotteries.

to Wailes for dividing Baldwin and Wilkinson counties and for laying them off into districts. Typical of instructions issued to surveyors for subsequent land lotteries, they are as follows:

To LEVIN WAILES Esq^{e}

Surveyor for District No.2 in the County of Baldwin

By virtue of the power in me vested by an act of the honorable the Legislature of this State, passed the 11th day of May 1803, entitled "an act to make distribution of the late Cession of lands, obtained of the Creek nation by the united States commissioners in a Treaty entered into at or near Fort Wilkinson on the sixteen day of June Eighteen hundred and two," You are hereby appointed and authorised to run and mark a dividing line, as contemplated in the above mentioned Act, between the Counties of Wilkinson & Baldwin, which are formed within that part of the said Cession, lying south of the Oconee river, and to lay out the said Counties into FIVE districts each, agreeably to the tenour of the law, and my Instrucutions hereunto annexed: for which you are to receive the same compensation for every Mile run and surveyed, as you are allowed by the aforesaid Act, as a district surveyor. Given under my hand and Seal this twenty first day of Feby 1804

Danl Sturges Sur^{y} Genl

Surveyor Gen^{ls} Office 21st Feby 1804

Instructions

To LEVIN WAILES Esq^{e}

For dividing the Counties of Wilkinson and Baldwin & for lay^{g} out the same into districts.

The temporary boundary line circumscribing that part of the late cession of Territory obtained of the Creek nation by the united states commissioners, lying south of the Oconee river, being now completed by the proper

As a result, he had a significant influence on the development of the state's land policies. His detailed and precise instructions issued to the lottery surveyors, written in his florid handwriting, remain on file today in the surveyor general records of the Georgia Department of Archives and History. Nor did he neglect his duties in administering headright grants, for when problems arose, he often went personally to inspect and resurvey the premises.

See Augusta Chronicle & Gazette of the State, Mar. 31, 1792, at 3, col. 3; Darien Gazette, Oct. 2, 1823, at 3, col. 5; Georgia Argus (Milledgeville), Jan. 30, 1810, at 3, col. 2; Georgia Journal (Milledgeville), June 16, 1818, at 3, col. 4; Louisville Gazette, Mar. 12, 1799, at 4, cols. 2–3; Id., May 7, 1799, at 3, col. 1; Executive Minutes (Feb. 6, 1817) (GDAH); Am. St. Papers, 2 Miscellaneous 72–79; and D. Pruett, Cousins by the Dozens, Supplement-I 12–31 (1978).

The author is indebted to Gordon B. Smith of Savannah and Robert S. Davis, Jr., of Jasper for providing the sources used to compile this sketch.

authorities; you are enjoined to proceed without delay to Fort Wilkinson for the purpose of effecting the objects of the foregoing appointment.

First. You are to commence the duties assigned you, by ascertaining with the utmost precision, the variation of your Compass or Needle from the true meridian; and to administer the oath prescribed by law to your Chain-men, with that seriousness and deliberation which is calculated to impress them wth. a due sence, of the solemn obligation they have entered into, to render a just and true acc^{t}. of the admeasurement of all lines as directed.

Second. You are required to divide the Territory above mentioned, by a line running from a point on the southern bank of the Oconee at Fort Wilkinson, South forty five deg^{s}. W^{t}.—according to the true meridian, to the Indian boundary line; leaving the Fort, with its Battlements to your left: which Line is to be considered a boundary between the Counties of Wilkinson and Baldwin, until altered by law.

Third. You are to mark the dividing line of the Counties of Wilkinson and Baldwin, in the manner following to wit. All Trees found standing precisely in the course thereof, are to be marked with four Notches; and at the termination of every quarter of a Mile from the beginning with the addition of the letters C.L. thus—

$$\underline{\underline{\overline{\overline{\text{CL}}}}}$$

—also the letter B to the right, and letter W to the left of Trees so found; otherwise Posts or large Stakes may suffice—and such trees as are by Surveyors called elbow trees, are to receive two Chops.

Fourth. In order therefore that you may attain the greatest possible accuracy in running and measuring this and all other lines required of you; you are to employ a man for the purpose of bearing a Pole before you—to become stationary at your command; and to remove such objects as may intervene, and obstruct your view—tending to render your observations less exact. In measuring you are to use no sticks which exceed twelve Inches in length. It will be your duty to give your Chain-Men a knowledge of the most approved method of Chaining—requiring them whilst in the act of measuring to conform thereto as near as circumstances will permit, by setting their sticks perpendicular with the plain of the Horizon: otherwise the meaning and intention of the Law, as respects the adjustment of your Chain by a standard in this Office, is intirely defeated.

Fifth. You are to lay out the Counties of Wilkinson and Baldwin into Five districts each, by lines running parallel with the divisional line of said Counties; and at such distances (as not only to include a certain number of parallel lines seperated by a space coinciding with the square of a Section of Two hundred two and an half acres) but to form areas equal in extent as near as circumstances will admit. This design you may accomplish by running a line transversely at right angles wth the County line, and in succession, the District lines; where you discover the ground most favorable—wch lines you are to blaze so plain as to be easily traced—and at the termination of every Forty five Chains, as you progress—mark that number on Trees so found—otherwise large stakes may be substituted. By pursuing this method in laying out the several districts, the number of fraction^{l} surveys will be greatly deminished, because confined to the two extremities of the said districts. In proportioning the districts agreeably to the extent of each County, I find district No. 1 in Baldwin County to include Fifteen parall. lines district No. 2 Fourteen parallel lines district No. 3 Thirteen parallels lines; district No. 4 Seventeen paral. Lines, and the rema[in]ing part of said County to form district No. 5. I also find district No 5 in the County of Wilkinson to include Thirteen parall lines, district No. 4 Thirteen paral: lines district No. 3 Fourteen paral: lines district No. 2 sixteen paral. lines, and the remaining part of said County in the same ratio, to form the district No. 1. This Scheme is fully displayed in a Map you will receive herewith; and to which you are enjoined to adhere—unless the County line, should intersect the temporary Indian line, considerably lower down than is therein represented—in the event of which, you are to proportion the districts accordingly.

SIXTH. The dividing lines of districts are to be marked and lettered by you in the manner following vizt All trees found standing precisely in the course thereof are to be marked with four Notches, and the letters D.L. thus

D.L.

fore and aft; and as often as convenient with the number of the adjoining districts, to the right and left—and all trees which are called elbow trees, are to receive two Chops.

SEVENTH. You are to transmit to this Office within fifteen days immediately after you have completed the running of the County and district lines, a fair transcript of your Field-Book: in which you are required to note down the course and distances of said lines; the names, situations, distances, and bearings of all Water Courses, lakes, Swamps, ponds, and other remarkable objects, crossed, or touched in the progress of that work: also a General Plan of the said lines, correctly delineated agreeably to your Notes—connected with the proportional lines; and exhibiting the whole extent of the Country, ceded south of the Oconee River, by the Creek nation, on the sixteenth day of June Eighteen hundred and two.

Danl Sturges Sur^{r} General

Surveyor Generals Office 28th Feby 1804

Sir

At the beginning and ending of the line dividing the Counties of Wilkinson & Baldwin, & wherever the dividing lines of Districts intersect the

Oconee or Appalachee rivers, and the temporary boundary line, you are to mark corners for the County line thus

O O O
X X X
O O O
B CL W

and for the district lines thus

O O O
X X X
O O O
2 DL 1

agreeably to the manner I have adopted for marking Corners of Tracts, to be laid out within the late cession of lands.

Yours

Danl. Sturges S. Genl.

Levin Wailes esq^{e}.

Surv^{or}. for District No.2 Baldwin County[18]

The following instructions were issued to the surveyors dividing the districts of Baldwin and Wilkinson counties and are typical of such instructions issued for subsequent land lotteries:

(Circular)—Sur^{o}. Gen^{ls}. Office 5th March 1804.

Instructions

To the District Surveyors for the Counties of Wilkinson and Baldwin

FIRST. You are to perform the duties of your respective appointments in your own proper persons, and to commence the same, by running lines parallel with the dividing lines of your Districts throughout the same, in regular progression; leaving a space of forty five Chains between each line. These lines being run you are to cross them by others at right angles, so as to form Tracts of forty-five Chains square, containing two hundred two and an half acres each; unless where the line of Limits between the Creek Nation, and your Districts; or the course of the River Oconee or appalachee may render the same impracticable: which method you are not to depart from, any further, than such particular circumstances may require.

SECOND. All lines run by you parallel with the dividing lines of your districts, are in the progress thereof to be blazed only, except, at their junction with the Oconee and Appalchee Rivers, and the Indian boundary line; where you are directed to mark Corners. But when running your Cross lines, you are to station (at any convenient distance, not less than five chains from their intersections with the first mentioned lines) so as to exhibit two Stations on every line of each survey, made, within the said Districts; save such lines of fractional surveys as fall short of Forty five chains—occasioned, by

18. Instructions of the Surveyor General 1–7 (GSGD).

the course of the Oconee and Appalachee Rivers, and the Indian boundary line. In such cases one station will suffice.

Third. You are required to mark the number of each survey contained in your Districts, on two Corner trees, or Stakes, of said surveys and on such stations thereof made, in the precise course of your cross lines; from the first to the last survey, in arithmetical progression; beginning with No. 1.—Which requisition you are to execute whilst running those lines; and to commence with them, at the distance of forty five chains, from the most extreme angle of your districts—Formed by the junction of a dividing line thereof, and the Indian boundary line.

Fourth. In closing the several surveys in your Districts, you are to run alternately from side to side, throughout the same, agreeably to the true meaning and design of the law; and not to deviate from this rule, unless for the purpose of ascertaining the areas of such fraction Tracts as are created by the course of the temporary boundary line, or the Oconee and Appalachee River.

Fifth. You are also to ascertain the contents of all Islands formed by the Oconee and Appalachee rivers lying south west of the middle, or main stream thereof within that part of the late cede[d] territory—to take as accurately as possible the meanderings of all water courses which may form boundaries to any of the surveys, and of all navigable Rivers or Creeks, whereby any of the surveys may happen to be divided—Which part of the duties required of you—you are to execute by running a line parallel with the general course of such streams; and by a sufficient number of offsets at right angles ther[e]with, to obtain the object sought.

Sixth. You are required to mark all corners with a chop, a blaze, a cross, a blase, and a chop thus,

—
o
x
o
—

and all stations with two chops, a blaze and two chops thus,

=
o
=

and all Trees standing precisely in the cou[r]se of lines, other than stations Trees, are to be blased; and elbow Trees are to receive one chop.

Seventh. The running and measuring of all lines, are to be performed by you with that mathematical exactness, which such skillful artists, as I presume you are would adhere to, in executing the plan of a Town or City—and in order that you may approach as near this required accuracy, as circumstances will permit; you are to employ a Man for the purpose [of] bearing a pole before you—to become stationary at your command, and to remove such objects as may intervene and obstruct your view; tending to render your observations less correct. In measuring you are to use no sticks which exceed twelve inches in length. It will be [your] duty [to] give [your] chain-men a knowledge of the most approved method of chaining—requiring them whilst in the act of measuring to conform thereto as near as possible; by setting their sticks perpendicular with the plain of the Horizon—otherwise the object of the law in adjusting your Chains, by a Standard in this Office, is intirely defeated.

EIGHTH. Your Chain-men are to be sworn, or affirmed to do justice in the admeasurement of all lines required of them.

NINTH. In your field Books, you are to note down the courses and distances of all lines run by you—the names situations, distances, and bearings, of all water courses, roads, or Indian paths—All large lakes, swamps, ponds, and other remarkable objects touched, or cross[ed], in running and measuring any of the aforesaid lines—the names of each corner and station tree, or Stake—the number of each survey, Islands if any, included—All boundaries, naturally or artificially created, either by Oconee, and Appalachee Rivers, &c. &c.—by the line of limits; your District lines; or by the numbers of adjoining Tracts.

TENTH. The respective surveys contained in your Districts, Islands included, are to be platted by a Scale of 20 Chains to an Inch: except such fractional surveys or Islands as contain less than one hundred and fifty acres—in which cases, you are to make use of a suitable scale, say 15 10 or 5—in order that your courses, distances, stations, &c. may not be crowded; as frequently happens in the returns of some surveyors, for the want of judgement to direct them otherwise.

ELEVENTH. You are required by law to return to this Office, within one hundred and fifty days from the running of the Indian boundary line, a fair transcript of your field-books, accompanied with the originals for examination—Also a Map of your respective districts handsomely executed—In which are to be correctly delineated, represented and numbered, agreeably to your field-Books, the several surveys made therein; together with a detached plat of each survey, certified as follows:

State of Georgia,

Scale of ______ Chains Per In:

The above plat exhibits the form and marks of Tract number ______ in District No. ______ in the County of ______ Surveyed the ______ day of ______ 1804, by ____________________

District Surveyor.

______ }
C.C.
______ }

which Plats are to be endorsed with the number of acres answering with the contents within—the number of survey corresponding with the arrangement in the General Plan, or Map—the number of your districts—and your own names with the initials of your Offices.

TWELFTH. Your District Maps are to be platted by a Scale of 40 Chains to an Inch—and no returns will be received which are not intirely free of erasures.

Although a portion of the time prescribed by law for the performance of the duties of your appointments, has through unavoidable circumstances elapsed, still, I am persuaded you have a sufficient space remaining if oeconomically disposed of to complete your surveys; and nearly all of your returns. I have authorised a person to designate (agreeably to the Law under which you are called upon to act) the division of the counties of Wilkinson and Baldwin; and to mark out within the same your respective districts. It will be adviseable in you to attend him throughout the progress of that work; or at least so far as concerns you individually—and as your Districts are formed to be in readiness to proceed to the duties you are to discharge therein. Let me advise you to compare your Instruments with that, used in laying out your districts; in order that you may thereby trace with the utmost precision the same lines: But if you are not skilled in the art of ascer[tain]ing the magnetic variation, [omission] you will find yourselves necessitated to adopt that measure, however inconvenient. I have directed the person appointed to lay out your Districts to proportion them by running a traverse line at right angles with the county line, in succession the District lines, where he discovers the surface of the country most favorable to the attainment of that object—on which lines, I have charged him to mark, at the end of every 45 Chains measured thereon; that number on trees, or stakes, thereby pointing out the number of lines to be run parallel with the dividing lines of Districts, by the respective surveyors, for districts No. 1, 2, 3, and 4 in Baldwin County; and districts No. 5, 4, 3, and 2 in Wilkinson. From which points marked 45, you are required to run the aforesaid lines; North East to the Oconee and Appalachee Rivers; and South West to the temporary boundary line.

I am charged by his Excellency the Governnor, to notify you, that in the event of any District Surveyors failing to prosecute the duties of his appointment, in his own proper person; or who by indolence, inattention, or otherwise, renders doubtful the finishing of his surveys within the time prescribed; or, who should be found deficient in the qualifications necessary to a due execution of the duties aforesaid—he, should not hesitate to declare vacant the appointment of such district Surveyor, and proceed in other respects as the law points out.

Dan[l] Sturges Sur. Gen[l] [19]

19. Id. 8–14.

Few of the maps and sketches that accompanied such instructions are to be found today.

Before proceeding, it may be helpful to imagine what these early surveying expeditions were like. The county and district lines were first run by the most proficient surveyors obtainable. The call then went out from the surveyor general for the district surveyors to begin. Winter was the preferred season for this work everywhere but in the mountains of north Georgia. In winter the weather was cooler, the crops had been gathered, and deciduous trees were shorn of their leaves. Consequently, there were fewer obstacles to penetrate. In addition, insects and snakes were less troublesome. The exigencies of the situation, however, rarely allowed the surveyors to choose their working season. Setting forth from their homes in wagons loaded down with supplies, tents, and an endless assortment of equipment, with perhaps an extra horse or two, these heralders of civilization first traveled to the state capital to be sworn in and to receive instructions. Then slowly they plodded off to their respective districts, over wretched roads studded with stumps, rilled with gullies, and, in the wet season, knee-deep in mud. The roads eventually turned to dim paths and then disappeared altogether on the fringe of civilization. Here they might stop at a lonely military outpost to purchase additional supplies and seek specific directions to their districts. For most of the remaining way they had to chop and hack a path through the wilderness, no doubt frequently getting lost in the process.

The first priority, upon reaching their districts, was to set up a base camp at a strategic location. This camp would probably be moved several times during the course of the work. The survey parties almost invariably consisted of the head surveyor, two chainmen, an "axman," a "marker," and one or two extra persons who assisted in cutting line and did odd chores.

The record is specific as to the procedure for laying out the districts—that can be gathered from the more than three hundred surviving volumes of field notes that resulted from these exertions. It was the simplest method that could be devised. Starting from any convenient corner of the district, the crew measured along one of the district lines for a distance equal to one side of a standard land lot. They then proceeded to chop lines parallel to the adjacent district line, crisscrossing the district with successive lines at the prescribed interval until they reached the far side of the district. Next, lines were chopped in like fashion at right angles to the lines previously run, and stakes were set or trees marked to

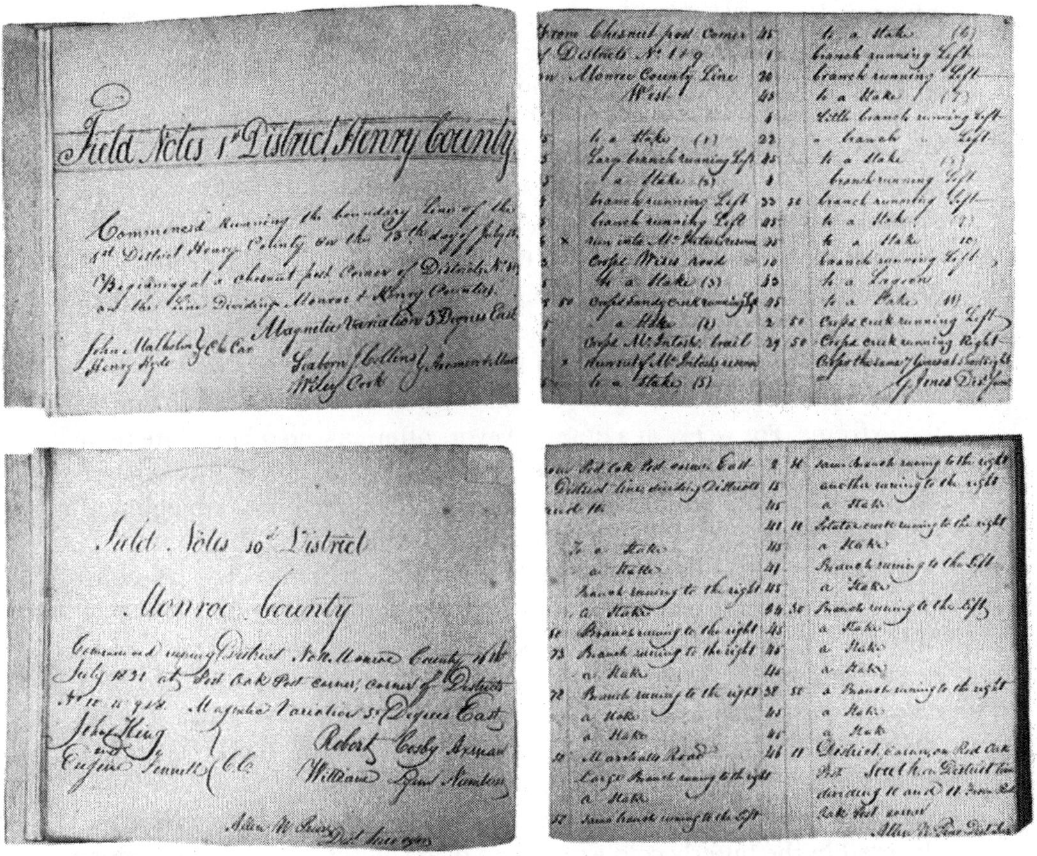

Land lottery surveyors' field notes. Courtesy of the Georgia Department of Archives and History.

designate land lot corners each time one of the previously run lines was crossed or a district line was intersected. This general procedure might be modified if natural obstacles such as large rivers, swamps, or mountains intervened or if a district was bounded by irregular objects such as rivers, trails, or ridges. Sometimes the meanders of irregular boundaries were traversed, but just as often they were not. Adjustments were made for magnetic declination, but no allowance was made for convergency of the meridians.

The corner trees, corner stakes, and line trees were incised with land lot numbers, which were vital in helping the settlers find their lots. (One of the early pioneers reported in his old age that more than eighty years afterward many of the original marked trees could still be found.)[20] An instrument resembling a branding iron and forced by the poll of an ax or hatchet was used to inscribe numbers.

Depending on the roughness of the terrain, a crew might run five to ten miles of line per day and on occasion even more. They often worked seven days a week, but some (like Daniel McDowell's crew) took Sundays off. McDowell, a Baptist preacher, in 1832 laid out the 25th District, 2d Section, of original Cherokee County (now in Gilmer, Gordon, and Murray counties). True to his calling, he spent his Sundays preaching sermons to gold diggers already in the region. No doubt his crew members were obliged to suffer through his fiery prelections also.[21]

The exertions involved in laying off the public domain were taxing to the extreme. For weeks at a time surveyors often saw no other white man and only an occasional bedraggled curious Indian. There was little to break the solitude but the singing of birds, the monotonous soughing of tall virgin trees, and sometimes the dull croaking of a bullfrog.

It is not too difficult to visualize these sojourners of the wilderness subsisting off the stale provisions they brought along, generally consisting of barrels of flour, biscuit, salt pork, bacon, beans, black-eyed peas, rice, dried fruit, coffee, sugar, tea, and perhaps beer, wine, or rum. On rare occasions this fare might be supplemented by creatures of the forest. There was little time to spare for hunting fresh meat, however, and in many areas very little was to be found.

The varying climes of the state had their effect. The surveyors often shivered in the bitterly cold wind of a north Georgia winter, their grimy fingers numb from manipulating the brazen compasses and links of steel. They found it difficult to gain a footing when traversing up steep, icy slopes and then floundered down on the other side, only to have to repeat the process over and over again. At night they slept on pine needles and hard, cold earth and perhaps heard the cry of a whippoorwill in the distance. In the sultry heat of a south Georgia summer, they endured day after weary day of hacking through nearly impenetrable baygalls and cypress swamps. Frequently they found themselves waist deep in muck, their clothes dripping with perspiration and their bodies plagued by rattlesnakes, redbugs, poison ivy, clouds of mosquitoes, and all sorts of other vermin. Sometimes they had to contend with days on end of seem-

20. W. Cotter, My Autobiography 24, 36 (1917).

21. Id.

ingly ceaseless rain, which smeared their field books and soaked them to the bone.

They hurried continually, lest supplies run out or the costs exceed their pay. Yet all the while they were confined to the line directed by their compasses and could veer neither left nor right to avoid obstacles; instead, they had to clamber over or through whatever lay in the path.

At length, their work accomplished, they could look upon land that was formerly a nondescript chunk of wilderness and see a district ordered into a tidy checkerboard pattern that would soon fill with settlers. Then they plodded back to the state capital to draw their maps, receive their hard-earned wages, and repeat many times over the story of their expedition and of strange sights encountered along the way.

Such were the adventures and privations the lottery surveyors experienced. In the course of a single generation, they subdivided more than twice the quantity of territory as was located in the old headright regions. This latter had been in the process of subdivision, piecemeal fashion, for the two preceding generations and the subdivision was not to be completed for another three generations.

The surveys for the counties of Baldwin, Wilkinson, and Wayne commenced in the spring of 1804 and were completed approximately a year and a half later.[22] The city of Milledgeville was staked off by Benjamin Easley during 1805 and 1806.[23]

The drawing for this first land lottery began in Louisville on July 22, 1805, and was completed on August 31 of that year.[24] Burke Chislom, chosen to draw the tickets from the wheel, was paid two dollars a day for his service.[25] A memo in the Executive Minutes for 1805 records that John Jacob Schley was paid forty dollars for making the wheel used in the drawing.[26]

22. Executive Minutes (Nov. 1, 1804, through Oct. 30, 1805) (GDAH); Instructions of the Surveyor General 6–53 (GSGD); 1804 Ga. Senate Journal 5.

23. Executive Minutes (Nov. 4, 1805) (GDAH); Journal [of the] Board [of] State Commissioners of Milledgeville, 1805 to 1812 passim (GSGD); Clayton 288.

24. Executive Minutes (Aug. 31, 1805) (GDAH); Augusta Herald, July 11, 1805, at 1, col. 3; Id., Aug. 1, 1805, at 1, col. 3; Id., at 3, col. 1.

25. Clayton 288.

26. Executive Minutes (Oct. 30 & Nov. 4, 1805) (GDAH).

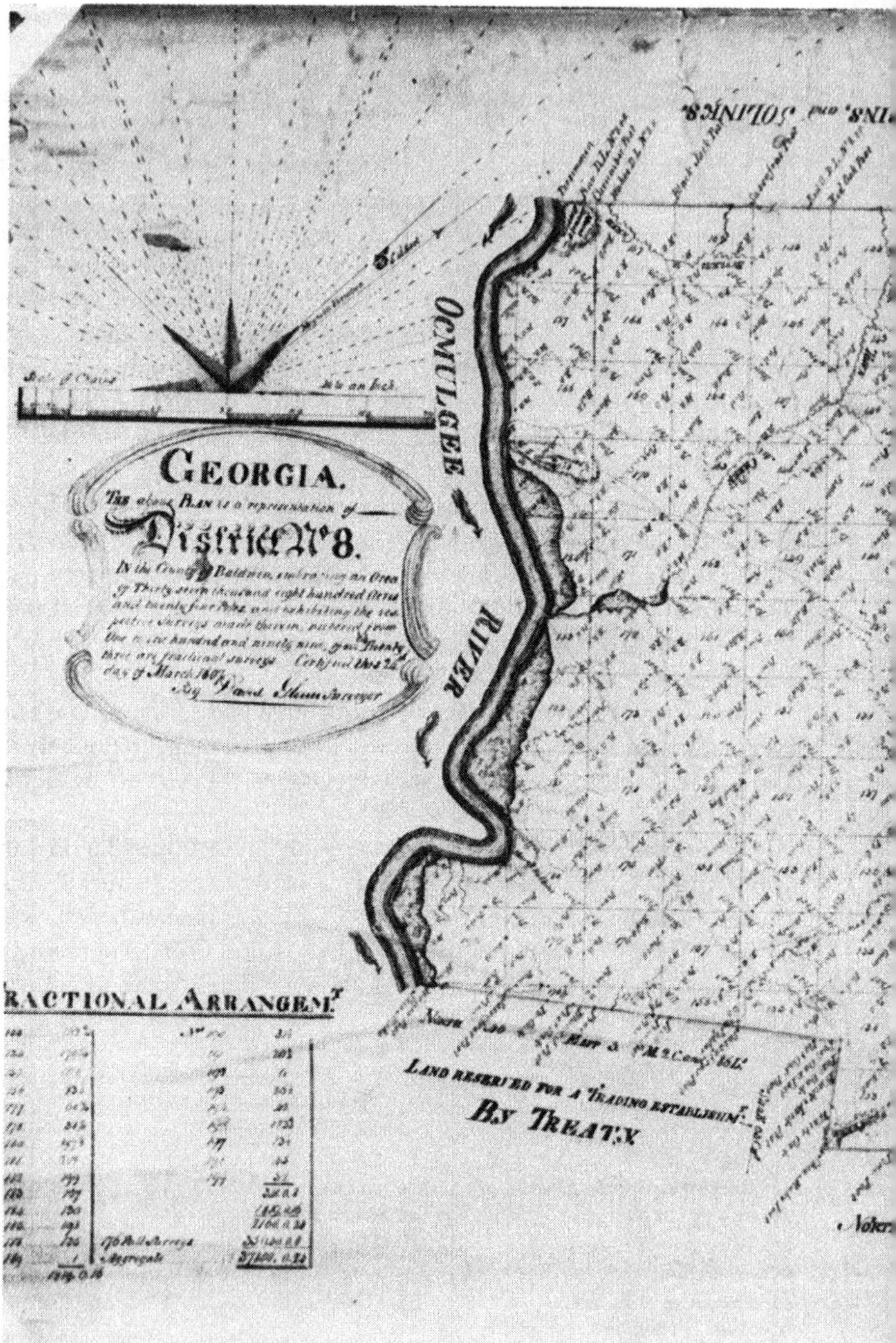

Map returned by David Glenn, surveyor of the 8th District of original Baldwin County. The district today is part of Jones and Bibb counties. Courtesy of the Georgia Department of Archives and History.

District N°6.
District N°9.
6. Miles
53 Chains and 50 Links
North 45 West
6. Miles and 60
South 45 West

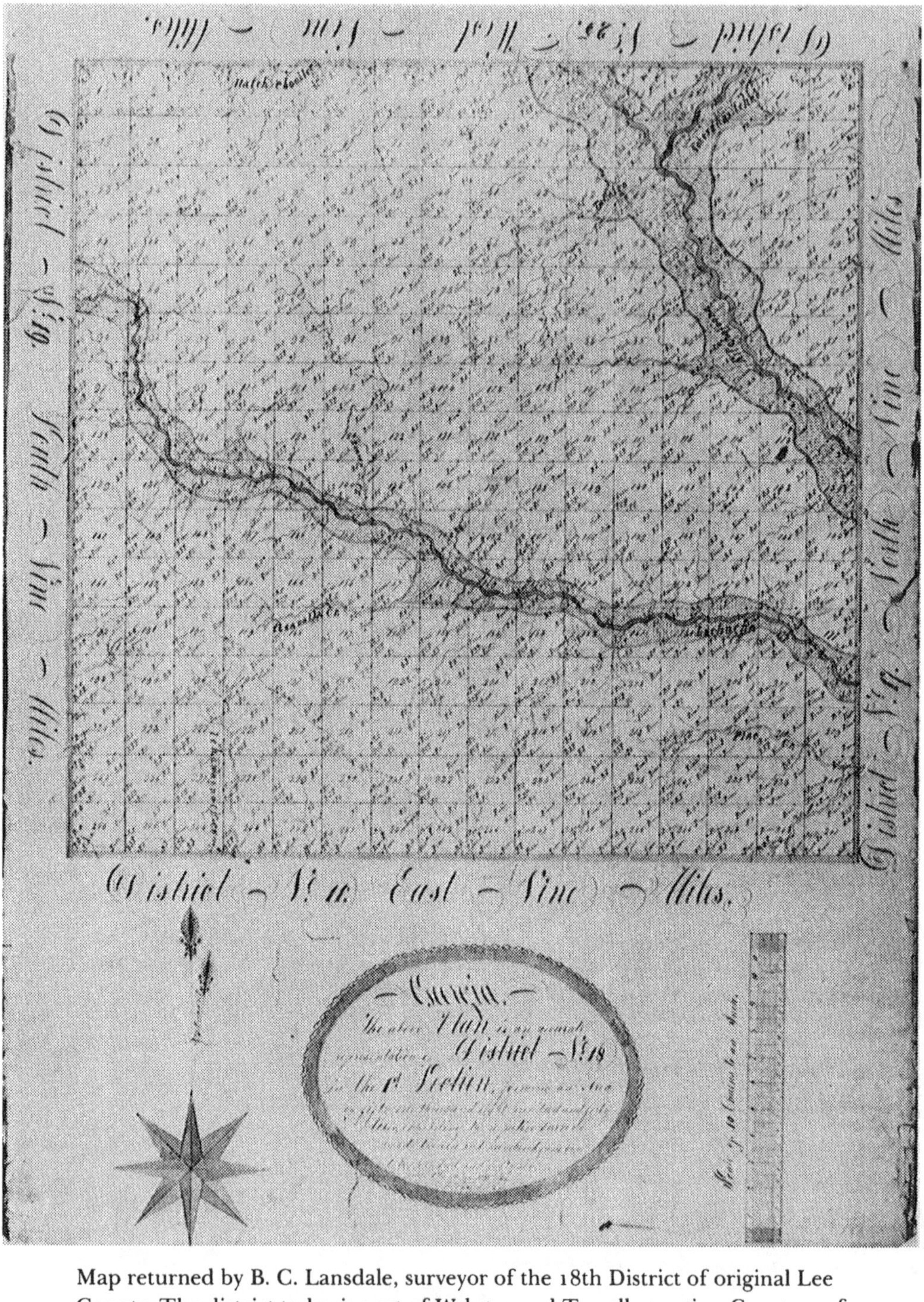

Map returned by B. C. Lansdale, surveyor of the 18th District of original Lee County. The district today is part of Webster and Terrell counties. Courtesy of the Georgia Department of Archives and History.

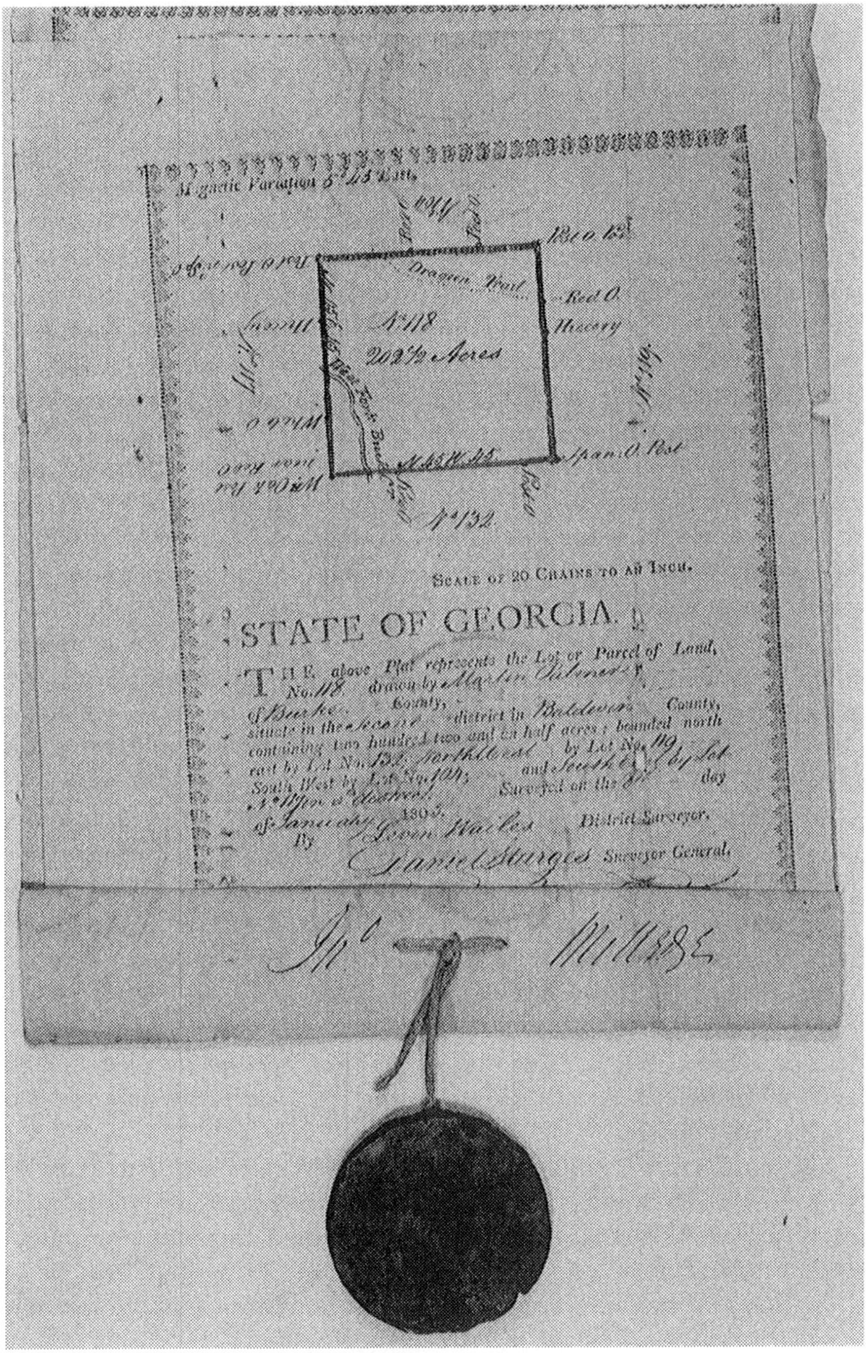

Grant of Land Lot 118 of the 2d District of original Baldwin County, signed by Governor John Milledge. Courtesy of Special Collections, Ina Dillard Russell Library, Georgia College, Milledgeville.

Land Lottery Act of 1806

Georgia had long cast its covetous eyes upon the fertile lands bordering the Ocmulgee River and had been disappointed that the treaty made at Fort Wilkinson did not extend its boundaries to that river. This disappointment was not to last long, however. In the year of the first lottery drawing, the state wrested from the Creeks all their remaining lands between the Oconee and Ocmulgee rivers. Excepted from this cession was a three- by five-mile reservation on the east side of the Ocmulgee River where Fort Hawkins was later built and east Macon stands today.[27] As consideration the United States paid the Creek Nation $206,000 in merchandise and agreed to furnish the nation with blacksmiths and strikers for eight years. The signature of the mixed-breed Creek chief William McIntosh, who will figure more prominently later in this narrative, appears on this treaty.

Governor Milledge called an extra session of the legislature for June 9, 1806, to make disposition of this latest cession. Needless to say, little prompting was required to induce the passage of an act providing for disposal of the lands in a manner almost identical to that of the previous cession. The act is summarized as follows:

Section 1. The dividing line between Baldwin and Wilkinson counties, as established by the land lottery act of 1803, to be extended "south forty-five degrees west, to the Ocmulgee river." That part of the latest cession lying north of this line to become part of Baldwin County and that part lying south of this line to become part of Wilkinson County.

Section 2. The cession to be divided into thirty-eight districts "as nearly equal" in area "as they can be conveniently made by running parallel lines south forty-five degrees west, and others intersecting them at right angles." District lines to be laid off by such persons as the surveyor general should appoint.

Section 3. Districts to be divided into lots by lines running "parallel" with the district lines and by other lines crossing "at right angles" so as to form tracts "forty-five chains square" and containing 202½ acres each, except for fractional lots.

Section 4. Fractional lots to be reserved from drawing and disposed of in such manner as a future legislature should direct.

Sections 5, 6, 7, 8. Surveyors to be elected by joint ballot of the legis-

27. Clayton 705; C. Kappler, Indian Affairs (Treaties), S. Doc. No. 319, 58th Cong., 2d Sess. 85 (1904).

lature. District surveyor with the highest number of votes to have first choice as to district, the one with the second highest number of votes to have second choice as to district, and so on in that order.

Section 9. Each surveyor to post a $10,000 bond.

Section 10. (Essentially identical to section 5 of the land lottery act of 1803.)

Sections 11, 12. (Essentially identical to sections 6 and 7 of the land lottery act of 1803, except surveyors to be advanced $150 instead of $300.)

Sections 13, 14, 15, 16. (Essentially identical to sections 8, 9, and 10 of the land lottery act of 1803, except as to qualifications of drawers.) Each unmarried white male citizen twenty-one years of age and older, each widow, and each spinster entitled to one draw. Each white male citizen with a family entitled to two draws. Each minor orphan or family of minor orphans entitled to one draw or two draws, depending on certain specified circumstances. Successful drawers in previous lottery excluded from participation.

Section 17. (Essentially identical to section 13 of the land lottery act of 1803.)

Section 18. Surveyors and chainmen to take an oath before beginning their work.

Section 19. Headright surveys that may have been made on lands embraced within the latest cession to be void.

Section 20. Prices of grants issued to successful drawers to be $6.00 per hundred acres. Any grants not taken out within one year of the drawing to revert to the state.

Section 21. Grants and applications for grants made contrary to this act to be void.[28]

Surveys for the Additions to Original Baldwin and Wilkinson Counties

The artificial boundary of the 1805 cession, running from the high shoals of the Appalachee to the headwaters of the Ocmulgee, was marked under the direction of Benjamin Hawkins in the summer of 1806.[29] Most of the

28. 1806 ex. sess. Ga. Laws 3; Clayton 290.

29. Letter from Benjamin Hawkins to John Milledge (July 29, 1806) (Telamon Cuyler Collection, UGL); Letters from John Milledge to Henry Dearborn (July 5, 1806), to Benjamin Hawkins (July 5, 1806), and to Levin Wailes (Aug. 12, 1806) (all in GLB, GDAH); 1806 Ga. House Journal 7.

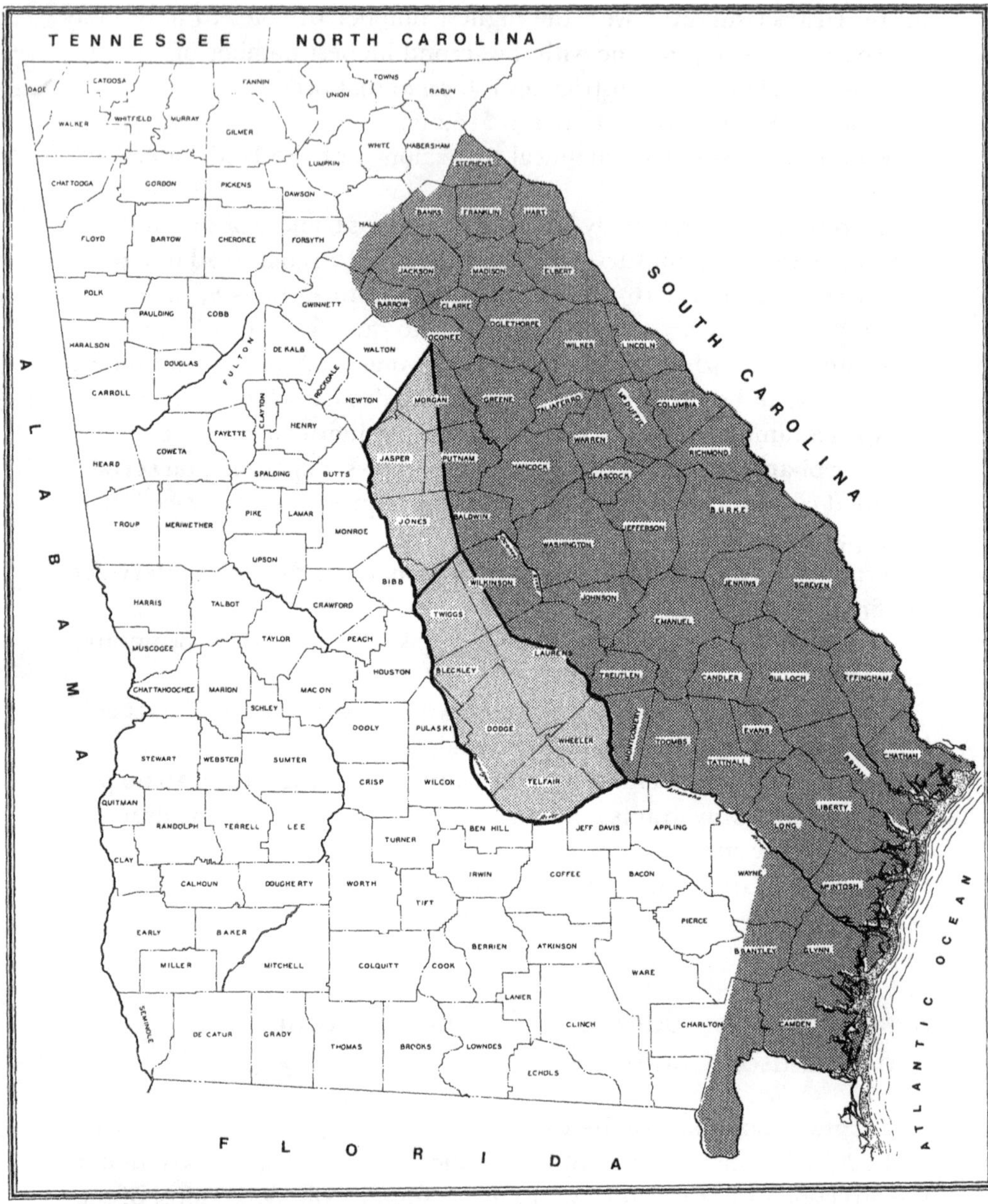

Lightly shaded area indicates lands distributed in the 1807 land lottery.

surveyors for this cession were sworn in during August 1806, and the surveys were finished about a year later.[30]

The drawing for the second land lottery began on August 10, 1807, and moved along very rapidly, being completed early the following month.[31] This was one of the last major operations of state government in Louisville. That same year the capital was moved to Milledgeville, which would be the scene of all later land lotteries.

Frauds and Mistakes in the 1805 and 1807 Land Lotteries

Under enactments of 1805 and 1806 the fractional lots were ordered to be sold at public auction and the proceeds devoted to worthwhile public purposes.[32] In November and December of 1808, Obadiah Echols, Reddick Simms, and Francis Flournoy, sworn commissioners of the fractional sales, were charged with conducting unscrupulous dealings. Upon trial by the state Senate, they were convicted of pocketing part of the proceeds and of rigging the sales so that they and their compatriots could purchase certain choice fractions at prices below the bid prices. For their "base, wicked, and corrupt designs" they were disqualified from holding any state office of trust or profit and were sentenced to be held in the Oglethorpe County jail until the costs of their trial had been paid into the state treasury.[33]

So many complaints about the surveys were made by grantees of the lots that a legislative committee directed the governor to look into the matter, and a resolution was adopted recommending that the complainants have recourse in the superior courts to the bonds of the surveyors.[34] Surveyor General Daniel Sturges informed the governor that he had received numerous reports that there were no lines to be found in the Ocmulgee swamps of the 15th, 25th, and 28th districts of Wilkinson

30. Executive Minutes (Mar. 18, 1807, through Sept. 21, 1807) (GDAH); Instructions of the Surveyor General 60–81 (GSGD); Letter from John Milledge to Levin Wailes (Aug. 12, 1806) (GLB, GDAH).

31. Executive Minutes (June 26, 1807) (GDAH); Augusta Chronicle, Aug. 1, 1807, at 4, col. 3; Id., Nov. 7, 1807, at 1, col. 4; Id. at 3, cols. 1–2.

32. 1805 Ga. Laws 56; 1806 Ga. Laws 64; Clayton 283, 341.

33. 1808 Ga. Senate Journal 119–72.

34. Executive Minutes (Jan. 9 & 18, 1808) (GDAH); 1809 Ga. Laws 103; Clayton 683, 689.

County and that many of the fractions contained more than double the quantity reported by the surveyors.[35] Skelton Standefer was sent to inspect the 28th District of Wilkinson (now in Twiggs) County, which had originally been surveyed by James Cooper. In May 1807 Standefer wrote Governor Jared Irwin from Ocmulgee Old Fields:

> Sir after Running the County line to the [Ocmulgee] River I began at Eleventh Corner and run South 45 East 15 Chains to the left of Capt Coopers line, that is his Corner Was 15 Chains nearer the River than it aut to have been and I found Capt Coopers first parrelal to only a few links over Thirtyseven Chains wide the next 45 Chs. from that and nearly in that order throughout and at the lower side of the District the parrelals were very wide. I then found that the Whole of the Work wou[l]d be to do over again and Immediately Commenced running the parrelals and there is not one Corner that wile answer and but two Stations in the hold District[.] I am now about half Done Myself and hands are all well tho verry tired.[36]

In June 1808 Governor Irwin appointed David McCord to resurvey the fractional lots in the 7th District of Baldwin (now Bibb) County, which had originally been surveyed by Benajah Smith. Three months later, McCord reported to the governor that according to his resurvey the fractions contained 1,895 acres, 1 rood, and 36 poles of land in excess of what had been represented by Smith. A joint committee in the legislature investigated the matter and came to the conclusion that Smith had intended no fraud on the state but had been prevented from making an accurate survey because of the high waters of the Ocmulgee River at the time of his survey. Smith was directed to pay for the resurvey ($520) but was absolved from all other liability.[37] As a result of McCord's resurvey, ten full-size and fractional lots were added to the district and fourteen of the previously existing fractional lots significantly increased in area. McCord's resurvey also accounts for the somewhat erratic arrangement of the numbering of the lots on the lower end of the district. Both Smith's and McCord's surveys are on file today in the surveyor general records of the Georgia Department of Archives and History.

Irregularities were also found in the fractional surveys of the 8th District of Wilkinson (now Telfair) County, which had been laid out by John

35. Letter from Daniel Sturges to David B. Mitchell (Nov. 22, 1809) (GSGD).

36. Letter from Skelton Standefer to Jared Irwin (May 13, 1807) (GSGD).

37. Letters from Jared Irwin to David McCord (Jan. 23 & June 20, 1808), and to Benajah Smith (Jan. 4, 1809) (all in GLB, GDAH); 1808 Ga. Senate Journal 6, 113–14; Clayton 687.

Coffee.[38] Coffee had been tardy in completing his work—a matter he explained by stating that soon after he and his men commenced, "the fall rains began to such excess that it was with great dificulty" that they were able to proceed. A month and a half into the survey they were "compeled to retreat on account of the great quantity of water . . . in the lower part" of the district. This had made it "impossible to conclude" the survey within the time allowed by law.[39] In June 1810 Daniel Sturges was sent to resurvey the fractions in Coffee's district and found an excess of 1,614 acres.[40] At the next legislative session a committee examined the two surveys and called as witnesses, among others, a Mr. Gatlin, who had acted as marker for Coffee. Gatlin testified that in running the lines along the Ocmulgee River he frequently had had "to wade to his breast in water, and at times to swim, in order as near as possible to ascertain the margin of the river." Moreover, in many instances the survey crew had been "at a loss and indeed found it impracticable assuredly to identify the same." Sturges, in his testimony, observed that when he made his resurvey "the river was uncommonly low, so that the dead river or cut-offs in the bends of the . . . river, were either dry or did not run."[41]

Based on the testimony, the committee concluded "that the inundation

38. John Coffee (1782–1836) was born in Prince Edward County, Virginia, and emigrated with his parents to Hancock County, Georgia, in 1800. He was serving as surveyor of Hancock County when he laid out the 8th District of Wilkinson, which later became part of Telfair County. He later moved to Telfair County and carved out a plantation in the 8th District, near Jacksonville. He directed the cutting of a road, known as the Old Coffee Road, which ran from Jacksonville to Mobile, Alabama. This road did much to open up southwestern Georgia to settlement. Coffee gained a reputation as an Indian fighter and served as an officer in campaigns against the Creeks and Seminoles. He represented Telfair County in the state House of Representatives and later in the state Senate. In 1832 he commanded a special militia force to guard surveyors laying out the Cherokee lands in northwestern Georgia (see Chapter 8). Later that year Coffee was elected to the U.S. House of Representatives and served until his death four years later. He is buried in Oakgrove Cemetery in McRae.

Coffee was a prominent figure in Georgia during the first third of the nineteenth century and a number of places are named for him: Coffee County; Coffee Bluff on the Ocmulgee River; the community of Coffee in Bacon County; the Old Coffee Road, part of which forms the boundary between present Berrien and Coffee counties; and Coffee Branch in Telfair County.

See Swilley, *John E. Coffee,* in 1 Dictionary of Georgia Biography 209 (K. Coleman & C. Gurr eds. 1983).

39. Letter from John Coffee to Jared Irwin (Feb. 3, 1807) (File II, GDAH).

40. Executive Minutes (June 15, Nov. 15, & Dec. 14, 1810) (GDAH); 1809 Ga. Laws 138; 1810 Ga. Senate Journal 18–19, 100–101.

41. 1810 Ga. Laws 170; Clayton 697–98.

of the waters on the low grounds, the dead rivers, lakes and lagoons" along the Ocmulgee had been the sole cause of the differences between the two surveys, that the margin of the river as exhibited by Coffee's survey and that of Sturges "must therefore be different," and that "in many instances arable land is found between their respective lines." The committee resolved that "no degree of blame" should be attached to Coffee, the rectitude of his conduct in making the survey was "indisputable," and "his reputation in that respect entirely unimpeachable."[42]

Because of the suspected inaccuracies in the surveys of the 7th District of Baldwin and the 8th District of Wilkinson, Governor Irwin in January 1808 had ordered the sale of the fractions held up.[43] In 1810 the legislature ordered these fractions, as resurveyed by McCord and Sturges, to be sold.[44]

Additional complaints about the surveys were filed. For example, Josiah Durden purchased a fractional lot reportedly containing 129¼ acres in the 1st District of Baldwin County adjoining the Milledgeville common and Fishing Creek. Upon resurvey, however, it was found to contain only 94 acres. The deficiency was "occasioned by a mistake in the District Surveyor who . . . , in laying down the distance from the corner to Fishing creek, . . . made the line eleven chain[s] and forty links longer than what appears to be by the re-survey." Jesse Sanford purchased a fractional lot in the 4th District of Wilkinson, which supposedly contained 161½ acres. After a resurvey by the surveyor general, it was found to contain only 94½ acres. Both men petitioned for redress, and the legislature recommended that the state treasurer refund to them a proportionate amount of what they had paid for the lots.[45] Even as late as 1955 an appeal came before the Georgia Supreme Court regarding Land Lot 103 of the 2d District of original Wayne County (now Brantley County). A resurvey of the land lot showed that it contained only 448 acres instead of the standard 490 acres.[46]

Although the above examples suggest that most of the surveying errors made in Georgia's first two land lotteries were caused by carelessness or by adverse field conditions, purportedly surveyors perpetrated a fraud

42. 1810 Ga. Laws 170; Clayton 697–98.

43. Executive Minutes (Jan. 18, 1808) (GDAH).

44. 1810 Ga. Laws 121; Clayton 657.

45. 1808 Ga. Senate Journal 67; 1813 Ga. Senate Journal 54.

46. Brunswick Pulp & Paper Co. v. Wilson, 211 Ga. 862 (1955).

in the second lottery. The fractional lots were reserved from the lotteries and sold at public auction. The story goes that some of the surveyors ran the lines of those fractions on the richest ground in such a way as to enclose 400 or 500 acres, but reported them as containing less than 202½ acres. The surveyors and their friends then attended the sale, bought the lots at the returned quantity, and reaped the profit.[47]

Indians, ever conscious that settlers invariably followed the work of the surveyor, also wrought mischief. On March 2, 1807, James Durham went before two justices of the inferior court of Wilkinson County and deposed that he saw some Uchee Indians destroy one corner of a lot in the 25th District of Wilkinson County, that many other corners had been destroyed as well as stations, and that, according to common report, Indians of the different tribes of the Creek Nation were actively canceling corners and numbers and killing beef cattle.[48]

47. King, *A Sketch of the History of Land Titles in Georgia,* in Report of the Second Annual Meeting of the Georgia Bar Association, 126, 156 (1886).

48. Affidavit of James Durham sworn before Wm. O'Neal and Thomas Gilbert (Mar. 2, 1807) (GDAH).

Chapter 5. The 1820 Land Lottery

Land Lottery Act of 1818

In 1813, while the United States was preoccupied with the war with Great Britain, the Upper Creeks, incited by the Shawnee leader Tecumseh, began a general uprising against the whites in Alabama and Tennessee. Many of the Lower Creeks sided with the Americans, however, and some of these, led by the Creek chief William McIntosh, actually fought with the Americans against the Upper Creeks. McIntosh was to play a pivotal role in Creek-American diplomacy for the next twelve years. In March 1814 the Upper Creeks were decisively defeated at the Horseshoe Bend on the Tallapoosa River in what is now Alabama by a force of Tennesseeans and Creek allies led by General Andrew Jackson.[1] In a treaty signed at Fort Jackson (near present-day Montgomery, Alabama), Jackson—acting as sole commissioner for the United States—exacted a terrible price from the Creeks for their inexpedience. He forced them to cede a huge expanse of territory in central and southern Alabama and a strip of land about seventy miles wide extending across southern Georgia from the Chattahoochee River to the western boundary of original Wayne County. This tract comprised about one-half of the total lands still remaining in Creek possession.[2]

Significantly, no responsible chief of the Upper Creeks signed the treaty and many of those who signed for the Lower Creeks were mixed-bloods. McIntosh signed in four places—once for himself, once as proxy for his uncle, and once each for two other chiefs.[3]

Jackson's principal object in securing the cession was a military one—to isolate the Creeks from the Chickasaws and Choctaws on the west and

1. Bonner, *William McIntosh,* in Georgians in Profile 114, 119–23 (H. Montgomery ed. 1958); 1 J. Parton, Life of Andrew Jackson 401–560 (New York 1861); U. Phillips, Georgia and State Rights 49–52 (1902).

2. Am. St. Papers, 1 Indian Affairs 826; C. Kappler, Indian Affairs (Treaties), S. Doc. No. 319, 58th Cong., 2d Sess. 107 (1904).

3. T. Abernethy, The South in the New Nation 1789–1819, at 371 (1961); Bonner, *William McIntosh,* in Georgians in Profile 114, 122–23 (H. Montgomery ed. 1958).

William McIntosh. From Thomas L. McKenney, *History of the Indian Tribes of North America* (1836 and 1844).

from the Spaniards and Seminoles in East and West Florida. Although Georgia acquired a considerable chunk of territory, Georgians were dissatisfied. In a memorial and remonstrance to the president, the Georgia legislature complained that not only were the lands in this cession sterile and unprofitable but the treaty left the Creeks in possession of the territory between the north boundary of the cession and the Ocmulgee and Altamaha rivers. It was feared that settlement would be deterred because any citizens who should move to the area would be hemmed in between the Creeks on the north and the Spaniards and Seminoles on the south. Indeed, the Indians had already warned Benjamin Hawkins that any attempt "to run out and settle the lands" would "be resisted with all their force."[4] Georgia's memorial and remonstrance concluded with a plea that immediate measures be taken to procure additional cessions from the Indians in conformity with the Articles of Agreement and Cession of 1802.[5]

In July 1817, apparently through well-placed bribes and circumspect negotiations with a small, unauthorized group of Cherokee chiefs, the United States was able to secure a treaty by which the Cherokees ceded two tracts of land—one in Tennessee and the other in Georgia.[6] The Georgia cession included the Cherokee lands that lay south and southeast of the Chattahoochee and Tallulah rivers. The United States, in return, agreed to give the Cherokees an equal area west of the Mississippi River, compensation for the improvements they had made on the ceded lands, and assistance to any who chose to emigrate to the West. Within two years about one-third of the Cherokee Nation removed to the Louisiana Territory.[7]

In spite of the treaty at Fort Jackson in 1814, hostilities broke out again between the Creeks and Americans. As a result, another treaty was signed in January 1818 at the Creek Agency on the Flint River in Georgia.[8] By its terms the Creeks were forced to cede two tracts. One was a tongue of land lying between the north boundary of the 1814 cession and the Ocmulgee and Altamaha rivers. The other lay about the head-

4. 2 Correspondence of Andrew Jackson 222 n. 1 (J. Bassett ed. 1927).

5. 1816 Ga. Laws 205, 206; Lamar 1168, 1169.

6. Am. St. Papers, 2 Indian Affairs 129; C. Kappler, Indian Affairs (Treaties), S. Doc. No. 319, 58th Cong., 2d Sess. 140 (1904). See G. Moulton, John Ross: Cherokee Chief 19 (1978).

7. U. Phillips, Georgia and State Rights 68 (1902).

8. Am. St. Papers, 2 Indian Affairs 152; C. Kappler, Indian Affairs (Treaties), S. Doc. No. 319, 58th Cong., 2d Sess. 155 (1904).

waters of the Ocmulgee River, bounded on the east and south by earlier cessions, on the west by the Ulcofauhachee River and Hightower Trail, and on the north by the Chattahoochee River and the Cherokee cession of 1817. For this territory the United States paid the sum of $120,000 and agreed to furnish the Creek Nation with blacksmiths and strikers for three years. One consequence of this treaty was that the irregular and ill-defined Hightower Trail became the line of demarcation between land grants issued in the lotteries of 1820 and 1821. This line was to confound the wits of surveyors working in the area thereafter.

In contrast to the previous cessions, Georgia moved slowly in disposing of these latest ones. It had been proposed in the House as early as 1815 that disposition be made of the lands obtained under the treaty at Fort Jackson,[9] and in 1817 a motion was considered in the Senate to dispose of the lands between the Chattahoochee and Flint rivers obtained in the Fort Jackson treaty.[10] Nothing came of either of these proposals, however. Delays in running the Indian boundaries, the general undesirability of the land, and the expectation of receiving additional cessions apparently caused Georgia's tardiness in granting the newly acquired territories to settlers.[11]

Finally, at its regular session in December 1818 the legislature passed an act, summarized as follows, to make disposition of the new territories:

Section 1. The Creek cessions of 1814 and 1818, embracing the area south and southwest of the Altamaha and Ocmulgee rivers, to be divided into three counties. That part lying west of a line beginning on the north boundary of the cessions two and three-quarter miles east of the Flint River and running south to the Florida line to form Early County. That part lying between Early County and a line beginning at Blackshear's Ford on the Ocmulgee River and running south to the Florida line to form Irwin County. The balance of the said territory to form Appling County. Early County to be divided into twenty-three districts, each "twelve miles and forty chains square, as near as convenience will admit, by running lines parallel with the dividing line" between Early and Irwin counties "and others crossing them at right angles." Early County districts to be divided into "squares of fifty chains, containing two hundred and fifty acres, by lines parallel with the district lines, and others crossing them at right angles," except for fractional lots. Irwin and

9. 1815 Ga. House Journal 11.
10. 1817 Ga. Senate Journal 39.
11. Id. 27.

Appling counties to be divided into districts "twenty miles and ten chains square, as near as can conveniently be, by lines running parallel with the line dividing said counties, and others crossing them at right angles." Irwin County districts to be numbered 1 to 16 inclusive. Appling County districts to be numbered 36 to 50 inclusive. Irwin and Appling county districts to be divided into "lots of seventy chains each way, containing four hundred and ninety acres each." Surveyor general to give district surveyors instructions for surveying, marking, and numbering lots.

Section 2. The Cherokee cession of 1817 and the Creek cession of 1818, embracing the area along the Chattahoochee, Tallulah, and Ulcofauhachee rivers and the Hightower Trail, to be divided into four counties. That part lying south of a line beginning at the point where the Yellow River crosses the cession line and running "north, sixty degrees east, to the Appalachee River" to form Walton County. That part lying between Walton County and a line beginning at the mouth of Big Creek and running "south thirty degrees west [east] to the top of the ridge and line of the county of Jackson, thence along said line to the head of Appalachee River, thence down the same to the corner of Walton county" to form Gwinnett County. That part lying between Gwinnett County and a line beginning "at a place where captain John Miller now lives, on the line of Franklin County, and running north, thirty degrees west, to the Chattahoochee river, and down the same to Gwinnett county line" to form Hall county. The balance of the said territory to form Habersham County. Walton, Gwinnett, Hall, and Habersham counties to be divided into thirteen districts "as equally as conveniently can be, by running lines parallel with the lines dividing the county of Walton and Gwinnett and others crossing them at right angles, and subdivided into tracts of fifty chains, containing two hundred and fifty acres each, by lines run as prescribed for subdividing the county of Early."

Section 3. County and district lines to be laid off by such persons as the surveyor general should appoint.

Section 4. Fractional lots to be reserved from drawing and disposed of in such manner as a future legislature should direct [but see sections 13, 14, 15, and 16 below].

Sections 5, 6, 7, 8. (Essentially identical to sections 5, 6, 7, and 8 of the land lottery act of 1806.)

Section 9. Each surveyor to post a $10,000 bond.

Section 10. (Essentially identical to section 5 of the land lottery act of 1803, except as to time for completing surveys.) Surveys to be completed within five months after the running of the district lines.

Section 11. District surveyors to receive $4 for each mile of line run, out of which they must defray all expenses, including compensation to chainmen and axmen. Governor to advance $400 to each district surveyor before the survey begins, and the balance to be paid upon completion.

Section 12. Surveyors of county and district lines to receive $5 for each mile of line run, out of which they must defray all expenses, including compensation to chainmen and axmen. Governor to advance $150 to each surveyor before the survey begins, and the balance to be paid upon completion.

Sections 13, 14, 15, 16. (Essentially identical to sections 8, 9, and 10 of the land lottery act of 1803, except as to the disposal of fractional lots above 160 acres in area and qualifications of drawers.) Fractional lots above 160 acres in area to constitute prizes and be drawn in the lottery along with full-size lots. Each unmarried white male citizen eighteen years of age and older, and each indigent or invalid veteran of the revolutionary war who was a successful drawer in either of the previous lotteries, entitled to one draw. Each white male citizen with a family and each indigent or invalid veteran of the revolutionary war who was not a successful drawer in previous lotteries entitled to two draws. Each widow and each minor orphan or family of minor orphans entitled to one draw or two draws, depending on certain specified circumstances. Successful drawers in previous lotteries, except as noted, and those persons who evaded military service, excluded from participation.

Section 17. (Essentially identical to section 13 of the land lottery act of 1803.)

Section 18. Surveyors, chainmen, and markers to take an oath before beginning their work.

Section 19. Headright surveys that may have been made on lands embraced within the latest cessions to be void.

Section 20. Prices of grants issued to successful drawers to be $18 each. Any grants not taken out within two years of the drawing to revert to the state.

Section 21. Grants and applications for grants made contrary to this act to be void.

Section 22. Lots 10 and 100 in each district to be reserved for the education of poor children.

Section 23. Proceeds from the sale of fractional lots to be set aside for the education of poor children.[12]

12. 1818 Ga. Laws 27; Lamar 416.

For some inexplicable reason, the numbering scheme for the districts of Appling County, as set out in the above act, was not followed. Also, the numbering of the districts for Early County varies considerably from the plan, for reasons that will be explained later.

Georgia-Florida Boundary

Before Appling, Irwin, and Early counties could be laid out, it was first necessary to establish the southern boundary of Georgia.[13] By the royal proclamation of 1763, the Treaty of Paris in 1783, and the treaty between Spain and the United States in 1795, Georgia's southern boundary had been fixed at a line beginning at the center of the Mississippi River on the thirty-first parallel, thence eastward along that parallel to the center of the Chattahoochee River, thence southward along the centerline of the Chattahoochee to its junction with the Flint River, thence in a straight line to the head of the St. Marys River, thence down the middle of that river to the Atlantic Ocean.

In accordance with the provisions of the treaty with Spain, Andrew Ellicott (on behalf of the United States) and Stephen Minor (on behalf of Spain) ran the line from the Mississippi to the Chattahoochee during 1798–99. Owing to the hostility of Indians in the area of the Chattahoochee, however, they abandoned the survey of the line east of that river and sailed around Florida and up the St. Marys. On February 27, 1800, they erected a large earth mound near the head of what they considered to be the main branch of the St. Marys.[14] This eastern terminus

13. Sources on the Georgia-Florida boundary are voluminous. They include M. McCall, The Georgia-Florida Boundary Dispute (1947) (unpub. M.A. thesis, University of Georgia, Athens, Ga.); S. McLendon, *History of the Georgia-Florida Boundary Line,* in Annual Report of the Secretary of State of the State of Georgia for the Year Ending December 31, 1922, at 5–18 (1923); and Justice Bradley's opinion in Coffee v. Groover, 123 U.S. 1, 31 L. Ed. 51 (1887). A list of Georgia acts and resolutions prior to 1851 regarding the boundary can be found in Cobb 152–53. A calendar of manuscript documents concerning the boundary, along with transcripts of some, can be found in P. Bryant, Georgia State Boundary Documents: Florida-Georgia Boundary, a Calendar (1978). In addition, more than eight hundred pages of correspondence, reports, and testimony concerning the Georgia-Florida boundary are set forth in the U.S. congressional documents. Space does not allow an itemized listing in this book, but for such a listing consult the C.I.S., U.S. Serial Set Index.

14. H.R. Doc. No. 96, 20th Cong., 2d Sess. passim (1829); H.R. Doc. No. 191, 21st Cong., 1st Sess. 2 (1830); A. Ellicott, The Journal of Andrew Ellicott, Late Commissioner on Behalf of the United States . . . for Determining the Boundary between the United States and

of the line running from the junction of the Chattahoochee and Flint rivers was to be the center of controversy for the next sixty years. Since the St. Marys has at least three large branches and all three lie in the lower reaches of the Okefenokee Swamp, identification of the true head of the river is difficult.

The branch Ellicott and Minor selected was the most northerly, but because the area was so swampy it was impossible to ascertain and mark a precise location. They therefore agreed that the actual head of the river was a point North 45° East, two miles from the mound. Ellicott and Minor determined the latitudes and longitudes of the eastern and western termini of the line and constructed a table showing the course and distance by the great circle by which the line could be run. The commissioners then submitted a joint report of their proceedings to their respective governments.[15] After ceding its western lands to the federal government in 1802, Georgia had no further interest in the portion of the line west of the Chattahoochee.

In contemplation of the disposal of the lands ceded in the treaty at Fort Jackson, Georgia attempted to get the federal government to have the south boundary of the state marked. Perhaps because it was negotiating the purchase of the Floridas and did not want to risk unsettling these deliberations, the federal government was slow to respond.[16] As an extemporaneous expedient, and apparently without consulting Spanish authorities, Governor William Rabun in April 1818 appointed William Green to mark a line that would serve as the southern boundary of Georgia. Green was also to run some of the county and district lines.[17]

---

the Possessions of His Catholic Majesty . . . 177–268, 278 (Philadelphia 1803); C. Mathews, Andrew Ellicott: His Life and Letters 127–97 (1908). The mound still exists, though much eroded.

15. H.R. Doc. No. 191, 21st Cong., 1st Sess. 2 (1830).

16. Letter from David B. Mitchell to James Monroe (Jan. 30, 1817) (GLB, GDAH); 1816 Ga. Laws 204; Lamar 1168; 1817 Ga. Senate Journal 45; 1819 Ga. Senate Journal 7–8.

17. Executive Minutes (Jan. 3, 1820) (GDAH); Letter from William Green to John Clark (Dec. 13, 1819) (Georgia-Florida Boundary File, Telamon Cuyler Collection, UGL); 1819 Ga. Senate Journal 8. William Montgomery Green was born in Dublin, Ireland, in 1767 of a family prominent in political circles. He received his education at Trinity College in Dublin. As a result of his participation in the ill-fated Irish rebellion of 1798, he immigrated to France and, in 1800, to the United States. He lived successively in New York, Beaufort, South Carolina, and Savannah, Georgia. He secured an appointment as professor of mathematics and astronomy at the University of Georgia in 1813 but was forced to resign three years later amid charges that he had failed to submit to the laws and regu-

Green's mission soon became urgent. Before he began his work, the lottery surveyors were already progressing in their surveys of the county and district lines under the act of December 1818.

After leaving Milledgeville, Green journeyed to Fort Scott in the southwest corner of Georgia. There he was to await the arrival of an Indian guard from the upper towns of the Creek Nation and then begin work at the junction of the Chattahoochee and Flint rivers. Because the Creeks had not yet withdrawn from south Georgia, Governor Rabun was attempting to hire friendly Creeks to escort Green and the other surveyors and guard them against marauding bands of unfriendly Creeks and Seminoles.[18]

The obstacles Green faced looked formidable indeed. "Located in one of the most remote and unsupplied points of the state," his guards were "to be sought among a people" who were "hostile to the very object about which" he was engaged. Moreover, the area had been the scene of several years of debilitating warfare with the Creeks. Only recently Andrew Jackson's army, in its devastating raids on the Seminoles in East Florida, had swept away everything in sight. Thus the entire country was destitute; "not an atom of subsistence" could be procured "nearer than 40 miles" of Fort Scott.[19]

The difficulty of obtaining guards and supplies delayed Green for a full month. His crew became "wearied and impatient" and the stock of provisions he had brought along seriously dwindled. Because he had "been furnished with no funds to meet" such contingencies, he could hardly afford to purchase additional supplies.[20]

---

lations of the college. Shortly thereafter he and his wife opened two private academies in Milledgeville—one for males and the other for females. Among the courses taught at the male academy was surveying. About 1830 he moved to Macon to become editor of the *Georgia Messenger.* Green died of stomach cancer on June 25, 1846, and is buried in Rose Hill Cemetery in Macon. See Bonner, *Journal of a Mission to Georgia in 1827,* 44 Ga. Hist. Q. 74, 82–85 (1960); J. Bonner, Milledgeville: Georgia's Antebellum Capital 96 (1978); and Georgia Messenger (Macon), July 2, 1846, at 3, col. 5.

18. Letters from William Green to William Rabun (May 13, 18, June 3, & 30, 1819) (all in Georgia-Florida Boundary File, GSGD); Letters from Robert Jackson to William Rabun (May 18, 22, 29, & June 24, 1819) (all in GDAH); Letter from Benjamin H. Sturges to William Rabun (May 20, 1819) (Telamon Cuyler Collection, UGL); Letters from William Rabun to Edmund P. Gaines (Mar. 18, 1819), to William S. Mitchell (Mar. 18 & April 16, 1819), to Benjamin Sturgis (June 1, 1819), to Robert Jackson (June 1, 1819), and to William Green (June 1, 1819) (all in GLB, GDAH); 1819 Ga. Senate Journal 7.

19. Letter from William Green to John Clark (Dec. 13, 1819) (Georgia-Florida Boundary File, Telamon Cuyler Collection, UGL).

20. Id.

The prospect that loomed before him did not appear inviting either. A "line of country was to be traversed, every where from 70 to 100 miles distant from any settlement and which from the large proportion of swamp, tangle, canebrake and miry pond" presented "difficulties to the passage of loaded horses, beyond the conception of any one who" had "not visited it." In addition, the "course was bordered by a ferocious & malignant population, restrained from hostilities . . . only by the dread of future consequences."[21]

With much difficulty Green was finally able to secure eleven Indian guards and got under way on May 31. After having carried the line only twenty miles from Fort Scott, however, his guards became so alarmed that it was only "with the utmost exertion of intreaty and authority" that he "forced them on as far" as he did.[22] Green sent several long, vivid dispatches to the governor about his activities—communications that provide valuable insights into the vicissitudes faced by early surveyors.

On June 3, 1819, Green wrote to the governor:

> I flattered myself, that I should several days earlier than this, have addressed you from the Eastern extremity of the line, of Irwin County.
>
> But notwithstanding the most unremitting & urgent exertions, I have not been able to commence the line untill within three days. The first days work was performed in Lagoons & exceedingly thick Cane brake & Swamp & reached only between two & three miles. yesterday & to day, tho' embarassed with the Swamps on Musquito & Atapulgus Creeks, we have increased the distance to 15 miles. We are now engaged in an extensive Swamp & Cane brake on head branches of Atapulgus. When this is cleared, I judge we shall have an open country to the Ochlockney, that is, for 40 or 50 miles, which we will probably clear, at the rate of ten or twelve miles per day.[23]

On June 8 Green wrote:

> Thus far, we have experienced no interruption. indeed . . . I have not seen an Indian for 5 days, Nor do know I of any danger to be reasonably feared, even where I am, except theft of horses, which if not prevented would be fatal to our expedition. I ground my hopes of security upon, what I know of the situation of the Red People they being totally dependent on the military posts for provisions & even Powder & Lead. Some few desperados . . . from the . . . love of mischief so natural to those people, or to plunder us of our

21. Id.

22. Id.

23. Letter from William Green to William Rabun (June 3, 1819) (GSGD).

> provisions, may be tempted to attack us, but trust we can repel any thing of this description. . . .
>
> Should I be permitted to finish the line & lay off the range of districts, bounded by it, in Irvine County, without hostile interruptions, no person engaged in the business will have much to fear. Three or four additional men, to their necessary attendants, will I believe, as a guard . . . be quite sufficient. Should anyone of the County Surveyors have run down to the line, I shall please God, finis[h] the lowest range in Irvine & then work up towards the Ocmulgee.
>
> If not, I will be obliged to run from the Eastern extremity, up to that river & work back again which I wish greatly to avoid.[24]

In the same letter, after explaining that rumors regarding the hostility of Indians toward the surveyors were greatly exaggerated, Green stated:

> Indeed the notions of people in the up country & even at Fort Gains, respecting the Indians in this section are altogether wild. The Surveyors have to encounter difficulties in subsisting their parties [; they] suffer the Plagues of Musquitos, without end [; and] The want of water & the danger of Sickness in this country [when] the Season is more advanced, but I believe they all carry their scalps themselves.[25]

Having dispensed with the running of the line, Green wrote the governor on June 30, 1819, from the upper part of the 12th District of Irwin:

> I have deemed it my duty to detach one of my party to apprise you of my movements & the circumstances that have occurred since my last communication of the 8th Inst. The morning after the express left me on the line, while working in a large tract of level land, partly pond or marsh & partly Prairie & interspersed at various distances, say from ¼ to 3 quarters of a mile, with hammocks of great richness & heavily wooded, the people noticed 15 or 16 Indians, apparently watching our party, on the left, at the distance of about a mile. We had just crossed a branch of the pond ½ mile wide and Knee deep in water part of the way. . . . our packs were perhaps 1½ mile behind, conducted by 6 of the guards. I judged the Indians, that were seen, to be merely a hunting party & ordered the men to proceed without seeming to notice them lest they should suppose us afraid. The Indians fell back behind the large hammock on the edge of which the packs were stationed, waiting the return of two men, who had been sent to seek a firm footing for them across the pond. A small party [of] 4 or 5 Indians, was seen soon after to pass round the left point of the hammock into which we were entering &

24. Letter from William Green to William Rabun (June 8, 1819) (GSGD).

25. Id.

one or two seemed passing back & forward with intelligence. Whether those near us had detached themselves from the first party, or were an additional one I could not determine. My people became greatly alarmed, begged me to stop & either wait the arrival of the Pack party, or turn back & join them. I despatched a man to inform those with the horses how matters stood & direct them to be on their guard & that we would join them in a few minutes. When I had completed the necessary marking, I moved across the marsh to join them as I saw the only grounds on which they could pass would bring them directly within reach of the Indians. I found on joining them they had seen the Indians & heard them yelling. One of the men, sent to find a trail returned. He fired twice at a large Crane. the 2nd time an Indian who was concealed in the long grass, sprung up & gave the war whoop, which was loudly & violently reiterated in various directions. The other man returned soon after, had seen the Indians moving to & fro, & conferring with each other & heard them whooping as the whole of the Pack guard had done. All the men were greatly alarmed & declared if we proceeded, we should necessarily be fired on, for that the Indians, never whooped so, but when they intended to fight. They begged I would return across the Ocklokany & send to the fort for a reinforcement. The men, with me, had been in all [of Andrew] Jackson's Indian campaigns, & stood high among their officers as good Soldiers. Believing we could not proceed without fighting, of which the Indians had given us fair warning, feeling it a sacred duty to avoid that, unless attacked & having no Indian, to hold any talk with them—I felt myself compelled with infinite reluctance & mortification to return upon my course a few miles & send to the Fort, for assistance & an Interpreter, without which, I found I could not preceed. Three days elapsed before my messenger returned, bringing a boy, who spoke some english; Conohatamico, or the Redground chief; & an Atapulgus warrior who knew the people settled about I-a-mo-ney, the name of the prairie & pond tract, in which we stopped. I forgot to mention before that the Indians had followed on our trail, to within 1½ m. of our camp and there turned off to the left. The Atapulgus man told us some of the Chehaws, a few of the Folemies' & Hoopaunies' & [some] of the Fowltown people, who were still hostile, were living there in March, to the number of 44 war[rs]. I sent the Indians & [s]ome of the soldiers, to the towns. They returned towards night, with Intelligence that no Indians would shew themselves & that they were gone to Suwanee. This seemed extremely suspicious to the guards & some of my own people, they believing they were Making a larger force & intending to waylay us in the neighborhood of Mikasukie to the northern extremity of which, the line would be near. The Indians would not go on with us—And I found, with feelings of distress I could scarce describe that I must either go on almost by myself, or return to the Fort & obtain an Indian or two known among the Seminoles & an interpreter, or an additional guard. The 13th Inst. started for the Fort with two men & 2 Indians, the third being left sick. Reached the Fort that night and proceeded next morning to the accomplishment of my business. An additional guard I could not obtain—Col[l] Arbuckel, not believing himself warranted in furnishing one. With great difficulty, loss of time, & exertion I obtained 2 Indians & Tom Perryman, half breed, interpreter—Packed a barrel [of] flour & ½ lb. of Pork & got back to camp the 19th in the morning. Obliged to stop, rest of the day, to rest the horses & in consequence of heavy rain. Indeed from the 10th to the 20th it rained almost incessantly & very heavily. We ran on without anything remarkable, till we neared the Mikasukie settlements. we began to find larger trails & fresher & more numerous marks of Indians & we had to stop twice by the advice of our Indians . . . but in no instance would the Indians shew. These appearances disturbed the people greatly. they went on with great ill will. on the 23[d] one of them Corporal Gray actually went off stating, he was sure of being Killed if he went on & 3 more declared they would follow his example. I found the corner tree of Irvine County, a little to west of the 50th mile on my line, about 11. A.M. same day—Marked as I suppose, by Col[n] [James C.] Watson. Pushed on to the end of 62¼ mls. found numerous mark of horse & footmen made within an hour or two. The Indians returned without being able to see any of the seminoles—but stated there must be a large town on the borders of a Swamp, to our right & thro' which our line passed. My people became so alarmed. I found it absolutely impracticable to proceed further on with them on the line. I therefore . . . turned up on a North Easterly course determining to commence running the districts lines—& sending off an Express acquainting your Excellency with the movement & requesting such assistance as might be necessary when I should again visit the line. I adopted this measure without much reluctance, as I believe it would facillitate & despatch the laying of[f] the land & that in 6 or at utmost 7 weeks after laying off my districts, every difficulty created by the Indians would be removed & I would have worked again down to line & could finish it without delaying the surveying work a moment. Travelling to reach the South Eastern corner of district No. 12 where I contemplated beginning, we were overtaken about two P.M. by a chief & 3 Seminole warriors. They came up two at a time—Proferred friendship. The chief Tustunuggee Cluco of Fowl Town addressed me with great fluency & dignity[.] Observed, he had followed me two days. the place pleasant for encamping & the [sun] hastening to the West [he] requested I should stop for the night, as he had much to say to me. I of course complied. He enquired after we had drunk together by whose authority I was running the line & why I had turned back. [He] expressed great uneasiness lest we should go into the State & propogate some evil report concerning the lower Indians. [He] Put me in mind we had not been interrupted, nor one of our horses stolen & assured me I might go all the way by myself. Mentioned we had frightened their women greatly & that [he] had followed me to set all these matters right. Said in reply to my request for a Pilot . . . that if any man of the town left his family now they must perish, depending entirely on his gun—But in 1½ month, the corn would be usable & I should have two or three. . . . Found in his talk a most ample proof of the truth of the statements I have made to you respecting the views of these people & the motives which would controll their conduct towards the surveyors. He urged me when I came down again, not to encumber myself with a guard as it frighted their people & was unnecessary. I learned from his discourse both himself & all the lower Indians were dreadfully apprehensive of incurring the resentment of [Chief William] McIntosh & a confirmation of intelligence I had rec[d] before, that McIntosh had sent a few warriors down from Broken Arrow, to the lower towns all the way to St. Marks, notifying them the line was to be run & that if any mischief or insult ever was committed he [would] come down & in his own phraseology, put hell upon them. Niamochla sent out for all the hunters & gave them their orders accordingly. I learned that the party at I-a-mo-ney were hunters from a village 18 miles south & had been greatly frightened. they are red club refugees from the upper towns. I shall please God lay off my districts from Ockmulgee downwards & with His blessing be finished about the 2nd or 3d week in August & shall not take a larger number of men than I have now, unless an Indian, or two. Deeming them unnecessary for safety & objectionable on the score of expense, I parted with 6 men yesterday. Indeed the country along the line is so much occupied with swamp & ponds & marsh that it is nearly impossible to get pack horses thro'. Of the 62¼ miles of the line run, ⅓[d] at least is swamp & canebrake. We lost one horse in Attapulgus Swamp & another in Tootulka. I gave them in charge of Indians & if they recover, they will be restored. No one can feel more deep regret than I do, at the unexpected & inconvenient length of time this business has occupied—And knowing the disposition of too many of our people, I should not be surprised, if, while I am anxiously, by every exertion & sacrifice in my power, discharging my duty & promoting the service of the state, I should be the object of censure & animadversion. from you, Sir, I expect nothing but fairness. . . . to others, I would suggest the difficulty of subsisting a party of men in so remote a section of the country—& a country so impracticable for loaded horses as this low & wet country is in many places—& furnishing now, no supply within it self. We have got but one deer & one Turkey since being out. . . . I commenced, the 31st Ult. & have lost by circumstances . . . already stated & heavy rains 18 days. In a remainder By the Blessing of God, I have run 62¼ miles, thro places, that must be passed in order to conceive their difficulty. . . . The line as far as it goes is run with an accuracy I scarcely expected to realize & it is abundantly marked. I regret greatly I was a few

> miles back when Col. [James C.] Watson's party came down. I shall meet [him] I hope as I move along. I forgot to state to you, Sir, that the chief was very anxious to persuade me the line was too far South—fearing least his town should be included within our state, he told lies & bullied a little, but finding me firm he gave it up. . . . With respect to the state's incurring any expense for additional guards—it is I again assure your Excellency absolutely unnecessary. the whole alarm has been propagated from Fort Gaines. the Indian who brought your express, came from that quarter & his stories principally occasioned the panic among my people, by which I have been so much embarrassed.[26]

After having completed sixty-two and one-quarter miles of his line, the depletion of provisions and the loss of packhorses compelled Green most reluctantly to turn northward to the Ocmulgee River to resupply his expedition. He notified the governor that he planned to return and finish late the following fall, but he never did.[27] For, in spite of all these exertions and adversities, the work came to naught. Surveyors who ran the district and land lot lines found that Green's line veered entirely too far south, just as the Seminole chief had maintained. As a consequence, in 1820 Governor John Clark appointed James C. Watson to run the line. Watson marked a line from the junction of the Flint and Chattahoochee to a point south of Ellicott's mound, and Georgia gave up its claims to the lands it had surveyed into land lots and districts south of this line.[28]

By the Adams-Onis Treaty of 1819 Spain ceded the Floridas to the United States, with actual possession commencing in 1821.[29] In 1825 the surveyor general of the territory of Florida, in preparation for a survey of the public lands therein, caused the boundary between Georgia and Florida to be run and marked by Daniel F. McNeil. McNeil's line (as it was called) lay north of Watson's line, at the midpoint by about fourteen chains. Nevertheless, the public land surveys for Florida were run to McNeil's line, thus overlapping Georgia's surveys and grants.[30]

Between 1825 and 1859 many long, vexing disputes transpired over

26. Letter from William Green to William Rabun (June 30, 1819) (GSGD).

27. Letter from William Green to John Clark (Dec. 13, 1819) (Georgia-Florida Boundary File, Telamon Cuyler Collection, UGL).

28. Executive Minutes (April 28 & June 12, 1820) (GDAH); Letter from John Clark to Wilson Lumpkin (April 15, 1820) (GLB, GDAH); Report from Daniel Sturges to John Clark (Mar. 6, 1820) (GSGD); 1820 Ga. Senate Journal 7.

29. 8 Stat. 252.

30. Coffee v. Groover, 123 U.S. 1, 31 L. Ed. 51 (1887).

the Georgia-Florida boundary—at first between Georgia and the United States government and then, after 1845, when Florida was admitted to the Union, between Georgia and Florida. The main point of contention in this border feud was the correct location of the eastern terminus of the line at the head of the St. Marys River. The western terminus at the junction of the Flint and Chattahoochee was fixed and easily ascertainable, but opinions varied as to the exact source of the St. Marys.

During this period three additional lines or partial lines were run between Georgia and Florida. In 1827 Thomas M. Randolph (as commissioner for the United States) and Thomas Spalding (as commissioner for Georgia), with John McBride as surveyor, ran a line that became known as McBride's line, but the survey was called off by Governor George M. Troup before its completion.[31] In 1831 Joel Crawford and J. Hamilton Couper as commissioners and Edward Lloyd Thomas as surveyor were hired by Georgia to run the line. Taking Lake Randolph (Ocean Pond, now conceded to be in Florida) as the source of the St. Marys, they ran a random line westward from the south edge of this lake and then ran the "true" line back to that point.[32] In 1854 Alexander A. Allen for Georgia and Benjamin F. Whitner for Florida attempted to run a line westerly from near Ellicott's mound, but this project quickly ended in fruitless contention.[33]

Neither Georgia nor Florida ever officially or unofficially recognized McBride's line, Crawford and Couper's line, or Whitner and Allen's line. In 1827 Georgia had prohibited headright surveys in the area of the head of the St. Marys[34] and had held up the sale of the fractional lots along the Florida boundary. In 1839, however, an act was passed permitting the sheriffs in the counties along the line to sell the fractions.[35]

31. H.R. Doc. No. 140, 20th Cong., 2d Sess. passim (1829); 4 Stat. 157; E. Coulter, Thomas Spalding of Sapelo 195–211 (1940).

32. S. Misc. Doc. 25, 33d Cong., 2d Sess. 360–61, 368–79 (1855); Goff, *Edward Lloyd Thomas, Surveyor,* 18 Emory U.Q. 107, 116 (1962).

33. Letter from James E. Broome to Herschel V. Johnson (May 14, 1854), and Agreement between Benjamin F. Whitner & Alexander A. Allen at Suwanee Springs (June 20, 1854) (both in Georgia-Florida Boundary File, Telamon Cuyler Collection, UGL); 1855–56 Ga. House Journal 252–53; Alexander A. Allen, Allen's Journal: A Trip along the Georgia-Florida Boundary June 14–July 22, 1854 passim (C. Trowell & F. Trowell eds., ca. 1985) (Occasional Paper from South Georgia College. No. 6). The John Macpherson Berrien Papers at the University of North Carolina, Chapel Hill, contain extensive correspondence concerning this survey.

34. 1827 Ga. Laws 127; 1827 Ga. Senate Journal 107.

35. 1839 Ga. Laws 40.

To settle the matter at last, both states agreed on a "compromise" line to be drawn from the junction of the Flint and Chattahoochee rivers to Ellicott's mound. This line would be regarded as the true boundary.[36] In 1859 Georgia appointed Professor Gustavus J. Orr of Emory College and Florida appointed Benjamin F. Whitner to run and mark the line.[37]

Orr and Whitner began their work at the junction of the Flint and Chattahoochee in November 1859. From there they reported having decided to run eastward along the arc of the great circle to Ellicott's mound according to the observations that had been made by Ellicott and Minor in 1800. The commissioners believed that they would very nearly strike Ellicott's mound but were concerned that any departure, however slight, from the mound would necessitate the running of a second or corrected line; otherwise the survey would not be in keeping with the resolutions of Georgia and Florida. The two commissioners recommended the adoption of amended resolutions providing that if the initial line departed no more than one-fourth of a mile from Ellicott's mound, it should be regarded as the true boundary, but if the line departed more than one-fourth of a mile from the mound, a corrected line should be run. The commissioners believed that such a resolution would save much time and expense, while the difference in territory involved would be negligible.[38] Laws to this effect were passed by Georgia[39] and Florida[40] in December 1859.

There is preserved today in the library at Emory University in Atlanta a long series of correspondence between Orr and his wife, written during the course of the survey. These letters provide a much more detailed and insightful account of the expedition than the official reports. In stark contrast to William Green's survey of the Georgia-Florida boundary some forty years before, Orr and Whitner had a relatively easy trek. The area was settled now so that supplies were readily obtainable, and the crew

36. 1857 Ga. Laws 326; 1858 Fla. Laws 162; 1851–52 Ga. House Journal 946–47.

37. Gustavus J. Orr (1819–87) achieved fame primarily in the field of public education. A graduate of Emory College at Oxford, he was professor of mathematics there at the time of his appointment to assist in running the Georgia-Florida boundary. Subsequently he became the first state school commissioner and later served as president of the National Education Association. See Coulter, *Gustavus J. Orr,* in 7 Dictionary of American Biography 57 (D. Malone ed. 1962); and Orr, *Gustavus J. Orr,* Atlanta Historical Bulletin, Fall–Winter 1971, at 1.

38. 1859 Ga. Senate Journal 316–17.

39. 1859 Ga. Laws 22.

40. 1859 Fla. Laws 28, 95.

often stayed at night in an inn or private home rather than camping in the open. It was possible to maintain almost continuous communication among the survey party, their respective governments, and their families. Most important, the area had long been cleared of troublesome Indians. The weather remained pleasant, and in spite of the swamps the trip proved to be a delightful excursion. Indeed, although up to his knees in water every day and soaked most of the time, Orr assured his wife that he was "fully convinced . . . camp life" was "more healthy than any other."[41]

When Orr and Whitner completed the line in January 1860, it passed just thirty-seven links north of Ellicott's mound.[42] The total length was 158 miles, 28 chains, and 3 links.[43] To make the line easily identifiable, the surveyors hacked trees all along the way. In addition, they set up lightwood posts surrounded with mounds of earth five feet high and ten feet in diameter, wherever practicable, each mile along the line. (There was no stone in the region from which to make more permanent markers.)[44] The Georgia-Florida boundary as thus marked was ratified by Florida,[45] Georgia,[46] and the United States,[47] and the question of its location was laid to rest.

The Orr and Whitner line, however, ran north of both the Watson line (to which the Georgia surveys had been extended) and the McNeil line (to which the Florida surveys had been run).[48] The Florida act of December 1859 recognizing the Orr and Whitner line provided that if any land granted by Georgia fell south of that line, such grants would be confirmed if they did not interfere with any Florida grant south of the McNeil line.

41. Letter from Gustavus J. Orr to Eliza C. Orr (Dec. 27, 1859) (Orr Family Papers, Robert W. Woodruff Library, Emory University, Atlanta, Ga.).

42. Field Notes of the Boundary Line between Florida and Georgia Surveyed by B. F. Whitner, on the Part of Florida and Gustavus J. Orr, on the part of Georgia (Division E [Surveys], Records of the General Land Office, Department of the Interior, Record Group 49, National Archives, Washington, D.C.).

43. Id.

44. Id.; Report on the Boundary Survey to his Excellency Joseph E. Brown from Gustavus J. Orr (1860?) (Georgia-Florida Boundary File, Telamon Cuyler Collection, UGL).

45. 1860 Fla. Laws 237.

46. Ga. Code of 1863 §§ 17, 21; 1866 Ga. Laws 221.

47. 17 Stat. 52.

48. All three lines are shown on township plats of the General Land Office in Florida Book 43.

The survey problem as well as the title problem thus presented with regard to lands along the boundary is illustrated in the U.S. Supreme Court case *Coffee v. Groover*,[49] which reversed a decision argued twice previously before the Florida Supreme Court.[50] In 1882 the heirs of Charles Groover filed suit to have A. J. Coffee ejected from certain lands in Madison County, Florida. Groover's heirs claimed under a chain of title extending back to grants by the state of Georgia in 1842, while Coffee claimed under grants by the United States government in 1857.

The heirs of Groover traced their title back to 1842 when the state of Georgia granted to James Groover fractional lots 199 and 200 in the 15th District of original Irwin County, Georgia. The plats of these two lots showed them bounded on the south by the Florida line. This line was shown to be the Watson line. James Groover conveyed his lands to Thomas A. Groover in 1855, who in turn conveyed them to Charles Groover in 1860. Charles Groover died in 1866.

Fourteen chains north of the Watson line in this area was the McNeil line. In 1857 the United States had granted to the state of Florida Fractional Section 29, Township 3 North, Range 9 East, which had as its north boundary the McNeil line and which overlapped the property that had been conveyed by Georgia to James Groover in 1842. Coffee derived title to Fractional Section 29 from the U.S. government grant to the state of Florida. Farther north than the McNeil line was the Orr and Whitner line, the one finally settled upon as the true boundary between Georgia and Florida.

In its decision the U.S. Supreme Court held that all Georgia grants of lands outside its own territory were invalid unless they had been confirmed by Florida. Inasmuch as Florida had confirmed only those grants between the McNeil and the Orr and Whitner lines, the grants by Georgia south of the McNeil line were void.

Surveys for Original Walton, Gwinnett, Hall, Habersham, Appling, Irwin, and Early Counties

After considerable delay, caused by sickness, death, and the resignation of the original commissioners, the line of the Creek cession of 1814 was

49. 123 U.S. 1, 31 L. Ed. 51 (1887). Justice Bradley, who delivered the opinion, once wielded the compass and chain himself. For a personal account of his surveying experience see Fairman, *The Education of a Justice*, 1 Stanford L. Rev. 217, 223–24 (1949).

50. Groover v. Coffee, 19 Fla. 61 (1882); Coffee v. Groover, 20 Fla. 64 (1883).

run by Captain William Barnett and Major General Edmund P. Gaines, both of the U.S. Army, and by John Coffee of Alabama during the latter part of 1815. The length of the Georgia portion of this line, running from the Chattahoochee River to the western line of original Wayne County, was 184 miles. The commissioners were able to proceed only after the mobilization of the Georgia militia and the showing of a strong body of regular soldiers convinced the Creeks that resistance would be futile.[51]

The artificial line of the Cherokee cession of 1817, running due north to the Tallulah River from the point where the Soque River empties into the Chattahoochee, was marked by James Blair and became known as Blair's Line.[52] The artificial boundaries of the Creek cessions of 1818 were marked under the direction of Wilson Lumpkin (who was later to serve as governor of Georgia) in the fall of 1818.[53]

The lottery surveys began in the spring of 1819 and dragged on for a full year because of molestations from Indians and tardiness in getting the south boundary of Georgia run.[54] Exacerbating these problems were the remoteness of many districts from bases of supplies, the debilitating heat in the summer of 1819, and the innumerable swamps of south Georgia. More formidable hardships were encountered in these surveys than in the surveys for any of the other land lotteries.

In accordance with a notice issued by the surveyor general on June 16, 1819, that the districts were ready to be divided into land lots, Thomas Cheely gathered his supplies, men, and equipment. They set out for their district—the 2d of Irwin, now primarily in Turner County. Crossing the Ocmulgee River eighteen miles below Hartford, they struck the Indian boundary two or three miles from the river. They then followed this

51. Executive Minutes (Oct. 24, 1815) (GDAH); 2 Correspondence of Andrew Jackson 210–11, 222 (J. Bassett ed. 1927); 6 Territorial Papers of the United States 506, 512–13, 632n, 776–77 (C. Carter ed. 1938); 3 Stat. 228; Letters from Edmund P. Gaines to John Coffee (Nov. 28, 1815), and to William H. Crawford (Mar. 30, 1816) (Letters Received by the Secretary of War, Registered Series, Record Group 107, National Archives, Washington, D.C.).

52. Instructions of the Surveyor General 41 (GSGD); Lamar 429; 1820 Ga. Senate Journal 6.

53. Letters from John C. Calhoun to Wilson Lumpkin (Aug. 4 & Oct. 3, 1818) (Letters Sent Relating to Indian Affairs, Volume D, Records of the Office of the Secretary of War, Record Group 75, National Archives, Washington, D.C.); Letters from Wilson Lumpkin to John C. Calhoun (Dec. 10, 1818, & Jan. 7, 1819) (Letters Received by the Secretary of War Relating to Indian Affairs, Record Group 75, National Archives, Washington, D.C.); Georgia Journal (Milledgeville), Dec. 15, 1818, at 3, cols. 2–3.

54. Field notes and district maps for Appling, Irwin, Early, Gwinnett, Habersham, Hall, and Walton counties in GSGD.

boundary in a southwest direction "in anxious expectation of finding" the corner of their district within fifteen or twenty miles. The country was so wild and rugged that a road had to be cut for this distance to get their wagon and horses through. Upon failing to find their corner, Cheely and two of his company proceeded on foot, following the Indian boundary to its intersection with the north boundary of the Creek cession of 1814. This point supposedly formed the southwest angle of their district. Finding no district corner, however, Cheely came to the mortifying realization that the district lines had not yet been run.

He and his men then returned home, where Cheely addressed a scathing letter to the governor. He blasted the surveyor general for "sporting" with their funds by ordering them "three & four hundred miles" from home without first receiving positive notice that the districts were ready. Cheely reported having met two other district surveyors on his return passage through Hartford who had had similar experiences.[55]

The Creeks still occupied the lands ceded under the Fort Jackson treaty. Indian guards were procured, with much difficulty, to protect surveyors marking the county and district lines.

Soon after beginning his survey of the district lines of Early County in June 1819, Robert Jackson detailed the difficulties he was encountering. Because of the intense heat and severe working conditions, some of his horses had died and several Indian guards had quit without giving notice. In addition, Jackson was apprehensive that the Spaniards were inciting the Indians to hinder the surveyors in their work.[56] A few weeks later Jackson had to postpone the continuance of his survey because the Florida line was not yet completed. In a letter to Governor Rabun, he reported having discovered a great error in the meandering of the Flint River as shown on the preliminary maps and notes furnished by the surveyor general. As a result, Jackson asserted, enough extra territory was available in Early County to lay off three or four additional districts.[57]

On November 24 A. B. Shehee reported that, having arrived within a few miles of his district (the 16th of Irwin), he fell in company with four Indians who made tedious inquiries concerning the surveyors' business in those parts: who had sent the surveyors, how far southward they were

55. Letter from Thomas Cheely to William Rabun (July 12, 1819) (GSGD).

56. Letter from Robert Jackson to William Rabun (June 24, 1819) (GSGD).

57. Letter from Robert Jackson to William Rabun (July 25, 1819) (GSGD). This accounts for the inconsecutive numbering of districts in original Early County and also for the fact that Early County was finally laid off into twenty-six districts instead of twenty-three, as specified in the 1818 land lottery act.

going, whether the Indians in the low country would be allowed to visit those in the up-country, and so on. These four Indians informed Shehee that other Indians were hunting along the Suwanee and Alapaha rivers and "would be troublesome" to the surveyors "if they did not Stop" their progress.[58]

Some weeks later, Shehee's party discovered that several of their corner posts had been destroyed and the numbers cut from the station trees—obviously the doings of mischievous Indians. A few days afterward Shehee and a member of his party went in search of their horses, which had either strayed from camp or been stolen. Upon returning late in the evening, they found one horse. To their surprise, however, the four members of the survey party who had remained at the camp were absent. Shehee and his companion then set out for the nearest white settlements. They overtook the other men, who explained that during Shehee's absence Indians appeared on each of the two horses belonging to the survey party and aimed guns toward the camp. As soon as the Indians realized they had been discovered, however, they dismounted, set one horse free, and disappeared into the woods with the other horse. Shehee's men were so frightened that they abandoned the camp.[59]

Having been threatened by Indians, Shehee ceased operations for the time being. His problems caused consternation among other district surveyors who were experiencing similar difficulties. Peter L. Jackson, surveyor of the 12th District of Irwin, jotted in his field notes that the Indians had manifested "a disposition to mar and deface" his work. Therefore, he had "thought it most advisable to corner in trees, having found Stakes pulled up by them." When the nearest tree "exceeded 5 steps" from the theoretical location of a corner he had "planted a stake and numbered also on a tree or trees, if any in 10 steps."[60]

Thomas E. Hardee, surveyor of the 7th District of Appling, wrote the governor that he had discovered the Indians busy displacing his corners, disfiguring numbers, and chopping the figures out of his stations. He could plainly perceive that the Indians kept close at his rear, but he could never get sight of the "horrid barbarians." If he had, he would have fired at them because his survey party was well armed. Hardee insisted that Governor Clark provide military protection for the surveyors.[61]

58. Letter from A. B. Shehee to John Clark (Nov. 24, 1819) (GSGD).

59. Id.

60. Field notes of Peter L. Jackson, surveyor of the 12th District of Irwin County 2 (GSGD).

61. Letter from Thomas E. Hardee to John Clark (Nov. 11, 1819) (GSGD); Letter from Thomas E. Hardee to John Clark (Dec. 18, 1819) (Telamon Cuyler Collection, UGL).

Clark presented Shehee's and Hardee's letters to the legislature, which was then in session, and a joint resolution was adopted recommending that the governor request federal troops or, if that failed, call out the state militia.[62] Clark requested help from Major General Edmund P. Gaines of the U.S. Army. Gaines agreed to send a battalion of infantry to guard the surveyors and to remove the Indians from the ceded lands. This movement commenced about the middle of December, and the surveyors apparently suffered no further interruptions.[63]

Problems of an entirely different character were experienced with the 10th District of Irwin (now largely in Berrien, Lanier, and Atkinson counties). As this district was to prove, mother nature could sometimes outwit the designs of surveyors even more effectively than could Indians. The 10th, straddling the Alapaha River, was almost entirely a tangle of impenetrable baygalls and sticky quagmires—unfeigned wilderness indeed. Certainly it was more than most surveyors were willing to contend with—even for the $4 a mile the state was willing to pay to have it surveyed. Robert Cunningham, who had the initial appointment of laying it out, was unaware of what he was getting into. After floundering around in this domain of the alligator, bullfrog, and water moccasin for a few days, Cunningham returned home. He claimed that his compass and field notes, which he had "very judiciously deposited . . . in a trunk" and concealed in a swamp, had been discovered by Indians, who promptly destroyed them. A year and a half later, however, after the area had been cleared of Indians, Cunningham was still at home, engaged in agricultural pursuits. Apparently he had no intention of returning to execute the survey.

Because the lottery tickets had already been drawn and applicants were coming forth almost daily for grants of land in the district, it was urgent that the work be discharged as quickly as possible. Governor Clark declared the post vacant and, after some indecision, collared Charles McKinnon, the surveyor of the 4th District of Irwin, to do the 10th. The surveyor general had recommended McKinnon to the governor, having found his work on the 4th District to be of superior quality.[64] In 1825, however, it was discovered that McKinnon had not run all the lines of

62. 1819 Ga. Senate Journal 46, 49, 52.

63. Letters from John Clark to Edmund P. Gaines (Dec. 23 & 30, 1819), and Letter from John Clark to Thomas E. Hardee (Dec. 30, 1819) (all in GLB, GDAH).

64. Executive Minutes (Feb. 14, 1821) (GDAH); Letters from Daniel Sturges to John Clark (Jan. 19, 1820, & Feb. 14, 1821) (GSGD).

the 10th. McKinnon sheepishly explained that he had been prevented by high waters from running some of the lines but had intended to return and finish as soon as conditions warranted. Nevertheless, not until the summer of 1827, seven years after the lottery tickets had been drawn, did McKinnon return to complete the job.[65]

Because of the vast impediment of the Okefenokee Swamp, only a few land lots were laid off in the 10th and 11th districts of Appling. In the 1850s the legislature had additional land lots laid off in these two districts, but most of the swamp was never subdivided.[66] Today the Okefenokee Swamp is a national wildlife refuge.[67]

The land lottery act of 1818 specified that land lots 10 and 100 in each district were to be set aside for the education of poor children, the only lottery act to make such a provision. Sales of these lots were attempted, beginning in 1823, and continued at sporadic intervals for slightly more than a decade.[68] Most of the lots were so poor or on such rugged terrain that little return was realized. Most remained unsold. A half-century

65. Letters from George M. Troup to Charles McKinnon (July 8, 1827), and to William Ashley (Sept. 27, 1827) (both in GLB, GDAH); Letters from Charles J. Shelton to George M. Troup (Nov. 15, 1825), from Charles McKinnon to George M. Troup (June 10 & July 18, 1827), and from Samuel Boyd to George M. Troup (Aug. 28 & Sept. 18, 1827) (all in GSGD); 1826 Ga. Laws 191; 1827 Ga. House Journal 201.

66. 1849–50 Ga. Laws 297; 1851–52 Ga. Laws 252; 1851–52 Ga. Senate Journal 28–29; 1820 Ga. House Journal 95; C. Trowell, Exploring the Okefenokee: The Mansfield Torrance Survey of 1850 (1989); Goff, *The Last Indian Town of Georgia,* 15 Georgia Mineral Newsletter 93 (1962).

67. Finding a worthwhile purpose for the Okefenokee Swamp posed a perplexing dilemma until the environmental movement of the twentieth century and offers a striking example of how attitudes toward the environment have changed. An 1854 act of the legislature providing for engineering studies to determine the feasibility of draining this vast quagmire referred to it as "unimproved and . . . worse than useless land." 1855–56 Ga. Laws 273. In 1891, after several decades of abortive proposals and indecision, the state conveyed its remaining 238,000 acres in the swamp, constituting the last large expanse of public domain, to the newly incorporated Suwanee Canal Company. The objective of this company was to drain the water off so as to reclaim the area for logging and eventual crop cultivation. This project proved too costly, however, and the company was forced into bankruptcy. In 1901 its assets were sold to the Hebard Lumber Company, which exploited the swamp for logging purposes through the 1920s. In 1936, following several years of lobbying from various quarters, the federal government purchased the landholdings of the Hebard Lumber Company, together with a substantial acreage of adjoining lands, and established the Okefenokee Swamp National Wildlife Refuge. See Coulter, *The Okefenokee Swamp, Its History and Legends,* 48 Ga. Hist. Q. 166, 291 (1964); and C. Trowell, The Suwanee Canal Company in the Okefenokee Swamp (1984).

68. 1823 Ga. Laws 96; 1832 Ga. Laws 118; 1833 Ga. Laws 6, 13–14; 1834 Ga. Laws 120; 1835 Ga. Laws 244.

later additional sales were attempted, but by then squatters had settled on many of the lots, making it difficult for the state to assert its title.[69] In 1903, to avoid protracted and expensive litigation, the legislature directed the governor and attorney general to compromise or submit to arbitration the claims to these lots.[70]

Land Lottery Act of 1819

In Washington on February 27, 1819, the Cherokees signed a treaty ceding lands in Tennessee and Georgia.[71] The Georgia lands lay just north and west of the lands the Cherokees had ceded in 1817. This cession came a mere two months after the passage of the land lottery act of 1818 and after the legislature had adjourned and gone home. At the meeting of the legislature in November and December 1819 an act, primarily to amend the 1818 act, was passed to dispose of these lands. This act is summarized as follows:

Section 1. That part of the cession lying between the fork of the Chattahoochee and Chestatee rivers and south of a line beginning on the Chattahoochee River where the dividing line between Hall and Habersham counties strikes the river and running "a due west course" to the Chestatee River, to become part of Hall County and to be divided into three districts "as nearly equal as practicable." That part lying between this line and "Blair's line, [extended] until the same strikes the top of the Blue Ridge," to become part of Habersham County and to be divided into six districts "as nearly equal as practicable." The balance of the territory to form Rabun County. Rabun County to be divided into five districts "as nearly equal as practicable."

Section 2. Districts 5 and 6 of the addition to Habersham County and Districts 1, 3, 4, and 5 of Rabun County to be divided into lots "of seventy chains each way, containing four hundred and ninety acres each, by lines running north and south, intersected by others at right angles," except for fractional lots. Districts 10, 11, and 12 of the addition to Hall County, districts 1, 2, 3, and 4 of the addition to Habersham County, and District 2 of Rabun County to be divided into lots "of fifty chains each way,

69. 1880–81 Ga. Laws 136.

70. 1890–91 Ga. Laws (vol. 1) 542, 577; 1903 Ga. Laws 86; 1903 Op. Atty. Gen. 8–10; 1905 Op. Atty. Gen. 12–13; 1912 Op. Atty. Gen. 53; 1914 Op. Atty. Gen. 52.

71. Am. St. Papers, 2 Indian Affairs 187; C. Kappler, Indian Affairs (Treaties), S. Doc. No. 319, 58th Cong., 2d Sess. 177 (1904).

containing two hundred and fifty acres, by lines running due north and south, intersecting others at right angles," except for fractional lots. Fractional lots above 160 acres in area to constitute prizes and be drawn in the lottery along with full-size lots.

Section 3. Surveyor general to issue instructions to surveyors.

Sections 4, 5, 6, 7, 8, 9, 10. Legislature to elect surveyors to lay off county and district lines and to divide districts into lots. Surveyors to receive $4 for each mile of line run. Each surveyor to take an oath and post a $5,000 bond before beginning his work. Governor to advance $400 to each surveyor before the survey begins.

Sections 11, 12, 13, 14, 15, 16. Lands contemplated in this act to be distributed in the same manner as prescribed in the land lottery act of 1818.

Section 17. Should there be more districts in the territory than contemplated by this act, then governor to appoint surveyors for such districts.

Section 18. This act in conformity with the land lottery act of 1818.[72]

Wilson Lumpkin was appointed commissioner to supervise the running of the artificial lines of the cession.[73] The surveys for Rabun County and the additions to Hall and Habersham counties were carried out during the spring of 1820, apparently without serious difficulty.[74] The drawing for the lottery began on September 1, 1820, and was completed early the following December.[75]

Frauds and Mistakes in the 1820 Land Lottery

In December 1824 John Loving, Samuel Jackson, and Fleming F. Adrian, who had been appointed to sell the fractional surveys in Walton, Gwinnett, Hall, Habersham, and Rabun counties, were ordered to appear before a "high court of impeachment"—the state Senate—to answer charges that they had embezzled funds from the sale. After a trial that was not completed until the following June, they were acquitted by one vote; a two-thirds majority was required to convict them.[76]

72. 1819 Ga. Laws 23; Lamar 429.

73. Am. St. Papers, 2 Indian Affairs 191; Letter from Wilson Lumpkin to John C. Calhoun (Oct. 12, 1819) (Letters Received by the Secretary of War Relating to Indian Affairs, Record Group 75, National Archives, Washington, D.C.).

74. Field notes and district maps for Rabun, Hall, and Habersham counties in GSGD.

75. Southern Recorder (Milledgeville), Dec. 5, 1820, at 3, col. 3; Id., Jan. 16, 1821 at 2, cols. 1–4, 1820 Ga. Senate Journal 7.

76. 1825 ex. sess. Ga. Senate Journal 77–128.

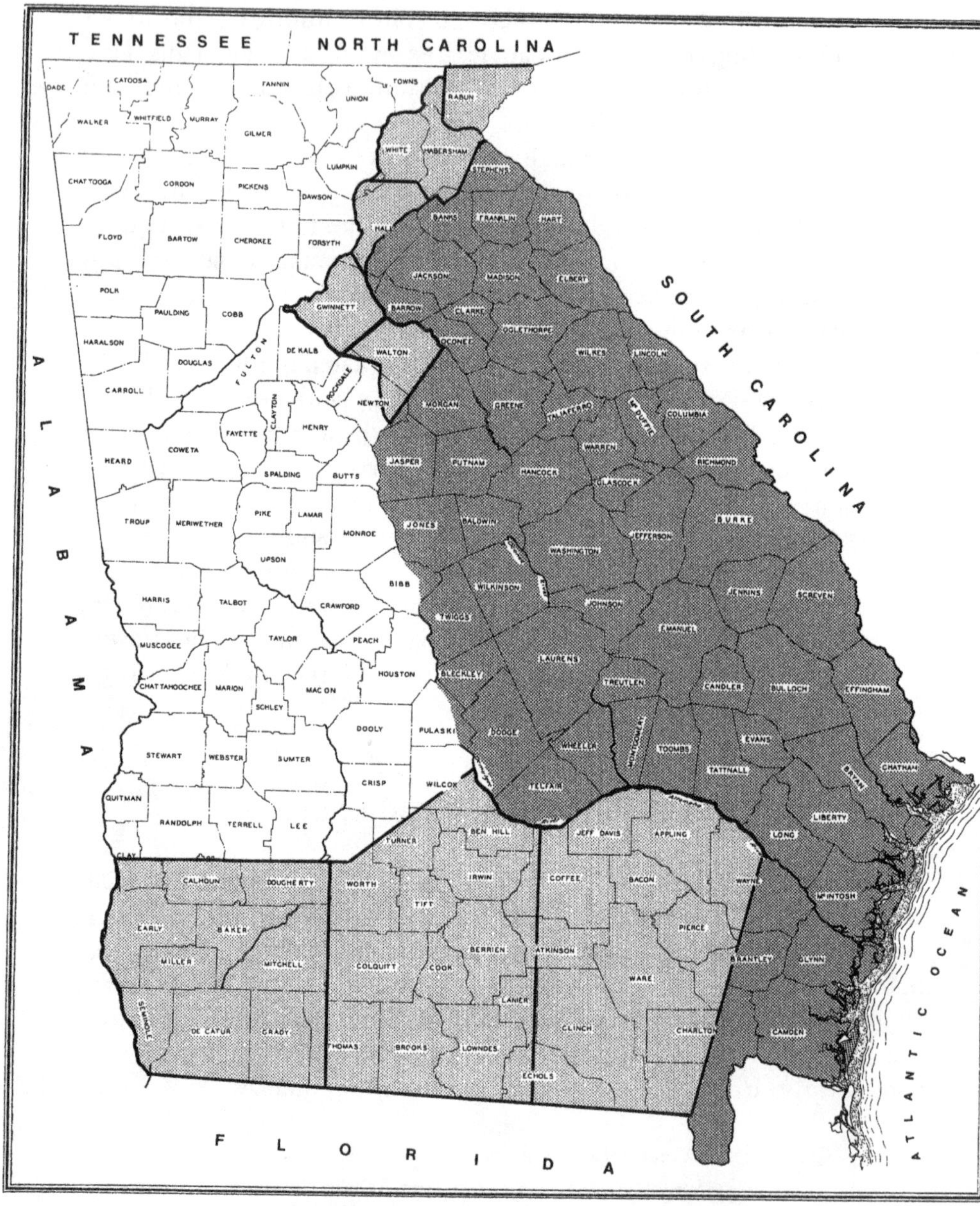

Lightly shaded area indicates lands distributed in the 1820 land lottery.

Great inaccuracies were found in John H. Brodnax's survey of the 11th District of Irwin County (later Lowndes County, now in Lowndes, Lanier, and Echols counties). The governor directed William B. Taylor to inspect and resurvey the district.[77] In April 1829 Taylor wrote Governor John Forsyth that he had started at the northwest corner of the district and run southerly along the district line. He found no corner at the end of the first 70 chains, found a corner at the end of the next 58.50 chains, and found the next corner at a distance of 65.56 chains. He reported that the district was similarly marked throughout, except that a considerable portion of its interior exhibited no signs of having been surveyed. Also, the line dividing the 11th and 12th districts varied 4° from the true meridian, and the line dividing the 11th and 10th districts varied 4° 40′ from the true parallel. Taylor noted that a number of swamps, branches, and creeks existed which had not been shown in Brodnax's survey and that the lots were very erroneously numbered. Sometimes as many as three lots bore the same number, and often two and three different numbers had been inscribed on the trees and posts for the same lot. According to Taylor, the method adopted by settlers to locate their particular lots was to employ a local surveyor "to run from the nearest marks at such distance as wou[l]d probably place them on or near their Lots had the same been regularly surveyed."[78]

In December 1829 the legislature took the matter under consideration. Taylor's attempts to rectify the chaotic condition of the district had been found impractical. As the legislature observed, "in making a resurvey" thereof "a derangement of the lots as originally marked, numbered, and granted" would "take place throughout the whole district." The result would be that "in many cases the drawers and owners" of the lots would "lose a part or the whole of their land, with the improvements thereon to their great injury." The following remedy was therefore enacted: if a "lot as originally marked and numbered" could "be identified as the land granted by the number, corner, and station trees, the drawer or owner of the . . . land" would "be entitled to the same without regard to the exact

77. Letter from William A. Knight & Jonathan Knight to George M. Troup (May 14, 1827) (GSGD); Letter from William B. Taylor to John Forsyth (Dec. 24, 1829) (GSGD); Letter from William B. Taylor to John Forsyth (Jan. 1, 1829) (Telamon Cuyler Collection, UGL); Letter from E. H. Pierce to William B. Taylor (Nov. 25, 1829) (GLB, GDAH); Letter from George R. Gilmer to Samuel A. Bailey (July 15, 1831) (GLB, GDAH); 1826 Ga. Laws 212; 1828 Ga. Laws 241; 1829 Ga. Laws 261; 1833 Ga. Laws 287; Dawson (pt. 2) 67, 112, 146; 1827 Ga. House Journal 191; 1829 Ga. Senate Journal 18.

78. Letter from William B. Taylor to John Forsyth (April 6, 1829) (GSGD).

quantity of acres" it contained "or precise distances of the lines." At the request and expense of the drawer or owner of any lot in the district, the county surveyor of Lowndes (of which the 11th District had become a part) was "to run round the" lot "according to the lines originally run" if the lines could "be ascertained." If the lines were open, they were to be closed "by running from corner to corner of the . . . lot, without regard to the course and distance of the . . . lines." If the lots could not be identified in this manner, the county surveyor was to number them "to correspond with the grants for the same . . . as nearly as he" could "according to the original plan of numbering the . . . district."[79]

Many problems were encountered with Beverly Spivy's survey of the 6th District of Habersham County (now in Habersham and White counties). Spivey was able to extricate himself from serious liability by submitting to the governor affidavits from his chainmen and axman. These averred that the course of most of his lines was "obstructed by impassable precipices, around many of which . . . considerable offsets" had to be taken. Frequently it then required "long & circuitous routs to obtain the true course again."[80] In 1868, however, nearly fifty years after Spivy's survey, the legislature hired J. J. Chitwood to resurvey the district. The reason given was that much uncertainty and confusion existed about the boundaries and that the owners were perplexed as to the exact locality of their lands.[81]

A medley of other vexatious problems appeared. In 1853 Obadiah T. Dickerson, county surveyor of Rabun, was appointed by the legislature to make a partial resurvey of the 1st and 2d districts of that county and to mark plainly all lines and corners that were unknown.[82] In 1856 an appeal came before the Georgia Supreme Court concerning lands in the 3d District of original Early County (now in Calhoun and Dougherty counties). The evidence indicated that some of the land lots were not situated in the same relative positions as shown in the plan of the district. In addition, several land lots were more than double the standard width and, although supposed to contain 250 acres, actually contained 400 or

79. 1829 Ga. Laws 127; Dawson 348.

80. Letter from Beverly Spivy to John Clark (June 29, 1821), Affidavit of Charles Wade & Hiram Noney (April 25, 1821), and Affidavit of William Lynn (April 27, 1821) (all in GSGD).

81. 1868 Ga. Laws 160.

82. 1851–52 Ga. Laws 452–53; Field notes and district map of Obadiah T. Dickerson, surveyor of the 1st and 2d districts of Rabun County (GSGD).

500 acres each.[83] It seems that many of the lines and corners shown in the field notes and plats of the southern part of Appling (later Clinch) County were not marked in the original survey because an act was passed in 1881 to have them marked.[84] Finally, a number of narrow, unsurveyed strips were discovered between districts in original Early County. The legislature ordered these strips sold by the inferior courts of the counties wherein they lay and the proceeds to go to the poor school fund. One such strip, found between the 4th and 6th districts, was attached to the 4th.[85]

A note of high praise was sounded, however, for some of the surveys for the 1820 land lottery. In 1913 certain fractional lots along the Tallulah River in the 4th and 5th districts of Rabun County were involved in litigation. Charles M. Strahan, professor of civil engineering at the University of Georgia, was called on to execute a resurvey. This was less than one hundred years after the initial survey, and many of the original corners were still intact. Strahan testified that, in going over the work of the old surveyors, he was impressed with their honest effort to carry out their instructions, that there were some errors as would have been expected in surveying through a rough virgin territory, but that the original surveys had been done conscientiously and with admirable accuracy for that time.[86]

83. Keel v. Pace, 20 Ga. 190 (1856), at 195–96.

84. 1880–81 Ga. Laws 655.

85. 1847 Ga. Laws 79; 1851–52 Ga. Laws 472; 1859 Ga. Laws 267.

86. State of Georgia v. Georgia Ry. & Pwr. Co., 141 Ga. 153 (1913), at 157. Strahan's field notes are filed with the original transcript of the case in the Georgia Department of Archives and History.

Chapter 6. The 1821 Land Lottery

Land Lottery Act of 1821

The rapid series of land cessions of 1814, 1817, 1818, and 1819 obtained by the United States for Georgia secured the former no long-lasting respite from the latter's vociferous agitation for a final execution of the terms of the 1802 compact. As a result, U.S. commissioners met a deputation of Creek chiefs at Indian Springs early in 1821. The conference received a jolt when Georgia presented trumped-up claims against the Creeks for nearly half a million dollars allegedly due the state under old treaties that dated back to the Revolution. McIntosh, a "white man's Indian" of the first sort and principal negotiator for the Creeks, was inclined to recognize the Georgia claims.[1]

In a treaty signed on January 8, the Creeks gave up the lands between the Ocmulgee and Flint rivers. The United States agreed to pay $200,000 and to assume the Georgia claims.[2] For his influence McIntosh received a thousand-acre reserve at Indian Springs and a one-square-mile reserve located on the west bank of the Ocmulgee River in what is now Butts County, where he maintained a home.

Only a small group of Creek chiefs attended the conference, and there is ample evidence that bribery and conspiracy occurred.[3] In addition to McIntosh's reserves, one-square-mile tracts were reserved for Michey Barnard, James Barnard, Buckey Barnard, Cussena Barnard, and Efau Mathlaw—all influential Creeks.[4] A one-and-a-half-by-three-mile tract on the Flint River, where the Creek Agency stood, was reserved for the use of the Creek Nation.

1. Am. St. Papers, 2 Indian Affairs 252–53; Bonner, *William McIntosh,* in Georgians in Profile 114, 128–29 (H. Montgomery ed. 1958); Lamar 1223.

2. Am. St. Papers, 2 Indian Affairs 248; C. Kappler, Indian Affairs (Treaties), S. Doc. No. 319, 58th Cong., 2d Sess. 195 (1904).

3. See Letter from David Adams & Daniel Newman to John Clark (Dec. 31, 1820) (GDAH); and Letter from David Adams, Daniel Newman, & John McIntosh to John Clark (Jan. 16, 1821) (Indian Affairs File, Telamon Cuyler Collection, UGL).

4. See Chapter 9.

An extra session of the legislature was convened in April 1821 by Governor Clark and an act, summarized as follows, was passed to dispose of the lands:

Section 1. The cession to be divided into five counties. That part lying south of a line beginning on the Ocmulgee River "opposite the town of Hartford" and running "due west" to the Flint River to form Dooly County. That part lying between Dooly County and a line beginning on the Ocmulgee River "opposite Fort Hawkins" and running "due west" to the Flint River to form Houston County. That part lying between Houston County, and a line beginning "at the Seven Islands" on the Ocmulgee River and running, "due west forty miles, thence due south" to the Flint River to form Monroe County. That part lying west of Monroe County and a line beginning at the northwest corner of Monroe County and running "north" to the Chattahoochee River to form Fayette County. The balance of the territory, lying north of Monroe County and east of Fayette County, to form Henry County.

Section 2. Each county to be divided into districts of "nine miles square, as near as practicable, the district lines running parallel to the lines dividing counties, and crossed by other lines at right angles, and said districts so laid out" to "be again subdivided by lines to be run in like directions into square tracts" containing 202½ acres each, "marked and numbered according to the plan heretofore pursued, under the instructions of the Surveyor General."

Section 3. Fractional lots to be reserved from drawing and disposed of in such manner as a future legislature should direct.

Sections 4, 5, 6. (Essentially identical to sections 5, 6, 7, and 8 of the land lottery act of 1806.)

Section 7. Each surveyor to post a $10,000 bond.

Section 8. (Essentially identical to section 5 of the land lottery act of 1803 except as to time for completing surveys.) Surveyors of county and district lines to complete their work within sixty days of the passage of this act. District surveyors to complete their work within ninety days of the time they are notified to begin.

Sections 9, 10. District surveyors to receive $2.50 for each mile of line run, out of which they must defray all expenses. Surveyors of county and district lines to receive $3.50 for each mile of line run, out of which they must defray all expenses. Governor to advance $300 to each surveyor before the survey begins.

Sections 11, 12, 13, 14, 15, 16, 17. (Essentially identical to sections 8,

9, and 10 of the land lottery act of 1803 except as to qualifications of drawers.) Each widow and each unmarried white male citizen eighteen years of age and older entitled to one draw. Each white male citizen with a family and each widow of a veteran of the revolutionary war, War of 1812, or Indian wars entitled to two draws. Each minor orphan or family of minor orphans and children of convicts entitled to one draw or two draws, depending on certain specified circumstances. Successful drawers in previous lotteries, those who evaded military service, and convicts excluded from participation.

Section 18. Governor to fill any vacancies caused by the death or resignation of surveyors, or by the failure of any surveyor to perform his appointed task, or by the presence of more districts in the territory than contemplated by this act.

Section 19. Surveyors, chainmen, and markers to take an oath before beginning their work.

Sections 20, 21. Prices of grants to successful drawers to be $19 each. Any grants not taken out within two years of the drawing to revert to the state.

Sections 22, 23. The land occupied by Fort Hawkins, along with an area of equal size across the Ocmulgee River therefrom, to be reserved and disposed of in such manner as a future legislature should direct.

Section 24. Indian reserves created by the Treaty at Indian Springs in 1821 (except those abandoned before the surveys begin) to be exempt from the operation of this act, and surveyors to make lots adjoining thereto fractional. Upon their abandonment by the Indians, the reserves to be sold and the proceeds used for the education of poor children.[5]

Fayette County, as finally laid out, varies substantially from the disposition implied in the above act. The distance between the Ocmulgee and Flint rivers, along the northerly line of Monroe County, was found to be considerably less than the forty miles called for in the act.

Surveys for Original Dooly, Houston, Monroe, Henry, and Fayette Counties

Wilson Lumpkin, commissioner for the United States, and Indian representatives marked the artifical lines of the cession in the spring of 1821.[6]

5. 1821 ex. sess. Ga. Laws 3; Dawson 246.

6. Letter from John C. Calhoun to Wilson Lumpkin (Mar. 14, 1821) (Letters Sent Relating to Indian Affairs, Volume E, Records of the Office of the Secretary of War, Record Group

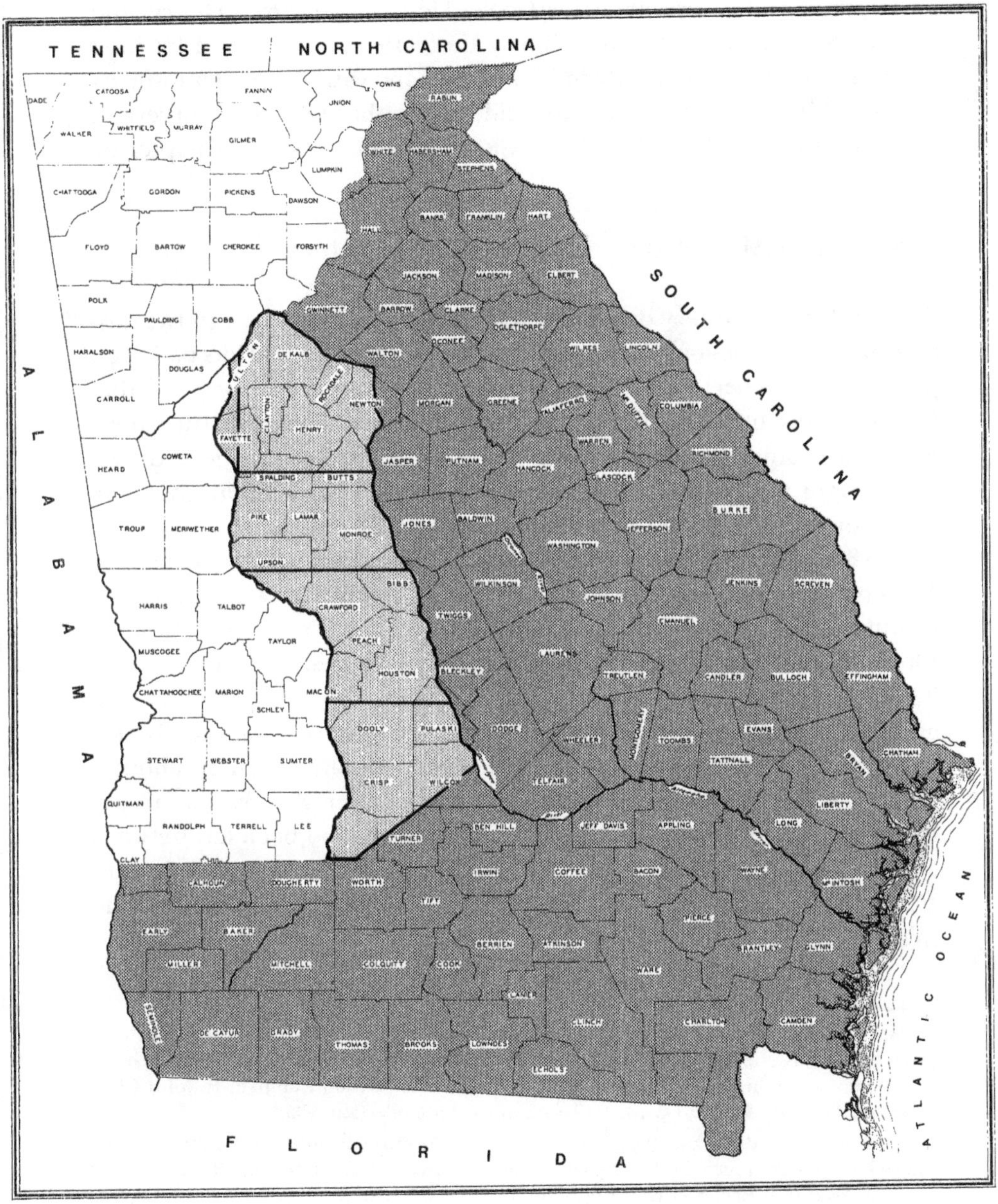

Lightly shaded area indicates lands distributed in the 1821 land lottery.

The county and district lines for Dooly, Houston, Monroe, Henry, and Fayette counties were marked shortly thereafter and the land lot lines run that summer. Apparently the work was done without incident worthy of note. The lottery drawing was held during that fall.[7] The surveys of the reserves set aside in the 1821 cession will be discussed in Chapter 9.

Frauds and Mistakes in the 1821 Land Lottery

During the 1821 lottery drawing it was discovered that, through the mistake of the commissioners, the numbers of several fractional lots had been placed in the wheel and drawn as prizes. The lottery act specified that fractions were not to be disposed of in this manner. The legislature, however, felt it unjust to deprive the persons whose names had been drawn of the right to take out grants on these lots and decreed that grants be issued on such draws in the usual manner.[8]

In 1824 word reached Milledgeville of inaccuracies in Murdock M. McLeod's survey of the 1st District of Dooly County (now south Macon County and northwest Dooly County). Governor Troup sent Edward Lloyd Thomas to inspect and, if necessary, to resurvey the district. Thomas discovered that the district had been only partially surveyed and that the line separating Buckey Barnard's Reserve from the district, which should have begun on the Flint River, actually began on another body of water nearly twenty chains east of the river. As a result, an unappropriated strip over a mile and a quarter long had been left between the river and the reserve. Needless to say, Thomas resurveyed the district and the legislature authorized the governor to put McLeod's bond in suit.[9]

75, National Archives, Washington, D.C.); Letter from Wilson Lumpkin to John C. Calhoun (Aug. 8, 1821) (Letters Received by the Secretary of War, Registered Series, Record Group 107, National Archives, Washington, D.C.); Letters from Wilson Lumpkin to John Clark (Mar. 31, April 28, & June 15, 1821) (Telamon Cuyler Collection, UGL).

7. Field notes and district maps for Dooly, Houston, Monroe, Henry, and Fayette counties in GSGD; Georgia Journal (Milledgeville), Sept. 11, 1821, at 2, col. 6; Id., Oct. 2, 1821, at 2, col. 4; Southern Recorder (Milledgeville), Dec. 11, 1821, at 3, col. 4.

8. 1823 Ga. Laws 92; Dawson 274.

9. Letter from George M. Troup to Edward Lloyd Thomas (Feb. 18, 1825) (GLB, GDAH); Letters from Edward Lloyd Thomas to George M. Troup (April 11, May 7, June 9, & Oct. 14, 1825) (all in GSGD); 1825 Ga. Laws 224; 1826 Ga. Laws 218; 1831 Ga. Laws

The grantees of land lots along the southernmost tier of the 16th District of Dooly County (now in Worth and Turner counties) discovered, to their delight, that they had received much more than they bargained for. Canneth Swain, who originally laid the district off, began by running the northern tier of land lots and then each successive tier southward. Either he accumulated considerable error in his survey or the district is too large in its north-south dimension, for the meridional dimensions of the most southerly tier of land lots in the district are considerably longer than the forty-five chains specified by law. Land Lot 112 of the district contains about 350 instead of 202½ acres, and all the other lots along the tier (except the fractions) have similar acreages. Swain's field notes, however, make no mention of this excess.[10] A surveyor who reran some of the lots in the 1880s believed that the original surveys had left an unappropriated strip between the south line of the 16th District of Dooly and the north line of the 7th District of Irwin. He believed that the owners to the north had appropriated this strip.[11] The north line of the 7th District of Irwin was the boundary of the Creek cession of 1814. It is therefore possible that attempts to resurvey this line in preparation for the lotteries caused a second line to be marked a few chains north of the old line, thus confounding the surveys in the region.

In the mid-twentieth century J. Burrus Smith was employed to resurvey Fractional Lot 269 in the 9th District of original Monroe (now Pike) County, which adjoins the Flint River. The state survey showed the lot as having the form depicted in the left-hand diagram on page 240 and containing an area of 152 acres, 3 roods, and 36 poles. Smith's resurvey revealed, however, that it was shaped as shown in the right-hand diagram and covered an area of over 280 acres—far larger than a standard, full-size land lot of the region. Smith's inspection of the land along the river indicated conclusively that the lot could not possibly have had the form shown in the state records at the time of the 1821 survey. Since the river had not been traversed in the original survey, Smith concluded that the error was caused by pure oversight. As a side note,

---

262; 1824 Ga. Senate Journal 69; 1825 Ga. Senate Journal 79; Goff, *Edward Lloyd Thomas, Surveyor,* 18 Emory U.Q. 107, 111 (1962).

10. Field notes and district map of Canneth Swain, surveyor of the 16th District of Dooly County (GSGD); Aerial photographs and tax maps in Worth County tax assessor's office, Sylvester, Ga.

11. Ford v. Williams, 73 Ga. 106 (1884) (original transcript of case in GDAH).

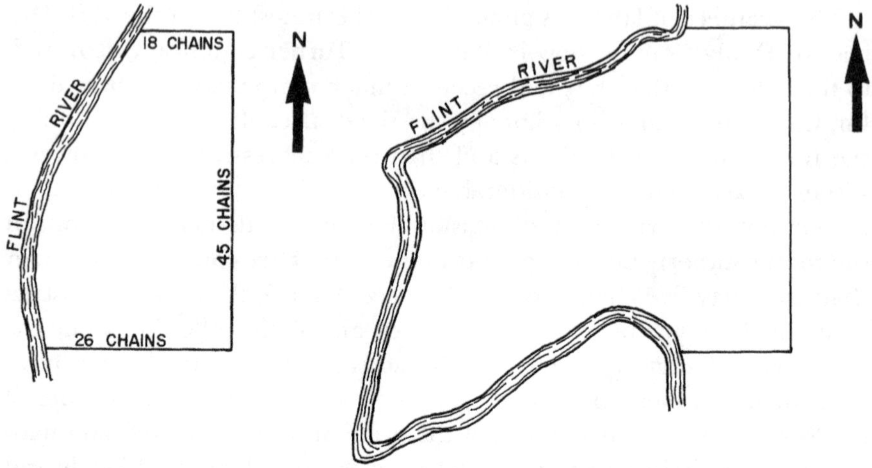

Left: Fractional Lot 269 as shown by the original survey. *Right:* Fractional Lot 269 as it actually exists.

the owner of the lot told Smith that once a party of surveyors working on a nearby tract had a barrel of peach brandy. After partaking liberally of these spirits and becoming rather intoxicated, they proceeded with their survey. In the process they lost their chain. Undaunted, they cut a grapevine with which to finish the job.[12]

Isidore Stouf, who performed the original survey of the 14th District of Monroe County (now in Monroe and Butts counties), noted in his field notes that the district contained nine extra chains in its north-south dimension. He had thrown this entire excess into the southernmost tier of land lots, thereby making each of the lots along this tier forty-five chains by fifty-four chains. Nevertheless, the state grants of lots along the southern tier show them as measuring the standard forty-five by forty-five chains and containing 202½ acres each.[13]

The problem of excess and deficiency was a phenomenon that occurred frequently because the county, district, and land lot lines were usually run by different surveyors using somewhat different methods and procedures. Because the law set forth no guidelines, surveyors generally followed their own whims and fancies in dealing with the problem. Rather than throwing the excess in the last tier of land lots as Stouf did,

12. Smith, *Surveying by Imagination,* 18 Surveying and Mapping 77 (1958).

13. Field notes of Isidore Stouf, surveyor of the 14th District of Monroe County, 5 (GSGD).

some district surveyors laid off a tier of fractional lots along one of the district lines to take up the excess. Personal discretion and convenience dictated which side of the district would receive the fractional lots. Other surveyors apportioned the excess throughout the district. Still other surveyors were indifferent to the problem and left long, narrow gores in their districts, which they did not mention in their field notes and reports. These, of course, became sources of confusion in later years.

Deficiencies too were often left in tiers of land lots along district lines, the surveyors in many instances making no note of their existence. Thus today in performing resurveys, "wide" ranges and "narrow" ranges of land lots are often encountered along district lines.[14]

14. See Heard v. Smith, 183 Ga. 725 (1937).

Chapter 7. The 1827 Land Lottery

Treaty at Indian Springs in 1825

The ink had hardly dried on the treaty at Indian Springs of 1821 when Georgians began again to clamor for the federal government to obtain further cessions of Indian territory. Having previously met success with the Creeks, President James Monroe directed United States commissioners at New Echota in July 1823 to attempt to obtain a cession from the Cherokees.[1] These overtures failed, even though bribes were offered to the Cherokee leaders through Chief McIntosh.[2] Subsequently McIntosh suggested to the commissioners that they might obtain a cession from the Creeks.

Years of debilitating warfare with the whites had reduced the Creeks to abject poverty and diminished their territory to only a fraction of its original extent.[3] Alarmed at seeing their lands slip away so rapidly, the head chiefs of the Creek Nation enacted a law in the spring and summer of 1824 that decreed death to anyone attempting to sell additional Creek lands. Moreover, they ostracized McIntosh from the council and banished him from the Creek Nation.[4] Nevertheless, a meeting was held between U.S. commissioners and the Creek national council in December 1824 at Broken Arrow, just west of the present site of Columbus. On December 14 the Creek council announced its firm refusal to make further land cessions.[5]

1. Abel, *The Cherokee Negotiations of 1822 and 1823,* 1 Smith College Studies in History 188 (1916); Bonner, *William McIntosh,* in Georgians in Profile 114, 133 (H. Montgomery ed. 1958). The commissioners were Duncan G. Campbell and David Meriwether, both native Georgians. They presided over the subsequent negotiations with the Creeks at Broken Arrow and at Indian Springs.

2. Bonner, *William McIntosh,* in Georgians in Profile 114, 133 (H. Montgomery ed. 1958); G. Moulton, John Ross: Cherokee Chief 25 (1978).

3. Niles Weekly Register (Baltimore), Dec. 4, 1824, at 222–23; U. Phillips, Georgia and State Rights 54 (1901).

4. Bonner, *William McIntosh,* in Georgians in Profile 114, 133–34 (H. Montgomery ed. 1958); Laws of the Creek Nation passim (A. Waring ed. 1960) (University of Georgia Libraries Miscellanea Publications, No. 1).

5. Bonner, *William McIntosh,* in Georgians in Profile 114, 135 (H. Montgomery ed. 1958).

Shortly afterward McIntosh secured the allegiance of a motley assortment of about forty more or less influential Creeks and from this group obtained authority to continue negotiations. As a result, a new conference was called for February 7, 1825, at McIntosh's tavern at Indian Springs.[6] Because of late arrivals, the proceedings did not get under way until the tenth. Many of the chiefs opposed to McIntosh did not show up at all, and some who did left on the night of the eleventh after having protested that this partial assembly had no authority to negotiate land cessions.[7]

On February 12, 1825, a treaty was signed between the U.S. commissioners and McIntosh and his followers.[8] This treaty purported, among other things, to cede all the remaining Creek lands in Georgia. For this the Creeks were to receive an equal quantity of land west of the Mississippi River, expenses for their removal, and further monetary considerations. The Creeks were given until September 1, 1826, to vacate the lands.

No doubt all sorts of bribery and corruption were used to obtain the treaty.[9] A supplemental article, signed two days later, after many of the Indians had left, provided that McIntosh was to receive a cash payment of $25,000 for the reserves he had acquired under the treaty in 1821—thus indicating his complicity in the matter. John Crowell, the Indian agent, alleged that the corruption of the U.S. commissioners was of so rank a character that it smelled to heaven.[10] The report of the select committee of the U.S. House of Representatives later charged with investigating the matter showed that of the fifty-one Indian representatives who signed, at least fourteen were known mixed-bloods and that representatives from only eight of the fifty-six towns in the Creek Nation signed the document. Of the Indians who signed, McIntosh was the only ranking chief, and he was merely fifth rank. Five were chiefs of inferior degree, and these five along with McIntosh were all from one town. Twenty-six of the Indians were officers called "law menders" or "law makers" but not

6. Id. 136–38; J. Bonner, Georgia's Last Frontier: The Development of Carroll County 12 (1971); E. Harden, The Life of George M. Troup 254 (Savannah 1859).

7. Am. St. Papers, 2 Indian Affairs 582–83; H.R. Rep. No. 98, 19th Cong., 2d Sess. 129–30 (1827).

8. Am. St. Papers, 2 Indian Affairs 563; C. Kappler, Indian Affairs (Treaties), S. Doc. No. 319, 58th Cong., 2d Sess. 214 (1904).

9. Letter from Benjamin Hawkins, Jr., to George M. Troup (April 10, 1825) (Telamon Cuyler Collection, UGL); J. Bonner, Georgia's Last Frontier: The Development of Carroll County 12–13 (1972); Bonner, *William McIntosh,* in Georgians in Profile 114, 138–40 (H. Montgomery ed. 1958).

10. H.R. Rep. No. 98, 19th Cong., 2d Sess. 502 (1827).

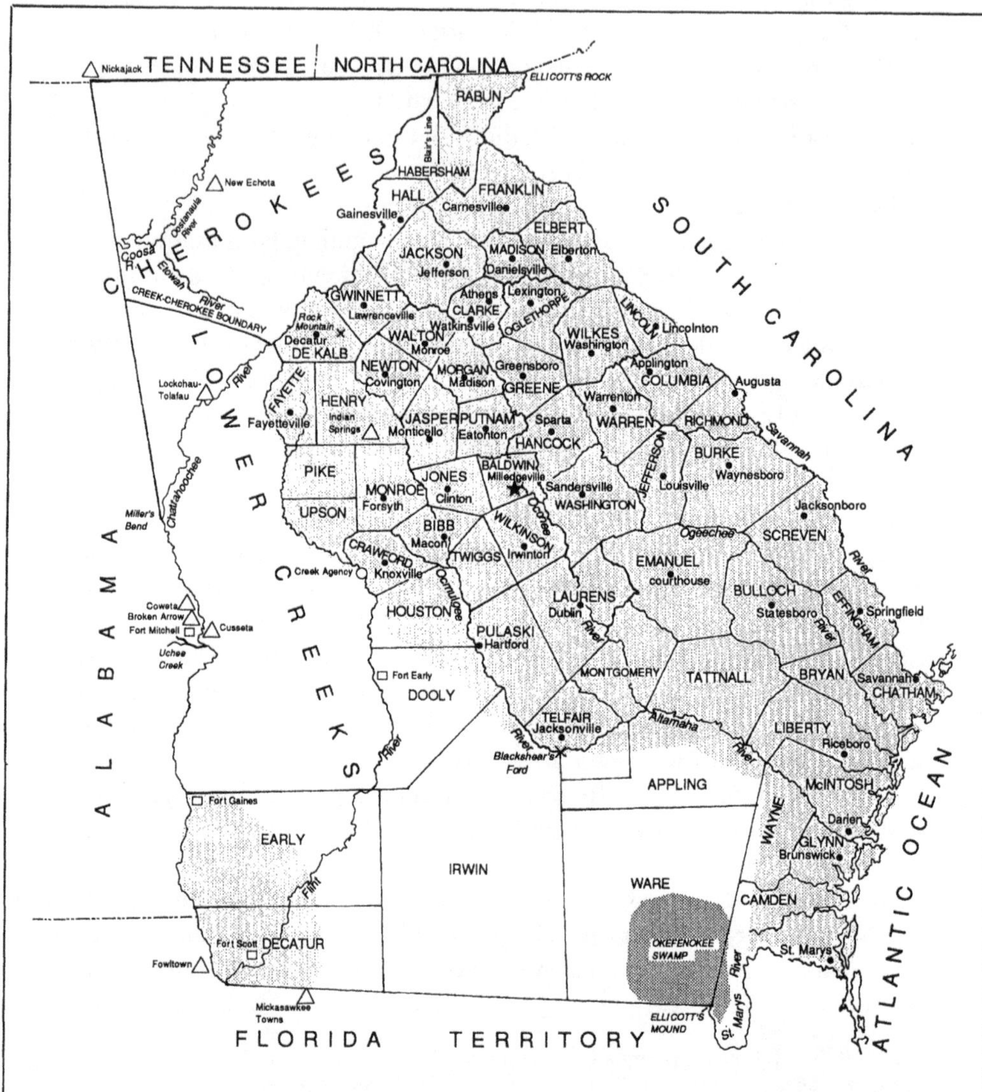

Georgia in early 1825. The shading indicates the approximate area of white settlement.

William McIntosh signing the Indian Springs Treaty in 1825. Drawing by George I. Parrish, Jr. Courtesy of the Atlanta Historical Society.

chiefs, fourteen were broken chiefs, three were Indians possessing no rank whatsoever, and two were persons wholly unknown to the council of the Creek Nation.[11]

The day after the signing Crowell dispatched a letter to the secretary of war stating his misgivings;[12] and several days later the Creek chiefs sent a letter to President Monroe declaring that the treaty was invalid.[13] These hasty protests did not make sufficient impression in Washington, however, until after the treaty had been ratified by the U.S. Senate and signed by President John Quincy Adams, who took office on March 4.[14] Given the slow communication of the day and the sluggishness of bureaucratic machinery, the fact that less than four weeks elapsed between the signing at Indian Springs and final approval by the president in Washington is further evidence of the treaty's perfidy.

11. Id. 5, 10, 254–56, 480, 821; Bonner, *William McIntosh,* in Georgians in Profile 114, 140 (H. Montgomery ed. 1958).

12. Am. St. Papers, 2 Indian Affairs 583; H.R. Rep. No. 98, 19th Cong., 2d Sess. 6, 132 (1827).

13. H.R. Rep. No. 98, 19th Cong., 2d Sess. 7, 133 (1827).

14. Id. 7.

Land Lottery Act of 1825

George M. Troup, the headstrong and fiery governor of Georgia and, ironically, first cousin of Chief McIntosh, called a special session of the legislature to meet May 23 to dispose of the lands.[15] He also requested McIntosh's permission to proceed immediately with the survey of the Creek lands rather than waiting until September 1, 1826.[16] After consulting with his followers, McIntosh granted Troup's request.[17] To this several Creek Indians vowed to kill the first surveyor who stretched a chain across their lands.[18]

Governor Troup's haste in this matter was dictated largely by political considerations. He was up for reelection in October and faced former governor John Clark, who had defeated Troup for the governorship on two previous occasions. The free distribution of the surveyed land to Georgia citizens through the lottery offered a prospect for patronage of gigantic proportions.[19]

McIntosh's reward for twelve years of pandering to white interests was death. On the afternoon of April 29, 1825, an execution party of about 170 Creek warriors concealed themselves in the woods near McIntosh's residence on the banks of the Chattahoochee River in what later became Carroll County. Shortly before dawn on the thirtieth they set fire to the house and outbuildings. McIntosh, awakened by the commotion, was forced by spreading flames to abandon the house. Upon making his appearance at the doorway, he was cut down by a hail of bullets. His bleeding body was dragged into the yard and there, in the glare of his burning home, he was knifed to death. McIntosh's son-in-law and Tome Tustunnugee, another Creek chief loyal to McIntosh, were also executed the same day.[20]

15. Id. 141–42; E. Harden, The Life of George M. Troup 263 (Savannah 1859).

16. Am. St. Papers, 2 Indian Affairs 757; H.R. Rep. No. 98, 19th Cong., 2d Sess. 8, 9, 273, 275, 519 (1827); E. Harden, The Life of George M. Troup 271 (Savannah 1859).

17. Am. St. Papers, 2 Indian Affairs 759, 761; H.R. Rep. No. 98, 19th Cong., 2d Sess. 8, 9, 142, 520, 540 (1827); E. Harden, The Life of George M. Troup 273–74 (Savannah 1859).

18. H.R. Rep. No. 98, 19th Cong., 2d Sess. 201, 352 (1827).

19. Bonner, *William McIntosh,* in Georgians in Profile, 114, 142 (H. Montgomery ed. 1958). Troup eventually won reelection by a small majority. E. Harden, The Life of George M. Troup 397–99 (Savannah 1859).

20. Bonner, *William McIntosh,* in Georgians in Profile, 114, 142–43 (H. Montgomery ed. 1958); E. Harden, The Life of George M. Troup 276–77 (Savannah 1859); Head, *Old Letters,* 14 Georgia Review 126 (1960); G. White, Historical Collections of Georgia 170–73 (New York 1855).

George M. Troup. From George White, *Historical Collections of Georgia* (1854).

In spite of the tumultuous events and blatant duplicity of the preceeding three months, the legislature met in May as planned. After two weeks of debate it passed an act, summarized as follows, to survey and distribute the lands obtained by the Indian Springs treaty:

Section 1. The cession to be divided into five "sections." That part lying south of a line beginning on the Flint River "opposite where the line dividing the counties of Houston and Dooly strikes said river and running due west" to the Chattahoochee River to form the first section. That part lying between the first section and a line beginning on the Flint River "opposite where the original line dividing the counties of Monroe and Houston [strikes said river] and running due west" to the Chattahoochee River to form the second section. That part lying between the second section and a line beginning on the Flint River "where the original line dividing the counties of Henry and Monroe strikes said river and running due west" to the Chattahoochee River to form the third section. That part lying north of the third section and east of the Chattahoochee River to form the fourth section. That part lying west of the Chattahoochee River to form the fifth section.

Section 2. Each section to be divided into districts of "nine miles square, as near as practicable, the district lines running parallel to the lines dividing sections, and crossed by other lines at right angles; and said districts" to "be again subdivided by lines to be run in like directions into square tracts" containing 202½ acres each "marked and numbered according to the plan heretofore pursued under the instructions of the Surveyor General."

Section 3. Fractional lots to be reserved from drawing and disposed of in such manner as a future legislature should direct.

Sections 4, 5, 6. (Essentially identical to sections 5, 6, 7, and 8 of the land lottery act of 1806.)

Section 7. Each surveyor to post a $10,000 bond.

Section 8. (Essentially identical to section 5 of the land lottery act of 1803 except as to time for completing surveys.) Surveyors of section, district, and reserve lines east of the Chattahoochee River to complete their work within sixty days of the time they are notified to begin. Surveyors of section, district, and reserve lines west of the Chattahoochee River to complete their work within sixty days of the running of the Georgia-Alabama boundary. District surveyors to complete their work within ninety days of the time they are notified to begin.

Sections 9, 10. District surveyors to receive $3.00 for each mile of line run, out of which they must defray all expenses. Surveyors of section, district, and reserve lines to receive $3.50 for each mile run, out of which they must defray all expenses. Governor to advance $300 to each surveyor before the survey begins.

Sections 11, 12, 13, 14, 15, 16, 17, 18. (Essentially identical to sections 8, 9, and 10 of the land lottery act of 1803 except as to qualifications of drawers.) The following entitled to one draw: each unmarried white male citizen eighteen years of age and older, each widow, each wife or child whose husband or father has been absent from the state for three years, and each veteran of the revolutionary war who was a successful drawer in the previous lottery. The following entitled to two draws: each white male citizen with a family; each indigent or invalid veteran of the revolutionary war, War of 1812, or Indian wars; each widow of a veteran of the revolutionary war, War of 1812, or Indian wars; and each veteran of the revolutionary war who was not a successful drawer in the previous lottery. Each minor orphan or family of minor orphans and children of convicts entitled to one draw or two draws, depending on certain specified circumstances. Successful drawers in previous lotteries (except as noted), evaders of military service, and convicts excluded from participation.

Section 19. (Essentially identical to section 18 of the land lottery act of 1821.)

Section 20. Surveyors, chainmen, and markers to take an oath before beginning their work.

Sections 21, 22, 23. (Essentially identical to section 20 of the land lottery act of 1818 except regarding payment of taxes.) Successful drawers to pay taxes on lands drawn, whether or not a grant has been taken out.

Section 24. The land occupied by the Old Creek Agency along with an area of equal size across the Flint River therefrom; "one mile square at Marshall's ferry on Flint river including the ferry; one mile square at McIntosh's on the Chattahoochie including the ferry; and a reserve of five miles square on the Chattahoochie river at the Coweta falls and including the same—the northern boundary to cross the river at a point one mile above the lower shoal" to be set aside for public purposes.[21]

The territory disposed of in the above act was soon organized into counties, the First Section becoming original Lee County; the Second, original Muscogee County; the Third, original Troup County; the Fourth, original Coweta County; and the Fifth, original Carroll County.[22]

21. 1825 ex. sess. Ga. Laws 3; Dawson 253.

22. 1826 Ga. Laws 57; Dawson 132.

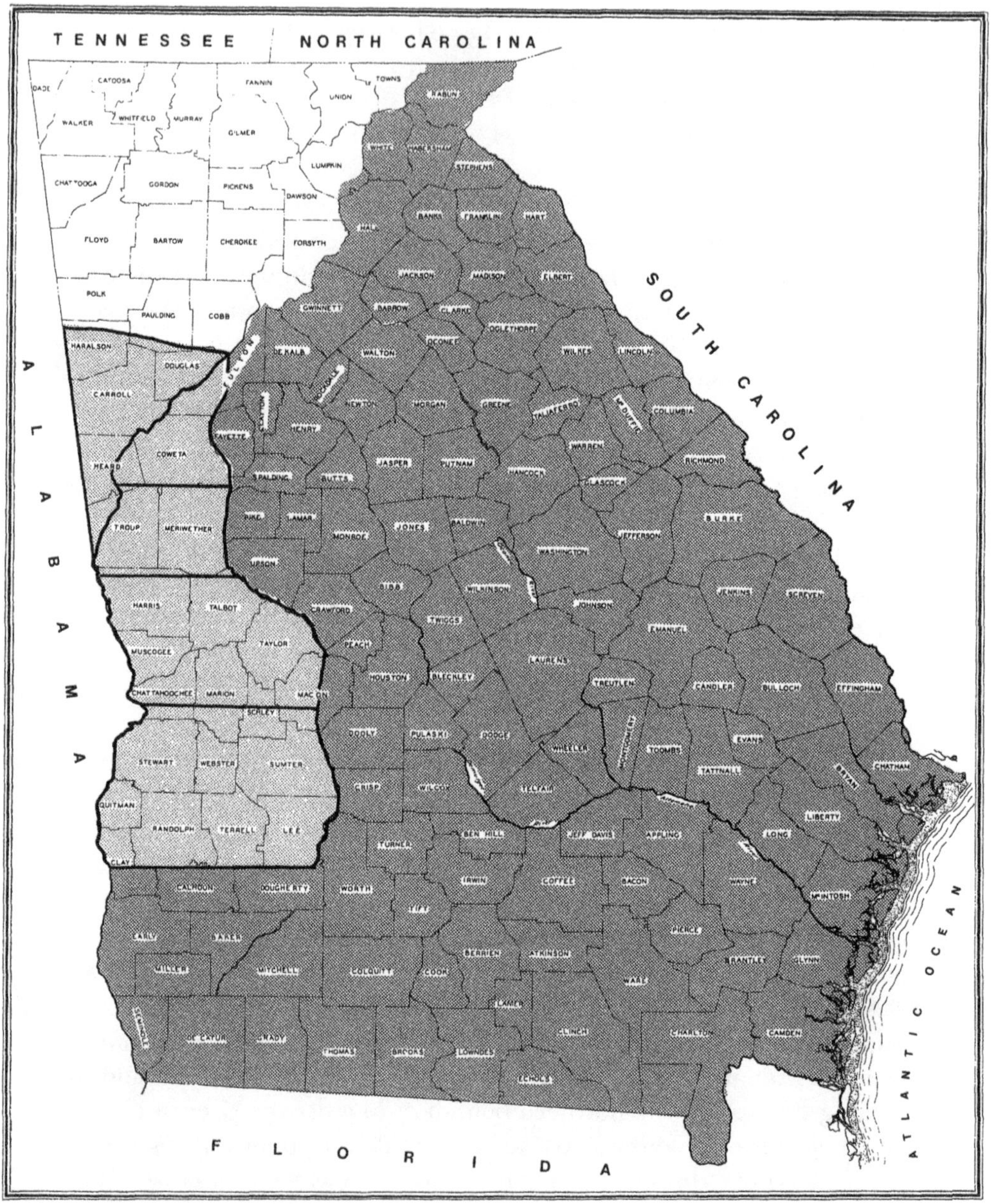

Lightly shaded area indicates lands distributed in the 1827 land lottery.

With the contemplated survey of the Creek lands at hand it now became necessary to run the boundary between Georgia and Alabama. In anticipation of the Creek cession, Governor Troup had, with characteristic impetuousness, communicated with Governor Israel Pickens of Alabama on the matter early in 1824. In July 1825 Pickens informed Troup that the Alabama legislature had made no provision as yet for running the line and that the project would have to be put off at least until the Alabama legislature met the following November. Thereupon Troup agreed to postpone the matter.[23]

Treaty at Washington in 1826

In the meantime the Creeks had raised a clamor of protest against the Indian Springs treaty, which, coupled with the execution of McIntosh, induced President Adams to refuse to execute the treaty. He directed the secretary of war to prohibit Troup from proceeding with the survey of the Creek lands.[24] In view of the president's actions, Troup put off the survey for the time being and fell back to the task of defending the treaty. A heated controversy followed between him and Adams that aroused national attention. The legislature and the people of Georgia lined up solidly behind Troup, but elsewhere the almost universal sentiment was in favor of the Creeks. To avoid a showdown, Adams began negotiations with the Creeks in November 1825 for an additional treaty. Georgia protested, but the negotiations continued. By the use of various forms of bribes and threats, Creek representatives were induced to agree to a new treaty of cession.[25] This treaty, signed in Washington on January 24, 1826, declared the old treaty void and ceded most of the balance of the Creek lands in Georgia.[26] In return the Creeks were to

23. Letter from Israel Pickens to George M. Troup (July 3, 1825) (GSGD); 1824 Ga. Senate Journal 20; Brantley, *Georgia Line—The Chain and Compass Line,* 4 Ala. Law. 282, 283 (1943); E. Harden, The Life of George M. Troup 221, 236–37, 328–29 (Savannah 1859).

24. Am. St. Papers, 2 Indian Affairs 809; E. Harden, The Life of George M. Troup 378 (Savannah 1859).

25. 2 Cong. Deb. 2670–78 (19th Cong., 1st Sess., 1826); 1 T. Benton, Thirty Years' View 59–64 (New York 1854); J. Bonner, Georgia's Last Frontier: The Development of Carroll County 16 (1971); Hryniewicki, *The Creek Treaty of Washington, 1826,* 48 Ga. Hist. Q. 425 (1964).

26. Am. St. Papers, 2 Indian Affairs 613, 662; C. Kappler, Indian Affairs (Treaties), S. Doc. No. 319, 58th Cong. 2d Sess. 264 (1904).

receive a tract west of the Mississippi River as well as certain monetary considerations and were to remove themselves from the ceded lands by January 1, 1827.

The treaty at Washington did not, however, deprive the Creeks of all their remaining lands in Georgia, for it left a small wedge-shaped strip of territory, located in the western part of original Carroll County, in Creek possession.[27] This omission set the stage for one of the most dramatic and far-reaching interruptions Georgia surveyors were ever to encounter in the survey of the state's public domain.

Georgia-Alabama Boundary

Meanwhile, Georgia and Alabama were ready to run and mark their common boundary. Governor Pickens believed the United States was a party in interest and should participate, but President Adams declined—perhaps because he did not wish further to incite the wrath of the Creeks.[28] Alabama appointed Arthur P. Bagby and Major Charles Lewis to act as its commissioners, while Governor Troup appointed Colonel Richard A. Blount, Major Joel Crawford, and Colonel Everard Hamilton as commissioners, James Camak as mathematician, and Edward Lloyd Thomas as surveyor.[29]

On July 2, 1826, these men met at Fort Mitchell in Alabama a few miles south of present-day Columbus and began a reconnaissance of the area.[30] The Articles of Agreement and Cession of 1802 defined the point at which the line would leave the Chattahoochee River: "running . . . up the said river Chattahoochee . . . to the great bend thereof, and next above the place where a certain creek or river called 'Uchee,' (being the first considerable stream on the western side, above the Cussetas, and Coweta towns,) empties into the said Chattahoochee."[31] Georgia contended that

27. J. Bonner, Georgia's Last Frontier: The Development of Carroll County 16–17 (1971); Hryniewicki, *The Creek Treaty of November 15, 1827*, 52 Ga. Hist. Q. 1 (1968). The Creeks also still possessed lands in northeastern Alabama during this period.

28. Brantley, *Georgia Line—The Chain and Compass Line*, 4 Ala. Law. 282, 284 (1943).

29. Id. 284–85; Dawson (pt. 2) 66.

30. Letter from Robert Blount to George M. Troup (July 4, 1826), and Journal of the Commissioners of Georgia, appointed to run the dividing line between the boundaries of that State & Alabama 1 (both in GSGD).

31. See page 114.

this site was the head of Miller's Bend at the lower edge of today's West Point, Georgia, whereas Alabama claimed that it was much farther up the river to the west of present-day LaGrange. When the disagreement became a stalemate, Alabama's commissioners respectfully withdrew from the proceedings and Georgia decided to continue with the survey.[32]

Since there were no established monuments to guide them, the commissioners and surveyors found it necessary first to run a random line to Nickajack and then, using this line as a guide, to run the "true" line back to the starting point. No incident worthy of note occurred until they reached what is now lower Chattooga County. They did, however, complain about the acute loneliness of the region, the scarcity of game, and the "noxious troublesome" insects.[33]

Since the line ran through the lands of both the Creeks and the Cherokees, Georgia had sought from these two tribes permission to allow its surveyors to pass through unmolested.[34] Little Prince, the head chief of the Creek Nation, supinely acceded to this request.[35] Charles Hicks, the principal chief of the Cherokee Nation, however, indignantly replied that there was no need for Georgia to run the line any farther north than the boundary between the Creeks and Cherokees.[36] When Georgia persisted with the survey, threats were made. When the surveyors reached the area of present lower Chattooga County, they were accosted by a small party of Cherokees, who stated that they had been sent by Hicks and Path-Killer "to forbid . . . marking them trees." The leader of the group informed the survey party that his name was Unakah Tehee (meaning "white-man killer") and, stated the commissioners, he "appeared angry enough to establish a good title to such a name." The commissioners and surveyors, however, evinced their courage by coolly continuing with their work, and they heard no more from the Indians.[37]

32. Brantley, *Georgia Line—The Chain and Compass Line,* 4 Ala. Law. 282, 285 (1943); Goff, *Edward Lloyd Thomas, Surveyor,* 18 Emory U.Q. 107, 111 (1962).

33. Journal of the Commissioners of Georgia, appointed to run the dividing line between the boundaries of that State & Alabama 21 (GSGD); E. Coulter, Joseph Vallence Bevan: Georgia's First Official Historian 91–92 (1964).

34. Letter from George M. Troup to Hugh Montgomery (Feb. 6, 1826) (GLB, GDAH); Letters from Hugh Montgomery to George M. Troup (Feb. 18 & May 17, 1826) (GSGD).

35. Journal of the Commissioners of Georgia, appointed to run the dividing line between the boundaries of that State & Alabama 2 (GSGD).

36. Letter from Hugh Montgomery to George M. Troup (May 17, 1826) (GSGD).

37. Report from J. Crawford, R. Blount, & E. Hamilton to George M. Troup (Oct. 17, 1826) (GSGD).

"Chiefs of the Creek Nation and a Georgian Squatter." The figure on the left is Little Prince. From Basil Hall, *Forty Etchings, from Sketches Made with the Camera Lucida, in North America, in 1827 and 1828* (1829). Courtesy of Special Collections, University of Georgia Libraries.

After reaching Nickajack and finding the rock that James Camak and the Georgia-Tennessee boundary commissioners had set up in 1818 as the northwest corner of Georgia, the survey party ran the "true" line back to the point of beginning—a distance of 145 miles, 58 chains, and 50 links, across mountains, over rivers, past swamps, and through great forests—arriving within one chain of the beginning point on September 19. No wonder Colonel Blount wrote Governor Troup that Edward Lloyd Thomas was justly entitled to stand at the head of his profession for accuracy and correctness.[38]

For several years the state of Alabama protested the line as run by the Georgia commissioners, but in 1840 it gave up the dispute and ratified the line. It felt that a fixed and known boundary was of greater importance than the acquisition of an inconsiderable amount of territory. Arthur P.

38. Letter from Robert Blount to George M. Troup (Sept. 26, 1826) (GSGD); Goff, *Edward Lloyd Thomas, Surveyor,* 18 Emory U.Q. 107, 112 (1962).

Bagby, one of Alabama's boundary commissioners of 1826, had become governor of that state and instigated the action.[39]

James Bright was appointed by the secretary of war to run the west boundary of the Creek cession in accordance with the treaty signed in Washington on January 24, 1826. The line he marked during the months of August, September, and October 1826 became known as Bright's Line. It lay just east of the line that had been marked as the Georgia-Alabama boundary.[40]

Surveys for Original Lee, Muscogee, Troup, Coweta, and Carroll Counties

From the very first intimation of a new Indian treaty, Governor Troup had denied the power of the federal government to annul the Indian Springs treaty. Instead of waiting for the deadline set in the Treaty of Washington for Creek removal, Troup ordered the surveyors to begin laying out the county and district lines on September 2, 1826. The Creeks protested to the federal government about this intrusion, and on September 16 the secretary of war fired off a letter of protest to Troup. The letter was delayed in its delivery, however, and when it did arrive Troup was able to reply that some of the surveyors of the county and district lines had already completed their work and that all would be finished in a very short time. He added that the surveyors had suffered no interruptions.[41]

William D. Conyers, surveyor of Muscogee County, reported to the governor on September 18 that the course of the district lines he was laying off had taken him through three Indian villages and near three or four others. Curiosity had "prompted" the Indians "to come out & follow" his crew "in one or two instances [at] some distance but no hostile disposition appeared." "On the contrary," Conyers added, "they appeared to [be] ready and willing to supply us with whatever we want that

39. 1839 Ala. Laws 182; Brantley, *Georgia Line—The Chain and Compass Line,* 4 Ala. Law. 282, 288–92 (1943).

40. Letter from James Bright to George M. Troup (June 26, 1826) (GSGD); Report and field notes of Bright's Line, run between August 21 and October 5, 1826 (National Archives, Washington, D.C.; microfilm copy in GDAH).

41. Am. St. Papers, 2 Indian Affairs 744; E. Harden, The Life of George M. Troup 428, 459–61 (Savannah 1859); Hryniewicki, *The Creek Treaty of November 15, 1827,* 52 Ga. Hist. Q. 1, 2 (1968).

they have."[42] John McBride, who ran the district lines of Troup County, informed the governor that his progress had "been . . . much impeded by bad swamps" and that both before and after his arrival in his section he had been told repeatedly that the Indians would oppose the survey. Yet he had not heard the first complaint from any Indian and those he encountered had manifested "no other dispositions but those of the most amicable nature."[43] Thomas Mitchell, surveyor of Carroll County, could find nothing more to complain about than the wet weather.[44]

The subdivision of the districts into land lots commenced late in December. While laying off the 18th District of Lee County, B. C. Lansdell had to contend with a very strong local attraction which played havoc with his compass needle. Upon examining the earth, he found deposits of iron rock.[45] Much more serious difficulties were to follow, however.

The Treaty of Washington had left a strip of Georgia's territory in western Carroll County in Creek possession. Governor Troup, not considering this treaty valid, had ordered the surveyors to extend their surveys to include all the Creek lands in Georgia. On December 29, 1826, James Kendrick, surveyor of the 12th District of Carroll County, sent the following message to Troup:

> at camp 12th District 5 Section, December 29th 1826
> Sir
>
> Yesterday we were visited by six Indians from the Burnt Villiage with their Linguister, who, demanded by whose authority we were surveying this side of Brights Line. they had with them Brights letter to Col. Crowell and a Plat of the survey made by him. I had a long & friendly talk with them. they requested us to desist from surveying any further. I refused to do so unless they had written orders from the head Chiefs. they replied they would return to their Town where the Chiefs were then in siting and get such orders as would authorise us to stop. They appeared to be much displeased at the survey said they never would consent to it and from their gestures & menances leaves no doubt in any mind but they will do it. Should they do so I will immediately come on to Milledgeville and leave my men here. Mr. Jordon who came to my camp last night informed me that six of the oakfusky Indians had painted themselves and had been hunting me for some time to prevent my surveying and [to] take away my instruments. I have not

42. Letter from William D. Conyers to George M. Troup (Sept. 18, 1826) (GDAH).
43. Letter from John McBride to George M. Troup (Sept. 28, 1826) (GDAH).
44. Letter from Thomas Mitchell to George M. Troup (Nov. 12, 1826) (GSGD).
45. Letter from B. C. Lansdale to George M. Troup (Nov. 20, 1826) (GSGD).

seen them yet. Their Linguister also observed that had those six came to our camp during our absence they would have done much mischief

Yours Respectfully
James Kendrick
District Surveyor[46]

The party of Indians obtained written orders from Little Prince, head chief of the Creek Nation, demanding that the surveyors desist from stretching a chain west of Bright's Line. A few days later they broke up Kendrick's operation and seized his compass.[47] Little Prince also dispatched a letter to Troup protesting the actions of the surveyors, but Troup ignored the letter.[48] On January 7, 1827, seven or eight defiant Indians confronted Elijah Byne and his men, who were running the lines of the 13th District of Carroll County. They presented Byne with Little Prince's order and, when Byne refused to discontinue his work, took his compass. At the same time they disdainfully brandished before him the compass they had seized from Kendrick.[49]

On January 21, 1827, James A. Rogers and his men had just marked a pine station thirty-one chains east of the southwest corner of Land Lot 167 of the 9th District of Carroll County when they collided with an angry party of Creek Indians.[50] Rogers sent the following urgent message to Governor Troup explaining the matter:

January 23, 1827 his Exelency Geo M Troup
Sir Inclosed you will find a copy of an Instrument of writing that was handed to me by A parcel of Indians on the 21st Inst and after I red the letter they demanded of me my compass which I had to Surrender to them but after a few minutes they agreed to give me back my compas and would come with me over the New Treaty line which they did and Threatened me very Sovere if I Should be caught over Brights line again a Surveying. I have come on to McIntoshes old place and have Stopt my hands until I hear from you[.] if you Intend to give me assistance I hope you will take the Earliest measure to get it to me as provision is scarce and my hands uneasy to go home. as to the Number of Min it will take to guard me I am unable to Say[.] my Situation is

46. Letter from James Kendrick to George M. Troup (Dec. 29, 1826) (GSGD).

47. Letter from George R. Clayton to George M. Troup (Jan. 12, 1827) (GSGD).

48. Letter from Little Prince to George M. Troup (Jan. 1, 1827) (GDAH).

49. Letter from George R. Clayton to George M. Troup (Jan. 12, 1827) (GSGD).

50. Field notes of James A. Rogers, surveyor of the 9th District of Carroll County 52 (GSGD); J. Bonner, Georgia's Last Frontier: The Development of Carroll County 17–18 (1971).

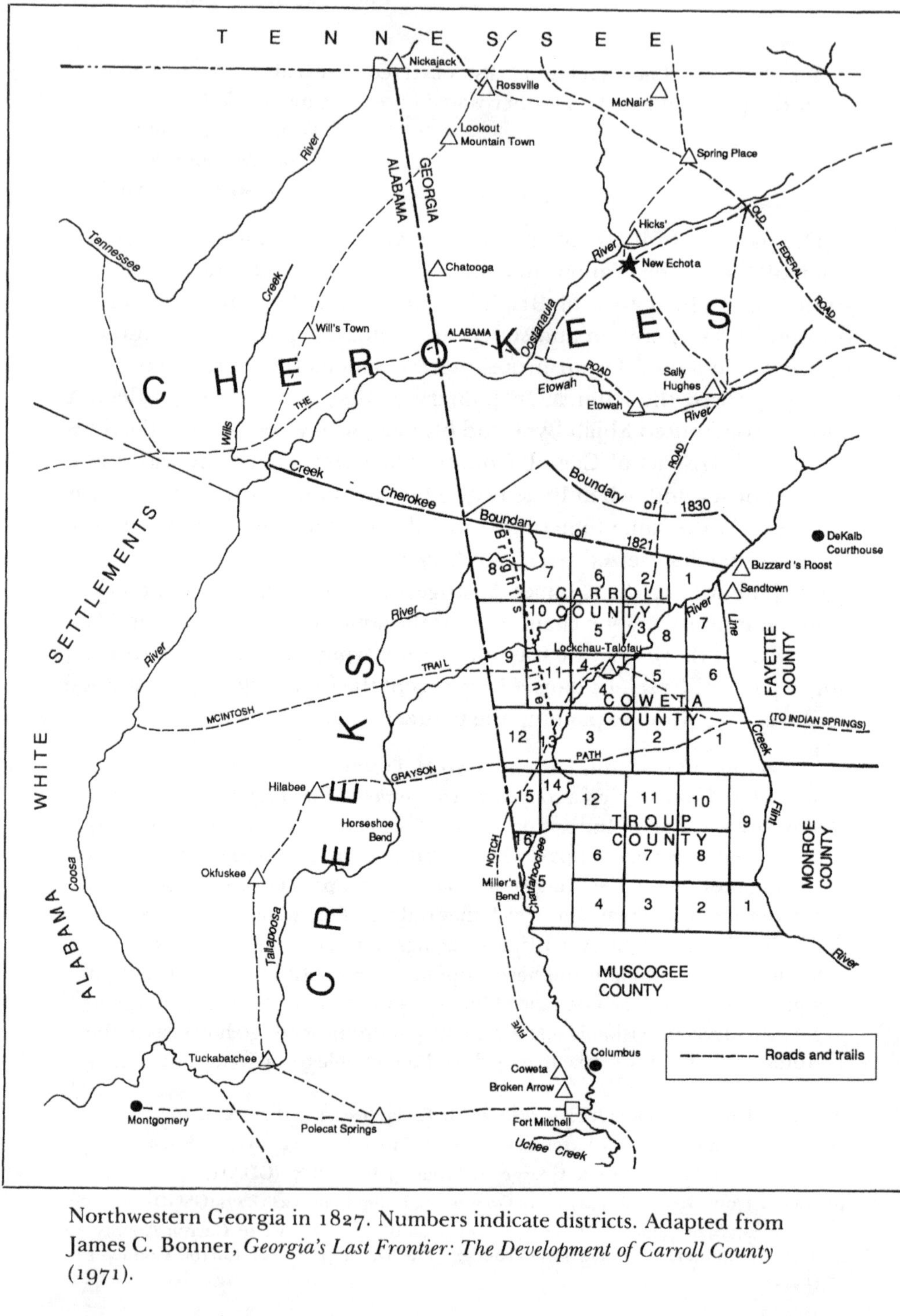

Northwestern Georgia in 1827. Numbers indicate districts. Adapted from James C. Bonner, *Georgia's Last Frontier: The Development of Carroll County* (1971).

this[.] there is three Settlements of Indians in my district that have in them about tin min and in two miles on the Alabama Side there is a Large Town that I am told have from forty to fifty wariors in it which is to be placed on the New Treaty line as Spies and [to] prevent the Land from being surveyed & as for further Information I will refer you to Maj[r] Parramore the bearur of this express and who was with me whin the Indians Stopt me

Your
James A. Rogers D.S

[enclosure]

To the Surveyors running the land west of the line of the late Creek Treaty

Creek Nation
January 12th 1827

Gentlemen We the undersigned Chiefs and head men of the Creek Nation having learnt with great regrett that you are engaged in Surveying the lands West of the line of the late Treaty and which was not Ceded by that Treaty we have again to request and demand of you in the Most friendly terms that you will desist from Stretching a Chain over any of our lands not Ceded by the Said Treaty[.] we flattered our Selves and we had a right to believe the Stipulation of that Treaty would have been friendly observ[d] by all parties concerned and that our former Differences with our Neghbours the Georgians would have been finally settl[d] and that we would in future live in the Strictest friendship in all our Intercourse[.] We are determined on our part not to do an act that can be Calculated to give offence but we [have] right[s] and we hope they will be respected

We are your friends
his
Little X Prince
mark

Signed in presence of
John Crowell
Agt. fo. I. A.

Oakfusky X Yoholo
Sparnee X Tustnugg Hargo
Eastichaeco X Chopco
Wicas X Hago
Coche X Hodgo
Carchus X Micue

James Istands Interpreter[51]

Wiley Williams, surveyor of the 11th District of Carroll, wrote the governor on January 15 that he had been informed of the action taken

51. Letter from James A. Rogers to George M. Troup (Jan. 23, 1827) (GSGD), *printed in* Am. St. Papers, 2 Indian Affairs 866.

Looking eastward along the south line of Land Lot 167, 9th District of Carroll County from the approximate spot where James A. Rogers and his survey crew were stopped by Creek Indians in 1827. The land lot line runs along the edge of the field, just left of the center of the picture. The area is located three miles southwest of Bowden. Photograph by the author.

against Byne and Kendrick. Williams crossed Bright's Line (which ran through his district) with every meridian he ran and was apprehensive that he might also be detained by the Indians since they had been in his area. Nevertheless, he offered to finish up Byne's district if Byne chose not to come back to complete the work.[52]

A week later Williams wrote that he had "flattered himself" that he would be able to finish his survey without trouble from the Indians, but in this he had been deceived. "Eight or ten lusty fellows" had ridden up to his camp the night before with a letter written by John Crowell and signed by several chiefs. The letter ordered him and his men to terminate their surveying activities west of Bright's Line. Williams asserted that the governor must provide protection for the surveyors that would be sufficient to guard the whole frontier line from the intersection of the Cherokee line with the state line, southward to Miller's Bend on the Chattahoochee. About one-fourth of Williams's district lay west of Bright's Line, and he had completed the running of the meridian lines at the time of this encounter with the Indians. Williams stated that he would commence running the parallels on the east side of Bright's Line and then attempt to survey the balance of the district, in the hope that a military force would be sent out in the meantime.[53] Frederick Williams, surveyor of the 16th District of Carroll, was apprehended by Indians on January 25 and reluctantly complied with their demand that he not proceed any farther.[54]

Upon being apprised of the Creeks' obstructionism, Governor Troup immediately ordered the state militia to be put on standby alert and requested the surveyors to continue their work. He vowed that a competent military force would be sent to protect them if necessary.[55] Troup also sent an angrily worded message to Little Prince denouncing him for having taken upon himself "to say what shall be the boundary line of Georgia" and for having "insulted my Surveyors—taken away their compasses and forced them to stop the survey." Demanding that he explain

52. Letter from Wiley Williams to George M. Troup (Jan. 15, 1827) (GSGD).

53. Letter from Wiley Williams to George M. Troup (Jan. 22, 1827) (GSGD), *printed in* Am. St. Papers, 2 Indian Affairs 865.

54. Letter from Frederick Williams to George M. Troup (Jan. 27, 1827) (GSGD). The General Assembly later compensated Frederick Williams $100 for losses he sustained in the survey. 1828 Ga. Laws 205.

55. Letters from George M. Troup to Col. Howan (Jan. 24, 1827), to Capt. Saffold (Jan. 24, 1827), to Lt. Nicholson (Jan. 24, 1827), to Wiley Williams (Jan. 26, 1827), and to Maj. Mitchell (Jan. 26, 1827) (all in GLB, GDAH).

these matters to his aide-de-camp "without delay," Troup concluded with a stern warning: "And now Little Prince I will talk plainly to you—you and your people have given me some trouble & if I catch you or your people on the land of Georgia interfering with our people I will give you some trouble too."[56] Both the chief executive of Georgia and Little Prince filed protests with the president.

In a special message to Congress on February 5, 1827, President Adams announced that under the federal act to regulate trade and intercourse with the Indian tribes[57] he had ordered U.S. marshals in Georgia to arrest any surveyors caught trespassing on Creek lands.[58] Without waiting for congressional action, Adams dispatched Lieutenant J. R. Vinton of the U.S. Army to Milledgeville with a message from the secretary of war stating that Troup must stop the survey of the lands beyond the treaty line or the United States would intervene with force.[59] The climax of the whole matter had now been reached. Never known for halfheartedness in word or deed, Troup fairly surpassed himself as a hotspur.[60] To the secretary of war's message, he replied:

> Sir: I received, this afternoon, from Lieut. Vinton, your letter of the 19th ult., and read, within the same hour, both it and the copy of it as published in the National Intelligencer of the 7th inst. No room was left to mistake the meaning of this dispatch. Lieut. Vinton announced himself . . . as the Aid of the Commanding General; and you are sufficiently explicit as to the means by which you propose to carry your resolution into effect. Thus the military character of the menace is established, and I am only at liberty to give it the defiance which it merits. You will distinctly understand, therefore, that I feel it to be my duty to resist, to the utmost, any military attack which the Government of the United States shall think proper to make on the territory, the people, or the sovereignty of Georgia; and all the measures necessary to the performance of this duty, according to our limited means, are in progress. From the first decisive act of hostility, you will be considered and treated as a public enemy; and, with the less repugnance, because you, to whom we

56. Letter from George M. Troup to Little Prince (Jan. 17, 1827) (GLB, GDAH).

57. This act provided that if any U.S. citizen should settle on Indian lands, attempt to survey such lands, designate boundaries by marking trees or otherwise, or perform any of various other specified acts, the offender would be subject to a fine or imprisonment or both. 2 Stat. 139.

58. Am. St. Papers, 2 Indian Affairs 862–63; H.R. Doc. No. 76, 19th Cong., 2d Sess. 1–5 (1827); 7 J. Adams, Memoirs of John Quincy Adams 219 (C. Adams ed. 1875).

59. Am. St. Papers, 2 Indian Affairs 865; E. Harden, The Life of George M. Troup 484 (Savannah 1859); Bonner, *Journal of a Mission to Georgia in 1827*, 44 Ga. Hist. Q. 74 (1960).

60. U. Phillips, Georgia and State Rights 62 (1901).

Drawing for the 1827 land lottery. Drawing by George I. Parrish, Jr. Courtesy of the Atlanta Historical Society.

> might constitutionally have appealed for our own defence against invasion, are yourselves the invaders, and, what is more, the unblushing allies of the savages whose cause you have adopted.[61]

The same day Troup ordered the attorney and solicitors general of the state to liberate any surveyors who might be arrested by United States officers and to make indictments against all such officers. Troup also ordered the mobilization of the militia and the collection of arms and ammunition to repel any hostile invasion of the state.[62] President Adams, apparently thoroughly intimidated by Troup's bombast, backed down and renewed negotiations with the Creeks for a cession of their lands in Georgia that had been left out of the 1826 treaty.[63]

61. Id. 63; H.R. Doc. No. 127, 19th Cong., 2d Sess. 5 (1827); E. Harden, The Life of George M. Troup 485–86 (Savannah 1859).

62. H.R. Doc. No. 127, 19th Cong., 2d Sess. 6 (1827); E. Harden, The Life of George M. Troup 486–87 (Savannah 1859).

63. H.R. Doc. No. 238, 20th Cong., 1st Sess. 7 (1828); E. Harden, The Life of George M. Troup 487–92 (Savannah 1859); Hryniewicki, *The Creek Treaty of November 15, 1827*, 52 Ga. Hist. Q. 1, 2 (1968).

Meanwhile, Little Prince, upon receiving Troup's insolent message of January 17 and perceiving his helplessness in the matter, had written Troup that the Creeks would not use force to prevent the survey beyond the treaty line and that the compasses his people had seized would be returned.[64] Needless to say, no surveyors were arrested,[65] and all were back at work in short order.[66]

Although some surveyors had not finished making their returns, the drawing of the names for the lottery commenced on March 6, 1827, and continued for sixty-nine days until all the prizes had been exhausted on May 25.[67] A final treaty was signed on November 15, 1827, in which the Creeks ceded the last of their lands in Georgia. Having relinquished the lands of their fathers, the Creeks departed for their new home beyond the Father of Waters.[68]

Frauds and Mistakes in the 1827 Land Lottery

Despite the whirlwind of controversy surrounding the survey of the Creek lands in 1826 and 1827, fewer problems or errors developed in these surveys than in those for previous lotteries. In April 1827, after the drawing for the lottery was well under way, it was discovered that the clerks who prepared the lottery tickets had designated a number of square lots as fractions and a number of fractions as squares. The latter had been placed in the wheel and some drawn as prizes. Also tickets had mistakenly been made for nonexistent square lots, some of which also had been drawn as prizes. Apparently the confusion had arisen because the tickets had been made before the completion of the surveys. Governor Troup instructed the surveyor general carefully to reexamine the returns of the surveyors to correct these errors. The names for which fractional and nonexistent lots had been drawn were placed back in the wheel to be drawn again.[69]

64. Letter from Little Prince to George M. Troup (Jan. 22, 1827) (GDAH).

65. E. Harden, The Life of George M. Troup 483n. (Savannah 1859); G. White, Historical Collections of Georgia 135 (New York 1855).

66. Letter from Samuel A. Bailey to George M. Troup (Feb. 13, 1827) (GDAH); Field notes of James A. Rogers, surveyor of the 9th District of Carroll County 53 (GSGD).

67. Executive Minutes (Mar. 3, 1827) (GDAH); M. Houston, Reprint of Official Register of Land Lottery of Georgia 1827 passim (1929).

68. C. Kappler, Indian Affairs (Treaties), S. Doc. No. 319, 58th Cong., 2d Sess. 284 (1904).

69. Executive Minutes (April 6, 1827) (GDAH); Letter from George M. Troup to the Commissioners of the Land Lottery (April 6, 1827) (GLB, GDAH).

Late in 1827 grantees of lots in the 1st District of Muscogee County (now in Macon County) complained of numerous errors. Examination proved that many of the lots reported by the surveyor as squares were actually fractions and vice versa. Also, some lots had been misnumbered on the trees and posts marking their bounds, and part of the district had not been run out at all. The legislature directed the governor to suspend the granting of lots in the district until the problem could be rectified and gave William P. Reid, the district surveyor, six months to resurvey the district. A year later, an act was passed to validate the grants of all fractions drawn in the lottery, to confirm the grants of those lots whose numbers had been changed by the resurvey, and to provide for the demarcation, by the county surveyor, of any lines in the district that still had not been surveyed.[70]

One other imperfection came to light: Samuel Torrence, surveyor of the 5th District of Muscogee, made a brief entry in his field notes stating that an overlap existed between the 5th and 6th districts of that county.[71] The area is today part of Chattahoochee County.

Creek-Cherokee Boundary

With the Creeks having ceded title to all their lands in Georgia, the state found it necessary to establish the south boundary of the Cherokee Nation—the line between the Cherokee possessions and the former Creek possessions. This boundary had originally been very nebulous, but in 1821 delegates from the Creek and Cherokee nations had agreed that the line should run from Buzzard's Roost on the Chattahoochee River (where Cobb and Douglas counties now corner) to the mouth of Wills Creek near present Gadsden, Alabama.[72] A white missionary living in the Cherokee territory at the time wrote: "Having no surveyor, a hunter, standing on the boundary at the Chattachoochee, pointed out the sup-

70. Letter from Henry Crowell to John Forsyth (Feb. 6, 1829) (GSGD); 1827 Ga. Laws 155; 1828 Ga. Laws 98; 1841 Ga. Laws 182; 1849–50 Ga. Laws 411–13; Cobb 692; 1827 Ga. House Journal 67–68; 1828 Ga. Senate Journal 89; Everett v. Towns, 17 Ga. 15 (1855).

71. Field notes of Samuel Torrence, surveyor of the 5th District of Muscogee County 3 (GSGD).

72. Laws of the Cherokee Nation: Adopted by the Council at Various Periods 70–72 (Tahlequah, Cherokee Nation 1852); Cherokee Phoenix, & Indians' Advocate (New Echota), Nov. 11, 1829, at 1, cols. 1–5; Extract of a Convention between the Creek and Cherokee Nations held 11 Decem. 1821, in letter from George M. Troup to John Quincy Adams (Feb. 11, 1826) (GLB, GDAH).

posed course to the mouth of Wills Creek. On this course a straight line was run by ranging, and they came out at the end of one hundred and forty miles within eighty rods of the intended place."[73] Evidently Georgia originally recognized this line as valid because its surveys under the 1825 land lottery act had been extended to it.

Nevertheless, in 1829 Georgia became embroiled in a controversy with the Cherokees over the location of the boundary. Georgia contended that the north boundary of the lands the Creeks had lately ceded ran from Suwanee Old Town on the Chattahoochee (in today's extreme northeast Fulton County) to Sixes Old Town on the Etowah (in present western Cherokee County), thence down that river to the Coosa.[74] Georgia employed Samuel A. Wales as commissioner and Edward Lloyd Thomas as surveyor to run this line.[75]

Because the Cherokees disputed the line run by Wales and Thomas, President Andrew Jackson directed General John Coffee of Alabama to investigate the matter. Coffee recommended that a third line, about midway between the lines claimed by Georgia and the Cherokee Nation, be adopted. The president approved Coffee's recommendation. The line was therefore drawn so that it began at the Shallowford on the Chattahoochee (below the bridge on present U.S. highway 19) and proceeded westerly along the divide between the waters of the Chattahoochee, Etowah, and Tallapoosa rivers.[76]

73. R. Walker, Torchlights to the Cherokees 191 (1931).

74. H.R. Doc. No. 2, 21st Cong., 1st Sess. 186–91 (1829); 1 The Papers of Chief John Ross 102–3, 169–76, 179–82 (G. Moulton ed. 1985); Goff, *Edward Lloyd Thomas, Surveyor,* 18 Emory U.Q. 107, 115 (1962); U.S. Bureau of Ethnology, Smithsonian Institution, Fifth Annual Report of the Bureau of Ethnology 269 (1883–84).

75. Letter from John Forsyth to Samuel A. Wales (Feb. 18, 1829) (GLB, GDAH); H.R. Ex. Doc. No. 89, 21st Cong., 1st Sess. 11–15 (1829); Athenian (Athens), June 2, 1829, at 2, col. 1, *quoted in* Savannah Georgian, June 6, 1829, at 2, col. 5; Cherokee Phoenix, & Indians' Advocate (New Echota), June 24, 1829, at 2, col. 1; Southern Recorder (Milledgeville), June 6, 1829, at 2, cols. 5–6, *quoted in* Savannah Georgian, June 11, 1829, at 2, col. 5; 1838 Ga. Senate Journal 96.

76. Goff, *Edward Lloyd Thomas, Surveyor,* 18 Emory U.Q. 107, 115 (1962).

Chapter 8. The 1832 Land and Gold Lottery

Land Lottery Act of 1830

With the Creeks expelled from their state, Georgians now turned their land-hungry eyes to the Cherokee lands that lay to the north. By the 1820s the Cherokees had made remarkable strides toward civilization. They had largely abandoned their former seminomadic state as hunters and fishermen and now farmed and raised livestock for a living. They had constructed schools, churches, sawmills, gristmills, roads, and ferries; had invented an alphabet; published a newspaper; and in many cases lived in houses as comfortable as those of their white neighbors.[1] Moreover, the Cherokees had set up a national capital at New Echota. In 1827 they adopted a written constitution asserting that they were a sovereign and independent nation with complete jurisdiction over their territory.[2]

Alarmed by these developments, the Georgia legislature in December 1828 passed an act that annexed the Cherokee country to Carroll, DeKalb, Gwinnett, Hall, and Habersham counties; declared all white persons in the Cherokee territory subject to Georgia law; and declared that after June 1, 1830, all Cherokee laws would be of no effect and all Indians residing in the Cherokee territory would become subject to Georgia law.[3] Although the United States Supreme Court eventually declared this statute void,[4] President Andrew Jackson, who was no friend of the red man, refused to enforce the decision and Georgia chose to ignore it.[5]

1. E. Boudinott, An Address to the Whites 8–9 (Philadelphia 1826); U.S. Bureau of Ethnology, Smithsonian Institution, Fifth Annual Report of the Bureau of Ethnology 260 (1883–84); Cherokee Nation v. State of Georgia, 30 U.S. (5 Pet.) 1, 8 L. Ed. 25 (1831).

2. H.R. Ex. Doc. No. 91, 23d Cong., 2d Sess. 10–19 (1835); Cherokee Phoenix (New Echota), Feb. 21, 1828, at 1.

3. 1828 Ga. Laws 88; Dawson 198.

4. Worcester v. State of Georgia, 31 U.S. (6 Pet.) 515, 8 L. Ed. 483 (1832).

5. 1831 Ga. Laws 259–61; 1831 Ga. Senate Journal 114–15; 1832 Ga. Senate Journal 10–13, 148–49; A. Johnson, Georgia as Colony and State, 278–79 (1938).

An unexpected development in 1829 hastened the course of events: gold was discovered in the Cherokee territory. Swarms of gold seekers poured into the region. Frontier lawlessness reigned supreme for many of the newcomers were vicious and profligate, bent on quick gain regardless of the hazards they might encounter. Freed from the restraints of a well-ordered society, they made life hideous with their gambling and drunken brawls.[6]

Anxious to restore order and to grab the region for white settlers, Governor George R. Gilmer called the legislature into session in October 1830, nearly a month ahead of schedule. Although the Cherokees had made no formal land cession, an act, summarized as follows, was passed the following December to survey and dispose of their lands to settlers:

Section 1. The territory to be divided into four "sections." That part lying east of a line beginning on the Georgia–North Carolina boundary "thirty six miles due West, from the North West corner of the first district" of Rabun County and running south to the Chattahoochee River to form the first section. That part lying between the first section and a line beginning "twenty-seven miles due West" from the last-mentioned corner and running "South to the Carroll [County] line, or to the boundary line dividing the organized, and unorganized parts of the State," to form the second section. That part lying between the second section and a line beginning "twenty-seven miles due West" from the last-mentioned corner and running "South until it strikes the Carroll [County] line or Alabama line" to form the third section. The balance of the territory to form the fourth section.

Section 2. Each section to be divided into districts "of nine miles square, as near as practicable, the district lines running parallel to the lines dividing sections, and crossed by other lines at right angles, and said districts so laid out, shall be again sub-divided by lines to be run in like directions into square tracts" containing 160 acres each, "marked and numbered, according to the plan heretofore pursued under the instructions of the Surveyor General."

Section 3. Fractional lots 100 acres in area and upward to constitute prizes to be drawn in the lottery along with full-size lots. Fractional lots less than 100 acres in area to be reserved from drawing and sold in such manner as a future legislature should direct.

6. A. Johnson, Georgia as Colony and State 275 (1938); W. McLoughlin, Cherokee Renascence in the New Republic 430–33 (1986); U. Phillips, Georgia and State Rights 72–73 (1901).

Section 4. Ninety-six district surveyors to be elected by the voters at large on the first Monday in February 1831. Surveyor general to determine by lot which district each surveyor should get. Members of the legislature disqualified as candidates to survey in the territory.

Section 5. Twelve persons to be elected by the legislature to run the section and district lines.

Section 6. Each surveyor to post a $10,000 bond.

Section 7. (Essentially identical to section 5 of the land lottery act of 1803 with the following exceptions and additions): Surveyors of section and district lines to complete their work within ninety days of the time they are notified to begin. District surveyors to complete their work within 120 days of the time they are notified to begin. District surveyors to note the extent of Indian improvements on each lot (see section 31 below).

Sections 8, 9. District surveyors to receive $2.50 for each mile of line run, out of which they must defray all expenses. Surveyors of section and district lines to receive $3.50 for each mile of line run, out of which they must defray all expenses. Governor to advance $300 to each surveyor before the work begins.

Section 10. (Essentially identical to section 18 of the land lottery act of 1821.)

Section 11. Surveyors, chainmen, and markers to take an oath before beginning their work.

Section 12. Surveyors to note the quality of each lot and to number the lots, "the enumeration to commence in all square districts in the Northwest corner and to run East."

Sections 13, 14, 15, 16, 18. The following entitled to one draw: each unmarried white male citizen eighteen years of age and older; each widow; each wife with children whose husband has been absent from the state for three years; each deaf and dumb or blind person; and each unmarried white female over the age of eighteen whose father was killed in the revolutionary war, War of 1812, or Indian wars. The following entitled to two draws: each white male citizen with a family; each veteran of the revolutionary war, War of 1812, or Indian wars who was not a successful drawer in previous lotteries; and each widow of such veterans. Each minor orphan or family of minor orphans, as well as children of convicts, entitled to one draw or two draws, depending on certain specified circumstances. Successful drawers in previous lotteries, convicts, residents of the Cherokee territory, and persons "either directly or indirectly concerned . . . with a certain horde of Thieves known as the Pony Club" excluded from participation.

Section 17. Right of navigation of streams in the Cherokee territory reserved to the state.

Sections 19, 20, 21, 22. (Essentially identical to sections 8, 9, and 10 of the land lottery act of 1803 except as to qualifications of drawers, which are given above.)

Section 23. Price of grants issued to successful drawers to be $18 each. Any grants not taken out within five years of the drawing to revert to the state. Successful drawers to pay taxes on lands drawn, whether or not a grant has been taken out (but see section 31 below).

Sections 24, 25, 26, 27, 28, 29. Grants and applications for grants made contrary to this act to be void.

Section 30. Governor to direct surveyors to proceed with their work as soon as possible. When surveys are completed, governor to direct lottery commissioners to proceed with the lottery (but see section 35 below).

Section 31. Until enacted to the contrary, no grant to issue for lots containing Indian improvements, nor shall successful drawers attempt to dispossess Indians from such lots. Successful drawers not required to pay taxes on such lots until entitled to obtain grants for them.

Section 32. Any person, whether Indians or others, who willfully cuts down or defaces marked trees, or removes landmarks made pursuant to this act, prior to the distribution of lands, to be guilty of a misdemeanor, and on conviction to be punished by confinement in jail or in the penitentiary for a term not exceeding four years.

Section 33. Any person who by force, menaces, or other means, prevents or attempts to prevent surveyors from running lines or performing any act required by this act, upon indictment and conviction thereof, to be sentenced to the penitentiary at hard labor for the term of five years.

Section 34. Governor to fill any vacancies of lottery commissioners.

Section 35. Governor to direct the election of justices of the peace and constables in the Cherokee territory upon the completion of section and district lines. Survey of districts suspended until Cherokees cede their territory to Georgia or until further enactment by the legislature.

Section 36. (Further details regarding selection of surveyors under sections 4 and 5 of this act.)

Section 37. Governor authorized and required to protect surveyors and to call out such portion of the militia as necessary for their protection should they be interrupted in the prosecution of their work.[7]

7. 1830 Ga. Laws 127.

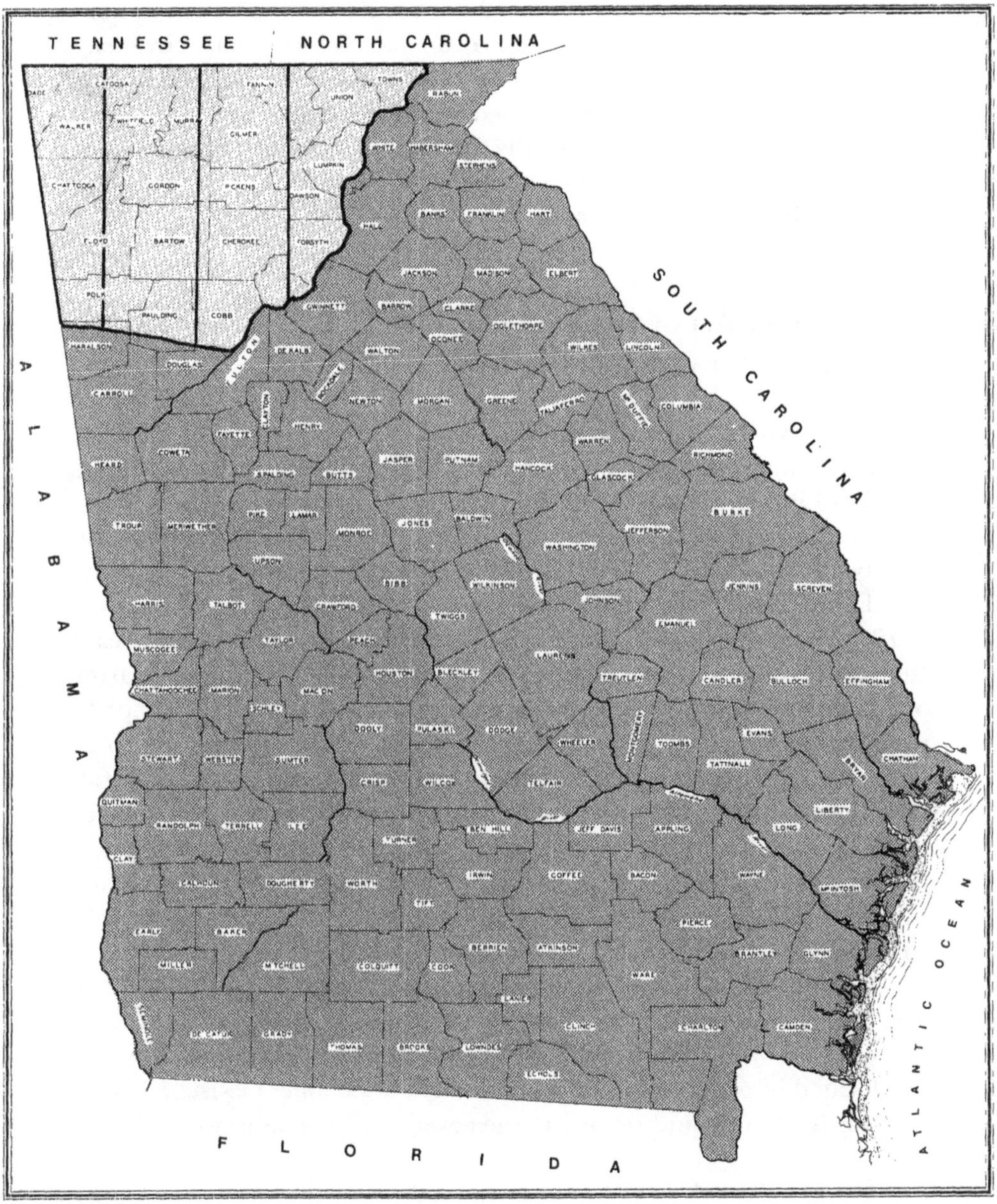

Lightly shaded area indicates lands distributed in the 1832 land and gold lottery.

The elaborate precautions in the above act for the protection of surveyors and their boundary marks are unique. Georgia had obviously benefited from experience gained four years before with the Creeks and was wary about surveying an unceded territory occupied by a semicivilized Indian nation. Undoubtedly the fact that the Jackson administration was now in power gave the state confidence that it could proceed in its designs without undue hindrance.

Gold Lottery Act of 1831 and Surveys for Original Cherokee County

Edward Lloyd Thomas, a Methodist preacher who was employed to run the section lines, commenced his work at the northwest corner of the 1st District of Rabun County on February 21, 1831.[8] Thomas had learned the art of surveying from his uncle Levin Wailes, who in 1804 had laid out the county and district lines of Baldwin and Wilkinson counties. Thomas assisted his uncle as chain carrier in the survey of the 2d District of Baldwin in 1805. When Wailes left Georgia for the Mississippi Territory two years later, he created a vacuum that Thomas largely filled. Within a quarter of a century Thomas's lines of demarcation ranged from the Oconee to the Chattahoochee rivers and from Tennessee to Florida; for, intermittently between his preaching, he found time to delineate the 4th District of Walton County, the county and district lines of Hall and Habersham counties, and the district lines of Henry County. In addition, he acted as troubleshooter when problems arose in the 1st District of Dooly County, surveyed the Georgia-Alabama boundary, and helped run the Creek-Cherokee boundary in 1829—a line later rejected by the government in Washington. Immediately upon completing the section lines of Cherokee, Thomas attempted to survey the troublesome Florida line. He also laid out the towns of Columbus, Oxford, and Salem (the latter of which was located in present-day Oconee County but no longer exists).[9]

In his field notes and reports to the governor, Thomas graphically de-

8. Field notes of Edward Lloyd Thomas, surveyor of the section lines of Cherokee County 2 (GSGD).

9. Goff, *Edward Lloyd Thomas, Surveyor,* 18 Emory U.Q. 107, passim (1962). Thomas's simple, two-story frame house, built in 1837, still stands on a beautiful knoll in Oxford, the town he designed and laid out. It has been lovingly restored by its present owners. Thomas is buried in the nearby public cemetery.

scribed the adversities that beset him and his men in the rugged north Georgia mountains. Initially, bitterly cold weather delayed the start of their work; then he and his party had difficulty finding the marked line that ran westward from the northwest corner of Rabun County to divide Georgia from North Carolina. In addition, they had to contend with rain, sleet, and snow. Strong local attraction deflected their compass needle. At one point they were halted by a steep, ice-covered mountain that lay directly on the line. Unable to climb it, they had to traverse around it. Another time they spent two nights without tents or cover when their packhorses could not ascend the precipitous mountain that Thomas and his companions were camped on. The party, however, enjoyed good health, found provisions easy to obtain, and suffered no interruptions from the Indians.[10]

Thomas completed the section lines early in April. He wrote the governor that because of an error in the 1818 survey of the north boundary of Georgia, the line was shorter than supposed. As a result, there was only enough land area in the Cherokee territory to make ninety-three districts instead of the ninety-six contemplated in the land lottery act.[11] The running of the district lines had commenced in March, and most of these lines were completed about a month and a half later.[12]

In November 1831 Governor Gilmer announced to the legislature that the law for surveying the Cherokee territory into sections and districts had been executed without difficulty, though one of the sectional surveyors—Benjamin H. Sturges—had failed to perform his job and to account for the money he had been advanced.[13] Sturges had subcon-

10. Field notes of Edward Lloyd Thomas, surveyor of the section lines of Cherokee County passim (GSGD); Letters from Edward Lloyd Thomas to George R. Gilmer (Feb. 28 & Mar. 5, 1831) (Georgia–North Carolina Boundary File, Telamon Cuyler Collection, UGL); Athenian (Athens), April 19, 1831, at 2, col. 3; *quoted in* Georgian (Savannah), April 25, 1831, at 2, col. 3.

11. Letter from Edward Lloyd Thomas to George R. Gilmer (Feb. 28, 1831) (Georgia–North Carolina Boundary File, Telamon Cuyler Collection, UGL); 1831 Ga. Laws 253–54; 1831 Ga. Senate Journal 169–70. It will be recalled (pages 117–18) that, when the Georgia-Tennessee boundary was run in 1818, the commissioners began at the northwest corner of Georgia and ran the line eastwardly. They terminated at a point considerably east of the Tennessee–North Carolina boundary—apparently believing they were at the point where that boundary intersects the north boundary of Georgia. When the distance they measured was subsequently scaled on maps in a westerly direction from the correct point, it gave an exaggerated impression of the length of Georgia's north boundary. This led to the error in calculating the number of districts that could be formed from the Cherokee territory.

12. Field notes of the district lines of Cherokee County in GSGD.

13. 1831 Ga. Senate Journal 14.

tracted the work out to Hardeman Owens, another sectional surveyor. Thomas surmised that Sturges had been "intimidated . . . by the difficulty of" surveying over "the mountains that nearly" covered his section. In addition, the gubernatorial election was approaching and Sturges apparently had felt it more expedient to spend his time stumping for one of the candidates than surveying his section. (The candidate eventually lost.)[14]

In December 1831 Wilson Lumpkin, who succeeded Gilmer as governor, recommended that the surveys of the Cherokee lands be completed and that the area be readied for white settlement.[15] Soon thereafter the legislature directed the district surveyors to begin their work on April 1, 1832, and the lottery commissioners to commence the drawing upon completion of the surveys.[16] The legislature also passed amendments, summarized as follows, to the land lottery act of 1830:

Section 1. Districts 1–5 and 11–15 of the First Section; Districts 1–3, 15–19, and 21 of the Second Section; Districts 1–4 and 17–21 of the Third Section; and Districts 1–3, 16, and 17 of the Fourth Section to be subdivided into "lots of forty acres each, by lines running parallel with the district lines, at the distance of twenty chains apart, and crossed by other lines, at right angles, and of the like distance from each other, marked and numbered according to the plan prescribed by the Surveyor General."

Section 2. Each white male citizen eighteen years of age and older, each widow, and each minor orphan or family of minor orphans entitled to one draw. Each head of a family entitled to two draws.

Sections 3, 4. (Essentially identical to sections 8, 9, and 10 of the land lottery act of 1803 except as to qualifications of drawers, which are given above.)

Sections 5, 6. The lands contemplated by this act to be drawn in a separate lottery from that contemplated by the land lottery act of 1830. The same commissioners to preside over both lotteries, and both lotteries to begin at the same time. Price of grants issued in this lottery to be $10 each.

Section 7. All laws militating against this act repealed.[17]

14. Letters from Edward Lloyd Thomas to George R. Gilmer (Feb. 28 & Mar. 5, 1831) (Georgia–North Carolina Boundary File, Telamon Cuyler Collection, UGL).

15. 1 W. Lumpkin, The Removal of the Cherokee Indians from Georgia 95–102 (1907).

16. 1831 Ga. Laws 141.

17. Id. 164.

Wilson Lumpkin. Courtesy of the University of Georgia School of Law and University of Georgia Libraries.

At the same session an act was passed to form into one county, called Cherokee, all Cherokee lands that had been divided between Georgia's five northern counties in 1828.[18] When the district surveyors began their work in April 1832, Governor Lumpkin gave assurances that they would be fully protected.[19] Before they arrived in their districts, he sent General John Coffee with a detachment of militia through the Cherokee lands as a show of force and a warning to the Indians of dire consequences if they should take any action against the surveyors.[20] Considering that the territory was inhabited by semicivilized Cherokees who had made no land cession, it is remarkable that there was not a mass outbreak of violence against the surveyors. Indeed, if one is to believe the dispatches of the surveyors themselves, they went to their respective districts with little fear of serious difficulties from the Indians.

General Coffee did report that surveyors from the 6th and 11th districts, 2d Section, had informed him that Indians were defacing numbers and "hiding the corners." Coffee was somewhat at a loss as to what to do but decided to send part of his militia force to demand that the chiefs reveal the names of the guilty parties. He realized that such information probably would not be divulged but hoped at least to alarm them sufficiently that they would desist.[21]

18. Id. 74.

19. Letters from Wilson Lumpkin to William S. Tarvin (April 3, 1832), and to John Coffee (April 5, 1832) (both in GLB, GDAH).

20. Letter from John Coffee to Wilson Lumpkin (May 6, 1832) (Telamon Cuyler Collection, UGL).

21. Letter from John Coffee to Wilson Lumpkin (June 17, 1832) (Telamon Cuyler Collection, UGL). An unidentified Cherokee described the arrogant indifference of the surveyors toward the Indians. In a letter to the editor of the *Cherokee Phoenix* he reported that on the preceding Monday morning (May 21, 1832) he and his family had been eating breakfast near the back door of their dwelling. Suddenly a force of Georgia militia came into view. He, "being at peace with all mankind," felt no cause for alarm and continued with his meal. Very shortly a "file of" surveyors "appeared at the door" and "signified something by a nod of the head." Then, "armed with an axe" and a "long chain," they drove a "spear" (clearly, a pole on which compass sights were taken) into the ground within three feet of his back door. On this "spear . . . a cap was placed and closely examined by the one who bore it." The surveyors next "marked deep into the wall" of his house "on each side of the door" large figures to designate the land lots—"93 and 124." Their "countenances . . . indicated the absence of justice and the bitter gnawing of conscience in trespassing in this manner on the possessions of a harmless Cherokee." A short distance away the "axe was . . . raised, and on each side of a beautiful shade tree" that he had "improved with some pains," they chipped away the bark and deeply inscribed the same figures. "This finished, the chain was resumed and a western course taken." They soon "disappeared in the heavy Indian

The only significant difficulty with Indians was suffered by Frederick A. Brown, who laid off the 10th District, 3d Section.[22] On June 22, 1832, a band of armed Cherokees arrested Brown while he was encamped on the farm of an Indian named Barnet. Brown's men were allowed to remain behind but were warned not to do any more surveying. Brown was taken to Athens, Tennessee, and charged with violating the federal act regulating trade and intercourse with the Indian tribes. This was the same act under which President Adams had attempted to have the Georgia surveyors arrested five and a half years before. Judge Keith of the U.S. District Court of Eastern Tennessee, however, ordered Brown released because he lacked jurisdiction in the matter. Brown returned to his district after a week's absence and continued with his survey.[23]

General Coffee obtained evidence that John Ross, one of the head chiefs of the Cherokee Nation, had ordered the arrest of Brown. Nevertheless, Coffee declined to take action against Ross because the United States was endeavoring to negotiate a treaty of cession with the Cherokees.[24] Governor Lumpkin agreed with this course of action.[25] Ross had apparently hoped that, by arresting Brown, he could bring before the federal courts the issue of the right of Georgia to survey the Cherokee lands.

Most of the surveys for Cherokee County were completed by early July 1832.[26] The tedious job of compiling the returns and making out lottery tickets was immediately begun. This lottery was in two parts—one a "land" lottery to dispose of the 160-acre lots and fractional lots of 100 acres or more and the other a "gold" lottery to dispose of the

---

forests with their calf skin leggings." Cherokee Phoenix & Indians' Advocate (New Echota), May 26, 1832, at 2, col. 5.

22. Letter from Thomas Porter to Wilson Lumpkin (May 10, 1832) (GDAH); Letter from Horace R. Ward to Wilson Lumpkin (June 28, 1832) (Telamon Cuyler Collection, UGL); Letters from Wilson Lumpkin to Wm. W. Williamson (May 3, 1832), to John Coffee (May 10 & 19, 1832), and to Thomas Porter (May 19, 1832) (all in GLB, GDAH).

23. Letter from John Coffee to Wilson Lumpkin (July 21, 1832) (Telamon Cuyler Collection, UGL); Letter from F. A. Brown to Wilson Lumpkin (Sept. 15, 1832) (GDAH); 1832 Ga. Laws 23; 1832 Ga. House Journal 107.

24. Letter from John Coffee to Wilson Lumpkin (July 7, 1832) (GDAH).

25. Letter from Wilson Lumpkin to John Coffee (July 14, 1832) (GLB, GDAH).

26. Field notes and district maps for Cherokee County in GSGD; Letters from Wilson Lumpkin to N. C. Barnett (Sept. 25, 1832), and to Josias Beall (Sept. 25, 1832) (both in GLB, GDAH).

40-acre lots. The 40-acre lots became popularly known as "gold lots," though very little gold was ever found on most of them. The drawings for both the land and gold lotteries began on October 22, 1832.[27] Eighty-five thousand names had been sent in for the land lottery, for which there were 18,309 prizes, or about three and a half blanks per prize. One hundred thirty-three thousand names had been sent in for the gold lottery, for which there were 35,000 prizes, or about four blanks per prize. The wheels were so large in circumference and so heavy that a strong man could hardly turn them. They had been built in the state penitentiary, reputedly using the labor of missionaries held there for preaching in the Cherokee Nation without a license and for agitating in favor of Cherokee rights. These rumors proved to be unfounded, however.[28]

As the lottery progressed, a veritable swarm of fortunate drawers descended upon the Cherokee Nation. The *Cherokee Phoenix* reported that they "have been passing and repassing single and in companies." The intruders asked questions such as "Ho, sir! where is the nearest line to this place, what district, number, corner, lot, station &c.?" The newspaper continued: "When we see the pale faces again, they are closely viewing the marked trees, and the carved posts. The gold drawers . . . are compared to the great flocks of pigeons that hasten to the ground in search of their food. Every lot has been viewed, and as many paths beaten, by the passing and the cross passing hunters."[29] James F. Smith even found it profitable to publish a book providing diagrams of each district containing 160-acre lots and the names of the successful drawers of each lot.[30]

All the land lots had been drawn by February 18, 1833, and all the gold lots by May 1, 1833.[31] The demand for the lots was so great that a special lottery was held later in the year to dispose of fractional lots in the Cherokee territory and the reverted and undrawn lots left over from previous lotteries. The tickets with names of applicants not drawn in the

27. Federal Union Office, Gold and Land Lottery Register passim (Milledgeville, Ga. 1833) (DeRenne Collection, University of Georgia Libraries, Athens, Ga.).

28. Niles' Weekly Register (Baltimore), Oct. 27, 1832, at 131, col. 2; Id., Nov. 24, 1832, at 196, col. 1.

29. Cherokee Phoenix & Indians' Advocate (New Echota), Dec. 8, 1832, at 2, col. 4; *quoted in* Niles' Weekly Register (Baltimore), Jan. 26, 1833, at 363, col. 2.

30. J. Smith, The Cherokee Land Lottery (New York 1838). This book, reprinted in 1968, is a handy reference for basic diagrams of land districts in original Cherokee County.

31. Federal Union Office, Gold and Land Lottery Register passim (Milledgeville, Ga. 1833) (DeRenne Collection, University of Georgia Libraries, Athens, Ga.).

lottery of 1832–33 were drawn in this lottery and the successful drawers allowed to take out grants in the usual manner.[32]

In 1832 the legislature divided Cherokee County into ten counties.[33] Any lots the surveyors reported as containing Indian improvements were temporarily withheld from granting. Nevertheless, many lots that the surveyors reported as unimproved were found to contain Indian improvements. When successful drawers attempted to dispossess the Indians from these lots, friction naturally resulted.[34] In 1835 and 1836 the legislature directed the governor to issue grants for all lots that had been drawn in the lottery.[35] The acts also authorized fortunate drawers of lots with Indian improvements to take possession of their land after May 25, 1838.

During all these events, negotiations for a cession had been going on in Washington and in the Cherokee Nation. Internal dissension among the Cherokees hampered their efforts to prevent the whites from occupying their lands.[36] After many pathetic but futile appeals to federal officials, the inevitable treaty was signed at New Echota on December 29, 1835. By its terms, the Cherokees ceded all their lands east of the Mississippi River. In return they received 7 million acres of land in what is now Oklahoma, expenses for their removal, liquidation of their debts to the United States, a cash payment of $5 million, and certain other considerations.[37] The Cherokees were given two years from ratification by the U.S. Senate to vacate the lands.

In the spring and summer of 1838 the U.S. Army under General Winfield Scott rounded the Cherokees up into camps, and that fall they were herded away to join their brethren in the West. Thus was consummated the final act of extirpating the aboriginal race from the soil of Georgia.[38]

32. Executive Minutes (Dec. 11, 1833) (GDAH); 1833 Ga. Laws 126, 392; Georgian (Savannah), Dec. 23, 1833, at 2, col. 3.

33. 1832 Ga. Laws 56.

34. Letters from Wilson Lumpkin to H. G. Clander (Jan. 3, 1833), to Isaac N. Johnson (Jan. 17, 1833), to Mark Castleberry (April 26, 1833), and to Charles H. Nelson (Mar. 21, 1835) (all in GLB, GDAH).

35. 1835 Ga. Laws 105; 1836 Ga. Laws 145.

36. G. Moulton, John Ross: Cherokee Chief 38–106 (1978).

37. C. Kappler, Indian Affairs (Treaties), S. Doc. No. 319, 58th Cong., 2d Sess. 439, 448 (1904).

38. S. Doc. No. 1, 25th Cong., 3d sess. 440–43, 459–66 (1838); Parks v. Ross, 52 U.S. (11 How.) 362, 13 L. Ed. 730 (1850); G. White, Historical Collections of Georgia 151–53 (New York 1855); T. Wilkins, Cherokee Tragedy 304–15 (1970); McLoughlin, *The Reverend Evan Jones and the Cherokee Trail of Tears,* 73 Ga. Hist. Q. 559 (1989).

Frauds and Mistakes in the 1832 Land and Gold Lottery

The lottery drawing was temporarily suspended in November and December of 1832 while state officials investigated charges of fraud against Shadrach Bogan, one of the lottery commissioners. The procedure was for a little boy to draw a ticket from the wheel containing the names of applicants and then for another little boy to draw a ticket from the wheel containing lot numbers. The tickets were then handed to a commissioner, who pasted them together back to back. Finally, they were distributed to clerks for entry into the books. It had been discovered that the handwriting on the tickets for five of the most valuable prizes was different from that of the clerks who had written the names on the other tickets.[39] Examination proved that Bogan had covertly pasted the names of certain of his friends and relatives, whose chances in the lottery he had purchased, on these five lottery tickets. The state Senate convicted Bogan of the charges, and he was barred from holding any state office or trust for twenty years. The five prize tickets were returned to the wheel to be drawn again.[40]

A novel method was employed in laying off the 13th District, 1st Section, of Cherokee (now in Lumpkin and Dawson counties). Since the Cherokee territory contained only ninety-three districts instead of the ninety-six specified in the land lottery acts, three more district surveyors had been chosen than were needed. Therefore, under an 1831 House resolution, the surveyor general was authorized to divide any three gold districts into two parts and appoint two surveyors to each of these districts.[41] Apparently under authority of this resolution, the 13th was divided into a northern and a southern half. Charles Smith took the

39. The five prizes in question were Land Lot 92, 14th District, 3d Section, situated at the junction of the Conasauga and Coosawattee rivers, containing the New Town Ferry; Land Lot 124, 14th District, 3d Section, containing the seat of government of the Cherokee Nation—New Echota; Land Lot 150, 15th District, 3d Section, containing a gristmill and sawmill; Land Lot 196, 23d District, 3d Section—the home of the Cherokee Chief Major Ridge, consisting of a ferry, a storehouse, and "a good two story building"; and Land Lot 244, 23d District, 3d Section, situated at the junction of the Hightower Trail and Oostanaula River. Estimates placed the value of each of these lots between $1,000 and $3,000. 1832 Ga. Senate Journal 302–3, 314–15.

40. 1832 Ga. Senate Journal 301–40.

41. 1831 Ga. Laws 253, 316.

northern portion and Walter Nunnelee took the southern. Smith began on the north line of the district and worked southward. In the end he left a narrow strip of land (varying in width from 1.25 to 4 chains) between his last range of lots and the north line of the southern portion of the district. This strip extended the full width of the district from east to west. Smith then went home, whereupon Governor Lumpkin scolded him and instructed him to return to his district and divide the strip into fractions. Walter Nunnelee, upon finishing the last range of lots in the southern portion of the 13th, discovered that he had inadvertently misnumbered the trees and posts from the sixth range southward. Although weary of his exertions, he nevertheless was obliged to go back and perform the tiresome task of chopping out the incorrect numbers and replacing them with correct numbers. The lots in the northern half of the 13th District (including fractions) are numbered 1 through 578; the lots in the southern half are numbered 1 through 554. Some confusion arose among successful drawers as to which of the two lots bearing the same number they could claim.[42]

In 1841 the legislature learned that, apparently because of the extreme ruggedness of the terrain, Horace R. Ward had left many of the lines and corners unmarked in his survey of the 19th District, 4th Section, of Cherokee (now Dade) County. Ward was instructed to return and finish the survey by April 1, 1842. If he failed to do so, the county surveyor of Dade was to complete the job. The governor was directed to put Ward's bond in suit.[43] Fifty-four years after Ward's original 1832 survey, an appeal was brought before the Georgia Supreme Court regarding certain boundaries in the district which had become confused because of the two surveys.[44]

For a number of districts, among other problems, an excess or deficiency in area was encountered which had to be dealt with. In April 1832, when James Sellers and his men began work on the 15th District, 3d Sec-

42. Field notes and district maps of Charles Smith & Walter Nunnelee, surveyors of the 13th District, 1st Section, of Cherokee County, in GSGD; Letter from B. L. Goodman to Wilson Lumpkin (Sept. 4, 1832), and Letter from Walter Nunnelee to Wilson Lumpkin (Sept. 25, 1832) (both in GSGD); Letters from Wilson Lumpkin to Charles Smith (Sept. 5, 1832), to B. L. Goodman (Sept. 20, 1832), to Walter Nunnelee (Sept. 20, 1832), and to Edmund B. Adams (April 25, 1833) (all in GLB, GDAH).

43. 1841 Ga. Laws 65.

44. Day v. Case, 78 Ga. 58 (1886).

tion (now in Gordon and Bartow counties), they found that the sectional surveyor had failed to place the northwest corner of the district and to mark most of the district's west line. They also found the district (which was a square) to be deficient in size in both its north-south and its east-west dimensions. As a result, a total of 225 full-size land lots were laid off instead of the usual 324, and a tier of fractional lots was laid off along the south and east lines of the district.[45]

An alternate method of dealing with excess and deficiency was employed in the 10th District, 3d Section (now in Whitfield and Murray counties). James Croker, who had assisted Frederick A. Brown in laying out this district, reported that he "frequently found a variation of from 5 to 8 & 10 Chains in closing on the district lines." In his district the error had been apportioned throughout, but, as far as Croker could learn, such errors generally had not been compensated for in other districts.[46]

One surveyor who apparently did follow the example of Croker was John Lawhon, who subdivided the 18th District, 2d Section (now in Cobb County). Seemingly this district was much larger than the prescribed size. As a result, Lawhon added a half-chain to each dimension of each land lot in the district.[47]

In districts where the procedures employed by Sellers, Croker, and Lawhon were not followed, the sizes of the lots often show wide discrepancies from the standard. For example, Land Lot 324, 9th District, 4th Section (now in Walker County), is found to contain 127.9 acres instead of 160 acres.[48] Land Lot 116, 4th District, 4th Section (now in Floyd County), by contrast, is found to contain 205¾ acres instead of 160 acres.[49]

Governor Wilson Lumpkin, who was well disciplined in the art of land measuring, recognized the problem of excess and deficiency. On the eve

45. Field notes and district map of James Sellers, surveyor of the 15th District, 3d Section, of Cherokee County, in GSGD.

46. Letter from James Croker to Wilson Lumpkin (July 30, 1832) (GSGD).

47. Interview with Roger L. Cordes, Georgia Registered Surveyor (Aug. 17, 1979); Plat of resurvey of Land Lots 53, 54, 55, 56, 57, 58, 60, 61, 62, 63, and 64 of the 18th District, 2d Section, of original Cherokee, now Cobb County, Georgia, by Roger L. Cordes (Oct. 19, 1976); 1973 Ga. Laws 2710.

48. Oliver v. Daniel, 202 Ga. 149 (1947). The lot, located at the corner of the district, sustained deficiencies from both directions that resulted from errors in closing on the district lines.

49. Ware v. McCall, 110 Ga. 574 (1900).

of the lottery drawing he wrote: "Many of the Surveyors, who had what were considered square districts, have made and returned fractions." Conceding that the law had not contemplated the existence of fractions in square districts, Lumpkin nevertheless declared that it was "right and proper that the whole country should be surveyed and an accurate return . . . made of the whole, whether it be in full lots or fractions."[50]

50. Letter from Wilson Lumpkin to Charles Smith (Sept. 5, 1832) (GLB, GDAH).

Chapter 9. The Reserves

As mentioned in the beginning of Chapter 1, several isolated tracts of land in Georgia are not laid out under either the headright system or the land lot and district system. Following is a brief description of each of these tracts.

Milledgeville

Milledgeville was reserved from granting under the land lottery act of 1803 and served as the state capital from 1807 until 1868. Commissioners were appointed, who, in turn, employed Benjamin Easley to lay out the town. Easley designed Milledgeville in checkerboard fashion with streets 100 feet wide, except Washington and Jefferson streets, which latter had parkways running down their centers and a total width of 120 feet. The original town was composed of eighty-four squares of four acres each and four public squares of twenty acres each. Each four-acre square was divided into four one-acre residential lots. A common encompassing the original town was reserved for the city's future expansion. Altogether Milledgeville with its common covers 3,290 acres.[1]

Macon and Fort Hawkins Reserves

By the Treaty of Washington in 1805, the Creeks ceded their lands between the Oconee and Ocmulgee rivers but reserved a tract, roughly three by five miles in area, on the east side of the Ocmulgee. By the same treaty the United States was given the right to build a military post and trading house on this reserved tract. Subsequently, Fort Hawkins

1. A Plan of Milledgeville the Capital and permanent seat, of the Government of the State of Georgia . . . Copied from the original Plan on the 2nd day of September 1808, by Daniel Sturges (GSGD); Letter from John Clark to John Milledge (Sept. 27, 1804), *printed in* H. Salley, Correspondence of Governor John Milledge 122 (1949); J. Bonner, Milledgeville: Georgia's Antebellum Capital 17 (1978).

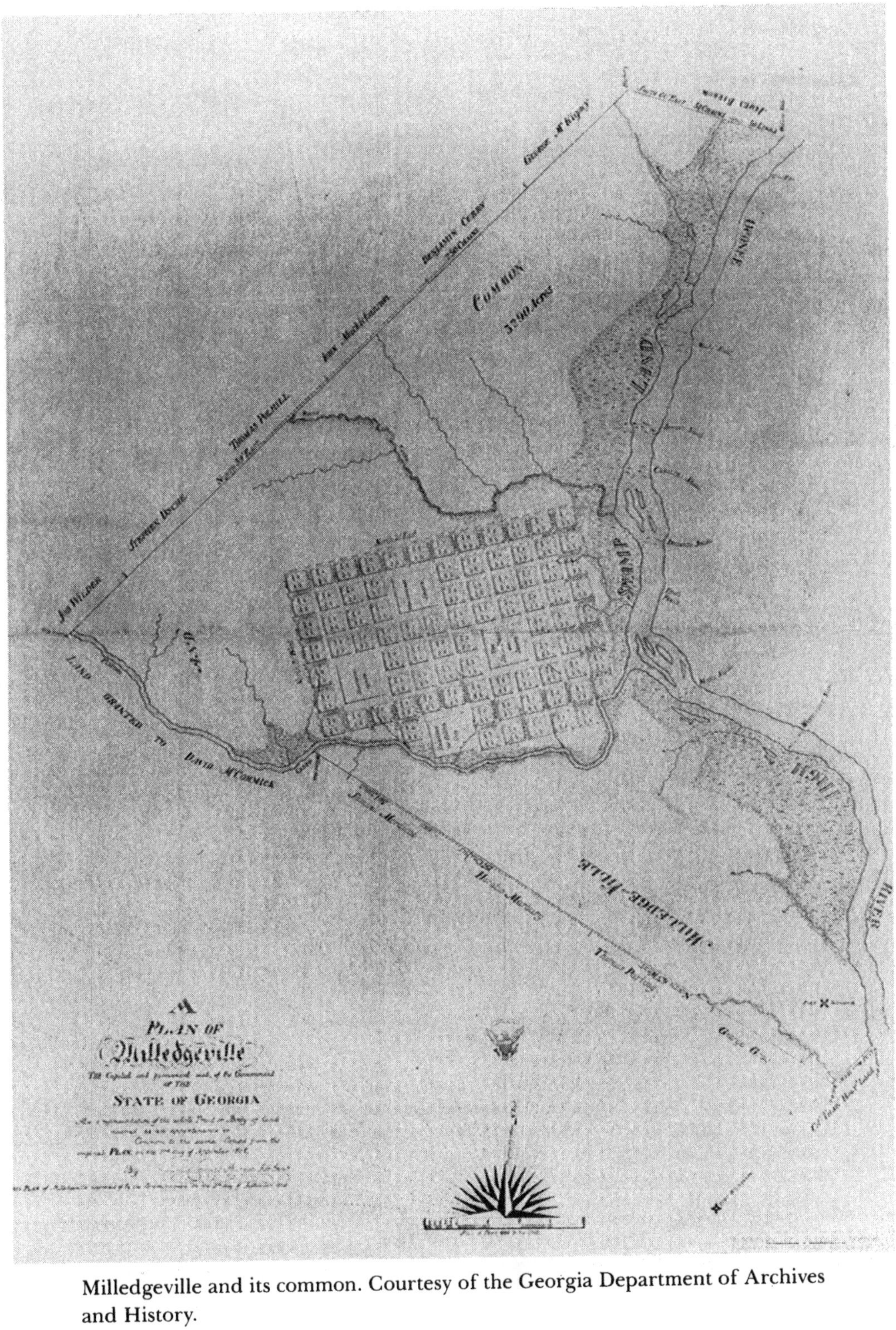

Milledgeville and its common. Courtesy of the Georgia Department of Archives and History.

was built on the site, and the tract became known as the Fort Hawkins Reserve.

In 1821, when the Creeks ceded their lands between the Ocmulgee and Flint rivers, they also relinquished their title to the Fort Hawkins Reserve. Realizing that this site at the head of navigation of the Ocmulgee was an ideal location for a town, the legislature exempted the Fort Hawkins Reserve and an equal quantity of land on the opposite side of the river from granting under the 1821 land lottery act.[2] An act of 1822 provided for the laying out of a town on the west side of the river, to be called Macon.[3] Five commissioners were appointed to oversee the surveying and auctioning of the lots.[4] They assembled at the site on January 20, 1823, and promptly appointed James Webb as surveyor. Work was begun a week later. Webb demarcated seven streets perpendicular to the river with alternating widths of 180 feet and 120 feet. Crossing these at right angles, he demarcated seven more streets with the same alternating widths. The wider streets apparently were designed to have tree-lined parkways running down their centers and along each side. One of the streets perpendicular to the river, Fifth, was carefully aligned with the site of a bridge that was to be built across the Ocmulgee. A four-acre square, interrupting the intersection of Fifth Street with one of its cross streets, Mulberry, was set aside for the erection of a courthouse. The simple gridiron of Macon was modified by Cotton Avenue, which cut diagonally across two of the blocks. Cotton Avenue had begun as an Indian trail, developed into a traders' path, and then became part of the Federal Road which ran through middle Georgia. It endured in the plan of Macon, as if to mock the usually effacing grid. As in the case of Milledgeville, most blocks in Macon contained four acres each. One entire block was reserved for an academy and another for a church or graveyard. Other blocks were divided by two alleys crossing at right angles. One alley was twenty feet wide and the other ten feet wide. Each lot ex-

2. 1821 ex. sess. Ga. Laws 12; Dawson 249.

3. 1822 Ga. Laws 112; Dawson 437.

4. Without completely itemizing sources, the following text concerning the survey of the Macon and Fort Hawkins reserves is derived from the documents filed under Records of the Town of Macon and the Public Reserve at Macon, in the GSGD; from Commissioners Journal and Ledger of Sale of Lots, Town of Macon, 1828 (GSGD); from Plat Book EEEE 2–4, 6, 8–11, 14–15, 107, 121–89 (GSGD); from Dawson 89, 262–63, 328, 344, 397, 437–38, 444–45, 448, 454, 457, 459–60, 466–67, 481; from Macon Advertiser, July 29, 1831, at 3; and from J. Sears, The First One Hundred Years of Town Planning in Georgia 172–77 (1979).

cept those along the river and along Cotton Avenue contained one-half acre. At the corners of each block was set "18 inches in the ground, a post of durable wood 8 inches square, three feet long, and marked with the number of the square." Intermediate points were designated by stakes. In all, 288 lots were laid off besides the three squares devoted to public purposes. Webb completed the survey on March 1, 1823.

Under an 1826 act the commissioners were authorized to enlarge the town by adding twenty lots of four acres each. Charles Philips arrived at Macon on January 15, 1827, to execute this task. At first, Philips encountered difficulties because many of the corner posts necessary to begin his measurements from were "either rotten down, or removed by mischevous persons," but once he was able to reestablish a sufficient number of corners, he found his compass bearings on the town lines to be the same as those of his predecessor. Philips finished setting off the additional lots in two and a half weeks.

A few months later, the legislature directed the governor to appoint three commissioners to oversee the subdividing of the remainder of the reserves. The commissioners selected Richard W. Ellis as surveyor, along with two chainmen and two axmen. They began work on January 21, 1828. Deeming it expedient to conform, as nearly as possible, to the regulations that had been followed in surveying the lottery lands, the commissioners consulted the surveyor general for advice. Among other things, the surveyor general recommended that "flags" be used "to enable the surveyor to take sights and back sights with greater ease and certainty." So large a portion of the reserves "consisted of close thickets, and the worst of Swamp" that the commissioners found this measure indispensable. Therefore, in addition to the surveyor, chainmen, and axmen, two other hands were appointed to carry the flags, to do the marking, and otherwise to assist. In cases where trees were unavailable for corners, the survey party set posts "of at least 6 inches in diameter, hewn towards the top on four sides to face the four tracts, the numbers of which" were marked thereon. The posts were made of decay-resistant woods such as sassafras, mulberry, locust, ash, and lightwood and were set at least twelve inches into the ground.

Because of the difficult terrain and exposure to swamps, Ellis took ill early in April. In little more than a month, he was too sick to continue and was replaced by William S. Norman, who finished the work eight days later.

The Macon and Fort Hawkins reserves as finally laid out consisted of

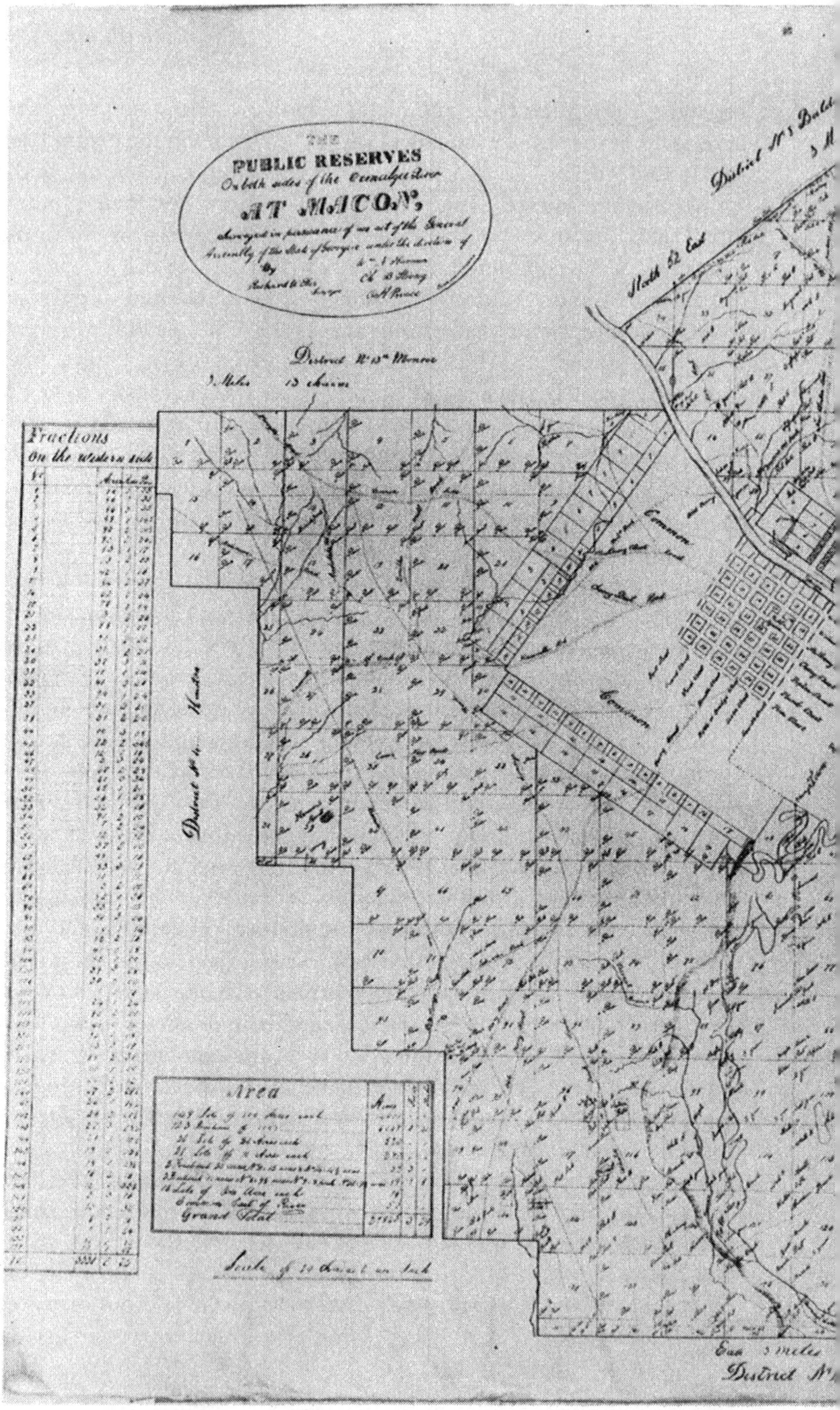
THE
PUBLIC RESERVES
AT MACON
Fractions
On the western side
Area
Grand Total

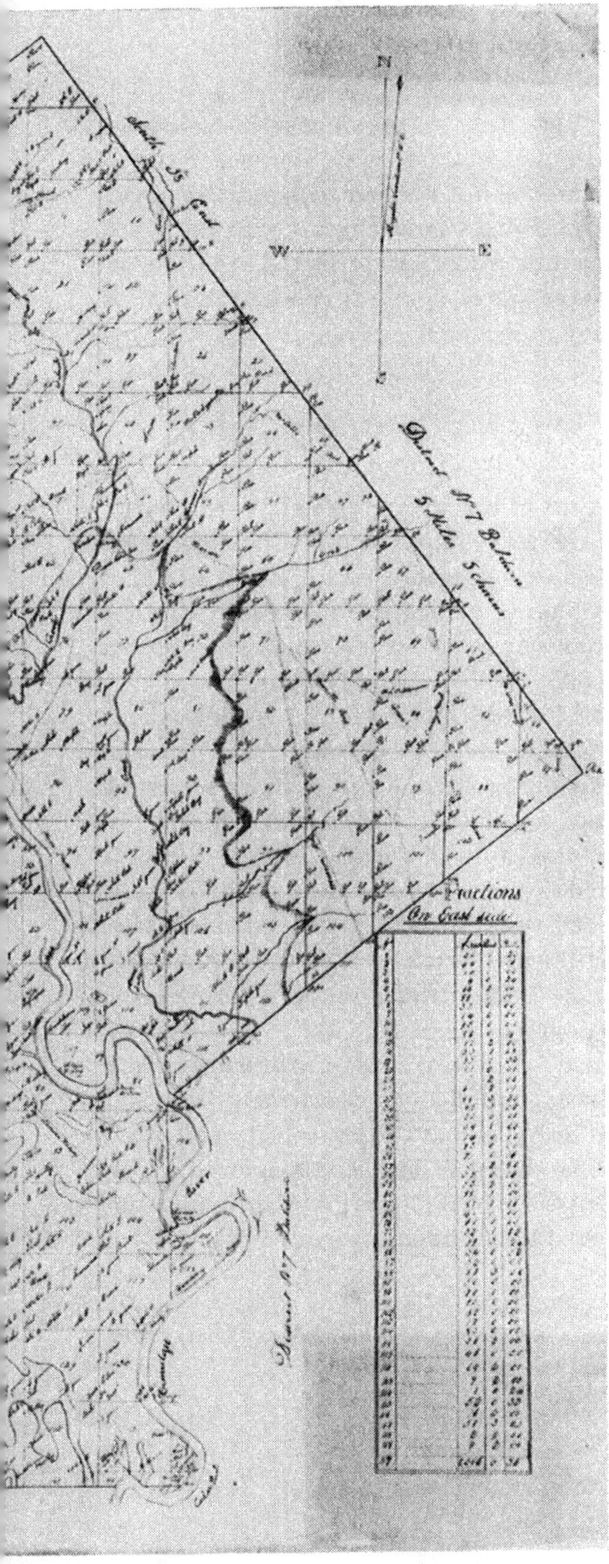

Original plan of the Macon and Fort Hawkins reserves. Courtesy of the Georgia Department of Archives and History.

the town proper, a common, a "Town Reserve" on which it was forbidden to cut trees, 16 one-acre tracts, 27 ten-acre tracts, 26 twenty-acre tracts, 149 hundred-acre "planting" tracts, and a number of fractional tracts. The town and its immediate appendixes were aligned with the river, while the large surrounding planting tracts were oriented in the cardinal directions. The total area of the Macon and Fort Hawkins reserves is approximately 22,000 acres.[5]

Columbus and Coweta Falls Reserve

By the 1825 land lottery act a tract of five square miles was ordered set aside at Coweta Falls (actually a series of rapids) on the Chattahoochee. These rapids constituted the head of navigation of the river.[6] In 1827, as it had for Milledgeville and Macon, the legislature provided for the appointment of a five-man commission to lay out a town on the reserved tract and to sell the lots so laid off.[7]

The commissioners employed Edward Lloyd Thomas to design and lay out Columbus. Thomas began his work on a cold, blustery day in January 1828 and finished the following April. As usual, he logged many day-to-day experiences, in addition to the necessary survey information, in his field notes. After only ten days' study of the site, Thomas drafted a proposed plan that was approved by the commissioners. A few days later the *Steubenville,* the first powered vessel to ascend the Chattahoochee, docked at Columbus. Thomas presented the captain of the vessel with a copy of his plan to carry to Pensacola and spread the word that Georgia was building a town on the Chattahoochee.

Thomas was compelled to discard the use of a compass because of very strong local attraction. The weather conditions were extremely damp and his crew was sick much of the time. Thomas's oldest son, Jacob (called "Truman"), succumbed to illness. Truman had been employed by his father as a chain carrier in the survey of the Georgia-Alabama boundary and again as a chain carrier in the Columbus survey. In late Febru-

5. Map of the Public Reserves on both sides of the Ocmulgee River at Macon, Surveyed . . . under the direction of Wm. N. Harmon, Ch. B. Strong, and O. H. Prince, Commissioners (ca. 1828) (GSGD).

6. 1825 ex. sess. Ga. Laws 16; Dawson 256–57.

7. 1827 Ga. Laws 183; Dawson 470.

ary Truman caught pneumonia from working in the rainy weather and died a few weeks later. He may have been the first person buried in the cemetery which his father provided in the plan of Columbus. Thomas continued the survey and had the sorrowful task of running a line by his son's grave. His field notes mention this and contain the simple comment that the site was a beautiful place for a graveyard.[8]

Columbus was oriented in the cardinal directions with half-acre town lots, a ten-acre square reserved for county purposes, a common, and a broad promenade extending along the river. This promenade afforded a scenic vista of the picturesque rapids below. The remainder of the reserve was divided into numbered ten-acre, twenty-acre, and one hundred-acre lots, as well as fractions of various sizes.[9]

Enterprising capitalists soon convinced the legislature that exploitation of the rapids for profit outweighed their value as scenery. In 1840 the promenade, as well as the adjacent riverbed, was divided into thirty-seven "water lots," which were devoted to industrial purposes and, to the consternation of their owners, became a source of recurring litigation.[10]

The Columbus and Coweta Falls Reserve covers a total of 10,440 acres.

Old Creek Agency Reserve

The Old Creek Agency Reserve is located on both sides of the Flint River in present Crawford and Taylor counties. It was the site of the Creek Agency from 1804 until 1827 and the site of the signing of the treaty of Creek cession in 1818.

By the terms of the 1821 treaty of cession, the Creeks gave up their lands between the Ocmulgee and Flint rivers but reserved a tract on the east side of the Flint to encompass the agency. Wilson Lumpkin surveyed this tract in May 1821.[11] Georgia acquired title to the site when the Creeks ceded their remaining lands in Georgia between 1825 and 1827.

Under the 1825 land lottery act, a tract equal in area to the agency tract

8. Field notes of Edward Lloyd Thomas, surveyor of the Columbus and Coweta Falls Reserve (GSGD); Goff, *Edward Lloyd Thomas, Surveyor,* 18 Emory U.Q. 107, 113–14 (1962).

9. Plan of that part of the Reserve, at the Cowetah falls, in Muscogee County . . . Certified 11th June 1828 by E. L. Thomas Surv^{r} (GSGD).

10. Muscogee Man. Co. v. Eagle & Phenix Mills, 126 Ga. 210 (1906). See page 477 note 13.

11. Plat of the Creek Agency by Wilson Lumpkin (May 29, 1821) (Telamon Cuyler Collection, UGL).

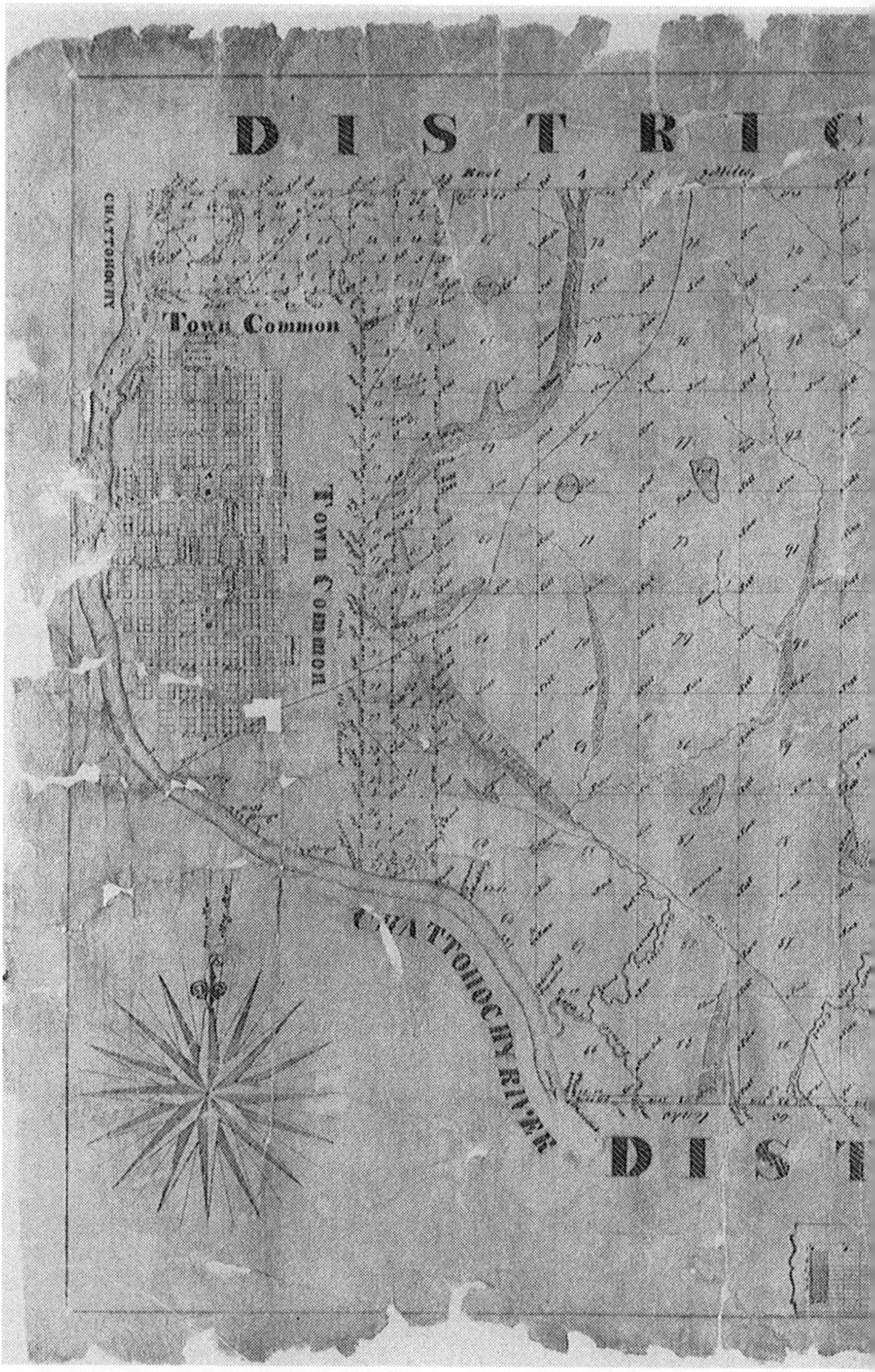

Original plan of the Columbus and Coweta Falls Reserve by Edward Lloyd Thomas. Courtesy of the Georgia Department of Archives and History.

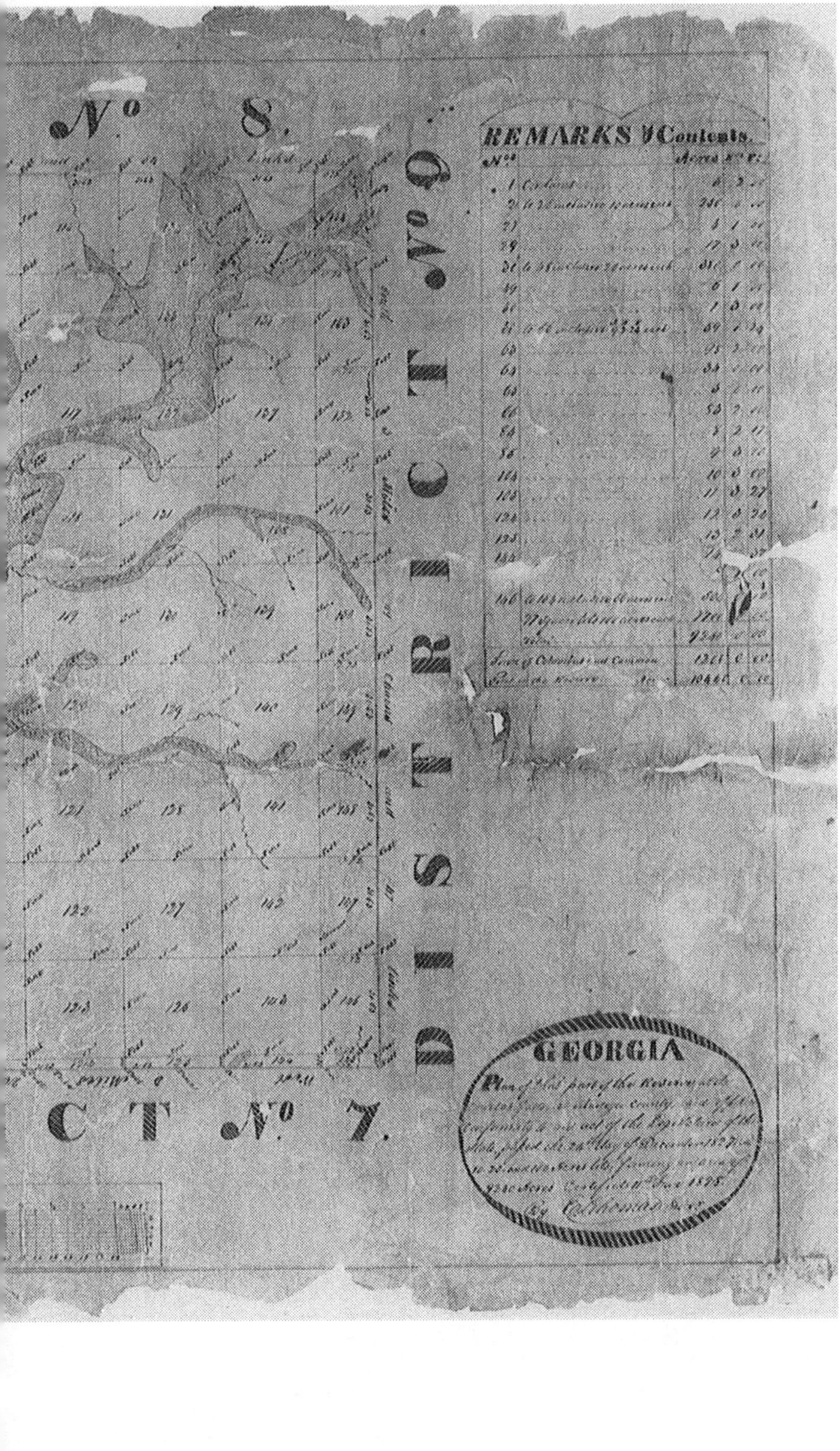

Nº 8.
REMARKS & Contents
DISTRICT Nº 9.
GEORGIA
CT Nº 7.

was laid off on the west side of the Flint River opposite the agency.[12] These two tracts make up what is known as the Old Creek Agency Reserve.

The reserve was subdivided into twenty-seven 202½-acre lots and twenty-one fractional lots by John Bethune in 1828.[13] The lots were subsequently granted under authority of the act of December 22, 1827.[14] Those on the east side of the river are numbered 1 through 20; those on the west side, 1 through 28. The reserve covers 8,055 acres.

Indian Springs Reserve

Located in present Butts County, the Indian Springs Reserve encompassed the tavern and other improvements made by Chief William McIntosh. It was the site of the signing of the Creek treaties of 1821 and 1825. By the former treaty, a one-thousand-acre tract around the spring was reserved, ostensibly for the use of the Creek Nation but in reality for the benefit of McIntosh. McIntosh relinquished his rights to the reserve in return for money by a supplemental article to the treaty of 1825.

The reserve was laid off by Wilson Lumpkin in May 1821[15] and subdivided by Hugh MacDonald in 1828.[16] The spring was left in the center of a ten-acre lot, and the remainder of the reserve was divided into lots ranging in size from 2 to 39.6 acres.[17] Streets running in the cardinal directions were laid off one-half chain wide. The lots were sold by commissioners appointed by the governor.

The Indian Springs Reserve did not grow to become the great resort area the legislature apparently thought it would, and the commissioners were granted authority to sell some of the streets and public passageways in 1838.[18] Today much of the Indian Springs Reserve is a state park.[19]

12. 1825 ex. sess. Ga. Laws 16; Dawson 256–57.

13. Plat Book EEEE 16, 52–64 (GSGD).

14. 1827 Ga. Laws 130; Dawson 266.

15. Plat of the Indian Spring by Wilson Lumpkin (May 23, 1821) (Letters Received by the Secretary of War, Registered Series, Record Group 107, National Archives, Washington, D.C.).

16. Plat Book EEEE 2, 4, 6, 13, 64–81 (GSGD).

17. Id.; 1827 Ga. Laws 119; Dawson 263.

18. 1838 Ga. Laws 142; Cobb 702.

19. O.C.G.A. § 50-16-1.

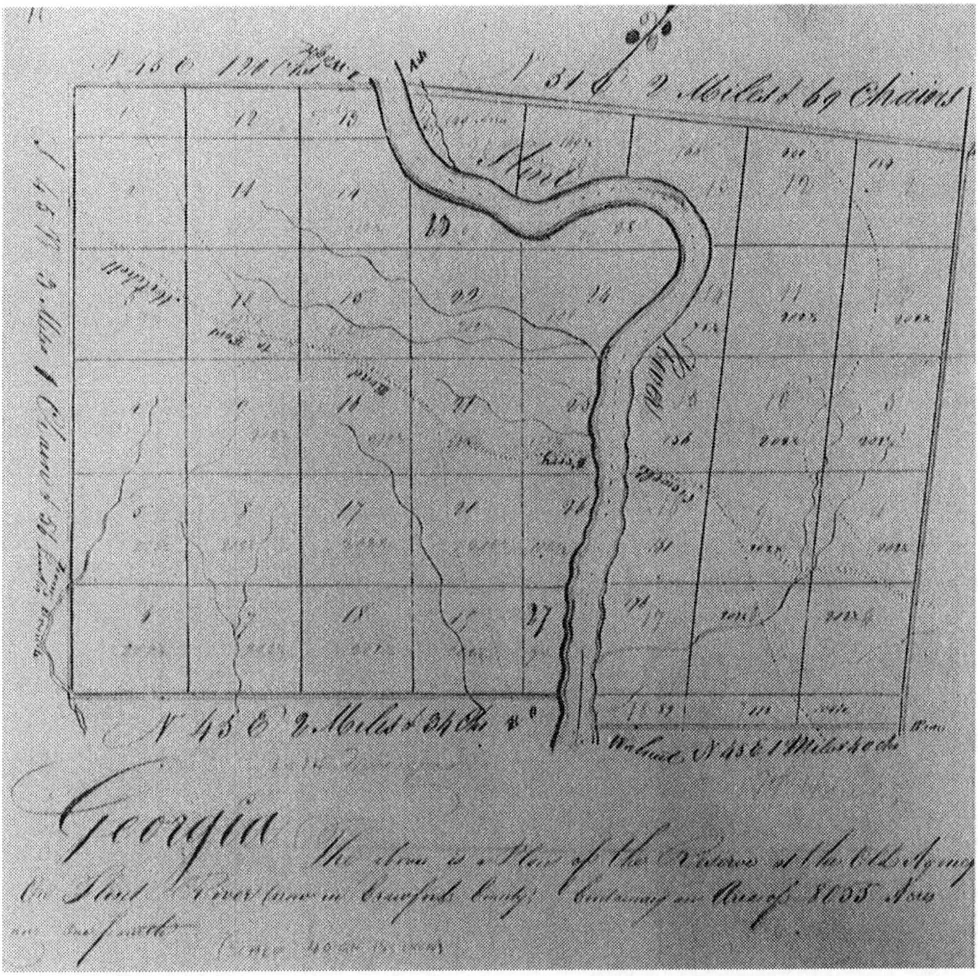

Old Creek Agency Reserve. Courtesy of the Georgia Department of Archives and History.

Indian Springs Reserve. Courtesy of the Georgia Department of Archives and History.

McIntosh's Lower Reserve

Located on the west side of the Ocmulgee River in present Butts County, McIntosh's Lower Reserve, like the Indian Springs Reserve, was set aside for the use of Chief McIntosh by the Creek treaty in 1821 and then sold by him to Georgia under a supplemental article to the Creek treaty of 1825. It was laid off by Wilson Lumpkin in May 1821. Lumpkin's survey shows it containing 640 acres, the amount specified in the 1821 treaty.[20] The same commissioners who superintended the Indian Springs Reserve were given authority to subdivide and sell McIntosh's Lower Reserve.[21] They employed Hugh MacDonald, who divided the tract into four lots of about equal size.[22] MacDonald's survey shows the total area of the reserve to be 529.1 acres. Lot 4 was granted to James H. Morrow in 1830 and lots 1, 2, and 3 to Benjamin F. Ward in 1834.[23]

McIntosh's Upper Reserve

McIntosh's Upper Reserve (also known as McIntosh's Ferry Reserve) straddles the Chattahoochee River in present Carroll and Coweta counties. It was the site of Chief William McIntosh's principal home, which he called Lockchau Talofau, and the scene of his bloody execution by Creek warriors in 1825. McIntosh lies buried on the site on a lonely hill overlooking the Chattahoochee River.

The reserve was surveyed by John Bethune in 1828 and sold to Peter B. Terrell in 1848 under authority of the act of December 22, 1827. The total area is 576 acres.[24]

Most of the tract was owned by the Georgia Power Company until 1978, when the company deeded its portion in Carroll County to the county for use as a park.[25] Subsequently, Chief Neal ("Tall Pine") McCormick, claiming to be the great-great-grandson and heir of McIntosh, interposed a

20. Plat of the plantation of the Indian Chief Gen! McIntosh by Wilson Lumpkin (May 24, 1821) (Letters Received by the Secretary of War, Registered Series, Record Group 107, National Archives, Washington, D.C.).

21. 1827 Ga. Laws 119; Dawson 263.

22. Plat Book EEEE 2, 7, 8, 12 (GSGD).

23. Grant Book "Butts Reserves" 1, 16–18 (GSGD).

24. Plat Book EEEE 17 (GSGD); 1827 Ga. Laws 130; Dawson 266.

25. Carroll County Georgian (Carrollton, Ga.), Jan. 21, 1978, at 1, cols. 4–5.

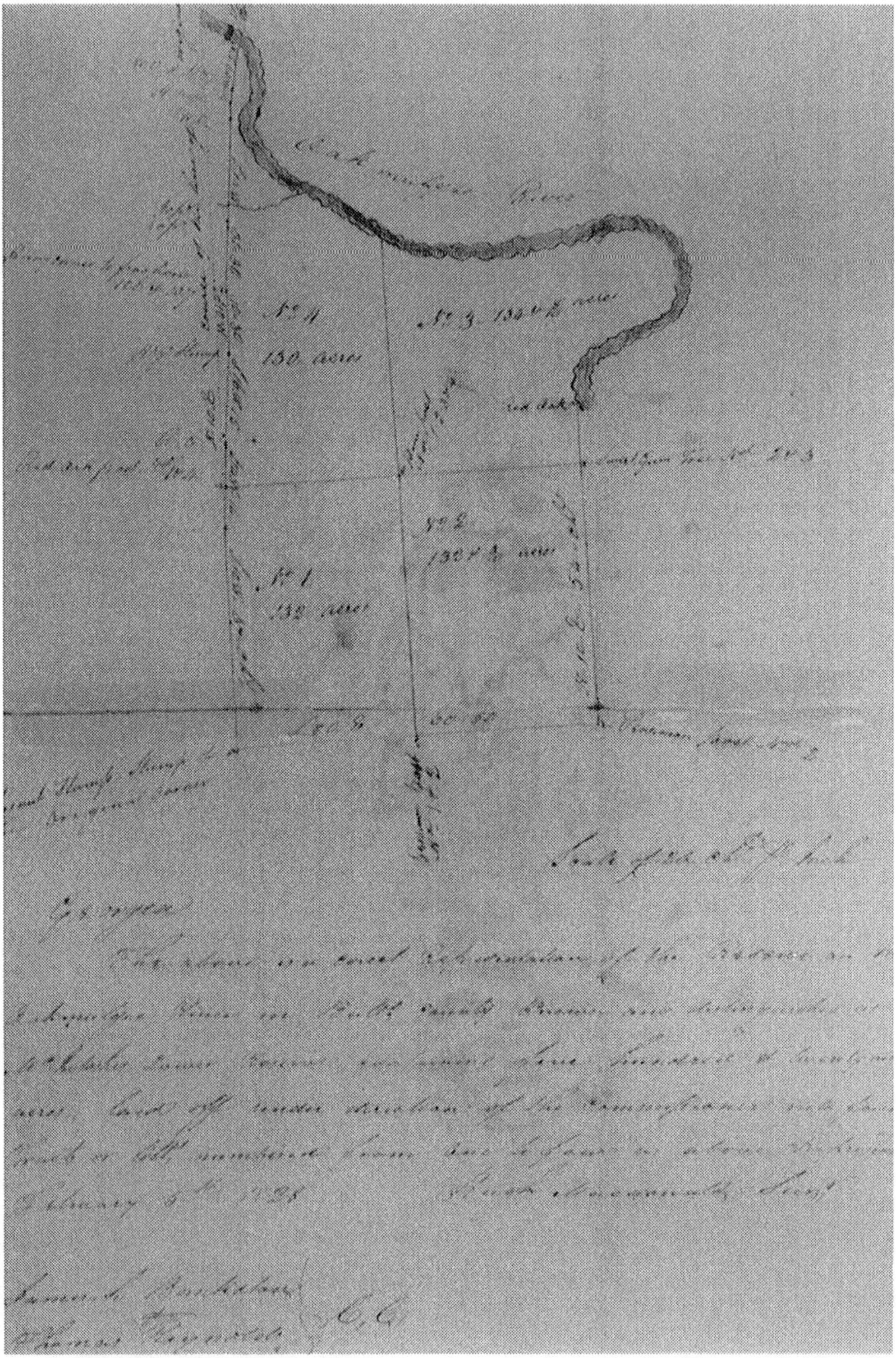

McIntosh's Lower Reserve. Courtesy of the Georgia Department of Archives and History.

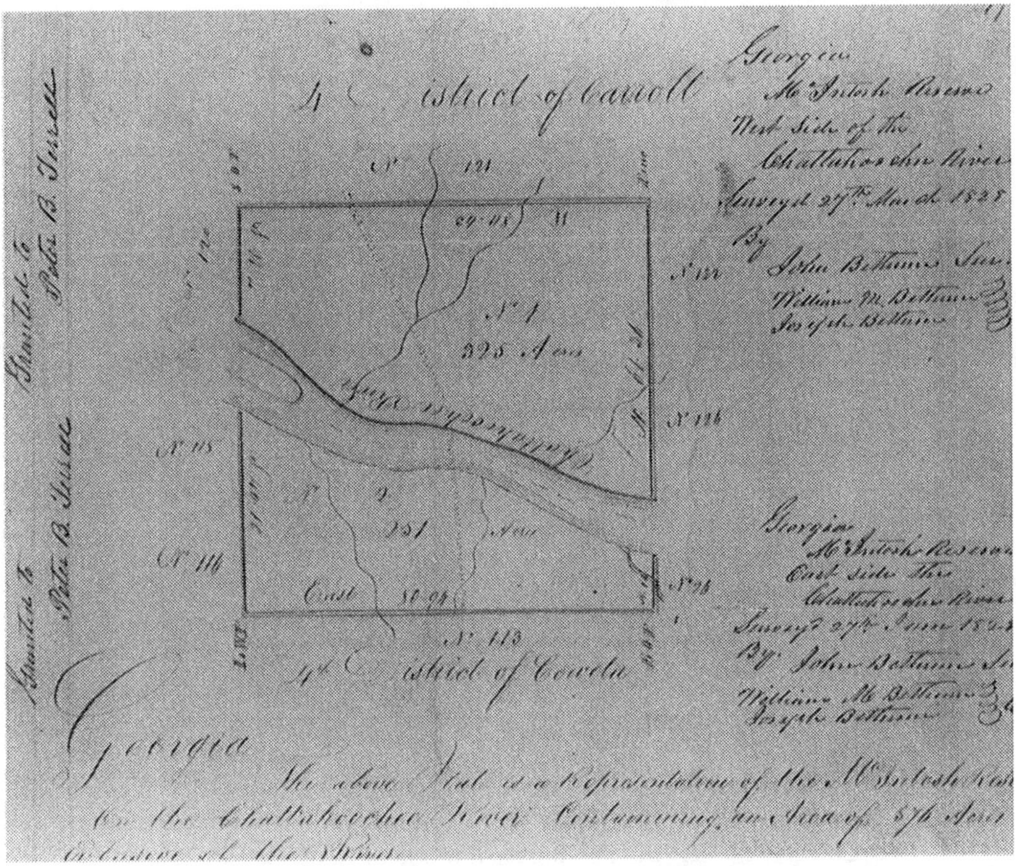

McIntosh's Upper Reserve. Courtesy of the Georgia Department of Archives and History.

claim for the reserve on the assertion that it had been left to McIntosh and his heirs forever.[26]

Buckey Barnard's Reserve

Located on the east side of the Flint River in present Macon County, this reserve today is covered by the town of Montezuma. It was reserved to

26. Atlanta Journal & Constitution, Feb. 26, 1978, at 8B; Carroll County Georgian (Carrollton, Ga.), Jan. 28, 1978, at 1, cols. 5–6.

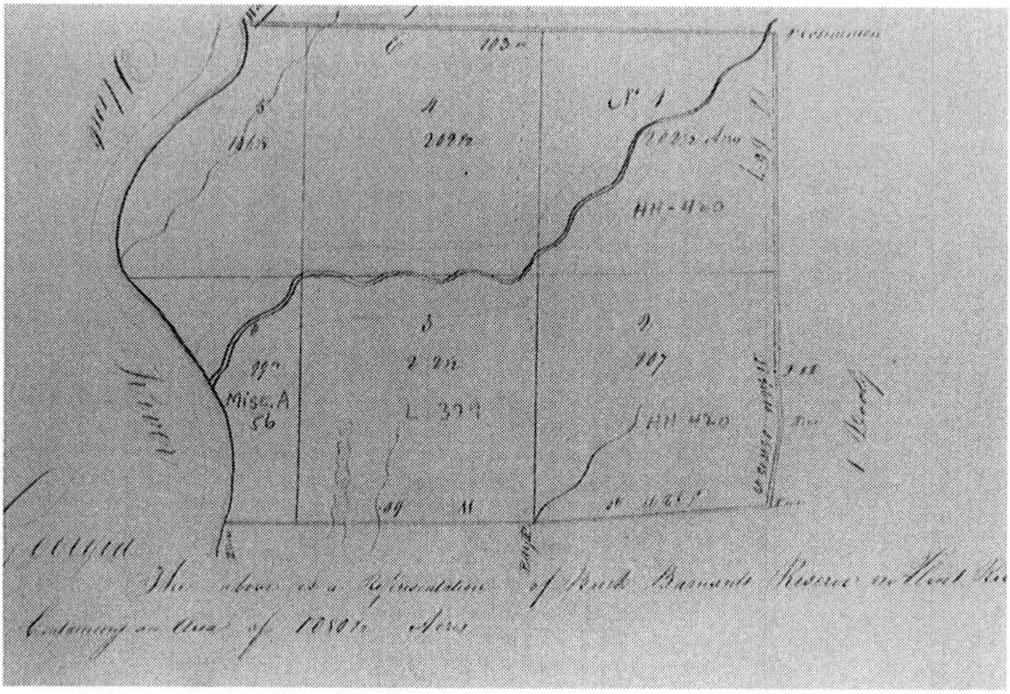

Buckey Barnard's Reserve. Courtesy of the Georgia Department of Archives and History.

Buckey Barnard, as were other tracts to his brothers and sister, by the Creek treaty of 1821. Buckey Barnard was the son of the half-breed chief of the Lower Creeks, Timpoochee Barnard. The elder Barnard held a commission as major in the U.S. Army and played a prominent role in relations between the United States and the Creek Nation during the early 1800s.[27]

By the terms of the treaty of 1821, Buckey Barnard was allotted a one-square-mile tract that included his farm and improvements. He would hold title to the property for as long as he should personally occupy it, and title would pass to the state upon his abandonment. Wilson Lumpkin laid the tract off in June 1821.[28] Situated a short distance east of the Flint River, it measured eighty chains by eighty chains.

27. L. Hays, History of Macon County Georgia 46–52 (1933).

28. Plat of the improvements of Buckey Barnard by Wilson Lumpkin (June 7, 1821) (Telamon Cuyler Collection, UGL).

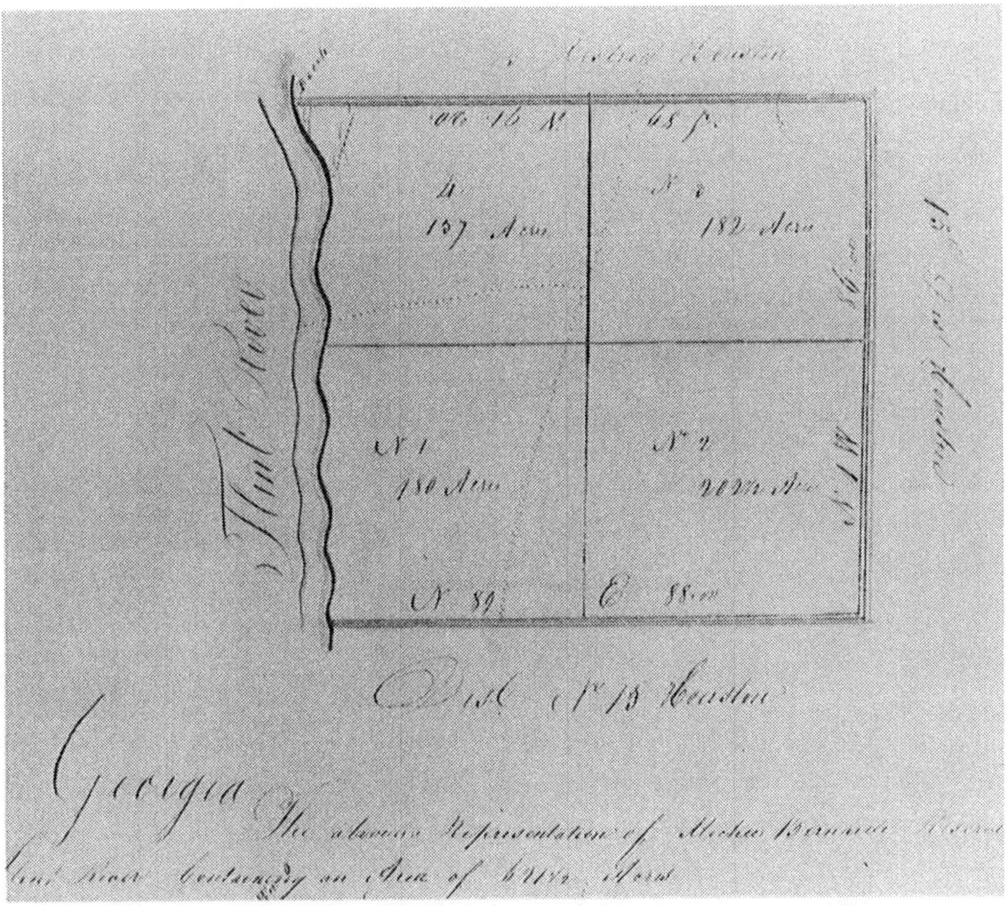

Michey Barnard's Reserve. Courtesy of the Georgia Department of Archives and History.

When the state surveys were conducted in the area in the spring and summer of 1821, the strip between the reserve and the river was carelessly left out. Edward Lloyd Thomas discovered this unappropriated parcel when he resurveyed the 1st District of Dooly County in 1824.[29] The addition of this strip to the reserve gave it a total area of 1,060.5 acres.

Buckey Barnard's Reserve was subdivided into six lots by John Bethune in 1828, and the lots were sold under the act of December 22, 1827.[30]

29. See page 238.

30. Plat Book EEEE 18, 46, 47 (GSGD); 1827 Ga. Laws 130; Dawson 266.

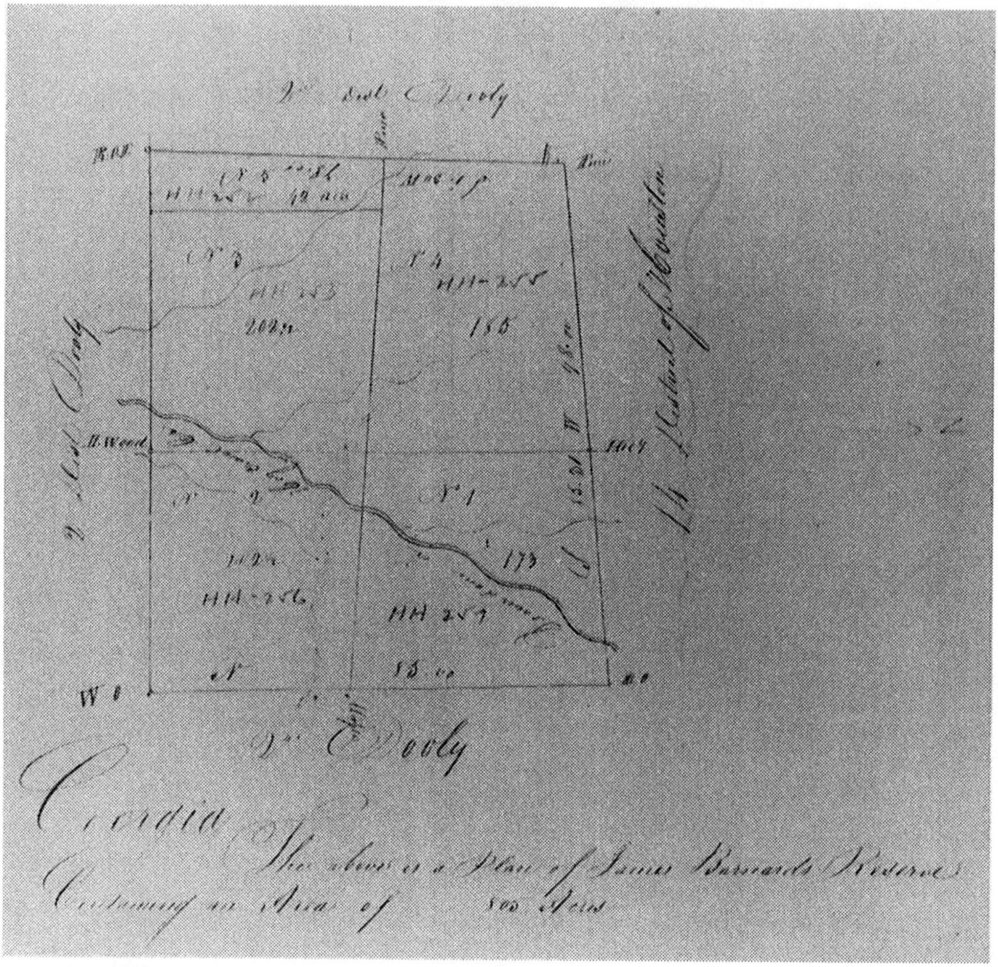

James Barnard's Reserve. Courtesy of the Georgia Department of Archives and History.

Michey Barnard's Reserve

Michey Barnard, a brother of Buckey, received this reserve under the Creek treaty of 1821 on the same terms as his brother. It is located in present Macon County on the northern edge of the town of Montezuma. Wilson Lumpkin laid the tract off in June 1821. His survey shows it to contain 640 acres.[31] Under the act of December 22, 1827, the reserve

31. Plat of the improvements of Michey Barnard by Wilson Lumpkin (June 1, 1821) (Telamon Cuyler Collection, UGL).

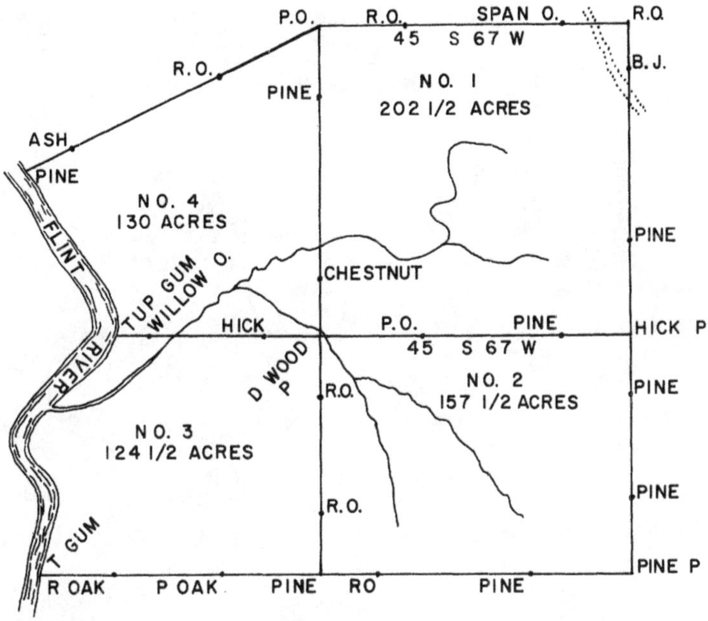

Diagram of Efau Mathlaw's Reserve. Compiled from plats in the Georgia Department of Archives and History.

was subdivided by John Bethune into four lots, all of which were sold to Alfred Clopton in 1834.[32] According to Bethune's survey, the four lots contain a total of 721.5 acres.

James Barnard's Reserve

James Barnard, a brother of Buckey and Michey, received this reserve under the Creek treaty of 1821 on the same terms as his brothers. It is located in present Macon and Dooly counties, straddling the line between the 14th District of original Houston County and the 2d District of original Dooly County. Wilson Lumpkin laid the tract off in June 1821. According to his survey, it contains 640 acres.[33] The reserve was subdivided into five lots by John Bethune under the act of December 22, 1827. The lots were sold to Cullen Horn and Solomon Folks in 1833.[34] Bethune's survey shows the five lots to contain a total of 805 acres.

32. Plat Book EEEE 18, 49, 50 (GSGD); 1827 Ga. Laws 130; Dawson 266.

33. Plat of the improvements of James Barnard by Wilson Lumpkin (June 4, 1821) (Telamon Cuyler Collection, UGL).

34. Plat Book EEEE 19, 48, 49 (GSGD); 1827 Ga. Laws 130; Dawson 266.

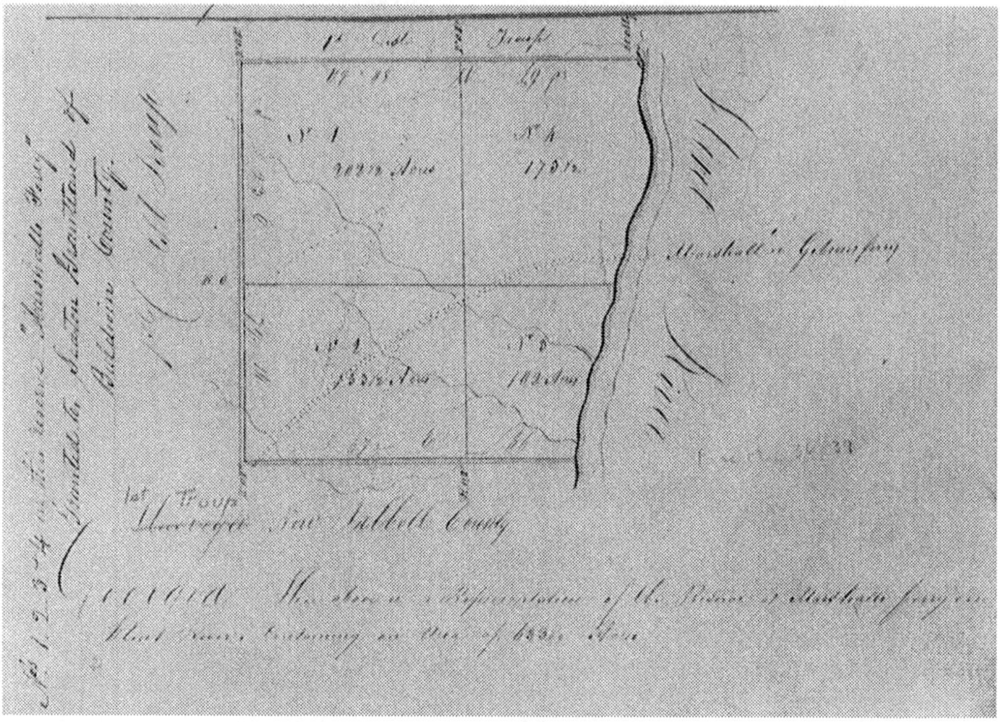

Marshall's Ferry Reserve. Courtesy of the Georgia Department of Archives and History.

Efau Mathlaw's Reserve

Also known as Cheneolly's Reserve, this tract is located on the east side of the Flint River in present Upson County. It was set aside for the full-blooded Creek chief Efau Mathlaw by the Creek treaty of 1821. Under the terms of the treaty, Mathlaw was to hold title for as long as he should personally occupy it, and title was to pass to the state if he should abandon it. Wilson Lumpkin laid the tract off in May 1821. His survey shows that it contains 640 acres.[35] The reserve was subdivided into four lots by

35. Plat of the plantation and improvements of Chief Efau-Emauthlaw by Wilson Lumpkin (May 26, 1821) (Telamon Cuyler Collection, UGL).

L. W. Harris, and the lots were sold under the act of December 22, 1827. According to Harris's survey, the four lots contain a total of 614.5 acres.[36]

Marshall's Ferry Reserve

Located on the west side of the Flint River in present Talbot County, this tract was reserved to the state by the 1825 land lottery act, which specified that a one-square-mile tract was to be set aside.[37] John McBride laid the reserve off in September 1826 and subdivided it into four lots.[38] The four lots were sold to Seaton Grantland of Baldwin County in 1848 under the terms of the act of December 22, 1827.[39]

36. Plat Book EEEE 51, 52 (GSGD); Executive Minutes (Jan. 14, 1829) (GDAH); 1827 Ga. Laws 130; Dawson 266. In addition to the reserves set aside for Chief McIntosh, Buckey Barnard, Michey Barnard, James Barnard, and Efau Mathlaw under the terms of the 1821 treaty, a one-square-mile tract was set aside for Cussena Barnard, sister of Buckey, Michey, and James. This tract was laid off by Wilson Lumpkin in June 1821. See Plat of the improvements of Cussena Barnard by Wilson Lumpkin (June 6, 1821) (Telamon Cuyler Collection, UGL). Evidently Cussena either died or abandoned the property before the state began surveying the area and, as specified by the treaty, title immediately passed to the state. This tract was subsequently overlapped by land lots and districts. Partial title abstracts of the reserves of Buckey, Michey, and James can be found in L. Hays, History of Macon County Georgia 55–58 (1933).

37. 1825 ex. sess. Ga. Laws 16; Dawson 257.

38. Plat Book EEEE 17 (GSGD); Letter from John McBride to George M. Troup (Sept. 28, 1826) (GDAH).

39. Grant Book "Frac. M" 36–39 (GSGD); 1827 Ga. Laws 130; Dawson 266.

Chapter 10. Some Closing Comments on the Land Lotteries

Results of the Land Lotteries

Clearly the land lottery system constituted a marked improvement over the headright system, both for getting the lands surveyed and for granting them to settlers in an efficient, expeditious manner. The numerous boundary complications and caveats that often accompanied headright grants were practically eliminated under the lottery system. Even today surveyors in the headright areas are plagued with problems unknown to those in the lottery areas.

There are several explanations for the errors in the lottery surveys: the cheapness of land during the period; the crude instruments used; the remoteness of the areas in which the surveys were conducted from supply bases; the lack of accurate geographic information about the new areas; the hostility of Indians; and the unskillfulness, indifference, carelessness, or fraudulent intentions of some surveyors. Often after an Indian cession, there was a sense of urgency that the lands be granted before the Indians and the federal government had time to reconsider and renege on the treaties. Thus great pressure was put on the surveyors to finish their work as rapidly as possible.

Numerous petty frauds and abuses occurred in the administration of the lottery grants. In addition to the instances previously described, many unqualified persons submitted fraudulent applications for draws, and the legislature was forced to pass numerous acts providing for their prosecution.[1] In 1845 the surveyor general and the secretary of the executive department apparently were involved in a collusive scheme to purchase certain choice reverted lots at nominal prices.[2] That same year Governor George W. Crawford, in his annual message to the legislature, disclosed that an examination of the books pertaining to lottery grants in

1. Clayton 340, 371, 452; Lamar 405, 425; Dawson 249–50, 256, 258, 261, 267, 273–75, 333, 337; Cobb 696–97, 699, 701.
2. See note 23 below.

the surveyor general's office had uncovered many alterations, forgeries, and erasures.[3] Throughout the lottery period, the surveyor general and other high government officials often speculated in reverted lots. Despite the endemic nature of such schemes, however, the lottery system itself—coupled with greater public vigilance—prevented the recurrence of abuses on the scale of the Pine Barrens Fraud and the Yazoo Fraud.

Numerous mistakes were made in the lotteries, both in the names of the fortunate drawers and in the records of the lots they drew. Time and remedial legislation have corrected most of these.[4]

A rough estimate of the number of lots and acreages made available in each of the lotteries is given in Table 3.

It is almost a cliché to extol the cheapness of land during the frontier period of this nation's development. Nevertheless, a few examples are appropriate in this context. In the 1820s Benjamin Beckman, the original grantee of Land Lot 77 of the 14th District of Henry County, is said to have swapped that land lot and a horse to Samuel F. Mitchell for a horse of superior quality.[5] Land Lot 78, lying just north of Land Lot 77, was sold for fifty dollars (or less than twenty-five cents an acre) in 1826.[6] Today the central portion of downtown Atlanta stands on these two lots and they, along with the improvements built on them, are worth hundreds of millions of dollars. An early innkeeper in Carrollton rejected a 202½-acre lot offered as payment for a night's lodging by one who had arrived the previous day to view the land he had drawn in the lottery.[7] It is estimated that roughly one-fourth of the lots drawn in the lotteries reverted.[8] In other words, about one-fourth of the successful drawers considered the lots they drew (ranging in size from 40 to 490 acres) not worth the cost of taking out grants (a minimum of $8.10 under the 1803 act to a maximum of $19.00 under the 1821 act). No wonder early surveys were so crude! The value of the lots often barely exceeded the cost of surveying them.

Over the long term, however, land values generally were somewhat

3. 1845 Ga. Senate Journal 20–21.

4. Sibley v. Haslam, 75 Ga. 490 (1885); Siles v. McRory, 32 Ga. 348 (1861); Sykes v. McRory, 10 Ga. 465 (1851). The acts of the General Assembly for the first half of the 1800s are replete with laws passed to correct bookkeeping errors made in the lotteries.

5. 1 F. Garrett, Atlanta and Environs 281 (1954); Kurtz, *The Story of Land Lot 77*, Atlanta Historical Bulletin, Dec. 1947, at 42.

6. 1 F. Garrett, Atlanta and Environs 60 (1954).

7. J. Bonner, Georgia's Last Frontier: The Development of Carroll County 21–22 (1971).

8. E. Banks, The Economics of Land Tenure in Georgia 19 n. 3 (1905).

Table 3. Number of Lots and Acreages Made Available in Each of the Land Lotteries

Year of lottery act	*Size of lots, in acres*	*Estimated number of lots made available, by size*	*Estimated number of acres distributed*
1803	202½	3,047	617,018
	490	829	406,210
1806	202½	10,476	2,121,390
1818	250	17,500	4,375,000
	490	10,000	4,900,000
1819	250	1,100	275,000
	490	500	245,000
1821	202½	12,200	2,470,500
1825	202½	21,000	4,252,500
1830	160	18,309	2,929,440
1831	40	35,000	1,400,000
Totals		129,961	23,992,058

SOURCES: E. Banks, The Economics of Land Tenure in Georgia 18 (1905); P. Starrett, Appraisal: The Creek Nation: Lands in Southern Georgia and Southeastern Alabama 198a–g (1957); The Georgian (Savannah), Oct. 4, 1832, at 2, col. 3; Niles Weekly Register (Baltimore), Oct. 27, 1832, at 131, col. 2.

higher than these examples suggest. Each successive lottery flooded the market with exploitable real estate and caused a momentary sharp drop in land prices, but the prices rebounded after a short time. An examination of deed records for the first half of the 1800s indicates that prices commonly paid for farmland ranged from $0.50 to $1.00 per acre. Even Governor Forsyth conceded in 1828 that many of the lots then being granted by the state at the rate specified in the 1825 lottery act—$18 apiece—were worth at least $200.[9] In July 1833 speculators drove the price of a single forty-acre gold lot to $30,000.[10]

Although Georgia remained overwhelmingly rural throughout the nineteenth century, more and more villages and towns began to dot the landscape. The huge counties created under the lottery acts were quickly

9. 1828 Ga. House Journal 380.

10. Western Herald (Auraria, Ga.), July 16, 1833, at 2, col. 4. It was Gold Lot 1052, 12th District, 1st Section, of original Cherokee, now Lumpkin, County.

and repeatedly sliced into smaller and smaller ones. By the time of the Civil War, 132 counties were delineated on the map of Georgia, necessitating the establishment of numerous seats of justice. The same general procedure for founding county seats that had prevailed in the headright areas was followed in the lottery counties. Either a commission (usually of five individuals) or, more often in the later years, the inferior court justices were empowered by a legislative act to fix upon the site of and lay out the county seat. Depending on the terms of the act, a tract usually ranging in size from fifty acres to a full land lot might be acquired by gift, purchase, or grant from the state. The town of Talbotton, laid out on the stage route from Macon to Columbus as the seat of Talbot County, was typical. Talbot County was cut from Muscogee County in December 1827.[11] The following March the inferior court justices purchased from William Gilbert Land Lot 266 in the 16th District of the county. Gilbert had drawn the lot in the lottery of the previous year. William McMurray, the county surveyor, was designated to delineate the streets and lots of Talbotton "agreeable to a plan recorded by the court." The town was laid out in a typical grid pattern, oriented in the cardinal directions. Pursuant to a special stipulation in the act organizing Talbot County, two one-acre lots were reserved for academies and four one-acre lots for future conveyance to religious denominations. McMurray was paid one dollar per lot for surveying the town.[12]

Of growing significance was the number of towns founded through private initiative. Oxford in Newton County has the most imaginative design. In 1837 the Georgia Methodist Conference purchased a 330-acre site for Emory College and the projected town of Oxford. Edward Lloyd Thomas laid out the college yard and circumjacent lands that same year. The town follows a grid pattern but, in addition, has streets radiating from the college square. There is little doubt that Thomas was influenced by the plan of Washington, D.C. The design of Oxford is unique among Georgia towns.[13]

In contriving the layout of Atlanta, the profit motive triumphed over all thought of utilitarian town planning. In 1842 Samuel F. Mitchell executed a deed of gift to the state for five acres of land and a railroad right-of-way, located in the northeastern portion of Land Lot 77 of the

11. 1827 Ga. Laws 69; Dawson 137.

12. Robins v. McGehee, 127 Ga. 431 (1906); 1827 Ga. Laws 65.

13. J. Sears, The First One Hundred Years of Town Planning in Georgia 29–30, 32 n. 6 (1979).

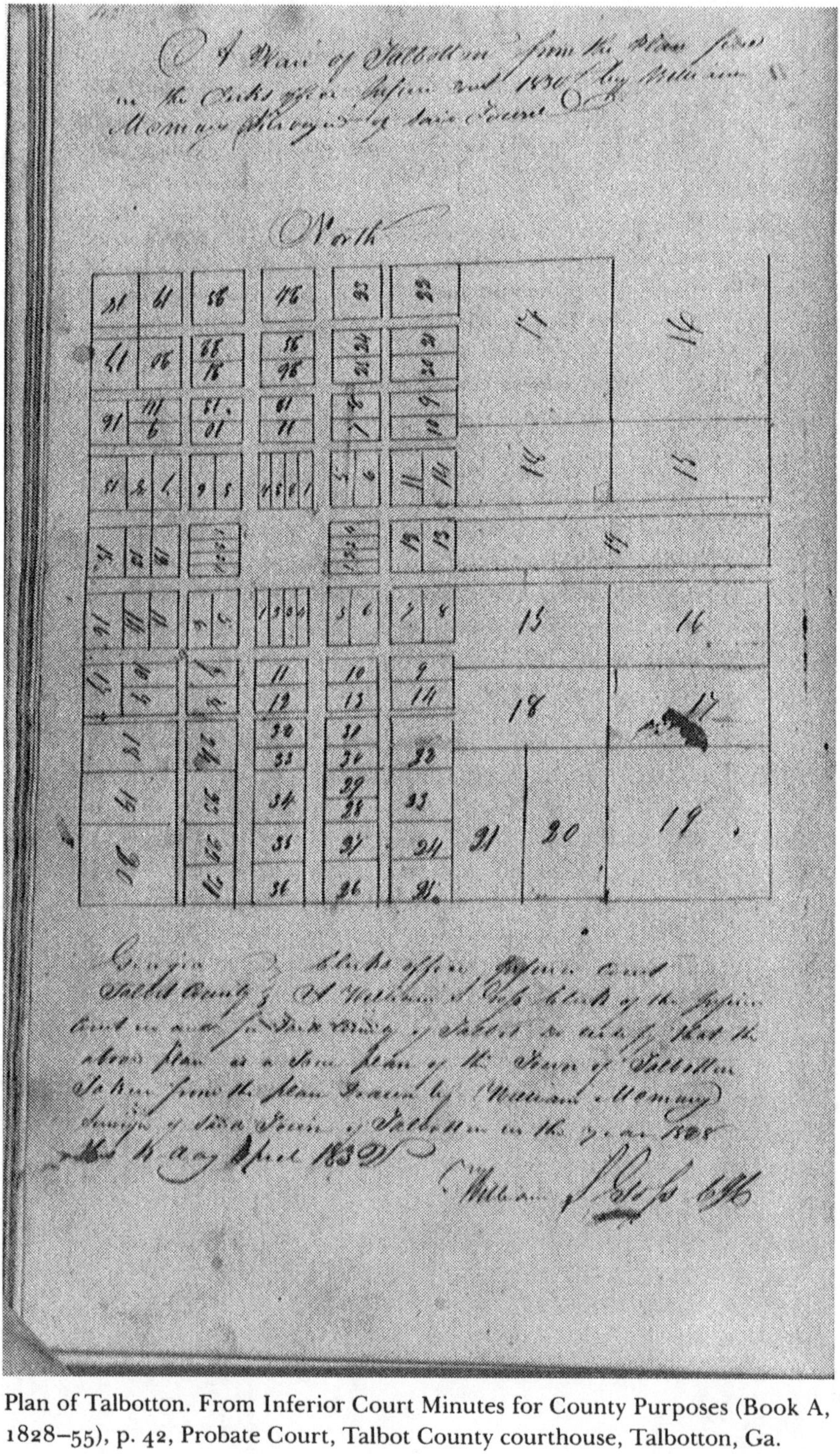

Plan of Talbotton. From Inferior Court Minutes for County Purposes (Book A, 1828–55), p. 42, Probate Court, Talbot County courthouse, Talbotton, Ga.

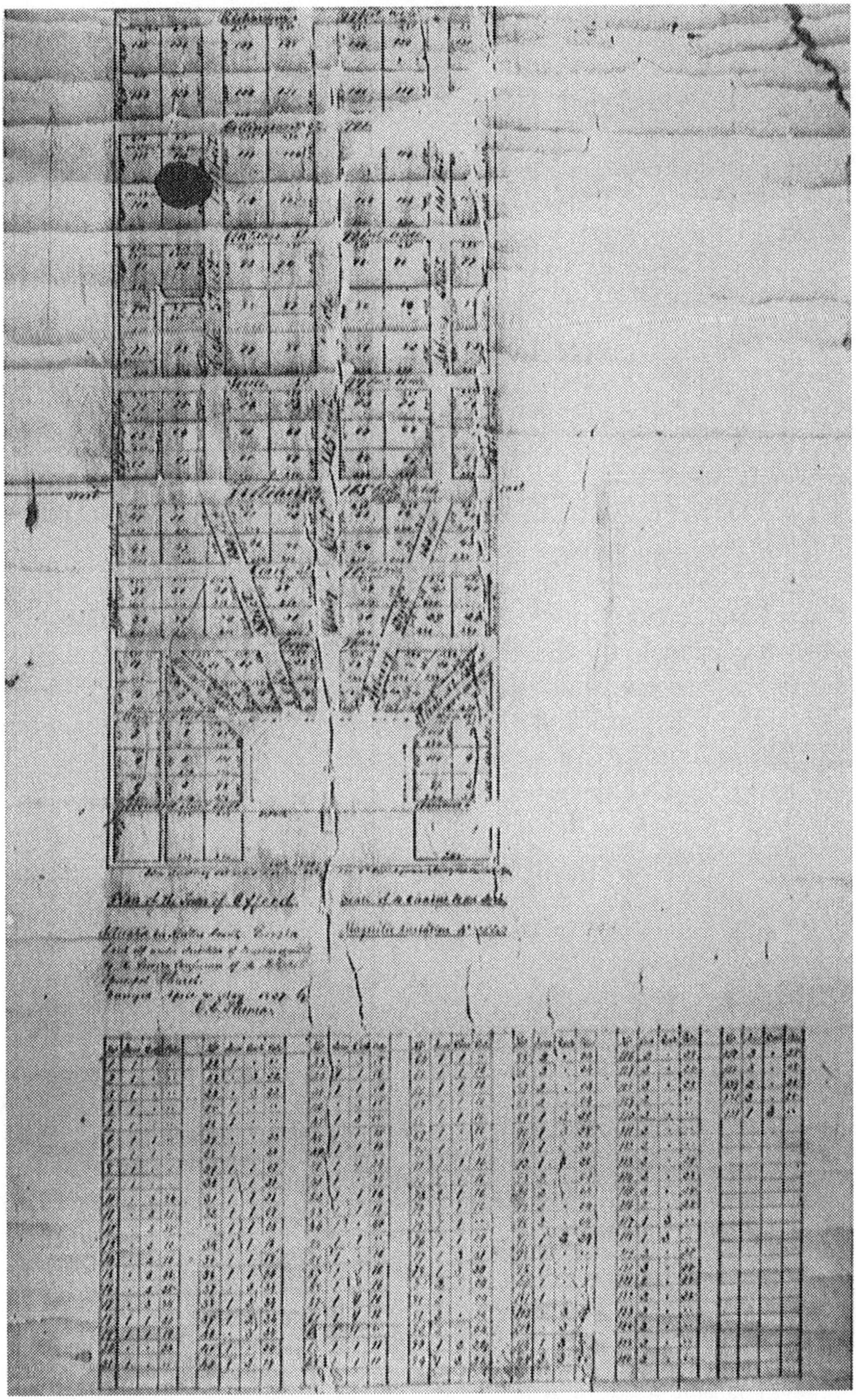

Original map of Oxford by Edward Lloyd Thomas. Thomas's house still stands on Lot 95. Courtesy of Special Collections, Robert W. Woodruff Library, Emory University.

14th District of original Henry (then DeKalb and now Fulton) County. The area at the time was almost a perfect wilderness. The five-acre tract (subsequently known as the State Square) had been selected as the terminal point of the Georgia-owned Western and Atlantic Railroad. Soon the Monroe Railroad (later the Central of Georgia) and the Georgia Railroad connected to the spot, causing real estate values to skyrocket. In 1844 Mitchell had Frederick C. Arms, assisted by former Governor Wilson Lumpkin, subdivide a portion of his land around the square into seventeen lots, each generally one-half acre in size. Lumpkin, then in his sixties, was fulfilling the last chore of his long and varied career—the superintendency of the Western and Atlantic Railroad. Arms was one of its engineers. Within two years Mitchell had Allen Pryor (for whom one of the streets was named) subdivide the remainder of Land Lot 77. The owners of the three adjacent land lots—51, 52, and 78—followed suit. The subdivisions of the four land lots—cornering beside the present Trust Company of Georgia building—form the basic plan of Atlanta. The principal object in these layouts was to cram a maximum number of lots into a fixed amount of space. Each delineation was aligned with the section of railroad track to which it lay adjacent, with the result that the central portion of Atlanta has five distinct orientations. The surveying was hastily and rather crudely executed. Thus Atlanta's configuration is a jumbled affair of narrow, twisting streets that cause traffic to snarl and create congestion. Barely three decades after Arms and Lumpkin's survey, however, this booming frontier town was the largest metropolis in the state—surpassing even urbane old Savannah, which had held the honor since the founding of Georgia.[14]

The large, elegantly drawn pictorial representations of districts that had been turned in by the lottery surveyors now provided the legislature with all the information it needed to delineate county boundaries with clarity and precision. Hardly any additional surveying of such boundaries was necessary in these areas because the acts creating successive counties usually designated their peripheries as existing land lot, district, and section lines or as natural features. So fast were new counties being created

14. Seignious v. MARTA, 252 Ga. 69 (1984); Samuel Mitchell Est. Inc. v. West. & Atl. R.R., 167 Ga. 728 (1928); Dougherty v. The West. & Atl. R.R. Co., 53 Ga. 304 (1874); 1 F. Garrett, Atlanta and Environs 188–90, 236, 238, 354 (1954); J. Johnston, Western and Atlantic Railroad of the State of Georgia 119–20 (1931); Kurtz, *The Story of Land Lot 77*, The Atlanta Historical Bulletin, Dec. 1947, at 42. Mitchell, *Atlanta's First Real Estate Subdivision*, Atlanta Historical Bulletin, May 1930, at 6.

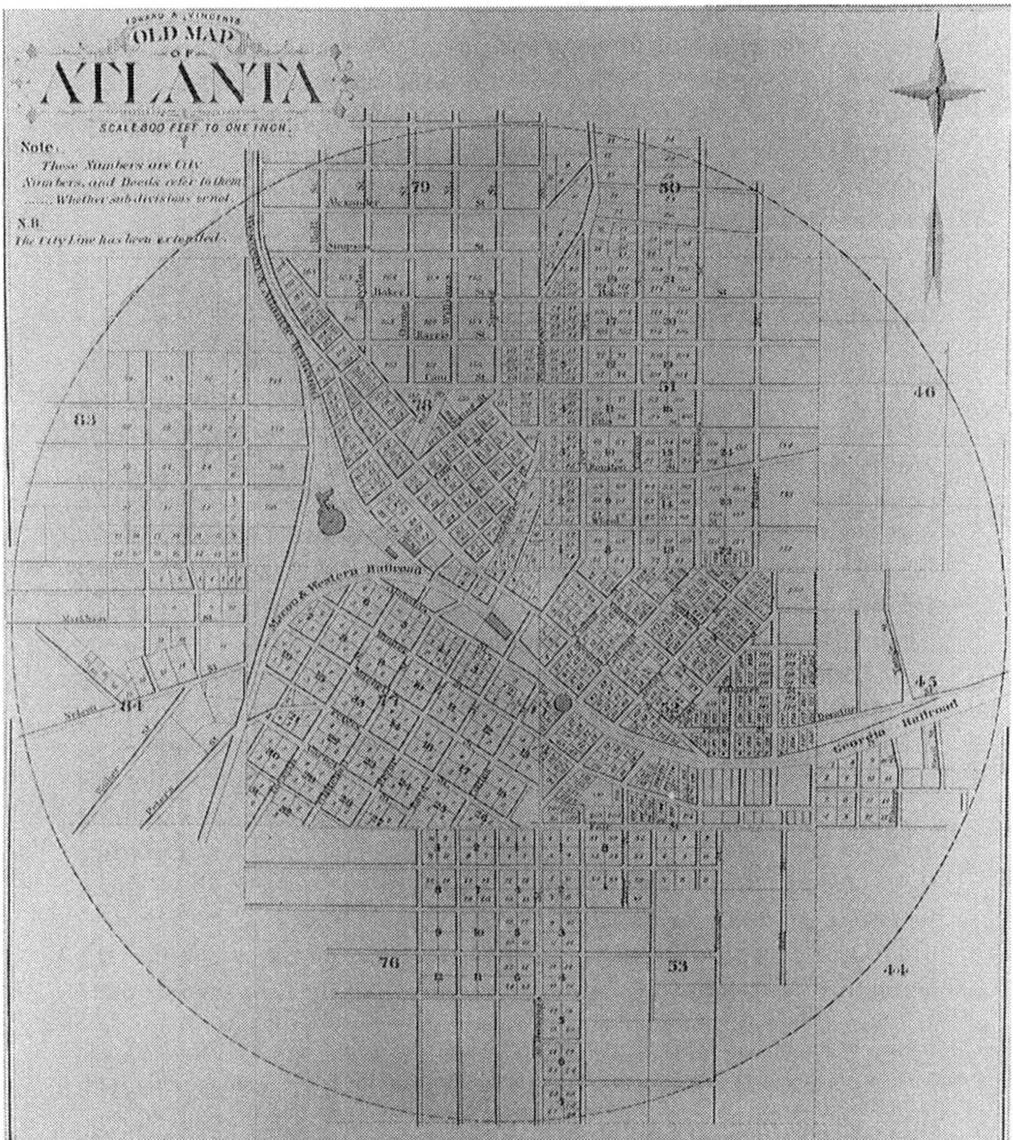

Early map of Atlanta. The State Square is in the center and the common corner of land lots 51, 52, 77, and 78 of the 14th District of original Henry County is a short distance to the northeast. From Griffith M. Hopkins, *City Atlas of Atlanta, Georgia* (1878).

and so numerous were the changes in county lines that the legislators often got vertigo poring too long and intently over the district maps filed in the surveyor general's office. But they repeated few of the absurdities that had been common in designating county lines in the headright regions. Glancing at a map of the state, one can see the general rectilinear shapes of the counties of western and southern Georgia, compared to those of eastern Georgia.[15]

The Passing of the Public Domain

The lottery system proved to be an efficient and popular method of land distribution, but successful drawers tended to be tardy in claiming their grants. Among the reasons were the relative cheapness of land, the usual percentage of poor lots, and a desire to avoid taxation for as long as possible.[16] Throughout the lottery period, headright grants could be obtained for roughly the same total cost as lottery grants. Although the amount of desirable ungranted headright lands diminished with each passing year, availability and proximity of such lands tended to suppress the incentive to move to wild, remote areas to take up unfamiliar lands. Successful lottery drawers often sold their draws before applying for a grant, and it was left to the transferees to secure the actual grants.[17] As a result, it generally required ten or twelve years for the bulk of the lots drawn in any lottery to be taken up.[18] Extending the time for obtaining grants soon became a chronic action of the legislature. To encourage those who had drawn poor lots to make application for their grants, grant fees were gradually reduced. By the mid-1830s fees on some lots were as low as $2.[19] In addition, beginning in 1814 for specified counties, the legislature discontinued the practice of extending the time for fortunate

15. For a penetrating analysis of county boundaries in the lottery areas see Vanderhill & Unger, *Georgia's Crenelated County Boundaries*, Studies in the Social Sciences, June 1977, at 59.

16. Beginning with the 1827 lottery, successful drawers were required to pay taxes on the lots they drew, whether or not they had taken out their grants. Otherwise they forfeited their rights thereto. 1825 ex. sess. Ga. Laws 14; 1830 Ga. Laws 139.

17. A. Powell, Actions for Land 144 (rev. ed. 1946).

18. M. Heath, Constructive Liberalism: The Role of the State in Economic Development in Georgia to 1860, at 147–48 (1954).

19. O. Prince, Digest of the Laws of the State of Georgia . . . to . . . 1837, at 568 (Athens, Ga. 1837).

drawers to claim their grants. Instead, when the time had expired, anyone could acquire any ungranted lot in these counties by paying certain stipulated fees.[20]

Despite considerable attrition to the public domain, a large number of lots remained in the state's possession. Finally, in 1843 an act was passed giving fortunate drawers until October 1, 1844, to take out their grants, after which time all lots remaining unclaimed or on which payments were in default would be deemed forfeited. The act provided that anyone could obtain any full-size lot remaining in the state's possession during October 1844 by paying $2,000; during November by paying $1,500; December, $1,000; January through April 1845, $500; May and June, $250; July and August, $100; September to December, $25; and after December, $5.[21]

This offer produced a response of unanticipated magnitude. By the time the $25 period rolled around, more than four thousand applications with cash payments enclosed had been received and several hundred people had arrived at the capital to purchase land. The $25 period was then extended to February 1, 1846. In addition, a drawing had to be held to determine who would get which lot because in many instances multiple applications had been made for the same land.[22] Contemporaneously (1845), an act was passed to dispose of lots the governor had ordered withheld from granting because of forgeries, alterations, and erasures found in the record books in the surveyor general's office. The act provided for the sale of the lots in a manner similar to that stipulated under the 1843 act.[23] An 1847 act required the sheriffs to advertise and sell the ungranted fractions in their respective counties.[24] In 1854

20. Id.; Lamar 407, 409; 1840 Ga. Laws 121.

21. 1843 Ga. Laws 67; Cobb 706; M. Heath, Constructive Liberalism: The Role of the State in Economic Development in Georgia to 1860, at 150 (1954).

22. Cobb 709; M. Heath, Constructive Liberalism: The Role of the State in Economic Development in Georgia to 1860, at 150 (1954).

23. 1845 Ga. Laws 31; Cobb 707. In the wake of the 1843 act, numerous persons journeyed to the state capital to examine the surveyor general's records to ascertain which lots were still available for taking out grants on. Evidently the surveyor general and the secretary of the executive department altered the records to make it appear that many choice ungranted lots had been granted and thus warded off attempts by others to obtain grants to these lots. As the prices diminished under the 1843 act, these two individuals purchased the lots for themselves at bargain prices. See Letter from S. W. Bloodworth to P. M. Compton (Mar. 26, 1845) (Georgia Surveyor General File, Telamon Cuyler Collection, UGL); Compton v. Killen, 60 Ga. 543 (1877); and 1855–56 Ga. Laws 556.

24. 1847 Ga. Laws 109; Cobb 709.

the governor was authorized to appoint an agent to auction off lots that remained unsold under the 1843 and 1845 acts.[25]

The public domain, which had captivated the attention of Georgians for more than a century, was now but a shadow of its former glory. It had been as much a part of frontier life as the log rollings, house raisings, and militia muster. That this era was now rapidly fading away forever was obvious. The achievement of the surveyors of the public domain had been little short of phenomenal. Since the day that William Bull and James Oglethorpe drove the first stake on the sandy white bluff at Yamacraw, over 37.5 million acres of wilderness had been transformed into private freeholds. Though the state land business was to survive an arrested existence for another half-century, there was little left to do except get all the records in order, compile some consolidated indexes to these records, rebind some musty, time-worn books, and, with diminishing frequency, issue an occasional grant.[26]

By the late 1840s, moves were afoot to merge the office of surveyor general with one of the other state offices. Bureaucratic inertia, however, delayed action until 1861. Then the exigencies of the Civil War caused the duties of the surveyor general to be assigned to the secretary of state.[27] The surveyor general thereafter ceased to function as a separate office.

An act of 1872 required the secretary of state to furnish the ordinaries in each county with a list of a few remaining lots in their respective counties that somehow had been overlooked in the granting process and the ordinaries to put the lots up for sale to the highest bidders.[28] This provision languished in the Georgia code until early in the twentieth century and then disappeared from the statute books. The state had finally been stripped of all public domain except some irregular and unknown tracts in the headright country and a few scattered lots and reservations that amounted to almost nothing.[29]

25. 1853–54 Ga. Laws 73.

26. 1825 ex. sess. Ga. Laws 34; 1835 Ga. Laws 320–21; 1837 Ga. Laws 278–79; 1851–52 Ga. Laws 17; 1855–56 Ga. Laws 556; 1861 Ga. Laws 136; 1865–66 Ga. Laws 325; 1866 Ga. Laws 219; 1868 Ga. Laws 186; 1870 Ga. Laws 504; 1897 Ga. Laws 14, 608.

27. 1861 Ga. Laws 72; Ga. Code of 1863 § 4966.

28. 1872 Ga. Laws 57.

29. Park's Ga. Code of 1914 §§1285, 3811–16. The 1872 act was carried in the official Georgia code of 1910 and in the unofficial Georgia code of 1926. It was omitted from the official code of 1933, the compilers branding it as "obsolete."

Part II. Georgia Land Surveying Law

Chapter 11. Nature of Law

Law is an extremely large field of study, for it deals with the entire range of human affairs. Illustrative of its bulk is the fact that there are at present over 3 million reported judicial opinions in the United States, and some forty thousand are added each year.[1] Beyond these are thousands of pages of statutory law, administrative rules and regulations, and the British case and statutory law applicable to this country.

Kinds of Law

Scholars have classified the legal systems of the Western world into two main categories: the civil law system and the common law system. Civil law is derived primarily from Roman law and from the French Civil Code. It is in force in Latin America and much of continental Europe. In civil law countries the primary source of law is legislation, and the courts are bound not by precedents but by the provisions of comprehensive, systematic codes.[2]

Common law is of Norman-English origin and was spread around the world by the expansion of the British Empire. It is in force throughout the United States (though civil law has much influence in Louisiana), Great Britain, most of Canada, Australia, New Zealand, Ireland, the Philippines, India, and to varying extents in many other countries.[3] Common law consists of those principles, maxims, usages, and rules of action which are based on tradition and immemorial custom.[4] It is evidenced chiefly in court decisions[5] and derives its force from the universal consent and long-standing practice of the people. The common law of any American jurisdiction is generally considered to be the English common law with

1. H. Bitner & M. Price, Effective Legal Research 189 (4th ed. 1979); J. Jacobstein & R. Mersky, Fundamentals of Legal Research 7, 17 (1977).
2. H. Berman & W. Greiner, The Nature and Functions of Law 476–77 (1966).
3. Id.
4. E. Morgan, Introduction to the Study of Law 27–28 (1926).
5. Tucker v. Adams, 14 Ga. 548, 569 (1854); 1 Bl. Comm. 71.

such modifications as are required by the different circumstances and conditions existing in that jurisdiction.[6]

As sources of law the United States also has the U.S. Constitution, state constitutions, enactments by Congress and the state legislatures, codes, treaties made with foreign powers, and local acts and ordinances. Collectively, these legal sources may be termed statutory law.

A third source—administrative law—has come into play especially during the last century. Administrative law encompasses those rules and regulations promulgated by the hundreds of boards and commissions existing at all levels of government.

Suits brought before the courts may be broadly divided into criminal cases and civil cases (the latter not to be confused with the civil law of continental Europe). Criminal cases involve prosecution for wrongs against the public or prosecution for violations of public rights and duties. Civil cases involve the settling of disputes between individuals and the enforcement of private rights and duties. Boundary disputes and most other actions concerning real property are civil in nature. In civil proceedings it is not necessary to prove "beyond the shadow of a reasonable doubt," as in criminal proceedings.[7] All that is necessary is to show a preponderance of evidence[8]—that superior weight of evidence which (though not enough to free the mind wholly from a reasonable doubt) is sufficient to incline a reasonable and impartial mind to one side of the issue rather than to the other.[9] The reason for this rule is obvious. If in civil actions it were necessary to prove beyond the shadow of a reasonable doubt, very few such cases could ever be settled. In resolving boundary disputes and other civil matters, slight evidence will prevail over no evidence at all.

Historical Background

The bulk of Anglo-American law is judge-made, rendered with reference to real-life disputes between individuals. An important aspect of this system is that it has been and is a continually evolving process of trial-and-error experiments, always attempting to retain the useful and practical and to discard the impractical.

6. E. Morgan, Introduction to the Study of Law 28 (1926).

7. O.C.G.A. § 24-4-3.

8. Id.

9. Id. § 24-1-1.

The basic English common law system had its beginnings with the Norman Conquest in 1066. Before this, disputes had been settled locally, according to local custom. One of the means used by the Norman conquerors to unify England under their rule was the establishment of a unified court system throughout the country. Judges appointed to these courts based their decisions on their knowledge of the customs and traditions of the people and on their own ideas of right and wrong. Eventually this system led to the practical replacement of a great mosaic of oral, local customs by a single system of law worked out by legal experts and "common" to all of England.[10]

Beginning in the late 1200s the decisions of judges began to be recorded in a series of books called the Year Books.[11] Thus when a judge was presented with a case, he could refer to similar cases for guidance in his decision. When a large number of judges had decided the same basic question in the same way, the decision became law. After a line of decisions had attained the status of law, it could be changed only when the customs and beliefs of the community changed. This is the origin of the doctrine of stare decisis, which will be more fully described below.

Modern court reporting began in the late 1500s with the publication, by various compilers, of sets of law reports containing a statement of the facts of each case and the judge's decision.[12] The reports were usually cited by the name of the compiler. Thus Plowden's Reports, compiled by Edmund Plowden, contain selected cases from the period 1549–80 and are contained in two volumes. Coke's Reports, compiled by Sir Edward Coke, contain selected cases from the period 1572–1616 and fill thirteen volumes.

When the English settlers came to America it was held, both here and in England, that they brought with them as a birthright and inheritance all English law then in existence, except those laws considered inapplicable or inconvenient to the new country.[13] Although reports of a few colonial cases exist, American law reporting did not really begin until after the Revolution.[14] At first, American jurists followed the custom of

10. A. K. R. Kiralfy, *English Law,* in An Introduction to Legal Systems 157, 158 (J. Derrett ed. 1968).

11. 2 W. Holdsworth, A History of English Law 525–56 (1966 reprint); 1 F. Pollock & F. Maitland, The History of English Law 216 (2d ed. 1898).

12. 5 W. Holdsworth, A History of English Law 355–78 (1966 reprint).

13. Helms v. May, 29 Ga. 121 (1859); State v. Campbell, T. U. P. Charlton 166 (1808); 1 Bl. Comm. 107–8.

14. L. Friedman, A History of American Law 282–85 (1973).

their English counterparts in citing cases by the name of the reporter. Great confusion soon arose, however, because of the large number of jurisdictions in this country and the prevalence of identical last names of reporters in different jurisdictions. As a result, by the mid-1800s most decisions came to be cited by a geographical or jurisdictional name rather than by reporter. Thus volume 1 of the United States Supreme Court Reports, originally cited "1 Dallas" (after Alexander J. Dallas, the reporter), came to be cited "1 U.S."; similarly, volume 1 of the Georgia Supreme Court Reports, originally cited "1 Kelley" (after James M. Kelley), is now cited "1 Ga."[15]

A few brief comments are necessary regarding statutory law. The great legislative body in Great Britain is, of course, Parliament, which has sat since the 1200s. In 1676 Parliament passed a statute entitled "An Act for Prevention of Frauds and Perjuries," more popularly known as the Statute of Frauds.[16] Before this date, when even much of the nobility was illiterate, it was usually necessary for land transfers to be oral. A transfer of title of freehold was generally effected by a symbolic ceremony called livery of seisin. In this ceremony the parties and their witnesses went upon the land and the grantor handed the grantee a twig, a piece of turf, a handful of earth, or a door latch as a token of delivery. As a further part of the ceremony, the grantor used proper and technical words to show that he intended to pass title. Afterward, the possession of the grantee was the only evidence of title, and the memory of the witnesses served as proof of the transfer of title.[17] This system worked well in an era in which land transfers were infrequent, when ownership usually passed from generation to generation, and when witnesses usually spent their lives in the villages and on the farms where they were born. As trade and manufacturing grew and society became more mobile, however, problems began to arise from the oral transfer of real property.

15. To avoid confusing readers who should be astute enough to check volume 1 of the U.S. Supreme Court reports, it will be pointed out that this volume does not contain any of the decisions of that court. Instead, it reports decisions rendered by the courts of Pennsylvania around the time of the Revolution. The decisions of the U.S. Supreme Court begin on page 399 in volume 2.

16. 29 Car. II, c. 3 (1677).

17. Sharp's Case, 6 Co. Rep. 26a, 77 Eng. Rep. 292 (K.B. 1601); Thoroughgood's Case, 9 Co. Rep. 136b, 77 Eng. Rep. 925 (K.B. 1612); 2 Bl. Comm. 315; Co. Litt. 48a, 49b; 1 R. Patton et al., Patton on Land Titles § 3 (2d ed. 1957). An ancient form of livery of seisin is mentioned in the Bible, where it is stated that a conveyance to Abraham was "made sure . . . in the presence of the children of Heth, before all that went in at the gate of his city." Genesis 23:17–18.

The Statute of Frauds made several provisions. One stipulated that any conveyance of an interest in land (other than one that passes by operation of law) has to be in writing and signed by the party to be charged if it is to be enforced in a court of law. The substance of the Statute of Frauds has been enacted in all fifty states, including Georgia.[18]

The statute does not require that a formal document be drawn up. Any writing, contemporaneous or subsequent, in which the party to be charged admits over his signature all the terms of the contract insisted upon is sufficient.[19] Thus a receipt,[20] an auctioneer's memorandum,[21] a personal letter,[22] the minutes of the board of directors of a private corporation,[23] or even mere hasty notes or memorandums scribbled down in books or on paper[24] may suffice to satisfy the terms of the statute.

Georgia's statutes date from the meeting of its first legislative body in 1755. After independence from Great Britain, each of the thirteen original states and many of the newer ones passed "revival of laws" acts. These declared that the English law operable in this country before the Revolution would continue in force. Georgia passed several such acts, the last one dated February 25, 1784. This act declared that all English common and statutory law in effect in the province as of May 14, 1776, and not contrary to the state's constitution, laws, or form of government would continue in full force and effect.[25] As several legal historians have pointed out, however, the passage of such laws was unnecessary because the English legal system was already ingrained in the American consciousness. Americans could no more cast aside the English legal system than they could cast aside the English language.[26]

A Georgia act of 1816 made it a penal offense maliciously or fraudulently to cut, fell, alter, or remove boundary trees and other landmarks. Anyone so doing was subject to a $500 fine and a one-year prison sentence.[27]

18. O.C.G.A. § 13-5-30.

19. Capital City Brick Co. v. Atlanta Ice & Coal Co., 5 Ga. App. 436 (1908).

20. Valdosta Machinery Co. v. Finley, 164 Ga. 706 (1927).

21. Arrendale v. Dockins, 166 Ga. 62 (1927); Ansley v. Green, 82 Ga. 181 (1888).

22. Star Laundry Co. v. May Dry Cleaning Co., 176 Ga. 34 (1932); F. & W. Grand Stores v. Eiseman, 160 Ga. 321 (1925).

23. 72 Am. Jur. 2d Statute of Frauds § 295.

24. Id.

25. Cobb 721.

26. L. Friedman, A History of American Law 95 (1973).

27. 1816 Ga. Laws 165; Lamar 582.

Georgia's first general code went into effect on January 1, 1863.[28] It was compiled under the direction of Richard H. Clark, Thomas R. R. Cobb, and David Irwin. Although revised and enlarged many times, it has never been completely rewritten.

Legal Citation

To the layman legal citation seems mysterious and complicated, but in reality the system is simple. In fact, because of the great mass of legal materials and the need to find specific sources with facility, law has its own unique classification and cross-referencing system that is more comprehensive and yet less complex than that found in any other depository of knowledge. This discussion will not attempt to cover all types of legal citation, but it should give the reader a working knowledge of most citations that will be encountered, as well as those used in this book.

Three types of materials need to be considered: (1) cases, (2) statutes, and (3) treatises and other general works. The general form for citing all of these sources is a number, followed by an abbreviation or group of words, which in turn is followed by another number. The first number is usually (but not always) the volume number. The abbreviation or group of words following it identifies the source. The number that follows this is usually the page number in that source. Very often a date or other explanatory material appears in parentheses. Consider the following example:

2 W. Blackstone, Commentaries on the
Laws of England 315 (Oxford 1766)

This is a reference to page 315 of volume 2 of the first edition of William Blackstone's *Commentaries on the Laws of England,* published at Oxford, England, in 1766.

A very few law treatises, such as Blackstone, have enjoyed such prestige that they practically have the force of law, and shortened forms are usually employed for their citation. For example, Blackstone's work is more commonly cited

2 Bl. Comm. 315

Now consider the following example:

Ga. Code of 1863 § 2208.

28. 1858 Ga. Laws 95; 1861 Ga. Laws 28.

This is a reference to section 2208 of the first Georgia code, which went into effect in 1863. Since it is a single-volume work, there is no need to designate a volume number. The symbol "§" indicates that the reference is to a section rather than to a page number.

Case citations usually give the names of the principal parties in the suit, followed by the information described above. Thus

Stewart v. Latimer, 197 Ga. 735 (1944)

indicates that the parties in this suit were Stewart and Latimer and that the report of the case can be found in volume 197 of the Georgia Supreme Court Reports, beginning on page 735. The case was decided in 1944.

Most cases are reported in "official" reports, published under the authority of the jurisdiction they cover, and also in "unofficial" reports put out by various private companies. The two best-known sets of modern unofficial reports are the "regional" reporters, published by West Publishing Company of St. Paul, Minnesota, and the American Law Reports, published jointly by Lawyers Co-operative Publishing Company of Rochester, New York, and Bancroft-Whitney Company of San Francisco, California.

The West Publishing Company has divided the United States into seven regions and publishes a set of reports for each region. Each set contains all the decisions of the highest courts of each state in that region. Georgia falls in the South Eastern region (abbreviated "S.E."). The other regions are North Eastern (N.E.), Atlantic (A.), Southern (So.), South Western (S.W.), North Western (N.W.), and Pacific (P.). Special sets of reports are also published for the states of New York and California.

The American Law Reports (abbreviated A.L.R.) do not attempt to reprint all decisions but instead contain leading cases selected from each jurisdiction. Following each case is a detailed survey of other cases that concern issues similar to the ones raised in the reported case. This survey serves as a cross-reference and is called an annotation.

A full citation for the case mentioned above would be

Stewart v. Latimer, 197 Ga. 735, 30 S.E.2d 633 (1944).

Some law reports are broken down into series. The "30 S.E.2d 633" in the above citation indicates that the report of the case can be found in volume 30 of the second series of South Eastern Reporter, beginning on page 633. If no series is indicated, then either the reference is to the first series of that set of reports or the set is not broken down into series. The above case is not reported in the A.L.R.

Some cases are not reported at all (that is, they have never been published), and some are reported in as many as six and even eight different sources. Since the Georgia Supreme Court reports and Georgia Court of Appeals reports (abbreviated "Ga. App.") should be readily accessible to the reader, other citations for Georgia cases are not given in this book. At least two citations, when available, will be given for cases decided in other jurisdictions.

As a final example of case citation, consider the following:

Fletcher v. Peck, 10 U.S. (6 Cranch) 87,
3 L. Ed. 162 (1810).

This indicates that the report of the United States Supreme Court case *Fletcher v. Peck* (the famous Yazoo Land Fraud case) can be found on page 87 of volume 10 of the United States Reports, on the same page in volume 6 of William Cranch's reports, and on page 162 of volume 3 of Lawyers' Edition. Lawyers' Edition is published by Lawyers Co-operative Publishing Company. Cranch's and the United States Reports are considered official reports; Lawyers' Edition is considered unofficial.

Several hundred different sets of court reports have been published in the United States and Great Britain over the past four hundred years. This large variety of reports should present little difficulty for the reader, however, because (excluding the state reports) a dozen or so sets contain almost all the cases that will need to be consulted. The state reports are usually cited with an abbreviation of the name of the state, so these likewise should present few problems.

A listing of abbreviations used in legal citation and their meanings can be found in the back of *Black's Law Dictionary.*

Reference to some of the above sets of reports as "unofficial" is not meant to discredit them. In fact, the unofficial are often better than the official reports. Many cases that go unreported in the official reports can be found in the unofficial ones. Moreover, several states have ceased publishing official reports altogether and now rely solely on the unofficial reports published by private companies. Lawyers' Edition of the decisions of the U.S. Supreme Court contains cases not found in the official reports, corrects errors in these reports, has its own internal cross-referencing system, and includes the briefs of the counsels for plaintiffs and defendants, which are usually left out of the official reports. For these reasons, it has become the most popular and widely used set of reports for the U.S. Supreme Court.

Two final examples of legal citation need to be mentioned.

1816 Ga. Laws 165

is a reference to page 165 of the acts of the General Assembly of Georgia for the year 1816.

9 C. J. Boundaries § 18

is a reference to section 18 under the topic "Boundaries" in volume 9 of the *Corpus Juris*. The *Corpus Juris* with its supplement, the *Corpus Juris Secundum* (C.J.S.), is a legal encyclopedia of over two hundred volumes that claims to cite all reported cases dealing with every aspect of American law.

To learn more about legal citation, the reader may consult Miles O. Price and Harry Bitner's *Effective Legal Research* and Frederick C. Hicks's *Materials and Methods of Legal Research.* Basic law collections can be found in county courthouses and in college libraries. Larger collections are located in law school libraries and in the state judicial building in Atlanta.

Some General Judicial Concepts

Law should be regarded more as a method of thinking and reasoning through problems than as a fixed body of definite rules. In other words, the ability to think clearly and to see the material issues of a given situation is of greater importance than the possession of an extensive knowledge of the laws.[29]

Because of its lack of any broad and rigid framework, the common law system is particularly responsive to the pressures of daily life. The rules laid down in cases are the product not solely of academic speculation but also of difficulties that have actually arisen. Thus they are practical rules in close touch with the needs of everyday life. The common law system is also capable of growth; for besides claiming authority to give binding pronouncements regarding the interpretation of legislative enactments, the courts authoritatively lay down new rules of law from time to time to meet new circumstances and the changing needs of society.[30]

One of the noteworthy features of the common law system is the strong lay contribution. Laymen devise a transaction or create a relationship; their legal advisers clothe it in legal form; and the court, in case of dispute, passes judgment on its legal validity. Laymen do not wait for judicial

29. R. Pound, The Spirit of the Common Law 1 (1921).

30. W. Geldart, Elements of English Law 16–17 (6th ed. 1963); A. K. R. Kiralfy, *English Law,* in An Introduction to Legal Systems 157, 159, 163 (J. Derrett ed. 1968).

approval before acting, nor is anyone concerned with the tidy place to be filled by their conduct in the legal pattern.[31]

An underlying assumption of the Anglo-American legal system is that all acts are lawful, all transactions duly formal, and all persons innocent of wrongdoing. Every transaction is given the effect the parties to it intend unless the transaction violates some paramount consideration of the legal order.[32]

An important legal principle that does much to check judicial caprice and has greatly strengthened the solidity and certainty of law is the doctrine of *stare decisis et non quieta movere* (literally, let it stand as decided and do not disturb what is settled). This self-imposed doctrine, usually termed simply stare decisis, requires courts to follow previously rendered decisions, provided that the principle so extracted is logically essential to the decision and is both reasonable and appropriate to the situation at hand. Thus stare decisis helps make law predictable so that people may judge with reasonable certainty the outcome of their actions.

The two highest state courts in Georgia are the Supreme Court and the Court of Appeals. The decisions of these two courts are binding on all the lower courts in the state and also on the federal courts in matters in which state law prevails.[33] Cases decided in other jurisdictions carry much persuasive authority but are not binding on the courts of Georgia.[34] Federal decisions are binding on the Georgia courts only so far as they relate to federal matters.[35]

The doctrine of stare decisis is not a straitjacket, however, for judges can overrule previous decisions if they feel that such decisions are erroneous or obsolete.[36] Thus the rule neither renders the courts impotent to correct past errors nor requires them to adhere blindly to rules that have lost their reason for being.[37] In addition, judges many times make

31. A. K. R. Kiralfy, *English Law,* in An Introduction to Legal Systems 157, 162 (J. Derrett ed. 1968).

32. Id.

33. Ga. Const. of 1983, art. 6, § 5, para. 3; Id., art. 6, § 6, para. 6.

34. Thompson v. Eastern Air Lines Inc., 200 Ga. 216 (1946); Miles v. State of Georgia, 96 Ga. App. 610 (1957).

35. The Georgia Railroad v. Cubbedge Hazelhurst & Co., 75 Ga. 321 (1885); Georgia R.R. & Banking Co. v. Stanley, 38 Ga. App. 773 (1928).

36. Ellison v. Georgia R.R. Co., 87 Ga. 691 (1891); Cain v. Monroe, 23 Ga. 82 (1857).

37. Fox v. Snow, 6 N.J. 12, 76 A.2d 877 (1950) (dissenting opinion); Flagiello v. Pennsylvania Hospital, 417 Pa. 486, 208 A.2d 193 (1965).

fine shades of distinction between previous cases and the case at hand. The rule of stare decisis is generally followed more rigidly with regard to property rights than other types of legal relationships.[38]

A term often confusing to the layman is the legal meaning of the word *intent*. From a legal standpoint, a person's intent is what is manifested by his words and actions, not what is hidden in his mind, since neither law nor science has ever discovered a method of reading a person's mind.[39] Thus what a person actually thinks is immaterial in determining his legal standing with others. In construing deeds and other instruments, the primary question is not what the parties meant to say but the meaning of what they did say.[40] A surveyor's intention is determined by what he actually does in making the survey, the description of the land he gives, and all the circumstances attending the transaction.[41]

The following chapters will examine several legal matters of interest to land surveyors. One word of caution is in order, however: law is not nearly as simple and clear-cut as it has been made to appear. This is because, in reality, law is not a set of disconnected principles but a single integrated mass. Its constituent parts are so completely intertwined that it is difficult to apply one segment of this mass to a given issue without examining in detail the ramifications of all the other segments. Ultimately each of these interacting segments will have an effect on the disposition of the matter under consideration. Unfortunately, to investigate this mass called law effectively, one must dismember it into a number of more or less arbitrarily selected categories and then place arbitrarily chosen limits on the categories selected. Thus treatises such as this sometime tend to give law the appearance of an oversimplified and inflexible method of resolving real-life issues. The reader is therefore urged, in addition to perusing this text, to study cases carefully and diligently contemplate their application to actual experience.

38. 20 Am. Jur. 2d Courts § 196.

39. Dunaway v. Windsor, 197 Ga. 705 (1944); Virgin v. Roe, 32 Ga. 572 (1861); Weniger v. Ripley, 134 Or. 265, 293 P. 425 (1930); 23 Am. Jur. 2d Deeds § 225; 13 C.J. Contracts §§ 49, 483; H. Elphinstone, Rules for the Interpretation of Deeds 36–41 (1889).

40. 23 Am. Jur. 2d Deeds § 225.

41. Masterson v. Ribble, 34 Tex. Civ. App. 270, 78 S.W. 358 (1904).

Chapter 12. Types of Real Property Interests

The topic of estates and other interests that make up landownership forms the heart of real property law. Because the surveyor is primarily concerned with defining the boundaries of a parcel of land, not with the title or interest the owner has in it, however, the subject is usually somewhat incidental to his professional function. Nevertheless, a general knowledge of the legal interests that exist in land is extremely desirable, and in some cases absolutely essential, if the surveyor is to perform his duties properly.

Some of the most intricate and involved principles of all law lie in the law of estates, and the very simplistic treatment given here cannot hope to impress on the reader the enormity and great complexity of the subject. Rather, this discussion is intended to do little more than briefly outline the major interests that constitute real property ownership.

Underlying Concepts

Property over which people exercise dominion naturally falls into two broad classes—personal and real.[1] Personal property (personalty) includes items that are movable in nature such as cars, furniture, tools, money, stocks, domesticated animals, and agricultural products, as well as rights in or issuing out of such items.[2]

Real estate or realty includes land, all things permanently attached to it, and anything intended to remain permanently in its place though not attached and any interests or rights in or issuing out of or dependent upon land.[3]

There is a special class of items of realty known as fixtures. A fixture

1. Fears v. State, 102 Ga. 274 (1897); O.C.G.A. § 44-1-1.
2. O.C.G.A. § 44-1-3.
3. Smith v. Odom, 63 Ga. 499 (1879); O.C.G.A. §§ 44-1-2, 44-1-6.

is an article or thing that was originally personalty but has become a part of the realty through the process of annexation or attachment.[4] For example, bricks are personal property when stacked on land and become real property if a building is constructed from them. If the building is destroyed by accident and the walls fall in, the bricks remain real property as long as they are left lying where they fell; but if the owner stacks them up or removes them, they will again become personal property.[5] Other examples of items usually held to constitute fixtures include wiring, counters, plant machinery, telephone switchboards, fences, door locks, water heaters, furnaces, and plumbing.[6]

Easements and licenses are rights in land in which the owner thereof has no title or ownership of the land itself.

Land has in its legal significance not only the earth's surface but also an indefinite extent upward as well as downward.[7] Thus landownership is actually three-dimensional: surface, subsurface, and air rights can be separately owned.

In most cases personalty is created specifically for use or sale by the creator and clearly belongs to its producer. The life of most items of personal property is relatively brief, and they are eventually consumed or pass into the rubbish heap. Consequently, the issue of original ownership of such items is usually easy to settle and does not become increasingly difficult over time.[8]

Land constitutes an interesting contrast to personalty. It is indestructible, immovable, and—with minute exceptions—not produced by people. Additionally, it is the source from which most other wealth must come.[9] As a result, a specialized and highly complex body of law has emerged concerning the ownership and disposition of land as compared to personal property.[10] Ownership in land is often characterized as a "bundle of sticks."[11] To say that a person "owns" a piece of land is really

4. Hargrove v. Jenkins, 192 Ga. App. 83 (1989); Ga. Real Est. Law § 10-1 (3d ed. 1986).

5. Guernsey v. Phinizy, 113 Ga. 898 (1901).

6. Ga. Real Est. Law §§ 10-18 to 10-36 (3d ed. 1986).

7. Thrasher v. City of Atlanta, 178 Ga. 514 (1933); Smith v. City of Atlanta, 92 Ga. 119 (1893); O.C.G.A. § 44-1-2.

8. J. Cribbet, Principles of the Law of Property 11 (3d ed. 1989).

9. Id.

10. One generally does not, for example, have an estate in personal property as in real property.

11. M. Harris, Origin of the Land Tenure System in the United States 4 (1953).

to assert that he is vested by law with a mass of rights, privileges, powers, and immunities with respect to the land, to the exclusion of others. Each of these rights, privileges, powers, and immunities constitutes a "stick" in the bundle.[12] The more sticks a person has, the greater his sum total of rights in the land. No one person ever owns all of the sticks. Consider, for example, a large office building that has been left to a widow for life, with her children succeeding to the interest upon her death. On this office building an outstanding mortgage exists in favor of a bank. In addition, it has been leased out for ninety-nine years, the lessee having in turn subleased parts of the building to various tenants. Obviously no one person "owns" the building, for there are many estates and interests in it parceled out among several persons.[13]

Even if all these rights and interests were vested in one person, that person still would not have absolute ownership. The state has certain sticks in the bundle, such as the power of taxation, condemnation, zoning, and police regulation; the adjoining landowners have the right to light, air, natural drainage, and lateral support; and the public at large has or may have the right to abate pollution and other nuisances, the right to enforce restrictive covenants, the right to use easements on the land, and riparian rights in the land.[14]

Before entering into a discussion of the specific types of interests that exist in land, it is necessary to describe briefly the historical background of the law of estates.[15]

Historical Background

Probably no other area of law is so firmly rooted in the past as the law of estates, and although it would be incorrect to characterize this system as feudal, it did grow out of the feudal system. For this reason some knowledge of the feudal background is essential to any real understanding of modern-day land law. This law has been a thousand years in the making,

12. Id.; W. Hohfeld, Fundamental Legal Conceptions 12 (1923).

13. Ga. Real Est. Law § 1-13 (3d ed. 1986); L. Friedman, A History of American Law 51 (1973).

14. Ga. Real Est. Law §§ 1-12, 1-17, 6-31, 6-39, 14-1, 14-4 to 14-7 (3d ed. 1986).

15. Most treatises on real property describe its historical background. The following is summarized from C. Moynihan, Introduction to the Law of Real Property 1–24 (2d ed. 1988), and H. Tiffany, A Treatise on the Modern Law of Real Property 13–23 (abr. ed. 1940).

during which time vast changes have taken place. The process has been one of evolution, not revolution.

Feudalism is a generic term that may be used to describe the social structure of western Europe during the Middle Ages. It arose out of the chaos that enveloped Europe after the fall of the Roman Empire, when almost all semblance of order and stability disappeared. During this period, trade and manufacturing were practically nonexistent; negotiable paper, stocks, and bonds unknown; and precious metals scarce. Land constituted almost all the known wealth. Under these circumstances the law regarding personal property was meager and comparatively simple, but the land law became one of the most technical and complex areas of human learning.

The fundamental theory of the feudal system was that all land was held either directly or indirectly of the king on the condition of rendering certain duties and services. A grantee of the crown could parcel out his grant to others without affecting his relationship with the crown, and these subgrantees could in turn parcel out their grants to others. In each case the grantee was a vassal to the person from whom he received his grant and was lord to those whom he made grants. The relationship between a lord and his vassal was termed tenure, and the land thus held was called a feud, fief, or fee. The word *feudal* and the *fee* of modern law come from these terms. This process of parceling out land to others who would then parcel it out to still others was called subinfeudation. It theoretically could go on indefinitely and in practice resulted in as many as six or seven lords (called mesne or middle lords) standing between the king (the lord paramount) and the lowest scale of tenants who actually worked the land (the tenants in demesne or tenants paravail). The concept of landownership as it exists today was unknown during this period. One "held" land of his lord rather than "owned" it.

In return for his feud each vassal was obligated to perform certain services for his lord, depending on the terms agreed upon at the time of the creation of the tenure. Tenures were either free tenures based on services considered proper for a free man to render, or were unfree tenures based on services of a menial nature, usually involving the cultivation of the lord's land. This is the origin of our classification of estates into freehold and nonfreehold categories.

There were three types of free tenure: (1) tenure in chivalry, (2) tenure in frankalmoin, and (3) socage tenure. Tenure in chivalry, originally the most prestigious form, usually involved the performance of military service by the tenant for his lord and frequently the furnishing of a specified quota of knights. The vassal obtained these additional knight services by subinfeudating his lands, as described above, on the condition that the grantees render the necessary service. Frankalmoin tenure arose from making a gift of lands to a church, monastery, or other religious body in return for services of a religious nature (such as saying masses or prayers). Socage tenure involved the rendition of fixed and definite agricultural services or a periodic payment of a definite amount of money to the lord. This quality of definiteness in the type and amount of services or money due distinguished socage tenure from unfree tenure.

Tenants in villeinage or serfs were originally bound to the soil under a form of tenure known as unfree, base, or villein tenure. They gradually acquired the status of free men, but the tenure itself continued to be classified as unfree.

Whatever the type of service required of a tenant, it was regarded as attaching to the land and running with it. Thus if a tenant should transfer his feud, the transferee was obligated to perform the service that had been required of the transferor.

Viewed as a whole, the system's structure was pyramidal in shape with the king at the apex, the mesne lords of varying degrees in the middle, and the serfs forming the broad base. Each person in the scale, except the king and the serfs, was tenant of the one above him and lord to those holding under him.

Certain "incidents" of tenure arose from the feudal relationship, independent of any express agreement between the lord and his vassal. These included homage, fealty, aids, reliefs, wardship, the right of marriage, fines, and escheat. Homage and fealty were ceremonial acts whereby the vassal acknowledged loyalty and obedience to his lord. Aids were financial contributions exacted from the vassal upon the knighting of the lord's eldest son, or for assistance in providing a dowry for the lord's eldest daughter, or for ransoming the lord himself if he were taken prisoner. Reliefs were sums paid to the lord by the heir of a deceased tenant for the privilege of succeeding to his ancestor's lands. During this period inheritance was a privilege that had to be paid for, not an unconditional right. Wardship was the right of the lord to the land as well as to the custody of the body of the minor heir of a deceased tenant in chivalry until such time as the heir should come of age. The right of marriage grew out of the right of wardship and was the right of the lord to dispose of the ward in marriage. Fines were payments exacted by the lord from his tenant for allowing the tenant to alienate or transfer his feud. Escheat was the return of a feud to the lord upon the death of a tenant without heirs or upon the tenant committing an act of treason or felony against the lord.

Such (in an all-too-brief summary) is how land tenures existed in England during the feudal period—its heyday running roughly from the time of the Norman Conquest in 1066 until the end of the thirteenth century. The feudal system was not just a system of land tenures; it was a way of life that pervaded both government and society. One's relationship to the land determined his status—hence the use of the word *estate* in describing the extent and character of a person's interest in land.

At first glance it may appear that all the advantages of the feudal system were on the side of the great lords, but one must remember that this was an age of violence and disorder with no strong central government. By placing himself under the protection of a more powerful lord, the vassal received a certain amount of security that could not be obtained otherwise.

The feudal system probably began to disintegrate almost from the time of its inception. It had arisen out of a need for stability and security, and the remedy it provided was a military society based on knight service. As the central government became stronger and society more orderly, tenures in chivalry declined in importance. Eventually tenants bound to perform military services were allowed to substitute a monetary payment. In 1290 an act was passed abolishing alienation by subinfeudation and providing that all alienations thereafter had to be by substitution.[16] Thus a tenant might transfer part or all of his feud without the payment of a fine to his lord, but the transferee would hold—not under the transferor—but under the transferor's lord. Since all growth of the feudal pyramid was thereby prevented, the death of the system was the inevitable consequence. By escheat mesne lords were gradually eliminated and the pyramid was progressively flattened until most lands came to be held directly of the king. The growth of commerce and industry made socage tenures more desirable than tenures in chivalry. In 1660 a parliamentary act converted all military tenures to tenures in "free and common socage."[17] By the same act most of the incidents of tenure were abolished, and the freeholder became practically the owner of the soil.

Although the settlement of America came late enough to escape most

16. Statute Quia Emptores, 18 Edw. I, c. 1 (1290).

17. Statute of Tenures, 12 Car. II, c. 24 (1660).

of the incidents of feudal tenure, there is no doubt that lands in the American colonies were to some extent held in tenure.[18] For example, Georgia's charter of 1732 provided that the grant made to the Trustees was "to be holden of us [the crown] . . . in free and common soccage" at a quitrent (service) of four shillings per year for every hundred acres granted.[19] After Georgia became a royal colony, land grants of the king to individuals bore similar language.[20]

The American Revolution destroyed any tenurial relationship between landowners in this country and the English king. By express statutory provision, lands today are held allodially (by a person in his or her own right) rather than by tenure.[21]

Classification

From its feudal origins, modern real property law has evolved into an extraordinary variety of forms of ownership in land. The forms that have survived to the present day did so mainly because they were found conducive to the needs of modern society. Their most notable feature is their great flexibility, which allows the creation of a diversity of rights and interests in land to meet a variety of human needs.

One of the fundamental aspects of real estate law is that property in land is made up of a fragmented scheme of ownership called estates. The principal concept of the theory of estates is that proprietary rights are projected on the plane of time. For a given parcel of land, there is one fee simple estate, which is of infinite duration. The fee simple may, however, be fragmented into interests of lesser duration such as life estates, estates for years, reversions, remainders, and so on. The total of these fragments make up the fee simple.[22] In addition, landownership may be viewed with reference to the time of enjoyment of the estate and with

18. 1 Am. Law of Prop. § 1.41; Barnes, *Land Tenure in English Colonial Charters of the Seventeenth Century*, in Essays in Colonial History Presented to Charles M. Andrews by His Students 4 (1931); H. Tiffany, A Treatise on the Modern Law of Real Property 22–23 (abr. ed. 1940).

19. 1 Col. Rec. Ga. 18. On the subject of quitrents see Beverley W. Bond, Jr., The Quit-Rent System in the American Colonies (1919).

20. 27 Col. Rec. Ga. 148–53.

21. O.C.G.A. § 44-5-2; Gibbons v. Gibbons, T. U. P. Charlton 113 (1807).

22. 1 Am. Law of Prop. §§ 1.7, 4.1.

reference to the number and connection of the owners of the estate. The general scheme of landownership as viewed from these standpoints may be outlined as follows:[23]

I. With reference to the duration and extent of the estate
   A. Freehold estates
      1. Estates of inheritance
         a. Fee simple
         b. Fee tail
         c. Defeasible fee
      2. Estates not of inheritance (Life estates)
         a. Estate for the life of the tenant
         b. Estate pur autre vie
   B. Nonfreehold estates (Leasehold estates)
      1. Estate for years
      2. Estate from period to period
      3. Estate at will
      4. Estate at sufferance
      (2–4: Classified as usufruct in Georgia)
II. With reference to the time of enjoyment of the estate
   A. Estates in possession
   B. Estates in expectancy (Future interests)
      1. Possibility of reverter
      2. Right of entry for condition broken
      3. Remainder
      4. Reversion
      5. Executory interest
III. With reference to the number and connection of the owners of the estate
   A. Estates in severalty
   B. Concurrent estates
      1. Joint tenancy
      2. Tenancy in common

In general, one who has an estate in land has the right of exclusive possession as well as the right to the profits thereof. There are other interests or rights in land, including easements and licenses, which fall outside the purview of estates. This is because these interests confer upon the

23. This outline, compiled by the author, is a modernized and greatly simplified version of that followed by Blackstone in his *Commentaries*. See 2 Bl. Comm. 103–95. Dower, curtesy, and several other interests extensively treated by Blackstone are omitted here because they are either obsolete or have been abolished in Georgia.

possessor merely a limited right of use of the land in conjunction with the fee owner and carry no rights to the profits. At law these interests are treated not as fragments of ownership in land but as burdens upon that ownership.

This discussion still has only hinted at the full meaning of the complex aggregate of legal incidents that make up landownership, for persons may enter into contractual relationships respecting the use of their lands, all interests in land carry with them a great number of rights and duties, and all lands are encumbered to some degree with a large assortment of legal and natural restrictions as to their use. These aspects are so diverse and variable that there is no conceivable way they can be cast into any general classification system. They lie well beyond the scope of this book and are mentioned merely to demonstrate the vastness and complexity of landownership.[24]

Using the general outlines that have been given, the major estates and interests in land will be briefly described.

Fee Simple

An estate in fee simple is one in which the owner is entitled to the entire property, with unconditional power of disposition during his life, and which descends to his heirs and legal representatives upon his death intestate.[25] It is the greatest and most extensive estate a person can have in land, every other estate being derivable therefrom and mergeable therein.[26] The term *fee simple absolute* is practically synonymous with the term *fee simple;* and the word *fee* standing alone, signifies a fee simple estate.[27] The fee simple is by far the most commonly conveyed class of estate in land and, in case of doubt, the law presumes a fee simple is intended.[28]

Originally, under the feudal system, estates were usually granted to a tenant for life only, with reversion to the lord upon the tenant's death. As time went on, grants were gradually extended to the sons and other issue of the tenant, who would stand in the tenant's place upon his death.

24. O. Browder et al., Basic Property Law 11–12, 341 (1966).
25. Houston v. Coram, 215 Ga. 101 (1959); O.C.G.A. § 44-6-20.
26. Gibbons v. Gibbons, T. U. P. Charlton 113 (1807).
27. 21 C.J. Estates § 8; 31 C.J.S. Estates § 8.
28. Montgomery v. Pierce, 212 Ga. 545 (1956); Brice v. Sheffield, 118 Ga. 128 (1903).

For the issue of the tenant to inherit, however, the grant had to be to the tenant "and his heirs," else the interest conveyed was a life estate only. Eventually estates were allowed to pass to collateral as well as to lineal heirs, and the tenant acquired complete freedom to alienate his property. Through this process the word *heirs* ceased to designate the person or persons who would take the place of the original grantee and came to mean, instead, that the grantee took an estate that would pass to his heirs or to the heirs of anyone to whom he transferred it. It was a rule at common law that the use of the word *heirs* was necessary to create a fee simple title.[29] This rule has been abolished in most jurisdictions, including Georgia. Thus although conveyances still frequently use words of inheritance, such terms are no longer necessary to create a fee simple.[30]

Although the fee simple represents the ultimate in landownership, the right of an owner to use his land as he pleases is considerably curtailed by the rights of adjacent landowners, by government regulation, and by any easements, restrictive covenants, or other encumbrances upon the land.[31]

Fee Tail

An estate limited to the grantee and his direct descendants and reverting to the grantor or his heirs or transferees if the grantee's line dies out is known as a fee tail.[32] A fee tail at common law could be limited to the bodily heirs of the grantee generally, or limited to the grantee's bodily heirs by a certain wife, or restricted to either his male or his female issue.[33] Such conveyances were popular in England during feudal times and later because they tended to prevent lands from escheating to the lord. They were also used to some extent in this country during its earliest years but quickly fell into disfavor because this country, in general, lacked a large landed aristocracy. One of the most unpopular acts of the Trustees in colonial Georgia was their policy of granting lands in tail male.[34] It was probably the result of this experience that Georgia, by its Constitution of

29. H. Tiffany, A Treatise on the Modern Law of Real Property 35–36 (abr. ed. 1940).

30. Grant v. Haymes, 164 Ga. 371 (1927); Craig v. Ambrose, 80 Ga. 134 (1887); O.C.G.A. § 44-6-21.

31. C. Moynihan, Introduction to the Law of Real Property 32 (2 ed. 1988).

32. Cole v. Ogg, 180 Ga. 343 (1935).

33. 2 Bl. Comm. 113–14; Co. Litt. 19b–25b.

34. See page 16.

1777, became one of the first states to prohibit the entailing of estates.[35] The constitution, however, failed to specify the effect of an attempted entailment, and conflicting opinions were issued by the courts. To resolve this confusion, the legislature enacted in 1821 that the use of words of entailment in a conveyance will vest an absolute unconditional fee simple title in the first taker of the estate.[36] Thus a grant "to A and then to his bodily heirs" vests fee simple title in A.[37]

There is one class of estates tail, however, that does not give the first taker a fee simple under the statute: the estate tail by implication. For reasons too lengthy and technical to discuss here, this class of estates tail is converted to either a life estate followed by alternative contingent remainders in fee simple or to a defeasible fee with an executory limitation over, depending on the exact language used.[38] The estate tail has been abolished (primarily by statutes converting it to some other type of estate) in almost all of the United States.

Defeasible Fee

A defeasible fee is an estate in fee that may endure forever, as would a fee simple, but is subject to termination if a specified contingency occurs.[39] A commonly cited example is a grant "to A but if she dies without children then to B."[40] Here A takes an estate in fee, which may run indefinitely as would a fee simple, but the estate is subject to termination if A dies without children. If this contingency occurs, the land passes to B. Other examples of defeasible fees include conveyances to someone until the happening of a certain event that may or may not occur,[41] or for as long as the property is used for a particular purpose,[42] or on the condition

35. Marbury & Crawford 12; 1 Rev. Rec. Ga. 295.

36. Raines v. Duskin, 247 Ga. 512 (1981); 1821 Ga. Laws 92; O.C.G.A. § 44-6-24.

37. Cole v. Ogg, 180 Ga. 343 (1935); Stamey v. McGinnis, 145 Ga. 226 (1916).

38. See Raines v. Duskin, 247 Ga. 512 (1981); Hertz v. Abrahams, 110 Ga. 707 (1900); Ga. Real Est. Law § 7-8 (3d ed. 1986); and Agnor, *Estates Tail in Georgia,* 13 Ga. B.J. 27 (1950).

39. Ga. Real Est. Law § 7-10 (3d ed. 1986).

40. Reynolds v. Rackley, 223 Ga. 586 (1967); Moody v. Baxley Turpentine Corp., 195 Ga. 482 (1943); McCoy v. Olive, 168 Ga. 492 (1929).

41. McDonald v. Suarez, 212 Ga. 360 (1956); Logan v. Hammond, 155 Ga. 514 (1923).

42. Franks v. Sparks, 217 Ga. 117 (1961); City of Barnesville v. Stafford, 161 Ga. 588 (1925); Moss v. Chappell, 126 Ga. 196 (1906); Atlanta Consolidated Street Ry. Co. v. Jackson, 108 Ga. 634 (1899).

that the property not be used for a particular purpose,[43] or for as long as the grantee performs certain specified acts or services.[44] Road rights-of-way and railroad rights-of-way are often conveyed as defeasible fees, the instrument providing that the property shall revert to the grantor if it ceases to be used for the purpose for which it was granted.

If the event on which the estate is to be determined occurs, then a fee simple vests in either the grantor or a third person, depending on the language used in the conveyance.[45] If the happening of the event becomes impossible of occurrence, then the estate in the grantee becomes fee simple.[46]

The owner of a defeasible fee has the same rights and privileges over his estate until the contingency occurs as if he were a tenant in fee simple.[47] For example, subject to the contingency, he may encumber, transfer, or sell the property and it may descend to his heirs.[48]

Life Estates

The duration of a life estate is limited to the life or lives of some particular person or persons.[49] The owner of such an estate is called the life tenant. Life estates may be subclassified into estates for the life of the tenant and estates pur autre vie.[50] As an example of the former, land is conveyed or devised to someone for life and at his death to go to his children or to some other specified person or persons.[51] To use another example, an owner in fee simple may convey his entire interest in a piece of land but reserve to himself a life estate thereto.[52] An estate pur autre vie is created when land is conveyed or devised to a person for the lifetime of some

43. Flaum v. Middlebury, Inc., 246 Ga. 682 (1980); 6 Am. Law of Prop. § 27.21; 1 R. Patton et al., Patton on Land Titles § 208 (2d ed. 1957).

44. Plemmons v. Sharp, 156 Ga. 571 (1923); Johnson v. Tullis, 152 Ga. 232 (1921).

45. 28 Am. Jur. 2d Estates § 27.

46. 31 C.J.S. Estates § 10.

47. Hill v. Alford, 46 Ga. 247 (1872).

48. 28 Am. Jur. 2d Estates § 26; 31 C.J.S. Estates § 10.

49. O.C.G.A. §§ 44-6-80, 44-6-81.

50. 21 C.J. Estates § 65; 31 C.J.S. Estates § 31.

51. Pearre v. McDonald, 168 Ga. 752 (1929); Green v. Driver, 143 Ga. 134 (1915); Taylor v. Kemp, 86 Ga. 181 (1890).

52. Kytle v. Kytle, 128 Ga. 387 (1907); Goodrich v. Pearce, 83 Ga. 781 (1889).

person other than the tenant himself.[53] For example, a conveyance may be made to someone for the lifetime of the grantor;[54] or a tenant holding an estate for his own life may convey his full interest, in which case the grantee takes an estate for the lifetime of his grantor.[55]

A life tenant is entitled to the full use and enjoyment of the property for the duration of his estate in it, provided he exercises ordinary care in the use of the property and does nothing to cause permanent injury to it.[56] He may, for example, lease the land out, thin timber in accordance with good forestry practices, mine or quarry it (in some cases), and convey up to his full interest in the property.[57] In general he is obligated to keep the property in good repair, to pay taxes and assessments, and to pay the interest on any outstanding loans on the property. He may also be obligated to keep the property insured and may not commit waste.[58]

Upon termination of a life estate, the residue is known as a reversion if it returns to the original grantor or his heirs and a remainder if it goes to third parties. Reversions and remainders along with certain other interests are known as future interests and will be discussed shortly.

Leasehold Estates

Leasehold estates include the estate for years, the estate from period to period, the estate at will, and the estate at sufferance. In Georgia leasehold interests other than the estate for years have generally been classified as usufructs.

An estate for years is limited in its duration to a period made fixed and certain by the instrument creating it.[59] Despite the implication of its name, it may be limited to a period of less than a year. The key is that it has a definite starting point and a definite ending point. It may originate by deed or will in which a remainder is granted or devised to someone

53. 21 C.J. Estates § 66; 31 C.J.S. Estates § 31.

54. Aycock v. Williams, 185 Ga. 585 (1938).

55. 1 Am. Law of Prop. § 2.17b; 31 C.J.S. Estates § 51.

56. Sutton v. Bennet, 215 Ga. 379 (1959).

57. Ga. Real Est. Law §§ 7-19 to 7-27 (3d ed. 1986).

58. Id. §§ 7-28 to 7-32, 7-51. Waste is the failure to exercise ordinary care for the preservation and protection of the property or the commission of acts tending to do permanent injury to it so as to reduce the value of the interest of the reversioner or remainderman. Id. § 7-51.

59. Hutcheson v. Hodnett, 115 Ga. 990 (1902); O.C.G.A. § 44-6-100.

else but is usually created by lease leaving a reversionary interest in the grantor.[60] Except for duration it is in most ways analogous to a life estate. The parties thereto, although often referred to as the landlord and tenant, are more properly called the grantor and grantee.[61] The grantee is regarded as the owner of the land during the continuance of his term. His interest may be bought and sold as any other real estate unless something in the grant prohibits it.[62] He is liable for taxes and assessments, and his interest may be made subject to a levy or lien.[63] A lease of land for one hundred years or more is substantially the equivalent of a conveyance of the fee.[64]

An estate from period to period is one that continues for successive similar periods (for example from year to year, month to month, week to week) until terminated by proper notice from either party at the end of a period.[65] An estate at will has no designated period of duration and is terminable at the will of either party to the transaction.[66] An estate at sufferance is really no estate at all. It arises when a person who had a possessory interest in land (such as an estate pur autre vie, an estate for years, an estate from period to period, or an estate at will) remains in possession after the termination of his estate without the consent of his landlord or other party who succeeds to the title.[67]

At common law and in most jurisdictions today, the estate from period to period, the estate at will, and the estate at sufferance are the classifications given to leasehold interests other than the estate for years. The parties are referred to as the landlord and tenant. By statute, how-

60. Ga. Real Est. Law § 7-110 (3d ed. 1986); 1 Am. Law of Prop. § 3.13.

61. Southland Inv. Corp. v. McIntosh, 137 Ga. App. 216 (1976); Evans Theatre Corp. v. DeGive Inv. Co., 79 Ga. App. 62 (1949).

62. Clark v. Herring, 43 Ga. 226 (1871); Shell Petroleum Corp. v. Jackson, 47 Ga. App. 667 (1933); Ga. Real Est. Law §§ 7-115, 7-121, 11-5, 11-27 (3d ed. 1986).

63. Henson v. Georgia Ind. Rlty. Co., 220 Ga. 857 (1965); Delta Air Lines, Inc. v. Coleman, 219 Ga. 12 (1963); Wilson Mfg. Co. v. Chamberlin, 140 Ga. 593 (1913); Evans Theatre Corp. v. DeGive Inv. Co., 79 Ga. App. 62 (1949); Harms v. Entelman, 21 Ga. App. 295 (1917).

64. Wright v. Central of Georgia R.R. Co., 146 Ga. 406 (1917); Atkinson v. Orr, 83 Ga. 34 (1889); Seaboard Coast Line R.R. v. Adcock, 119 Ga. App. 812 (1969); S. Mitchell, Real Property under the Code of Georgia and the Georgia Decisions 208–9 (5th ed. 1960).

65. 1 Am. Law of Prop. § 3.23.

66. Id. § 3.28.

67. Morgan v. Maddox, 216 Ga. 816 (1961); Taylor v. West, 142 Ga. 193 (1914); Willis v. Harrell, 118 Ga. 906 (1903); Williams v. Durham, 77 Ga. App. 840 (1948); Carruth v. Carruth, 77 Ga. App. 131 (1948).

ever, Georgia has departed from this traditional breakdown by classifying leasehold interests other than the estate for years as usufructs.[68] A usufruct is a grant under which the tenant acquires merely the right to possess and enjoy real estate for a fixed period of time or at the will of the grantor.[69] No estate passes from the owner of the land under this arrangement.[70] Any renting or leasing of land for a period of less than five years conveys only a usufruct, unless the lease specifically provides that a greater interest is conveyed.[71] Georgia law presumes that a lease of land for five years or more conveys an estate for years.[72]

Estates in Possession

Up to this point the classes of estates that have been discussed are what may broadly be termed estates in possession, that is, the owner has a present right to the possession and profits of the land and does not have to wait for some future circumstance or contingency to occur before he can take advantage of the possession and profits. Contrasted to the estates in possession are the estates in expectancy or, to use the more modern term, the future interests.[73]

Future Interests

A future interest or future estate is one in which a person owns an interest in land but is not entitled to the present possession and enjoyment of the land because of the presence of a present possessory estate therein in favor of someone else.[74] The future interest vests in possession and enjoyment at the end of the preceding estate, at which point it changes from a future interest to a present possessory estate.[75] It is important to

68. Southland Inv. Corp. v. McIntosh, 137 Ga. App. 216 (1976).
69. O.C.G.A. § 44-7-1.
70. Id.
71. Id.
72. Ginsberg v. Wade, 95 Ga. App. 475 (1957).
73. 2 Bl. Comm. 163; 31 C.J.S. Estates §§ 13–15.
74. W. Agnor, Future Interests: The Law in Georgia §§ 1-1, 1-5 (1979).
75. Id.

understand that a future interest is not an interest to arise in the future.[76] It is subject to present ownership and, with certain practical exceptions, is salable, devisable, descendable, leviable, and taxable.[77] Future interests are most often created by a will or trust agreement but can be created by deed or lease or may arise by operation of the law. Future interests include the possibility of reverter, the right of entry for condition broken, the remainder, the reversion, and the executory interest.[78]

If the owner of land in fee simple conveys a defeasible fee in the land to someone and does not otherwise make disposition of the fee if the interest conveyed should terminate, then the interest he retains is either a possibility of reverter or a right of entry for condition broken. The distinction between these two types of future interests is technical and will not be discussed here.[79] Some eminent authorities deny that these interests can truly be considered estates.[80]

A remainder is the remnant of an estate, depending on a particular prior estate created at the same time and by the same instrument, and limited to vest in possession immediately upon the expiration of the particular estate and not in abridgment of it.[81] The particular estate that supports a remainder is most often a life estate but may be an estate for years. An example is a conveyance "to A for life and at her death to be equally divided among her children."[82] Here A's life estate is the particular estate and her children hold the remainder.

A reversion is the remnant of an estate left in the grantor and his heirs, to commence in possession after the expiration of a lesser estate which he has granted from the estate he owns.[83] It usually arises when the owner of land in fee simple conveys a life estate or an estate for years in the

76. Id.

77. Id. §§ 1-1, 3-1 to 3-4; Ga. Real Est. Law § 7-53 (3d ed. 1986).

78. W. Agnor, Future Interests: The Law in Georgia § 1-1 (1979); 1 Am. Law of Prop. §§ 4.3, 4.5.

79. See W. Agnor, Future Interests: The Law in Georgia §§ 1-10, 1-11 (1979); 1 Am. Law of Prop. §§ 4.6–4.15.

80. 1 Am. Law of Prop. § 4.6; S. Mitchell, Real Property under the Code of Georgia and the Georgia Decisions 217, 235 (5th ed. 1960); H. Tiffany, A Treatise on the Modern Law of Real Property 203 (abr. ed. 1940).

81. Booth v. Terrell, 16 Ga. 20 (1854); 21 C.J. Estates § 130; 31 C.J.S. Estates § 68.

82. Pearre v. McDonald, 168 Ga. 752 (1929); Pearson v. Cochran, 152 Ga. 276 (1921); Green v. Driver, 143 Ga. 134 (1915); Bull v. Walker, 71 Ga. 195 (1883).

83. Guess v. Morgan, 196 Ga. 265 (1943).

land to someone and does not make express disposition as to the residue of the fee.[84] In such cases the fee returns to the grantor or his heirs or assignees upon the expiration of the lesser estate. This is because it is a rule of law that the fee simple must exist in some person or persons. If the owner of such an estate grants a smaller estate, then whatever is not granted remains in him.[85] Reversions may also arise when the owner of a life estate or an estate for years conveys an estate the duration of which is less than he owns.

The distinction between a remainder and a reversion should be carefully noted. If the owner of a reversion transfers his interest, then the interest his grantee receives is still a reversion, not a remainder.

An executory interest (or an executory devise, as it is called if created by will) is an estate created in a third person to arise upon the defeasance of a prior estate in the same property.[86] For example, in a grant "to A but if she dies without issue then to B," A has a defeasible fee and B has an executory interest.[87] The executory interest is distinguished from the remainder and the reversion in that it always displaces a preceding fee interest. The remainder and the reversion, on the other hand, take effect in possession upon the natural expiration of a preceding estate and are regarded as a part of the total fee.[88] Executory interests are useful in estate planning but otherwise are only infrequently created.

Estates in Severalty

Heretofore this discussion has considered estates in land with reference to their duration and extent and to the time of their enjoyment. Another way of viewing estates is with reference to the number and connection of the owners. When land is held by one person alone, his ownership is known as an estate in severalty. This is the more common and usual form of landholding. Contrasted to the estate in severalty are the estates in plurality, also known as the concurrent estates.[89]

84. Haralson v. Redd, 15 Ga. 148 (1854).

85. 28 Am. Jur. 2d Estates § 171; 2 Bl. Comm. 175.

86. 28 Am. Jur. 2d Estates §§ 333, 334.

87. Reynolds v. Rackley, 223 Ga. 586 (1967); Murrelle v. Broughton, 142 Ga. 41 (1914); Payne v. Rosser, 53 Ga. 662 (1875); Harris v. Smith, 16 Ga. 545 (1855).

88. 28 Am. Jur. 2d Estates § 197; 1 Am. L. of Prop. § 4.53.

89. 2 Bl. Comm. 179; 31 C.J.S. Estates §§ 16–18.

Concurrent Estates

When two or more persons own simultaneous possessory interests in the same land, they have a concurrent estate thereto.[90] At common law and in some jurisdictions today several varieties of concurrent estates are recognized, but in Georgia only two types have any significance: the joint tenancy and the tenancy in common. Except for matters involving the right of survivorship, these two types of estates have almost exactly the same characteristics. When one of the tenants in a joint tenancy dies, his entire interest goes to the surviving tenants and this process is repeated until there is only one survivor left, who receives the whole estate in severalty.[91] In a tenancy in common, when one tenant dies his interest descends to his heirs or passes under his will.[92] Although joint tenancies are not prohibited, the law does not favor them, and a conveyance or devise of the same land to two or more persons will create a tenancy in common, unless the instrument clearly provides that a joint tenancy is intended.[93] Likewise, when land passes to two or more persons by descent, they take as tenants in common.[94]

Joint tenants have equal shares in the land they possess. Tenants in common may have equal or unequal shares in their land. If an instrument creating a tenancy in common does not fix the respective shares of the cotenants, they are presumed to take equal shares.[95] A cotenant (whether he holds as a joint tenant or as a tenant in common) may dispose of his share at pleasure and without the knowledge or consent of the other tenants, and the purchaser becomes a cotenant with the remaining tenants.[96] Such disposition by a joint tenant severs the joint tenancy and the transferee becomes a tenant in common with the other cotenants.[97]

Tenants in common may voluntarily partition their land between them-

90. H. Tiffany, A Treatise on the Modern Law of Real Property 281–82 (abr. ed. 1940).

91. Sams v. McDonald, 117 Ga. App. 336 (1968); 20 Am. Jur. 2d Contenancy and Joint Ownership § 3; 48A C.J.S. Joint Tenancies § 3.

92. 23 Am. Jur. 2d Descent and Distribution § 34; 26A C.J.S. Descent & Distribution § 8; 86 C.J.S. Tenancy in Common § 9.

93. Wilson v. Brown, 221 Ga. 273 (1965); Ga. Real Est. Law § 7-81 (3d ed. 1986); O.C.G.A. § 44-6-190.

94. Brown v. Brown, 209 Ga. 620 (1953); Zeagler v. Zeagler, 190 Ga. 220 (1940); O.C.G.A. § 44-6-190.

95. Ga. Real Est. Law §§ 7-82, 19-141 (3d ed. 1986).

96. Motor Aid, Inc. v. Ray, 53 Ga. App. 772 (1936).

97. 20 Am. Jur. 2d Contenancy and Joint Ownership §§ 15, 16; 2 Am. Law of Prop. § 6.2.

selves so that each takes a portion thereof in severalty.[98] If they cannot agree on how to manage their property and cannot make a mutually satisfactory partition, then any one of them may initiate legal proceedings to have a compulsory partition or sale of the property made.[99]

Easements

In its broadest context, an easement may be defined as a right which one person has to use the land of another for a specific purpose not inconsistent with the general property in the owner.[100] It is an interest in land but carries no title to the land over which it is exercised, nor does it carry any right to participate in the profits arising from the land.[101] In other words, the right to the fee and the right to an easement in the same realty are independent of each other and may coexist, though vested in different persons.[102] In general, easements are limited to the purpose for which they were created,[103] and the grantor retains full right and dominion over the land except so far as is reasonably necessary for the use and enjoyment of the easement.[104] A person cannot possess an easement on his own land because all uses of an easement are comprehended within his general right of ownership.[105] Easements almost invariably involve an incidental or indirect use of land for purposes such as roads, railroads, power lines, pipelines, private ways, canals, party walls, or navigation.[106]

Easements may be classified as appurtenant or in gross and as affirmative or negative. An appurtenant easement is an easement that an owner of real property may, for the benefit of his property, enjoy in the real

98. Ga. Real Est. Law §§ 7-93 to 7-97 (3d ed. 1986).

99. Id. §§ 7-98 to 7-107; Stone v. Benton, 258 Ga. 539 (1988).

100. Brown v. Tomlinson, 246 Ga. 513 (1980); 25 Am. Jur. 2d Easements and Licenses § 1. It must be conceded, however, that for many forms of easements the fee owner is excluded from making any practical use of the property over which the easement extends.

101. 25 Am. Jur. 2d Easements and Licenses § 2; 19 C.J. Easements § 200; 28 C.J.S. Easements § 73.

102. Southern Ry. Co. v. Wages, 203 Ga. 502 (1948).

103. 19 C.J. Easements § 200; 28 C.J.S. Easements § 73.

104. Georgia Power Co. v. Sullivan, 217 Ga. 699 (1962); State Hwy. Dept. v. Noble, 114 Ga. App. 3 (1966).

105. Burk v. Tyrrell, 212 Ga. 239 (1956); Grizzard v. Broom, 136 Ga. 297 (1911).

106. Ga. Real Est. Law § 8-1 (3d ed. 1986).

property of another.[107] The land burdened by the easement is called the servient tenement because it serves the other tract, and the land benefited by the easement is called the dominant tenement because of its dominant position with respect to the use being made of the servient tenement.[108] A common example of an appurtenant easement is a private way extending across one parcel of land (the servient tenement) for access to another parcel (the dominant tenement). An appurtenant easement is regarded as attaching to the dominant tenement. Thus a conveyance of the dominant tenement will carry with it the appurtenant easement even though the latter is not expressly mentioned in the instrument.[109]

An easement in gross is a right to use the lands of another with no dominant tenement involved.[110] Public roadways, railroad rights-of-way, and utility line rights-of-way are usually easements in gross.[111]

An affirmative easement authorizes its owner to do acts on the land of another which, without the easement, would be a trespass.[112] A negative easement is one that precludes the owner of the land subject to the easement from doing an act that, without the easement, he would be entitled to do.[113] An example of a negative easement is an easement for light and air over one tract for the benefit of an adjoining tract. In this case the owner of the land subject to the easement is precluded from erecting structures or barriers on his land which would shut off the flow of light and air to the adjoining tract.[114] The difference between a negative easement and a restrictive covenant that runs with the land is often very tenuous, but basically an easement constitutes an interest in the land while a covenant creates only a contractual right.[115]

In general, easements originate by the same methods that other rights or title to land originate. They may be created by express grant,[116] by

107. 28 C.J.S. Easements § 4.

108. Ga. Real Est. Law § 8-2 (3d ed. 1986).

109. Id. § 8-3; Stovall v. Coggins Granite Co., 116 Ga. 376 (1902); Atkinson v. Drake, 101 Ga. App. 485 (1960).

110. Ga. Real Est. Law § 8-4 (3d ed. 1986).

111. Id.

112. 25 Am. Jur. 2d Easements and Licenses § 8; 28 C.J.S. Easements § 3.

113. 25 Am. Jur. 2d Easements and Licenses § 8; 28 C.J.S. Easements § 3.

114. Ga. Real Est. Law § 8-5 (3d ed. 1986).

115. Id. See pages 490–94.

116. See pages 354–62. For discussion of the wording necessary in a conveyance to create an easement see pages 452–54.

dedication,[117] by prescription,[118] by condemnation,[119] by reservation or exception in a deed,[120] by reference to a plat showing the existence of the easement,[121] or by implication.[122] Since easements constitute an interest in land, under the statute of frauds they must usually be created in writing to be legally enforceable.[123] They are binding on grantees of the land over which they exist, except bona fide purchasers without notice.[124]

A parol license is usually revocable at any time. When the licensee, upon the faith of the license, has relied upon the license by expending money and making improvements, however, it becomes an easement running with the land if revocation would be harmful to the licensee.[125] Thus when an owner gives someone oral permission to construct a road,[126] a railroad,[127] a sewer line,[128] a ditch,[129] a dam,[130] or a party wall[131] through or upon his land, the license becomes an easement in the property so occupied when the licensee has constructed such facilities. Also, if such facilities are constructed without any agreement with the landowner, yet the owner, knowing of the construction, remains inactive until large sums of money have been expended or until the project is completed, he will be regarded as having acquiesced in the construction and cannot at a later date maintain an action for trespass.[132]

117. Atlanta & West Point R.R. Co. v. City of Atlanta, 156 Ga. 251 (1923). See pages 367–74.

118. See pages 374–88.

119. See pages 365–67.

120. For example, an owner in fee may sell a portion of his lands, subject to an easement of access across the portion sold, to reach lands he retains. McElwaney v. MacDiarmid, 131 Ga. 97 (1908).

121. See pages 459–61.

122. Martin v. First State Bank & Trust Co., 231 Ga. 511 (1973). With reference to private ways see pages 448–49.

123. Cook v. Pridgen, 45 Ga. 331 (1872).

124. Calhoun v. Ozburn, 186 Ga. 569 (1938).

125. Bell Industries, Inc. v. Jones, 220 Ga. 684 (1965); Waters v. Baker, 190 Ga. 186 (1940); Cherokee Mills v. Standard Cotton Mills, 138 Ga. 856 (1912).

126. Woodruff v. Bowers, 165 Ga. 408 (1927).

127. See page 454.

128. Bell Industries, Inc. v. Jones, 220 Ga. 684 (1965); Waters v. Pervis, 153 Ga. App. 71 (1980).

129. Brantley v. Perry, 120 Ga. 760 (1904).

130. Southwestern Railroad v. Mitchell, 69 Ga. 114 (1882); Winham v. McGuire, 51 Ga. 578 (1874).

131. Rawson v. Bell, 46 Ga. 19 (1872).

132. Central of Georgia Ry. Co. v. Standard Fuel Supply Co., 144 Ga. 92 (1915); Green v.

Easements may be terminated or extinguished by the terms of the instrument under which they were created,[133] by agreement or release between the parties,[134] by cessation of purpose or necessity,[135] by abandonment,[136] by adverse possession or prescription,[137] or by merger.[138] Private ways, under Georgia law, may be terminated upon the failure of the owner to maintain and repair the way.[139]

Licenses

With respect to real property, a license is a temporary, revocable privilege to do an act or series of acts on the land of another without possessing any estate or interest in the land itself.[140] Its only really significant legal attribute is to relieve the licensee of any charge of trespass.[141] A license may authorize an almost endless variety of acts on land such as passing over it, storing items on it, flooding it, cutting wood, or entering to watch a sports spectacle or theatrical performance.[142]

A license is distinguished from an easement primarily in that the latter carries with it an irrevocable interest in the land. It is distinguished from a lease in that a licensee has neither possession of nor the right to the profits of the land, while a tenant has the exclusive right of possession along with the profits.[143]

A license may be created by writing but is more often created orally or by implication.[144] By opening a place of business, for example, a person

South Bound R.R. Co., 112 Ga. 849 (1901); Ga. Real Est. Law § 23-64 (3d ed. 1986).

133. For example, they may be created to endure for a specified length of time or until the happening of a certain contingency. 25 Am. Jur. 2d Easements and Licenses §§ 99, 102.

134. 25 Am. Jur. 2d Easements and Licenses § 109; 28 C.J.S. Easements § 61.

135. S. A. Lynch Corp. v. Stone, 211 Ga. 516 (1955); Russell v. Napier, 82 Ga. 770 (1889); O.C.G.A. § 44-9-5.

136. See pages 390–91.

137. Wimpey v. Smart, 137 Ga. 325 (1911). See pages 374–88.

138. See pages 352–53.

139. See page 449.

140. Henson v. Airways Service, Inc., 220 Ga. 44 (1964); Ga. Real Est. Law § 8-45 (3d ed. 1986); 53 C.J.S. Licenses § 79.

141. 25 Am. Jur. 2d Easements and Licenses § 125; 53 C.J.S. Licenses § 84.

142. H. Tiffany, A Treatise on the Modern Law of Real Property 524–25 (abr. ed. 1940).

143. Berolzheimer v. Taylor, 230 Ga. 595 (1973); Ga. Real Est. Law § 8-45 (3d ed. 1986); 25 Am. Jur. 2d Easements and Licenses §§ 3, 123.

144. Ga. Real Est. Law § 8-46 (3d ed. 1986).

impliedly licenses the public to enter for the purpose of transacting such business.[145]

A license is not generally assignable and may be revoked at the will of the licensor.[146] As stated previously, however, when a licensee, upon the faith of the license, has expended money in constructing facilities, the license may become an easement running with the land.[147]

Merger of Interests

If several estates or interests in the same land vest in the same person and there is no intermediate estate or interest in favor of someone else, the lesser estates and interests are merged in the greater.[148] For example, if remaindermen convey their interest to the life tenant, then the life tenant has, not a life estate plus a remainder, but full fee simple title to the land.[149] When a testator devises land to his wife for life and the wife is his sole heir, the wife's life estate merges with her inheritance of the reversion upon the death of the testator, giving her fee simple title.[150] If the owner of an easement later acquires the fee to the land over which the easement exists, then the easement is extinguished by being merged into the fee.[151] Generally liens, leases, restrictive covenants, mortgages, and security deeds in favor of one who later acquires the fee are eliminated by merger.[152]

For merger to operate, the ownership of the various estates and interests must be coextensive and equal in validity, quality, and all other circumstances of right.[153] Thus a half-interest for life will not merge with a five-sixths remainder interest,[154] and a tenancy in common in favor of a

145. H. Tiffany, A Treatise on the Modern Law of Real Property 525 (abr. ed. 1940).

146. Ga. Real Est. Law § 8-48 (3d ed. 1986); 25 Am. Jur. 2d Easements and Licenses § 126; 53 C.J.S. Licenses § 86.

147. See preceding section.

148. Seaboard Air-Line Ry. Co. v. Holliday, 165 Ga. 200 (1927); O.C.G.A. § 44-6-2.

149. Ga. Real Est. Law § 7-129 n. 1 (3d ed. 1986).

150. Bowen v. Driggers, 138 Ga. 398 (1912); Smith v. Moore, 129 Ga. 644 (1907); Wilder v. Holland, 102 Ga. 44 (1897).

151. 25 Am. Jur. 2d Easements and Licenses § 108; 28 C.J.S. Easements § 57.

152. Ga. Real Est. Law § 7-126 (3d ed. 1986).

153. Seaboard Air-Line Ry. Co. v. Holliday, 165 Ga. 200 (1927); 25 Am. Jur. 2d Easements and Licenses § 108.

154. Seaboard Air-Line Ry. Co. v. Holliday, 165 Ga. 200 (1927).

widow and her children for the life of the widow will not merge with a remainder interest to vest in the children upon the death of the widow.[155] If land is conveyed "to A for life, then to B for life, then to C in fee simple," the acquisition by A of C's interest will not result in merger (unless B dies before A). Instead A will have a life estate and A's heirs will have a remainder, to vest in possession upon the death of B.[156]

155. Luquire v. Lee, 121 Ga. 624 (1904).

156. 28 Am. Jur. 2d Estates § 378; 1 H. Tiffany, The Law of Real Property 101 (3d ed. 1939).

Chapter 13. Transfer of Real Property

This chapter presents the various methods modern law recognizes for transferring real property from one person or legal entity to another.

Deed

General

With reference to real property, a deed is most often understood to be a written instrument passing title to or an interest in land.[1] It is by far the most commonly used method of transferring such rights or title. An essential characteristic of a deed is that it conveys a present estate or interest, although the possession or enjoyment of that estate or interest may be deferred until later.[2] Thus a writing executed in the form of a deed but evidencing the intention that the estate not vest until the death of the maker is a will.[3] On the other hand, an instrument presently conveying an estate in land but postponing the enjoyment of the estate until a future date is a deed.[4] For example, if a person conveys land in fee simple but reserves a life estate in the land, then the instrument immediately conveys an interest (a remainder) to the grantee, with his enjoyment of that interest postponed until the death of the grantor.[5]

A security deed is distinguished from a mortgage in that a mortgage does not convey title but merely provides a security for debt.[6] A sales contract, contract for deed, or bond for title, as distinguished from a deed, is an agreement to convey title in the future upon payment of consider-

1. 26 C.J.S. Deeds § 1.

2. Caldwell v. Caldwell, 140 Ga. 736 (1913).

3. Gardner v. Thames, 223 Ga. 378 (1967); Tyson v. Hutchinson, 164 Ga. 661 (1927); Hester v. Young, 2 Ga. 31 (1847).

4. Martin v. Smith, 211 Ga. 600 (1955); Crawford v. Thomas, 150 Ga. 435 (1920).

5. Patterson v. Patterson, 210 Ga. 359 (1954); Smaha v. George, 195 Ga. 412 (1943).

6. Hirsch v. Northwestern Mut. Life Ins. Co., 191 Ga. 524 (1941); Dewit v. Bozeman, 17 Ga. App. 666 (1916); O.C.G.A. § 44-14-30.

ation.[7] A deed is usually executed at a later date to confirm that the sale was actually completed.[8] A lease transfers temporary possession of land rather than its title.[9]

No particular form is necessary for a deed to be valid, and the cardinal rule of construction is always to ascertain the intention of the parties as gathered from the instrument as a whole.[10] There are thousands of pages of case law dealing with the construction of deeds in such matters as the identity of the grantor and grantee; their capacity to execute a deed; the nature of the estate granted; the description of the property conveyed; the time of taking effect; proper consideration, signing, attestation, and delivery; recording and notice; appurtenant and incidental rights; convenants; conditions; restrictions; excèptions; and reservations. Surveyors are primarily concerned with rules regarding the description of the property conveyed. These will be discussed in some detail in Chapters 14 through 17. The remainder of this section will define the different kinds of deeds and describe some of their formal requisites.

Kinds of Deeds

The various types of deeds include warranty deeds, limited warranty deeds, quitclaim deeds, voluntary deeds, security deeds, executor's deeds, administrator's deeds, and fiduciary deeds.

A warranty deed is one by which the grantor warrants the title to the land conveyed against the valid claims of all persons. It includes covenants of the right to sell, covenants of quiet enjoyment, and assurances that the land is freed from encumbrances.[11]

A limited or special warranty deed limits the grantor's liability to possible defects created in the title after his acquisition of the land. It does not, therefore, cover defects and encumbrances created prior to the grantor's ownership.[12]

A quitclaim deed merely releases to the grantee, without warranty, whatever right, title, or interest the grantor has in the land described and

7. Ga. Real Est. §§ 18-2, 20-70 (3d ed. 1986).

8. Ingram v. Smith, 62 Ga. App. 335 (1940).

9. Warehouses, Inc. v. Wetherbee, 203 Ga. 483 (1948); 23 Am. Jur. 2d Deeds § 4.

10. See pages 395–96.

11. Ga. Real Est. Law §§ 19-6, 19-168 (3d ed. 1986). See text at notes 38–44 below.

12. Id. §§ 19-7, 19-168.

more or less excludes any implication that the grantor has a good title or any title at all.[13] If the grantor has any title or interest in the land, then his quitclaim deed, if properly drawn and executed, is just as effective as a warranty deed to pass that title or interest.[14] If he has no interest in the land, the grantee acquires nothing by quitclaim deed except color of title to the land and the grantor is relieved of the liability of warranting title.[15] Quitclaim deeds are most commonly used to clear up titles to land in which the grantor has or may have some residual interest, rather than as a general method of conveying land.

A voluntary deed conveys land as a gift or without valuable consideration.[16]

A security deed, also known as a deed to secure debt or debt deed, is a deed conveying land to secure a specific indebtedness.[17] It usually contains a power of attorney authorizing the grantee or his assigns to sell the property upon default of payment and a statement that, upon full payment of the debt, the grantee or his assigns will cancel the deed or reconvey the property to the grantor.[18]

Executor's deeds are deeds from the executor conveying lands of a testate decedent, and administrator's deeds are those from the administrator conveying lands of an intestate decedent.[19]

Fiduciary deeds include deeds of guardians, trustees, receivers, sheriffs, attorneys in fact, as well as executors, administrators, and other persons acting on behalf of the true owner of the land.[20]

Occasionally one will encounter what is known as a transfer or an assignment. This is a writing, usually on the back of a deed, in which the grantor purports to convey to the grantee "the within deed and the land therein described."[21] The object is generally to avoid the trouble and expense of drawing up a new deed, but the use of transfers is not considered good title practice.

13. Id. § 19-5; Morrison v. Whiteside, 116 Ga. 459 (1902).

14. McDonald v. Dabney, 161 Ga. 711 (1926).

15. Patterson v. Fla. Rlty. & Fin. Corp., 212 Ga. 440 (1956).

16. Ga. Real Est. Law § 19-8 (3d ed. 1986).

17. Id. §§ 19-10, 21-2.

18. Id. § 21-2.

19. See next section.

20. Ga. Real Est. Law § 19-19 (3d ed. 1986).

21. Edwards v. Decatur Bank & Trust Co., 176 Ga. 194 (1932); Cross v. Citizens Bank & Trust Co., 160 Ga. 647 (1925); Hightower v. Haddock, 153 Ga. 160 (1922).

Parts of a Deed

No particular form is necessary for a deed to be valid. An instrument containing merely an identification of the parties, words of conveyance, and a description of the land, if duly executed and delivered, may be sufficient to constitute a valid deed.[22] There are, however, certain formal parts that a well-drawn conveyance usually includes. These parts, in the order that they usually appear in a deed, are known as the caption, premises, granting clause, description, habendum, warranty, testimonium, signature, and attestation.[23]

The caption, or heading, indicates the place of execution.[24] If the description of the property in the deed does not disclose the county and state in which the land is located, then the property is presumed to be located in the place indicated in the caption.[25]

The premises set forth the parties to the deed, usually give the date of execution, and include the granting clause and property description.[26]

The granting clause, a part of the premises, states the consideration and contains language that has the effect of conveying title.[27] To be operative, all deeds, whatever else they may lack, must contain words of conveyance.[28] The fact that the consideration was not actually paid in no way affects the validity of the deed.[29] The grantor's remedy in such cases is a legal action to enforce the obligation, not a cancellation of the deed.[30] A recital of the full consideration,[31] nominal consideration,[32] a consideration of goods,[33] services,[34] "love and affection,"[35] or "benefits

22. O.C.G.A. § 44-5-33; H. Tiffany, A Treatise on the Modern Law of Real Property 671 (abr. ed. 1940).

23. Ga. Real Est. Law § 19-22 (3d ed. 1986).

24. Id. § 19-23.

25. Id. §§ 13-56, 19-23.

26. 23 Am. Jur. 2d Deeds § 18; 26 C.J.S. Deeds § 22; 2 R. Patton et al., Patton on Land Titles § 332 (2d ed. 1957).

27. Ga. Real Est. Law § 19-25 (3d ed. 1986).

28. 23 Am. Jur. 2d Deeds § 19; 26 C.J.S. Deeds § 28.

29. Barret v. Simmons, 235 Ga. 600 (1975); Morris v. Johnson, 219 Ga. 81 (1963).

30. Morris v. Johnson, 219 Ga. 81 (1963); Harry v. Griffin, 210 Ga. 133 (1953); Whidby v. Willis, 151 Ga. 43 (1920).

31. O.C.G.A. § 13-3-40.

32. Stein v. Maddox, 234 Ga. 164 (1975); Morris v. Johnson, 219 Ga. 81 (1963).

33. 17 Am. Jur. 2d Contracts § 95; 17 C.J.S. Contracts §§ 76, 87.

34. Boddie v. Ridley, 197 Ga. 221 (1944).

35. Warren v. Warren, 221 Ga. 227 (1965); Dunn v. Evans, 139 Ga. 741 (1913).

and advantages" accruing to the grantor[36] is usually sufficient.

The description of the property conveyed, also a part of the premises, follows the granting clause.

The habendum clause, which follows the premises, is a holdover from ancient forms of conveyancing. It serves no useful purpose other than further to describe the nature of the estate being conveyed. It may usually be omitted entirely without affecting the meaning or validity of the deed.[37]

The warranty clause is a covenant by the grantor, for himself and his heirs, warranting to the grantee the estate granted.[38] Typically it also includes covenants of the right to sell, the right of quiet enjoyment, and freedom from encumbrances.[39] It may appear as a continuation of the habendum or as a separate paragraph.[40] A limited or special warranty is one in which the grantor warrants the title only against his own acts and against persons claiming by, through, or under him. It does not cover possible defects created prior to the grantor's acquisition of title.[41]

The existence of visible easements on the land conveyed, such as roads, railroads, or transmission lines, does not constitute a breach of warranty where an inspection of the premises by the grantee would have readily revealed physical evidences of such easements;[42] and a deficiency in area does not constitute a breach of warranty where the conveyance is by the tract rather than by the acre.[43] Thus when the area is mentioned only by way of description and not by way of covenant, there can be no recovery due to a deficiency.[44]

The final paragraph of a deed is called the testimonium.[45]

The signature of the grantor and the attestation follow the testimonium.

36. 17 Am. Jur. 2d Contracts § 96; 17 C.J.S. Contracts § 74.

37. Ga. Real Est. Law § 19-27 (3d ed. 1986).

38. 18 C.J. Deeds § 74; 26 C.J.S. Deeds § 22.

39. O.C.G.A. § 44-5-62; Ga. Real Est. Law § 19-168 (3d ed. 1986).

40. Ga. Real Est. Law § 19-28 (3d ed. 1986).

41. Id. §§ 19-7, 19-168.

42. Calhoun v. Ozburn, 186 Ga. 569 (1938); Desvergers v. Willis, 56 Ga. 515 (1876); Annotation, 48 L.R.A.(N.S.) 619 (1914).

43. Ga. Real Est. Law § 19-175 (3d ed. 1986).

44. Id.

45. Id. § 19-29.

Execution and Delivery

To be valid a deed must be in writing, must contain the signature of the grantor, and must be delivered to the grantee.[46] In addition, to provide sufficient record notice of title, it must bear the signatures of at least two witnesses (one of whom is vested with notarial powers, unless the deed contains an acknowledgment).[47] As between the parties, however, a deed is valid though not attested.[48] Any writing adopted by the grantor as his signature is sufficient. He may simply make a mark such as an "X,"[49] his name may be printed or stamped on the deed,[50] or he may have someone else sign his name to the deed in his presence.[51]

Although it is customary to place seals after the signatures, they are actually a quaint relic from times when people as a general rule could not write. Seals, therefore, are of little importance today.[52] To render an instrument sealed, it must be stated in the body of the instrument that it is under seal, and some sort of seal must be attached.[53] The seal may be stamped on wax affixed to the instrument or it may be stamped on the paper itself, or it may be a printed or scribbled mark.[54] A printed "L.S." (*locus sigilli,* the place of the seal) is commonly used.[55] A deed is perfectly valid without a seal and many judges have ridiculed their continued use.[56]

Delivery is the actual placing by the grantor of the deed in the possession of the grantee with the mutual intention of passing title.[57] A deed

46. Id. §§ 19-15, 19-34.

47. Id. §§ 19-53, 19-60, 19-61. In practice acknowledgments are rarely used.

48. Bailey v. Williams, 215 Ga. 395 (1959); Budget Charge Accts., Inc. v. Peters, 213 Ga. 17 (1957); Blue Ridge Apt. Co. v. Telfair Stockton & Co., 205 Ga. 552 (1949).

49. Horton v. Murden, 117 Ga. 72 (1903).

50. 18 C.J. Deeds § 80; 26 C.J.S. Deeds § 34.

51. Guthrie v. Gaskins, 171 Ga. 303 (1930); Brown v. Wilcox, 150 Ga. 88 (1920).

52. Atlanta, Knoxville, & Northern Ry. Co. v. McKinney, 124 Ga. 929 (1905); Ga. Real Est. Law § 19-44 (3d ed. 1986).

53. Baxley Hdwe. Co. v. Morris, 165 Ga. 359 (1927); Chastain v. L. Moss Music Co., 83 Ga. App. 570 (1951).

54. Stansell v. Corley, 81 Ga. 453 (1888); Williams v. Greer, 12 Ga. 459 (1853); O.C.G.A. § 1-3-3.

55. Cannon v. Gorham, 136 Ga. 167 (1911); Lowe v. Morris, 13 Ga. 147, 153 (1853).

56. "The original common-law rule that a seal is essential to the validity of a writing is medieval nonsense, and should have no place in modern jurisprudence." Allen v. Montgomery, 25 Ga. App. 817 (1920). For eloquent discourses on the matter see opinions by Justice Lumpkin in Drumright v. Philpot, 16 Ga. 424 (1854) and Lowe v. Morris, 13 Ga. 147 (1853).

57. O'Neal v. Brown, 67 Ga. 707 (1881).

that is not delivered is invalid, even though all other formal requisites for its validity are met.[58] Delivery may be by hand,[59] the grantor may mail it to the grantee,[60] or the grantor may give it to a third person with instructions that it be delivered to the grantee.[61] If a deed is made out to an infant child, delivery to the parent is sufficient.[62] The recording of a deed raises the presumption of its delivery.[63] If the grantee repudiates a deed upon tender of delivery by the grantor, then there is no effective delivery.[64]

After an effective delivery of a valid deed, title cannot subsequently be divested by the mere alteration, mutilation, loss, or destruction of the deed, or by its surrender or redelivery to the grantor.[65]

Recording and Notice

The sole purpose of recording deeds and other instruments affecting title to land is to impart notice, and it is only creditors, mortgagees, and subsequent purchasers for value that are protected against unrecorded instruments.[66] Therefore, failure to record a deed does not in itself affect the passing of title between the parties to the deed.[67]

In legal contemplation there are two kinds of notice: actual and constructive. Notice is actual when it is given directly and personally to the person to be charged with it.[68] Constructive notice is information or knowledge of a fact imputed by law because the fact could have been discovered by proper diligence and the situation was such as to cast upon a person the duty to inquire into it.[69] Constructive notice, once established,

58. Giuffrida v. Knight, 210 Ga. 128 (1953); Childs v. Mitchell, 204 Ga. 542 (1948).

59. Ga. Real Est. Law § 19-85 (3d ed. 1986).

60. Allgood v. Allgood, 230 Ga. 312 (1973).

61. Preston v. Ham, 156 Ga. 223 (1923); Baxter v. Chapman, 147 Ga. 438 (1917).

62. Whitworth v. Whitworth, 233 Ga. 53 (1974); First Nat'l Bank of Cornelia v. Kelly, 190 Ga. 603 (1940).

63. W. L. Schautz Co. v. Duncan Hosiery Mills, 218 Ga. 729 (1963); Dollar v. Thompson, 212 Ga. 831 (1957).

64. Stallings v. Newton, 110 Ga. 875 (1900); Hilson v. Beardsley, 94 Ga. 50 (1894).

65. Latham v. Fowler, 192 Ga. 686 (1941); Jordan v. Pollock, 14 Ga. 145 (1853); 23 Am. Jur. 2d Deeds § 359; 26 C.J.S. Deeds §§ 173-77.

66. 26 C.J.S. Deeds § 74.

67. Whittington v. Wright, 9 Ga. 23 (1850).

68. Picklesimer v. Smith, 164 Ga. 600 (1927).

69. Hamilton v. Edwards, 245 Ga. 810 (1980).

is as binding as actual notice upon the person or persons to whom it applies.[70] Examples of information for which the law imputes constructive notice are legislative acts, legal notices in newspapers, and matters of public record such as documents recorded in the courthouse.[71] In addition, actual possession of land is constructive notice to the world of whatever right or title the occupant has in the land.[72]

A duly recorded deed is constructive notice of its contents, and the law presumes that a purchaser has examined every deed and instrument affecting his title.[73] He is therefore charged with notice of every fact that an examination of the records would have shown, whether or not he actually searched the records.[74] In the case of two deeds from the same grantor conveying overlapping interests in the same land, the first to be recorded generally takes precedence, even if it was executed after the other deed.[75] A recorded deed outside a purchaser's chain of title does not constitute constructive notice.[76]

The rules regarding notice are the basis for determining priorities between overlapping conveyances of the same land, which will be discussed in Chapter 14.

Deeds are recorded in the office of clerk of court in the county where the land being conveyed lies.[77] A deed to land lying in two or more counties should be recorded in each county.[78] The creation of a new county from previously existing ones does not necessitate the re-recording of deeds to land in the new county that are already recorded in the older counties.[79]

Deeds are indexed in books in the clerk's office by the name of the grantor and grantee. The grantor indexes are usually entitled "Grantor"

70. Jordan v. Pollock, 14 Ga. 145 (1853).

71. 58 Am. Jur. 2d Notice §§ 17-21; 66 C.J.S. Notice § 13.

72. Pierce v. Thomas, 258 Ga. 469 (1988); Ammons v. Cent. of Georgia Ry. Co., 215 Ga. 758 (1960); Toms v. Knighton, 199 Ga. 858 (1945); Wren v. Wren, 199 Ga. 851 (1945).

73. Harper v. Durden, 177 Ga. 216 (1933); Talmadge v. Interstate Bldg. & Loan Asso., 105 Ga. 550 (1898).

74. Harper v. Durden, 177 Ga. 216 (1933); Talmadge v. Interstate Bldg. & Loan Asso., 105 Ga. 550 (1898).

75. Michael v. Poss, 209 Ga. 559 (1953).

76. Real Estate Operators v. McMahon, 171 Ga. 454 (1930); Hancock v. Gumm, 151 Ga. 667 (1921).

77. O.C.G.A. § 44-2-1.

78. Bagley v. Kennedy, 94 Ga. 651 (1894); Kennedy v. Harden, 92 Ga. 230 (1893).

79. Sapp v. Cline, 131 Ga. 433 (1908).

or "Direct," and the grantee indexes are usually entitled "Grantee" or "Reverse." Occasionally other instruments affecting title to land such as mortgages, affidavits, liens, powers of attorney, condemnation proceedings, and court judgments will be found recorded in the deed books.

The quality of public record keeping varies greatly over the state, and the present system as a whole is rather antiquated. The use of electronic data-processing equipment and other modern devices will likely work a revolution in record keeping in the near future.

Will or Descent

The body of law dealing with the inheritance of property through will or descent is large and technical, but the surveyor will rarely need to have more than a very superficial knowledge of this law. Modern-day practice calls for the execution of executor's deeds or administrator's deeds conveying the lands of a deceased, thus usually eliminating the need for a surveyor to consult probate records to ascertain links in the chain of title to land.

A will is the legal declaration of a person's intention regarding the disposition of his property after his death.[80] When a person dies testate (with a valid will), his will is probated (proved that it is genuine and meets the requirements of law) in the probate court of the county of his domicile.[81] Except for noncupative (oral) wills, all wills must be in writing, must be signed by the testator or by someone acting under his direction, and must be attested by at least two witnesses in order to be admitted to probate.[82] A will is not legally operative until probated.[83] Land devised in a will passes first to the executor, who holds it for administrative purposes.[84] After matters of debt and distribution are cleared up, the executor conveys the land by deed or other instrument, or by implication, to the beneficiaries of the will or to purchasers for value.[85]

In general the same rules for construing deeds are followed in construing wills, the primary object in the case of wills being to accomplish

80. O.C.G.A. § 53-2-1.

81. Id. §§ 53-3-1 to 53-3-49.

82. Id. § 53-2-40.

83. 1 D. Redfearn, Wills and Administration in Georgia § 111 (5th ed. 1988).

84. Ga. Real Est. Law § 16-43 (3d ed. 1986).

85. Id.

the purpose intended by the testator.[86] There are a few significant distinctions, however, in the construction of wills as compared to deeds. A testator may revoke his will at any time before death. Furthermore, a will takes effect only at the instant of the testator's death, not at the time of delivery as in the case of deeds.[87] Thus a devise of land in a will is valid if owned at the time of death, even though it was not owned at the time of execution;[88] and, of course, a devise of land owned at the time of execution but not owned at the time of death is void.[89] Intestacies are not favored in construing wills. If a will is equally capable of two or more constructions, one of which will leave no intestacy and the others which will leave a partial or complete intestacy, then the construction that leaves no intestacy is favored.[90] In the case of conflict between different provisions in a will, the latter provision is usually favored.[91] Generally greater flexibility is permitted in the construction of wills than in deeds.[92]

If a person dies intestate (without a valid will), his estate is distributed in the manner prescribed by the laws of inheritance. The probate judge appoints an administrator to collect the assets of the decedent, to pay any debts or claims against the estate, to distribute the remainder of the estate to the heirs, and to perform certain other functions.[93] Title to real property owned by the decedent immediately vests in his heirs,[94] but the administrator has the power to sell part or all of the land if there are debts that cannot otherwise be paid or if no equitable division can be worked out between the heirs.[95]

Because of the cost and relative complexity of estate administration, a quick and simple method has been devised for winding up the affairs of an intestate estate. Under this procedure, an heir of the deceased files

86. Id. § 16-7.

87. Id. §§ 16-7, 16-21; O.C.G.A. §§ 53-2-3, 53-2-70; 1 D. Redfearn, Wills and Administration in Georgia §§ 8, 142 (5th ed. 1988).

88. 1 D. Redfearn, Wills and Administration in Georgia § 12 (5th ed. 1988).

89. 94 C.J.S. Wills § 76; 95 C.J.S. Wills § 631.

90. Ga. Real Est. Law § 16-7 (3d ed. 1986); 1 D. Redfearn, Wills and Administration in Georgia § 142 (5th ed. 1988).

91. Rigdon v. Cooper, 203 Ga. 547 (1948); Watts v. Finley, 187 Ga. 629 (1939); O.C.G.A. § 53-2-92.

92. Ga. Real Est. Law § 19-153 (3d ed. 1986); Harriss v. Howard, 126 Ga. 325 (1906).

93. O.C.G.A. §§ 53-6-23 et seq.; V. Jenkins, Handbook for Ordinaries of Georgia 75–125 (1965); 2 D. Redfearn, Wills and Administration in Georgia §§ 248 et seq. (5th ed. 1988).

94. O.C.G.A. § 53-4-8.

95. Id. § 53-8-23.

a petition with the probate judge for an order that no administration of the estate is necessary.[96] Several conditions must exist for such an order to be issued: all the heirs must be competent adults; they must have agreed upon an amicable division of the estate; all debts of the estate must have been paid; and a sufficient length of time must have elapsed since the intestate died to make the existence of undisclosed debts highly improbable.[97]

The records pertaining to estate administration proceedings are filed in the probate court in the county where the deceased had assets at the time of his death. The probate court is often the sole depository of plats and other title instruments drawn up during the administration of estates, a point surveyors should bear in mind when searching for such records.

Often, either through ignorance or indifference, the heirs simply take possession of the assets of a decedent without going through the legal formalities for settling the affairs of the estate. The consequences of such actions are the concern of the title attorney, not the surveyor.

Public Grant

Grants of land by the crown or state have been extensively treated in Part 1 of this book. The subject today is primarily of historical interest, and even attorneys rarely find it necessary to run titles back to the original grant to assure good title. Accordingly, only a few additional comments will be made.

The title to all privately owned lands in Georgia originates by grant, either from Great Britain or, after independence, from the state.[98]

Sometimes it may be helpful, when preparing to resurvey a tract of land, to examine the original land grant plats to the tract. Copies of almost all such plats, except for those lost in the Revolution, are in the surveyor general records of the Georgia Department of Archives and History in Atlanta. In addition, for the headright areas, such plats are often recorded in the county where the tract was located at the time of the grant. These locally recorded headright grant plats were entered into books kept especially for that purpose or, after 1867, were fre-

96. Id. §§ 53-10-1 to 53-10-4.

97. Id.

98. O.C.G.A. § 44-5-1; Clark, *Public Lands,* in 25 Encyclopedia of Georgia Law §§ 2 to 4, 6 to 7, 11 (1974).

quently entered in the ordinary's minute books. Today these books are usually deposited in the probate judge's office. Sometimes the plats were re-recorded in the deed books in the office of the clerk of court.

To find the original plats in the surveyor general records for the land lottery areas, one simply needs to know the land lot number, the district number (and section number for original Cherokee County), and the present or original county the land is located in. To find the headright grant plats for a particular area, it is usually necessary to run the title back to the original grant. Such searches are often hampered by the loss of records and the failure by many owners in the past to record their documents of title. The problem of finding headright plats could be alleviated if maps were compiled showing the relative positions of all original headright grants, but such a project would be extremely costly.

Eminent Domain

Eminent domain is the right of the nation or state or of those to whom the power has been lawfully delegated to appropriate private property for public use.[99] Condemnation is the implementation of the power of eminent domain.[100]

The right to take private property for public uses is an obvious necessity for the proper functioning of government as well as for the public welfare, but just and adequate compensation must be paid to the individual whose property is taken or damaged.[101] In addition, except in the case of private ways, private property can be taken only for a public use or purpose; a taking for strictly private uses or purposes is unauthorized.[102]

Although the power of eminent domain is mentioned in the state and federal constitutions, it does not depend for its existence on a specific grant in the constitutions. It is, rather, an inherent attribute of sovereignty that exists in a sovereign state independent of any constitutional recognition thereof. The constitutional provisions are merely for the purpose of limiting a power that otherwise would be without limit.[103]

Eminent domain may be exercised by the federal government, by

99. 29A C.J.S. Eminent Doman § 1.

100. Id.

101. Thomas v. City of Cairo, 206 Ga. 336 (1950).

102. City of Atlanta v. Atlanta Gas Light Co., 144 Ga. App. 157 (1977).

103. State Hwy. Dept. v. Smith, 219 Ga. 800 (1964); 26 Am. Jur. 2d Eminent Domain § 2.

the state government, by counties and municipalities, by agencies of these governing bodies, and by corporate bodies and individuals (such as railroads and utility companies) upon whom the power has been conferred by the legislature or Congress.[104] Property may be condemned for any use for which the general public may derive benefit—for example, roads, railroads, utility lines, subways, harbor facilities, canals, schools, airports, hydroelectric projects, parks, historic sites, wildlife preserves, public housing, and slum clearance.[105]

A party having the power of eminent domain may condemn only as much land as is necessary for the particular public use to which the land is to be put.[106] It is recognized, however, that the decision to condemn always involves a certain amount of discretion on the part of the condemnor as to what and how much property is to be taken, and the courts will not interfere with this discretion unless it is abused or exercised in bad faith.[107]

It would be pointless here to enter into a detailed discussion of the procedure for condemnation, which is controlled primarily by statute. In general, there must first be a bona fide effort to negotiate a voluntary purchase.[108] When the condemning authority must resort to litigation, the venue of the proceeding is the county where the land lies. If it lies in more than one county, the venue is any county in which a part of the land lies.[109] The owner of the property and all persons having an interest in it must be served with a notice or petition which, among other things, describes the land sought with the same exactitude as is required for a deed and which also describes the interest to be acquired.[110] The compensation paid to the condemnee is the actual value of the land taken, plus an award for any damages to his remaining lands, and all expenses caused by the proceeding.[111]

104. O.C.G.A. §§ 22-1-1 to 22-1-4, 22-3-1 to 22-3-122, 50-2-22.

105. Id. §§ 6-3-22, 20-2-521, 20-2-522, 46-4-57, 52-3-4; Ga. Real Est. Law §§ 2-8, 2-17 (3d ed. 1986).

106. Piedmont Cotton Mills v. Georgia Ry. & Elec. Co., 131 Ga. 129 (1908).

107. City of Atlanta v. First Nat'l Bank of Atlanta, 246 Ga. 424 (1980); Harwell v. Georgia Pwr. Co., 154 Ga. App. 142 (1980).

108. Ga. Real Est. Law § 2-22 (3d ed. 1986).

109. Id. § 2-26.

110. Gunn v. Georgia Pwr. Co., 205 Ga. 85 (1949); Ga. Real Est. Law §§ 2-25, 2-34 (3d ed. 1986); C. Pursley, Georgia Eminent Domain § 2-1 (1982).

111. Ga. Real Est. Law § 2-42 (3d ed. 1986).

A condemnor is authorized to acquire no greater interest in the land sought than is reasonably necessary for the use intended.[112] In the past an easement or defeasible fee was usually acquired for transportation, communication, and utility rights-of-way, but the fee simple has been gaining in frequency for these purposes. If an easement or defeasible fee is acquired, then full title reverts to the owner or his successors if the condemnor ceases to use the property for the purpose for which it was taken.[113] If fee simple title is acquired and the condemnor later ceases to use the property for the stated purpose, no reversion takes place and the condemnor may put the property to other uses.[114]

Upon the completion of a condemnation proceeding, a deed conveying the property from the condemnee to the condemnor is placed on record in the county or counties where the land is located.

Dedication

General

In a dedication the owner of land consents, either expressly or by action, that his land may be used by the public for a particular purpose.[115] To constitute a valid and complete dedication there must be an intention by the owner to dedicate and an acceptance on the part of the public of the property so dedicated.[116] Both the intent to dedicate and the acceptance of the dedication may be express or implied.[117]

Among the uses to which land may be dedicated are highways, streets, and roads;[118] alleys;[119] sidewalks;[120] parks, public squares, and

112. Id. § 2-37; C. Pursley, Georgia Eminent Domain § 2-6 (1982).

113. City of Atlanta v. Fulton County, 210 Ga. 784 (1954); O.C.G.A. § 22-2-85.

114. Galloway v. Banks County, 246 Ga. 472 (1980); Sadtler v. City of Atlanta, 236 Ga. 396 (1976).

115. Mayor & Council of Macon v. Franklin, 12 Ga. 239 (1852).

116. Nelson v. Girard, 215 Ga. 518 (1959); Maddox v. Willis, 205 Ga. 596 (1949).

117. Lowry v. Rosenfeld, 213 Ga. 60 (1957).

118. Waldrep v. Hall County, 227 Ga. 554 (1971); R. G. Foster & Co. v. Fountain, 216 Ga. 113 (1960); Johnson v. State, 1 Ga. App. 195 (1907); O.C.G.A. § 32-3-3.

119. Tribble v. Mayor, etc., of Forsyth, 225 Ga. 204 (1969); Perry v. Weaver, 11 Ga. App. 186 (1912).

120. R. G. Foster & Co. v. Fountain, 216 Ga. 113 (1960).

commons;[121] parking lots;[122] schools;[123] wharves and landing places;[124] religious and charitable purposes;[125] and cemeteries.[126] Land may be dedicated to any use of a nature which the general public can enjoy. Furthermore, it is not essential that the right of use be vested in a corporate body.[127]

Where land has been dedicated to public uses only, the underlying fee remains unaffected. The owner of the fee is not deprived of his title and retains exclusive rights for every purpose of user or profit not inconsistent with the public easement.[128]

When a dedication, whether express or implied, has been made for a specific purpose, the public authorities have no right to use the property for other purposes except by right of purchase or eminent domain.[129] The public authorities may, however, put dedicated property to all customary uses within the definition of the use.[130] For example, land dedicated for streets may be used for purposes other than travel, such as the construction of sewers and drains, the laying of gas and water pipes, the erection of telegraph and telephone wires, and other purposes not inconsistent with the use of the land for streets.[131]

121. Stanfield v. Brewton, 228 Ga. 92 (1971); City of Abbeville v. Jay, 205 Ga. 743 (1949); Caffey v. Parris, 186 Ga. 303 (1938); City of LaFayette v. Walker County, 151 Ga. 786 (1921); Mayor & Council of Macon v. Franklin, 12 Ga. 239 (1852); O.C.G.A. § 36-64-2.

122. Lowry v. Rosenfeld, 213 Ga. 60 (1957).

123. Gartrell v. McCravey, 144 Ga. 688 (1915); Chapman v. Floyd, 68 Ga. 455 (1882).

124. 23 Am. Jur. 2d Dedication § 4; 26 C.J.S. Dedication § 8; H. Tiffany, A Treatise on the Modern Law of Real Property 757 (abr. ed. 1940).

125. 26 C.J.S. Dedication, § 8.

126. Floral Hills Mem. Gardens, Inc. v. Robb, 227 Ga. 470 (1971); Haslerig v. Watson, 205 Ga. 668 (1949).

127. Chatham Motorcycle Club, Inc. v. Blount, 214 Ga. 770 (1959); Haslerig v. Watson, 205 Ga. 668 (1949).

128. Smith v. Lemon, 166 Ga. 93 (1927); Seaboard Air-Line Ry. Co. v. Greenfield, 160 Ga. 407 (1925); Long v. Faulkner, 151 Ga. 837 (1921); Brown v. City of East Point, 148 Ga. 85 (1918).

129. State Highway Dept. v. Alexander, 222 Ga. 354 (1966); Harris v. Powell, 177 Ga. 15 (1933).

130. Faulkner v. Georgia Pwr. Co., 243 Ga. 649 (1979); Franklin v. Board of Lights, 212 Ga. 757 (1956); Norton v. City of Gainesville, 211 Ga. 387 (1955); Huff v. City of Macon, 117 Ga. 428 (1903).

131. City of Albany v. Lippitt, 191 Ga. 756 (1941).

Intent to Dedicate

An intention on the part of the owner to dedicate the property is absolutely essential to a dedication.[132] Also, there is no power of dedication when there is no power to alienate.[133] Therefore, a valid dedication cannot be made by a life tenant; a mortgagor; one who owns a mere easement in the land; a trespasser or squatter; a mere stranger; one who has parted with his title; or the holder of record title to land, title to which has vested in an occupant by adverse possession. Finally, a person cannot dedicate to public uses land that belongs to his spouse and children.[134]

No particular form is necessary to make a dedication.[135] It may be done by written conveyance,[136] orally,[137] or by acquiescence.[138] An intention on the part of the owner to dedicate must be manifested by his conduct from the facts and circumstances of the particular case, based on the owner's acts, and not on what is secreted in his heart.[139] A dedicator may specify obligations or conditions in his offer of dedication, and if the dedication is accepted, the public will be bound to fulfill the obligations.[140]

Maintenance and use by the public of a piece of land, with the knowledge of and without objection from the owner, to the extent that the public accommodation or private rights of nearby owners may be materially affected by interruption is sufficient to show dedication.[141] The element of time is of no consequence under such circumstances and the public immediately acquires rights in the land so dedicated, if the public uses it.[142] But the mere use of one's property for travel purposes by a small portion of the public, even for an extended period of time, and

132. Swift v. Mayor of Lithonia, 101 Ga. 706 (1897); Collins v. Mayor, etc., of Macon, 69 Ga. 542 (1882).

133. Jacobs Pharmacy Co. v. Luckie, 143 Ga. 457 (1915); 23 Am. Jur. 2d Dedication §§ 11 to 19; 26 C.J.S. Dedication § 7.

134. 18 C.J. Dedication § 16; 26 C.J.S. Dedication § 7.

135. Mayor & Council of Macon v. Franklin, 12 Ga. 239 (1852).

136. City of Albany v. Lippitt, 191 Ga. 756 (1941); County of Gordon v. Mayor etc. of Calhoun, 128 Ga. 781 (1907); O.C.G.A. §§ 32-3-2, 32-3-3.

137. Ellis v. Mayor, etc., of Hazlehurst, 138 Ga. 181 (1912).

138. Hyde v. Chappell, 194 Ga. 536 (1942); Shirley v. Morgan, 170 Ga. 324 (1930).

139. Dunaway v. Windsor, 197 Ga. 705 (1944).

140. Jenkins County v. Dickey, 139 Ga. 91 (1912).

141. Lowry v. Rosenfeld, 213 Ga. 60 (1957); Haslerig v. Watson, 205 Ga. 668 (1949); Morgan v. Shirley, 172 Ga. 727 (1931); O.C.G.A. § 44-5-230.

142. Fountain v. Bryan, 229 Ga. 120 (1972).

an occasional maintenance by the public authorities, will not constitute a dedication of such property to a public use without a manifest intent to dedicate.[143] After such use for twenty years or more, however, the element of prescription may set in.[144]

Allowing public authorities to pave and maintain an area for use by the public for parking purposes is implied dedication;[145] but merely allowing the public to park on one's property in the conduct of business will not amount to dedication, even if this practice continues for twenty years or more.[146]

If an owner subdivides a tract of land into lots and makes a plat showing such lots with designated streets, alleys, parks, and other public areas, then sells the lots with reference to the plat, the owner is presumed to have irrevocably dedicated the streets and public areas for the use of all lot owners in the subdivision, whether or not there has ever been an acceptance of the dedication by the public authorities or the public generally.[147] The name of the owner or developer does not have to appear on the plat,[148] nor does the plat have to be recorded[149] to constitute a valid dedication under such circumstances. The mere survey and platting of a body of land into lots and streets, however, is insufficient to lay off and open such streets.[150] When the lots are sold and the streets put into condition for travel, it is then that they in fact become streets.[151] Merely leaving a narrow strip of land between two lots does not in itself create a street or alley.[152] On the other hand, if a person subdivides a tract into lots and leaves a vacant strip of land abutting the lots, and if the deeds describe

143. Tribble v. Mayor, etc., of Forsyth, 225 Ga. 204 (1969); Hudspeth v. County of Early, 210 Ga. 386 (1954); Dunaway v. Windsor, 197 Ga. 705 (1944); Mayor & Council of Madison v. Booth, 53 Ga. 609 (1875).

144. Shirley v. Morgan, 170 Ga. 324 (1930); Swift v. Mayor of Lithonia, 101 Ga. 706 (1897).

145. Lowry v. Rosenfeld, 213 Ga. 60 (1957).

146. Thrash v. Wood, 215 Ga. 609 (1960).

147. Higgins v. Odom, 246 Ga. 309 (1980); Smith v. Bruce, 241 Ga. 133 (1978); Lee v. Warren, 230 Ga. 165 (1973); Ford v. Harris, 95 Ga. 97 (1894); Bayard v. Hargrove, 45 Ga. 343 (1872).

148. Hobbs v. Ware County, 247 Ga. 385 (1981).

149. Dept. of Transportation v. Ladson Investments, 158 Ga. App. 687 (1981).

150. Robins v. McGehee, 127 Ga. 431 (1906); Parsons v. Trustees of Atlanta University, 44 Ga. 529 (1871).

151. Robins v. McGehee, 127 Ga. 431 (1906).

152. Ga. Real Est. Law § 8-15 (3d ed. 1986). Compare Hardigree v. Hardigree, 244 Ga. 830 (1979); and Miller v. Wells, 235 Ga. 411 (1975).

the lots as bounded by a street, the owner is estopped from denying the use of the strip as a street.[153]

The requirements necessary to show an intent to dedicate are more stringent for property owned by utility companies and other providers of quasi-public services.[154] For example, the fact that a railroad company allows the public to use the approaches to its stations and depots does not amount to a dedication, even though many who use the approaches do not come there to do business with the company.[155] When the owner of wharf property permitted the use of the area by the public, allowed the city to pave it and build a railroad track over it, and this went on for more than twenty years and the city continued to accept payment of taxes, there was no implied dedication.[156]

It is generally held that use by the public of unenclosed wild lands, for whatever period of time, raises no presumption of dedication and confers no rights in the public.[157]

When dedication results from mere use and acquiescence, generally no dedication is implied beyond what is actually used.[158] If a person dedicates, by implication, a strip of land between the center of a road and the curb for road purposes and also dedicates, by implication, a strip of land several feet back from the curb for sidewalk purposes, no dedication is implied as to the grassy strip between the curb and the sidewalk.[159]

Acceptance

To be complete, a dedication must be accepted by the public[160] because the public may not need the land.[161] In fact, it may be an inconvenience and burden instead of a benefit.[162] Even if land is conveyed by deed to

153. Hamil v. Pone, 160 Ga. 774 (1925).

154. 26 C.J.S. Dedication § 19.

155. Seaboard Air-Line Ry. Co. v. Greenfield, 160 Ga. 407 (1925); Georgia R.R. & Banking Co. v. City of Atlanta, 118 Ga. 486 (1903).

156. Mayor, etc., of Savannah v. Stand. Fuel Supply Co., 140 Ga. 353 (1913).

157. Lines v. State of Georgia, 245 Ga. 390 (1980); 26 C.J.S. Dedication § 20.

158. R. G. Foster & Co. v. Fountain, 216 Ga. 113 (1960); Ellis v. Mayor, etc., of Hazlehurst, 138 Ga. 181 (1912).

159. R. G. Foster & Co. v. Fountain, 216 Ga. 113 (1960).

160. Hutchinson v. Clark, 169 Ga. 511 (1929).

161. Jackson v. Chatham County, 225 Ga. 641 (1969).

162. Penick v. Morgan, 131 Ga. 385 (1908); Kelsoe v. Town of Oglethorpe, 120 Ga. 951

a public body, there must still be an acceptance by the body before the tract will vest in that body.[163] But failure of acceptance by a public body will not allow the owner to appropriate dedicated property to his own use if there has been a continuous and notorious use by the public so as to constitute acceptance by the public.[164]

Express acceptance by a public body may be shown by the official minutes of that body,[165] by a resolution of the body recognizing the dedication,[166] or by recognition of a dedicated street on the official maps of the city or county.[167] A county map showing a road as a public road but not a county road is insufficient to show acceptance.[168] Refusal by the state highway department to designate a road as a rural road for county road mileage purposes does not determine acceptance.[169] Merely recording a subdivision plat bearing the statement that certain designated public areas shown thereon are dedicated to the public does not amount to acceptance;[170] nor does approval of plats by a zoning board necessarily constitute the county's acceptance of public areas designated thereon.[171]

Maintenance and use by public authorities is usually sufficient to show implied acceptance,[172] but isolated and occasional acts of maintenance or use by the public authorities is not sufficient to show acceptance.[173] For example, use of a way by garbage trucks to collect refuse and by the fire department to reach property for fire inspections does not constitute a dedication and acceptance.[174] Likewise, if the county fills in an area with

---

(1904); Georgia R.R. & Banking Co. v. City of Atlanta, 118 Ga. 486 (1903); Parsons v. Trustees of Atlanta University, 44 Ga. 529 (1871).

163. Jackson v. Chatham County, 225 Ga. 641 (1969); Mayor, etc., of Athens v. Gamma Delta Chapter House Corp., 86 Ga. App. 53 (1952).

164. Chatham Motorcycle Club, Inc. v. Blount, 214 Ga. 770 (1959); City of Abbeville v. Jay, 205 Ga. 743 (1949); Haslerig v. Watson, 205 Ga. 668 (1949).

165. Tillman v. Mayor of Athens, 206 Ga. 289 (1949); Parsons v. Trustees of Atlanta University, 44 Ga. 529 (1871); Mayor & Council of Americus v. Johnson, 2 Ga. App. 378 (1907).

166. Dennis v. Howell, 164 Ga. 657 (1927).

167. Young v. Sweetbriar, 222 Ga. 262 (1966); Adams v. Richmond County, 193 Ga. 42 (1941).

168. Chatham Motorcycle Club v. Blount, 214 Ga. 770 (1959).

169. Fountain v. Bryan, 229 Ga. 120 (1972).

170. Lewis v. DeKalb County, 251 Ga. 100 (1983).

171. Smith v. Gwinnett County, 248 Ga. 882 (1982).

172. Horton v. Wayne County, 243 Ga. 789 (1979); Ross v. Hall County Bd. of Comm'rs., 235 Ga. 309 (1975); Adams v. Richmond County, 193 Ga. 42 (1941).

173. Chatham Motorcycle Club, Inc. v. Blount, 214 Ga. 770 (1959).

174. Hasty v. Wilson, 223 Ga. 739 (1967).

dirt and cuts and removes weeds on two occasions, these actions alone are inadequate to show acceptance.[175]

Continuing to accept payment of taxes raises the inference of non-acceptance.[176]

When a dedication is beneficial, greatly convenient, or necessary to the public, an acceptance will be implied from slight circumstances.[177]

When an owner subdivides a tract of land into lots and records a plat showing streets, parks, and other public areas, this dedication is irrevocable when third parties have acquired interests in the subdivision by buying lots. In such cases it is immaterial whether the dedication has been accepted by a public body.[178]

Public maintenance and use of a strip of land of definite width dedicated for road purposes is acceptance of the entire width of the strip, even though such maintenance and use do not cover the entire width.[179]

Acceptance of a dedication need not be immediate but may be made when the public necessity or convenience arises.[180] An offer of dedication may be withdrawn at any time before acceptance;[181] and an offer of dedication will be presumed to have been declined if a long period of time has elapsed without any acts of acceptance on the part of the public.[182] In one case, for example, a person subdivided a tract into lots and streets, made a plat of the land so subdivided, sold the lots with reference to the plat, and dedicated all the streets to a town. The town accepted, by implication, some of the streets, by opening and maintaining them, but failed expressly to accept the other streets and these remaining streets were never opened and used as thoroughfares. The court held that there was no valid acceptance of the streets that had not been opened.[183]

When land located in an incorporated town is dedicated, it is usually considered vested in the town as trustee for the public.[184] If land is dedi-

175. Raines v. Petty, 170 Ga. 53 (1929).

176. Id.; Mayor, etc., of Savannah v. Stand. Fuel Supply Co., 140 Ga. 353 (1913).

177. Arlington Cem. Corp. v. Bindig, 212 Ga. 698 (1956).

178. Lee v. Warren, 230 Ga. 165 (1973); East Atlanta Land Co. v. Mower, 138 Ga. 380 (1912).

179. Muggridge v. State Hwy. Dept., 229 Ga. 670 (1972); Carroll v. DeKalb County, 216 Ga. 663 (1961); Stringer v. Willingham, 86 Ga. App. 217 (1952).

180. East Atlanta Land Co. v. Mower, 138 Ga. 380 (1912).

181. 23 Am. Jur. 2d Dedication § 49; 26 C.J.S. Dedication §§ 35, 58.

182. Mayor, etc., of Athens v. Gamma Delta, 86 Ga. App. 53 (1952); Ga. Real Est. Law § 5-4 (3d ed. 1986).

183. Kelsoe v. Town of Oglethorpe, 120 Ga. 951 (1904).

184. County of Gordon v. Mayor, etc., of Calhoun, 128 Ga. 781 (1907).

cated to a public use before the existence of a municipal corporation and later a town is incorporated whose limits include the land so dedicated, that land at once vests in the town, in trust for the public.[185]

Duties of the Surveyor with Respect to Unwritten Dedication

The surveyor has no authority to determine whether there has been an unwritten dedication. He merely reports the conditions of use as found. For this reason it is important that he understand some of the legal aspects of dedication so that he can place his client on notice when necessary in the execution of property surveys.[186]

Adverse Possession or Prescription

General

Title by adverse possession or prescription is the right to property which a possessor acquires by reason of the continuance of his possession for a period of time fixed by the law.[187] More specifically, it is a method of acquiring ownership of another's land by exercising possession over it for a requisite length of time without having any agreement with the true owner and without paying him any compensation.

The terms *adverse possession* and *prescription* are often used interchangeably. Technically, however, adverse possession applies to the acquisition of an estate in fee simple absolute, while prescription is concerned with the acquisition of incorporeal hereditaments or rights in the lands of another (for example, easements).[188] To avoid confusion, this discussion will use the two terms interchangeably.

The primary reason for allowing title to land to pass by adverse possession is to protect people in the enjoyment and possession of property that they have held and improved at great expense and labor for the better part of their lives.[189] Also underlying the system is a sound public policy

185. Id.; City of LaFayette v. Walker County, 151 Ga. 786 (1921); Lastinger v. Town of Adel, 142 Ga. 321 (1914).

186. C. Brown & W. Eldridge, Evidence and Procedures for Boundary Location 158 (1962).

187. O.C.G.A. § 44-5-160.

188. 3 Am. Jur. 2d Adverse Possession § 4; 2 C.J.S. Adverse Possession § 3.

189. Conyers v. Kenans, 4 Ga. 308 (1848).

of encouraging those who diligently develop and improve the land, as against those who are content to hold bare legal title inactively for many years.[190] Then too, there is the effect of the law as a statute of repose, to compel litigants to bring their claims to court before time obliterates documentary and parol evidence.[191]

A title that is perfected by the requisite adverse possession extinguishes all other titles, and the possessor becomes the true owner.[192] Such a title is inheritable, devisable, and transferable according to all ordinary methods.[193]

Under Georgia law, for adverse possession to ripen into a prescriptive title, the possession must be continued for seven years or more under color of title[194] or twenty years or more without color of title.[195] The possession cannot be permissive; must not have originated in fraud; must be open, continuous, exclusive, uninterrupted, and peaceable; and must be accompanied by a claim of right.[196]

Acts of Possession

To constitute adverse possession, the possessor must either remain permanently on the land or occupy it in such a way as to leave no doubt in the mind of the true owner as to the adverse claim.[197] The particular acts or conditions that may constitute acts of ownership and give rise to adverse

190. Ga. Real Est. Law § 12-5 (3d ed. 1986).

191. Id. "So general in this state has been the loss or destruction of records by the burning of courthouses, carelessness of clerks, and the reluctance of rural landowners to spreading their muniments of title upon the public rolls, that it may be safely asserted that a majority of our land titles are dependent at one point or another upon the rules of prescription for their validity. Chains of title stemming back in unbroken links to the state grant are so rare that the legislature has mercifully permitted proof of a forty-year period of ownership to stand as a prima facie showing of title." Id. Quotation used with the permission of the copyright owner, The Harrison Company, Norcross, Georgia.

192. Danielly v. Lowe, 161 Ga. 279 (1925).

193. A. Powell, Actions for Land 413 (rev. ed. 1946).

194. Barfield v. Vickers, 200 Ga. 279 (1946); O.C.G.A. § 44-5-164. However, adverse possession under a tax deed executed on or after July 1, 1989, gives the prescriber title after only four years. 1989 Ga. Laws 1393; O.C.G.A. § 48-4-48.

195. Jordan v. Way, 235 Ga. 496 (1975); O.C.G.A. § 44-5-163. An exception is roads and private ways which under certain circumstances may be obtained by seven years' user without color of title. See pages 438, 447–48.

196. O.C.G.A. § 44-5-161.

197. Denham v. Holeman, 26 Ga. 182 (1858).

possession are almost without limit.[198] Examples include cultivation;[199] enclosure with a fence;[200] construction and maintenance of buildings and other structures;[201] residency on the land;[202] keeping of livestock, farm equipment, and other items on the property;[203] leasing of the land to a tenant;[204] use of the land for church purposes;[205] clear-cutting of the land and replanting it in trees;[206] working the land for turpentine production (in some cases);[207] mining;[208] and the construction and use of roads,[209] railroads,[210] and transmission lines.[211]

The following acts, by themselves, will not constitute actual and notorious possession: sporadic acts of trespass;[212] enclosure by a fence in disrepair and partially broken down;[213] occasional cutting of timber;[214] cutting of firewood and occasionally putting out fires;[215] use of an unenclosed area as range for livestock;[216] splitting of rails;[217] boxing of

198. Ga. Real Est. Law § 12-28 (3d ed. 1986).

199. Cheek v. Wainwright, 246 Ga. 171 (1980); Pressly v. Jennings, 227 Ga. 366 (1971); Key v. Stringer, 204 Ga. 869 (1949).

200. Dyal v. Sanders, 194 Ga. 228 (1942); Roumillot v. Gardner, 113 Ga. 60 (1901).

201. Everett v. Culberson, 215 Ga. 577 (1960); Hearn v. Leverette, 213 Ga. 286 (1957).

202. Ga. Real Est. Law § 12-30 (3d ed. 1986); 3 Am. Law of Prop. § 15.3; 2 C.J. Adverse Possession § 8; 2 C.J.S. Adverse Possession, § 49.

203. Hughes v. Heard, 215 Ga. 156 (1959); Hearn v. Leverette, 213 Ga. 286 (1957).

204. Baxter & Co. v. Wetherington, 128 Ga. 801 (1907); McMullin v. Erwin, 58 Ga. 428 (1877).

205. Henson v. Bridges, 218 Ga. 6 (1962); Slaughter v. Land, 194 Ga. 156 (1942).

206. Brown v. Williams, 259 Ga. 6 (1989); Cheek v. Wainwright, 246 Ga. 171 (1980); Pannell v. Continental Can Co., 554 F.2d 216 (C.C.A. Ga. 1977).

207. Rogers v. Manning, 200 Ga. 844 (1946); Dyal v. McLean, 188 Ga. 229 (1939); Baxter & Co. v. Wetherington, 128 Ga. 801 (1907).

208. House v. Palmer, 9 Ga. 497 (1851).

209. Davis v. The State, 9 Ga. App. 430 (1911).

210. Seignious v. MARTA, 252 Ga. 69 (1984); Murphey v. Cent. of Georgia Ry. Co., 135 Ga. 194 (1910); Georgia Pac. Ry. Co. v. Strickland, 80 Ga. 776 (1888).

211. Kerlin v. Sou. Bell Tel. & Tel. Co., 191 Ga. 663 (1941).

212. Denham v. Holeman, 26 Ga. 182 (1858).

213. Turner v. McKee, 97 Ga. App. 531 (1958).

214. Shahan v. Watkins, 194 Ga. 164 (1942); Robertson v. Abernathy, 192 Ga. 694 (1941); Bussy v. Jackson, 104 Ga. 151 (1898).

215. Bridges v. Brackett, 208 Ga. 774 (1952); Long v. Young, 28 Ga. 130 (1859).

216. Gordon v. Ransom & Lomax Lbr. Co., 147 Ga. 55 (1917); McCook v. Crawford, 114 Ga. 337 (1901).

217. Joiner v. Borders, 32 Ga. 239 (1861); Grimes v. Ragland, 28 Ga. 123 (1859).

timber for turpentine;[218] payment of taxes;[219] prospecting for gold and occasional mining of ore;[220] posting of signs forbidding trespassing and driving away hunters;[221] and having the land surveyed.[222]

An adverse possession need not be held by the person who claims it.[223] He may hold by a tenant.[224] In such a case the tenant's possession is that of his landlord, and the landlord is the true possessor.[225] The title that ripens is in favor of the landlord.[226] However, an independent adverse occupation by a tenant of another's land that does not purport to be covered by the terms of the lease and is not based on any authorization purported to be conferred therein by the landlord does not give the landlord prescriptive title to the land so held.[227]

Color of Title

Color of title has been defined as "a writing, upon its face professing to pass title, but which does not do it, either from a want of title in the person making it, or from the defective conveyance that is used—a title that is imperfect, but not so obviously so that it would be apparent to one not skilled in the law."[228] It is wholly immaterial how imperfect or defective the instrument may be, so long as it is in writing and defines the extent of the claim.[229] A valid deed conveying title is never color of title.[230] An instrument relied on as color of title does not have to be recorded to constitute color of title. If it is not recorded, however, adverse possession will not extend beyond the land actually occupied, although the length

218. Hilton v. Singletary, 107 Ga. 821 (1899).
219. See text at notes 280–84 below.
220. Graham v. Mitchell, 78 Ga. 310 (1886).
221. Fitzpatrick v. Massee-Felton Lbr. Co., 188 Ga. 80 (1939).
222. Dillon v. Mattox, 21 Ga. 113 (1857).
223. McMullin v. Erwin, 58 Ga. 428 (1877).
224. Id.
225. Id.
226. Id.
227. Olsen v. Noble, 208 Ga. 899 (1953).
228. Capers v. Camp, 244 Ga. 7 (1979); Beverly v. Burke, 9 Ga. 440 (1851).
229. Veal v. Robinson, 70 Ga. 809 (1883); Field v. Boynton, 33 Ga. 239 (1862).
230. Gooch v. Cit. & Sou. Nat'l Bank, 196 Ga. 322 (1943).

of time necessary to perfect title by adverse possession will be reduced from twenty to seven years.[231]

Examples of color of title include a devise in a will,[232] a quitclaim deed,[233] a deed from one who had never owned the land,[234] a deed from one who has already conveyed the land to another,[235] a deed made by one of several cotenants without the consent of the other cotenants,[236] a deed made to defraud creditors,[237] a letter purporting to convey title,[238] a void tax deed,[239] a forged bond for title,[240] and a void grant from the state.[241]

The same certainty of description necessary in an instrument constituting a conveyance of title is required in an instrument relied on as color of title.[242] Therefore, a writing that purports to convey land but is so indefinite in description that the land cannot be identified is inoperative as color of title.[243] Also, a writing that purports to convey a certain tract of land cannot serve as color of title over another tract of land which a party took possession of honestly believing that the color of title covered that tract.[244] A security deed cannot operate as color of title until the right of possession accrues under it by foreclosure or otherwise.[245] A mortgage cannot operate as color of title because the mortgagee has no title to the

231. Tucker v. Long, 207 Ga. 730 (1951); Carstarphen v. Holt, 96 Ga. 703 (1895); O.C.G.A. § 44-5-167.

232. Blalock v. Webb, 190 Ga. 769 (1940).

233. Johnson v. Girtman, 115 Ga. 794 (1902).

234. Smart v. Miller, 260 Ga. 88 (1990); McMullin v. Erwin, 58 Ga. 427 (1877); Gittens v. Lowry, 15 Ga. 336 (1854).

235. Wright v. Smith, 43 Ga. 291 (1871).

236. Street v. Collier, 118 Ga. 470 (1903).

237. Moore v. Mobley, 123 Ga. 424 (1905).

238. Wooding v. Blanton, 112 Ga. 509 (1900).

239. Pannell v. Continental Can Co., 554 F.2d 216 (C.C.A. Ga. 1977); Smith v. Powers, 208 Ga. 768 (1952).

240. Millen v. Stines, 81 Ga. 655 (1888).

241. Moody v. Fleming, 4 Ga. 115 (1848). For other examples of color of title see Ga. Real Est. Law § 12-47 (3d ed. 1986); S. Mitchell, Real Property under the Code of Georgia and the Georgia Decisions 151–53 (5th ed. 1960); and A. Powell, Actions for Land 343, n. 14 (rev. ed. 1946).

242. Crawford v. Verner, 122 Ga. 814 (1905).

243. Laws v. Oakey, 216 Ga. 408 (1960); Bennett v. Rewis, 212 Ga. 800 (1957); May v. Sorrell, 149 Ga. 610 (1919).

244. Walton v. Sikes, 165 Ga. 422 (1927); LaRoche v. Falligant, 130 Ga. 596 (1908); Luttrell v. Whitehead, 121 Ga. 699 (1904).

245. Lawhorn v. Steele, 232 Ga. 857 (1974); Thomas v. Stedham, 208 Ga. 603 (1952).

land but only a lien or security on the land for debt;[246] and a bond for title is not sufficient color of title until the vendor is paid.[247] A writing that does not on its face profess to pass title but expressly states that title will be conveyed at a future time and upon certain conditions is not sufficient to constitute color of title.[248]

The Possession Cannot Be Permissive

Except in the case of prescriptive easements, permissive possession cannot be the foundation of prescription unless an adverse claim is made, together with actual notice of such claim to the owner.[249] The rule applies equally to prescription under color of title for seven years and prescription for twenty years without color of title.[250] Thus a tenant cannot prescribe against his landlord;[251] a grantor remaining on the land after he has sold it cannot prescribe against his grantee;[252] one cotenant cannot prescribe against the other cotenants until actual ouster, exclusive possession after demand, or express notice of adverse possession;[253] the holder of a security deed cannot prescribe against his grantor;[254] and a

246. Fudge v. Bailey, 182 Ga. 119 (1935); Phillips v. Bond, 132 Ga. 413 (1909).

247. Brown v. Huey, 103 Ga. 448 (1897); Paxson v. Bailey, 17 Ga. 600 (1855).

248. Shippen v. Cloer, 213 Ga. 172 (1957).

249. Harris v. Mandeville, 195 Ga. 251 (1943); Kesler v. Verner, 161 Ga. 118 (1925); Rucker v. Rucker, 136 Ga. 830 (1911). "The doctrine that prescriptive titles to the fee in real estate by seven years' possession cannot originate in consent, because the possession there must be adverse all the time, does not . . . apply to a right-of-way . . . for the reason that knowledge and acquiescence of the owner of the lands is of the very essence of the right-of-way against the owner. . . . 'Steps must be taken to prevent the enjoyment of the way' by the owner, to stop the continuance of the use, or it will ripen into a prescription, though he acquiesce silently or consent openly all the time." Everedge v. Alexander, 75 Ga. 858 (1885). "Possession must be adverse in order to form the basis for prescription. A notable exception exists, however, in the case of private ways. The use may originate in permission, and yet may ripen by prescription. . . . When a way once begins to be traveled by the people of a neighborhood, they adjust themselves thereto. Land is bought, farms are opened, and houses are built on the faith that such property can be reached over the existing road. The longer the owner remains inactive the greater will be the inconvenience. The code therefore requires him to act promptly." Kirkland v. Pitman, 122 Ga. 256 (1904).

250. Woods v. Brannen, 208 Ga. 495 (1951).

251. Dickson v. Davis, 237 Ga. 883 (1976); Coates v. Jones, 142 Ga. 237 (1914).

252. Johnson v. Hume, 163 Ga. 867 (1926).

253. Fuller v. McBurrows, 229 Ga. 422 (1972); Erwin v. Miller, 203 Ga. 58 (1947); O.C.G.A. § 44-6-123. Compare Roumillot v. Gardner, 113 Ga. 60 (1901).

254. Lawhorn v. Steele, 232 Ga. 857 (1974); Thomas v. Stedham, 208 Ga. 603 (1952).

party who otherwise holds permissive possession of land cannot prescribe against the true owner.[255]

The Possession Must Not Have Originated in Fraud

For prescription to run, the claimant must have entered into possession of the property in good faith.[256] Good faith, as contemplated by the law, has relation to one's actual state of mind.[257] If one takes possession of land which he knows does not belong to him or to the person from whom he purchases, no prescription will run in his favor no matter how long he holds possession.[258] For example, the devisee to a tract of land could not hold the land adversely where she knew that her testator, subsequent to the execution of his will, had divested himself of title to the property.[259]

The Possession Must Be Open

The acts of possession must be of such open, public, and notorious character as to attract the attention of the true owner in the exercise of ordinary vigilance.[260] There can be nothing clandestine or stealthy about the possession.[261] In other words, the possessor must hold the property in open sight of the whole community[262] so that the owner may be warned and put on notice that his title is contested and in danger.[263]

Tunneling under a piece of property from an adjoining tract[264] or maintaining underground pipes and other structures[265] are not acts of a

255. Woods v. Brannen, 208 Ga. 495 (1951); Mack v. Felt, 166 Ga. 705 (1928).

256. Lee v. Ogden, 83 Ga. 325 (1889).

257. Fraser v. Dolvin, 199 Ga. 638 (1945); Kelly v. Tucker, 175 Ga. 796 (1932); McMullin v. Erwin, 58 Ga. 427 (1877).

258. Harrison v. Holsenbeck, 208 Ga. 410 (1951); Parker v. Salmons, 101 Ga. 160 (1897); Cowart v. Young, 74 Ga. 694 (1885).

259. Erwin v. Miller, 203 Ga. 58 (1947).

260. Ga. Real Est. Law § 12-22 (3d ed. 1986).

261. 2 C.J. Adverse Possession § 57; 2 C.J.S. Adverse Possession § 50.

262. S. Mitchell, Real Property under the Code of Georgia and the Georgia Decisions 148 (5th ed. 1960).

263. 2 C.J.S. Adverse Possession § 51; H. Tiffany, A Treatise on the Modern Law of Real Property 786 (abr. ed. 1940).

264. Marengo Cave Co. v. Ross, 212 Ind. 624, 10 N.E.2d 917 (1937), commented on in 72 U.S. L. Rev. 186–92 (1938); Pierce v. Barney, 209 Pa. 132, 58 A. 152 (1904).

265. Ga. Real Est. Law § 12-22 (3d ed. 1986); 25 Am. Jur. 2d Easements and Licenses § 60; Kodzoff v. Paul, 108 F. Supp. 265 (1952).

sufficiently public nature to amount to adverse possession. The fact that it may be difficult or impossible to occupy a tract of land openly and notoriously does not alter the rule that the possession must be public.[266]

The Possession Must Be Continuous (Uninterrupted)

The rule requiring continuity of possession is one of substance and not of absolute mathematical continuity, provided there is no break so as to make a severance of two possessions.[267] There may be slight intervals during which the prescriber or his agent or tenant is not actually on the land, as in the case of changing tenants, or when the nature or character of the business does not require his presence every day.[268] The bringing of a suit against an adverse possessor, which is dismissed and not prosecuted to a successful termination, constitutes no disturbance of adverse possession.[269]

Continuity of adverse possession is not, however, established by dis-

266. "Plaintiff further says that his possession . . . was evidenced by his continuous use and occupancy of said lands for the purpose of range for cattle and hogs; that he constantly and continuously so used and occupied said premises to the exclusion of all other persons, except with his consent; that he repeatedly occupied said lands by himself and his agents for the purpose of cutting timber for different purposes; that said lands, being on the swamp of the Oconee river, were unfit for actual, physical residence thereon, and were unfit for cultivation; that they were useful and valuable only for the purposes stated, and that his occupancy thereof and possession thereof for said purposes was as complete and exclusive as was possible . . . . All this may be true, and still there is no reason alleged why the property could not be in some manner inclosed so as to make this evidence of possession notorious. . . . The fact that the land was situated in a swamp of a river would not necessarily prevent it from being inclosed in such a manner as to attract the attention of adverse claimants; and while a fence erected for the purpose of inclosing the land might be destroyed or washed away in times of high water, it could be promptly replaced so as to make the possession continuous within the meaning of the law. That it would be expensive and troublesome to keep such property inclosed would not alter the law in reference to the subject of what amounts to actual possession, but would be simply a necessary burden to be borne by an owner of property of this character. But let it be conceded, for the sake of the argument, that it is impossible to inclose the land, that it is impossible to cultivate the land, that it is impossible to put the same to any use or occupy it in any manner which would be so notorious as to attract the attention of adverse claimants; still the plaintiff fails to make a case by his allegations, for the reason that, if this state of affairs exists, title to the property can not be acquired by prescription, because, when the character of property is such that it is impossible to be in actual possession thereof, title thereto can pass from one to another only by written evidence of title." McCook v. Crawford, 114 Ga. 337 (1901).

267. Chamblee v. Johnson, 200 Ga. 838 (1946); Clark v. White, 120 Ga. 957 (1904).

268. Walker v. Steffes, 139 Ga. 520 (1912).

269. Kile v. Fleming, 78 Ga. 1 (1886).

connected periods of occupancy, by casual or incidental acts of dominion or trespass, or by sporadic acts of ownership.[270]

What is required is that there be such complete continuity of possession that for each day out of the entire period during which the adverse claimant has control of the land a cause for legal action will lie for the true owner.[271]

The Possession Must Be Exclusive

The acts of possession must be sufficient to prevent others from occupying the land.[272] It is not necessary, however, that the claimant prohibit any and all use of the property by the general public. The use of the property by another person concurrently with the claimant—by the permission of the latter or in subordination of his claim—does not mitigate against the exclusiveness of his possession.[273]

The Possession Must Be Peaceable

The words *peaceable possession* mean about the same thing as quiet enjoyment.[274] Peaceable possession is possession unbroken by an ouster and undisturbed by successful litigation.[275]

The Possession Must Be Accompanied by a Claim of Right

Before prescription can arise under an asserted claim of right, the claim must be honestly entertained.[276] A person entering upon lands, not claiming in good faith the right to do so by virtue of any title of his own or by virtue of an agreement with someone whom he believes to hold title, is a mere squatter.[277] A squatter can never gain prescriptive title to land,

270. Clark v. White, 120 Ga. 957 (1904); Bryne v. Lowry, 19 Ga. 27 (1855); 2 C.J.S. Adverse Possession § 150.

271. 2 Am. Jur. 2d Adverse Possession § 54; 2 C.J.S. Adverse Possession § 150.

272. Ga. Real Est. Law § 12-23 (3d ed. 1986).

273. 2 C.J.S. Adverse Possession § 54.

274. Gittens v. Lowry, 15 Ga. 336 (1854).

275. Ga. Real Est. Law § 12-24 (3d ed. 1986).

276. Mayor & Council of Forsyth v. Hooks, 182 Ga. 78 (1935).

277. Id.

no matter how long he holds possession.[278] Nor can a squatter defeat the title of the true owner by conveying the land to purchasers who have full knowledge of the nature and character of the squatter's title.[279]

Payment of Taxes

In some jurisdictions an adverse claimant must pay taxes on the property in order for prescription to run.[280] Such, however, is not the case in Georgia,[281] though payment of taxes is some evidence of adverse possession[282] and nonpayment may imply lack of good faith.[283] Payment of taxes, without more, is insufficient to support an adverse claim.[284]

Tacking

The possessions of successive adverse claimants of property may be tacked (that is, added together) to fill out the necessary number of years to obtain a prescriptive title.[285] There must, however, be privity between the successive occupants; that is, each occupant must have received his possession through some act of the preceding occupant or by operation of the law.[286] Otherwise continuous possession by different tenants is regarded as successive trespasses.[287] The transfer may be in writing or by a parol agreement or understanding.[288]

Intermittent possession by a person cannot be tacked so as to make

278. Id.; Halpern v. The Lacy Inv. Corp., 259 Ga. 264 (1989); Hannah v. Kenny, 210 Ga. 824 (1954).

279. Hannah v. Kenny, 210 Ga. 824 (1954); McCamy v. Higdon, 50 Ga. 629 (1874).

280. 3 Am. Jur. 2d Adverse Possession § 125; 3 Am. Law of Prop. § 15.1; 2 C.J.S. Adverse Possession § 213.

281. Ga. Real Est. Law § 12-36 (3d ed. 1986). On the other hand, the owner of outstanding mineral interests may lose his rights to them by failure, for seven years, to pay taxes and to exercise mining rights. Hayes v. Howell, 251 Ga. 580 (1983).

282. Hughes v. Heard, 215 Ga. 156 (1959); Chamblee v. Johnson, 200 Ga. 838 (1946); Baxter & Co. v. Wetherington, 128 Ga. 801 (1907).

283. Mitchell v. Crummey, 134 Ga. 383 (1910).

284. Talmadge v. Adams, 240 Ga. 193 (1977); Watts v. Griswold, 20 Ga. 732 (1856).

285. Fraser v. Dolvin, 199 Ga. 638 (1945); O.C.G.A. § 44-5-172.

286. Cliett v. Met. Life Ins. Co., 195 Ga. 257 (1943).

287. Hays v. Morrison, 30 Ga. 971 (1860).

288. Blalock v. Redwine, 191 Ga. 169 (1940).

out the prescriptive period.[289] An innocent purchaser cannot tack his possession to that of his grantor if the grantor's possession originated in fraud.[290]

Persons against Whom Prescription Does Not Run

Generally prescription will not run against those who are physically or legally incapable of bringing suit to recover lands adversely held. Therefore, prescription will not run if the true owner is a minor,[291] is confined in prison,[292] or is insane.[293] Prescription does not run against a remainderman until the death of the life tenant;[294] or against an unrepresented estate until representation, provided the lapse does not exceed five years.[295] Prescription does not run against the state or any of its political subdivisions.[296]

It is also a well-settled rule that a person cannot hold his own land adversely to himself.[297] Thus in a case in which a person used a private driveway across his land for several years and then sold off a portion of the land that included part of the drive, he could not convert the part of the drive on the property he had sold into a prescriptive easement in his favor.[298]

289. Walker v. Steffes, 139 Ga. 520 (1912); Clark v. White, 120 Ga. 957 (1904).

290. Ellis v. Dasher, 101 Ga. 5 (1897); Hunt v. Dunn, 74 Ga. 120 (1884).

291. Ellis v. Dasher, 101 Ga. 5 (1897); Whittington v. Wright, 9 Ga. 23 (1850); O.C.G.A. § 44-5-170.

292. O.C.G.A. § 44-5-170.

293. Id.; Lawrence v. Boswell, 155 Ga. 690 (1923); Verdery v. Sav., Fla., & West. Ry. Co., 82 Ga. 675 (1889).

294. Harper v. Paradise, 233 Ga. 194 (1974); Futch v. Jarrard, 203 Ga. 47 (1947).

295. Dozier v. Parker, 219 Ga. 725 (1964); O.C.G.A. § 44-5-173.

296. See text at notes 299–307 below.

297. Ga. Real Est. Law § 10-13 (3d ed. 1986); 3 Am. Jur. 2d Adverse Possession § 129.

298. Burk v. Tyrrell, 212 Ga. 239 (1956). "Nor can the plaintiffs tack to their prescriptive period the time when Swanson, between 1923 and 1950, owned both tracts and used this private way over both tracts. Where two parcels of land are owned and occupied by the same person, there cannot be such use of one of them as to create an easement by prescription over the other tract. Unity of possession of both the dominant and servient tenements is inconsistent with adverse use, and such use could not begin until after a severance of the two estates." Id. See also Hogan v. Cowart, 182 Ga. 145 (1935).

Adverse Possession against the State

Adverse possession does not run against the United States,[299] the state,[300] a county,[301] a municipal corporation [302] or any public corporation created by the state or the United States.[303] There seem to be few exceptions to this rule. When land that was subject to entry and grant has been held under a private claim of right for twenty years or longer, the land is presumed to have been granted by the state.[304] This provision is more a form of estoppel, however, than a method of acquiring adverse rights to land. In the absence of better evidence, old boundary fences and other improvements may determine the location of boundaries between public and private property, such as a public street and the abutting, privately owned lots.[305] The courts may refuse to grant an injunction against the accidental encroachment of a building of less than one foot into a highway right-of-way if the effect of such an injunction would be to force the demolition of the building, thus incurring considerable and needless expense for the owner.[306] But, of course, the owner obtains no prescriptive title.

The reason adverse possession does not run against the state is because the state cannot possibly watch and protect its countless possessions. The business of government is transmitted entirely through agents, who are numerous and scattered. The utmost vigilance could not guard against losses and schemes to defraud the public. The government is therefore

299. Texas v. Louisiana, 410 U.S. 702, 35 L. Ed. 2d 646 (1973); Gibson v. Chouteau, 80 U.S. (13 Wall.) 92, 20 L. Ed. 534 (1872); 2 C.J.S. Adverse Possession § 10.

300. State of Georgia v. Paxson, 119 Ga. 730 (1904); Brinsfield v. Carter, 2 Ga. 143 (1847).

301. Adams v. Richmond County, 193 Ga. 42 (1941).

302. Odd Fellows v. City of Thomasville, 226 Ga. 4 (1970); Austin v. Mayor & Council of Madison, 153 Ga. 494 (1922).

303. Ga. Real Est. Law § 12-10 (3d ed. 1986).

304. Montgomery v. Trustees of the Masonic Hall, 70 Ga. 38 (1883); O.C.G.A. § 44-5-14. For many tracts of land it is impossible to trace the title back to the original state or colonial grant because of the loss and destruction of instruments in the chain of title. Because of the difficulty under such circumstances of proving that lands have been granted, it will be presumed that they were granted if they have been in private possession for a long period of time and if other existing elements of title are consistent with private ownership of such lands.

305. City of Mt. Carmel v. McClintock, 155 Ill. 608, 40 N.E. 829 (1895); Bridges v. Town of Grand View, 158 Ia. 402, 139 N.W. 917 (1913); City of Madison v. Mayers, 97 Wis. 399, 73 N.W. 43 (1897).

306. Stephens v. State Highway Dept., 223 Ga. 713 (1967).

exempt from the operation of prescription upon grounds of public policy and not because of any extraordinary prerogative.[307]

Extent of an Adverse Claim

If one goes into possession of a tract of land without color of title,[308] or under color of title that is not recorded,[309] he cannot claim beyond the extent of his actual possession. If one goes into actual possession of a portion of a lot under a recorded color of title that describes the entire lot, he has constructive possession of the entire lot and his possession will be adverse as to the entire lot.[310] If, however, adjoining adverse claimants are in constructive possession of the same land because of an overlap in their descriptions, no prescription can arise in favor of either as to the overlap.[311]

Actual possession of a portion of a lot under a recorded deed that purports to convey that lot and several contiguous lots is constructive possession of all the lots, if all are described as being in one body.[312] The tracts must be adjacent to or corner with one another.[313] A statement of the total acreage may satisfy the rule that the lots must be described as constituting a single tract.[314] Division of a tract by a road, stream, or swamp does not make it two tracts.[315]

When land is described in a color of title as bounded by the land of an adjacent owner, there can be no prescription against such owner further than the actual possession of the grantee in the color of title extends.[316]

If a person goes into possession of land beyond the boundaries re-

307. J. Wait, The Law of Operations in Engineering and Architecture 365 (1901).

308. Hall v. Gay, 68 Ga. 442 (1882); Cook v. Long, 27 Ga. 280 (1859).

309. Tucker v. Long, 207 Ga. 730 (1951).

310. Mincey v. Anderson, 206 Ga. 572 (1950); Knighton v. Hasty, 200 Ga. 507 (1946); Shahan v. Watkins, 194 Ga. 164 (1942); Roberson v. The Downing Co., 120 Ga. 833 (1904).

311. Tucker v. Long, 207 Ga. 730 (1951); Robertson v. Abernathy, 192 Ga. 694 (1941).

312. Campbell v. Gregory, 200 Ga. 684 (1946); Georgia Minerals Co. v. Cox, 154 Ga. 861 (1922).

313. Georgia Pine Inv. & Mfg. Co. v. Holton, 94 Ga. 551 (1894); Morris v. Gibson, 35 Ga. App. 689 (1926).

314. Tucker v. Wimpey, 158 Ga. 820 (1924).

315. Atlanta & Charlotte Air-Line Ry. Co. v. Colbert, 171 Ga. 196 (1930); Dodge v. Cowart, 131 Ga. 549 (1908).

316. Holloway v. Woods, 195 Ga. 55 (1942).

cited in his deed or color of title under the mistaken idea that the recited boundaries include the land, he can obtain title to the land beyond the boundaries only by twenty years' adverse possession.[317] Casual possession and use of land along the boundary line of adjoining owners, however, is construed to be permissive.[318]

Where land is conveyed and the grantor reserves the mineral interests, he forfeits his rights to the minerals by failure, for seven years, to pay taxes and to exercise mining rights.[319]

Title Obtained by Adverse Possession

The nature and extent of the estate acquired by adverse possession are determined by the claim of the possessor, coupled with the fact of possession under the circumstances contemplated by law.[320] In general, when a person fulfills the requirements to obtain a prescriptive title to land, he obtains the fee to the land.[321] On the other hand, adverse use of land for such purposes as roads,[322] alleys,[323] private ways,[324] railroads,[325] transmission lines,[326] cemeteries,[327] and party walls[328] merely confers an easement unless held under color of title purporting to convey the fee.

Prescriptive easements can also be obtained to maintain building cornices and other projections over an adjoining owner's land,[329] to pollute

317. Bridges v. Brackett, 205 Ga. 637 (1949); Walton v. Sikes, 165 Ga. 422 (1927).

318. Godley v. Hinely, 165 Ga. 717 (1927).

319. Hayes v. Howell, 251 Ga. 580 (1983).

320. 3 Am. Jur. 2d Adverse Possession § 102.

321. Danielly v. Lowe, 161 Ga. 279 (1925); A. Powell, Actions for Land 412–13 (2d ed. 1946).

322. Annotations: 104 Am. St. Rep. 679 (1904); 57 Am. St. Rep. 744 (1896).

323. Henderson v. Ezzard, 75 Ga. App. 724 (1947).

324. See pages 447–48.

325. Maryland & Pennsylvania R.R. Co. v. Mercantile-Safe Deposit, 224 Md. 34, 166 A.2d 247 (1960); 65 Am. Jur. 2d Railroads § 74; 74 C.J.S. Railroads § 85; Annotation, 127 A.L.R. 517 (1940).

326. Chancey v. Georgia Pwr. Co., 238 Ga. 397 (1977); Kerlin v. Sou. Bell Tel. & Tel. Co., 191 Ga. 663 (1941).

327. See page 496.

328. Forsyth Corp. v. Rich's, Inc., 215 Ga. 333 (1959); Joel v. Publix-Lucas Theatres Inc., 193 Ga. 531 (1942).

329. 19 C.J. Easements § 86; 28 C.J.S. Easements § 20; 2 G. Thompson, Commentaries on the Modern Law of Real Property § 336 (5th ed. 1980).

a stream,[330] to overflow another's land,[331] and for rights of passage over nonnavigable streams.[332] Prescriptive rights cannot be obtained as to lateral support,[333] ancient lights,[334] or for maintaining awnings over public sidewalks.[335]

If the owner of land allows the public to acquire a prescriptive easement across his lands subject to implied conditions, the public takes subject to those conditions. Thus, in one case, a landowner allowed the public to use a road through his land but kept a gate across it at each end to protect his property. The public acquired a restricted prescriptive right and, with that qualification, was entitled to use the road.[336]

If a person goes into adverse possession of land under color of title, he obtains the type of estate specified in the color of title.[337] For example, if the color of title purports to convey a life estate, the prescriber can obtain only a life estate to the land.[338]

Surveyor's Duties with Respect to Adverse Possession

Making judgments as to whether adverse rights have ripened into title is beyond the scope of the land surveyor's rights or duties.[339] The surveyor merely gathers evidence of possession and relates that possession to written title lines.[340] Because the land surveyor is often the first to discover encroachments and other evidence of adverse possession, it is important that he have a working knowledge of its legal aspects. He can then recognize it when he comes upon it and put his client and other interested parties on notice.

330. Anneberg v. Kurtz, 197 Ga. 188 (1944).
331. Terrell v. Terrell, 144 Ga. 32 (1915).
332. Seaboard Air-Line Ry. v. Sikes, 4 Ga. App. 7 (1908).
333. Mitchell v. Rome, 49 Ga. 19 (1873).
334. O.C.G.A. § 44-9-2.
335. City Council of Augusta v. Burum, 93 Ga. 68 (1893).
336. Green v. Bethea, 30 Ga. 896 (1860).
337. Ga. Real Est. Law § 12-46 (3d ed. 1986).
338. Id.
339. Holloway v. Woods, 195 Ga. 55 (1942); C. Brown & W. Eldridge, Evidence and Procedures for Boundary Location 171–73 (1962).
340. C. Brown & W. Eldridge, Evidence and Procedures for Boundary Location 171 (1962).

Year's Support Proceeding

The year's support, or twelve months' support,[341] is real or personal property set aside by the probate court out of the estate of a deceased husband or wife for the maintenance and support of the surviving spouse and their minor children for twelve months. It is practically a unique creation of Georgia law, almost unknown in other jurisdictions. Its purpose is to prevent the spouse and children from being turned out without home or means of support at the death of the person responsible for their support. It may be awarded whether the deceased died testate or intestate. In general, a year's support award is superior to the debts of the deceased, materialmen's liens, back taxes, and certain other obligations (including even express terms in the deceased's will).

The code provides that the spouse and minor children of a deceased person, or someone acting in their behalf, shall make application to the probate judge, who in turn shall appoint five appraisers to set apart, for the spouse and children, real and personal property from the decedent's estate, sufficient for their support and maintenance for twelve months. Additional awards may be made in succeeding years. The spouse and children receive fee simple title to the property so awarded and may subsequently sell or encumber it like any other property.

If real property is set aside in a year's support proceeding, then it must be described in a manner legally sufficient to pass title, or else the proceeding is void. The appraisers, therefore, have authority to employ a surveyor to survey and make a plat of the land to be set aside.

Tax Execution

Delinquent ad valorem taxes on real or personal property are collected through the sale of the property under tax executions issued by the tax collector or tax commissioner.[342] The execution is issued against the owner of the property, if known, or against the property itself if the

341. O.C.G.A. §§ 53-5-1 to 53-5-21; Ga. Real Est. Law §§ 17-1 to 17-20 (3d ed. 1986); 3 D. Redfearn, Wills and Administration in Georgia §§ 323–42 (5th ed. 1988).

342. O.C.G.A. §§ 9-13-140, 48-4-1 to 48-4-48; Matthews, *Executions,* in 11A Encyclopedia of Georgia Law §§ 307, 313 (1979).

owner is unknown. The sheriff or constable advertises the property and sells it at public auction at the courthouse in the county where the tax was levied. The levying officer makes out a deed to the purchaser, who acquires whatever right, title, or interest the debtor had in the property. Until a bar of redemption is filed, the debtor may reacquire the property by paying to the purchaser the amount paid for the property at the tax sale, plus interest and certain other fees. The purchaser at the tax sale may file a bar of redemption one year after the sale, thus closing off the rights of the original owner.

Judicial Sale

Under the general term *judicial sale*[343] is included the sale of real and personal property in bankruptcy proceedings, foreclosure of mortgages or trust deeds, enforcement of liens of various kinds, partition proceedings, guardian's sales, sales by executors and administrators, and other types of sales made under a judgment or order of court. In Georgia the term includes the sale of property by the sheriff under a court decree to satisfy personal debts. Such sales are conducted in the same manner as sales under tax executions.

Abandonment

Abandonment of property or of a right is the voluntary relinquishment thereof by its owner or holder, with the intention of terminating his ownership and without concern as to who may subsequently take possession.[344] Title to a corporeal hereditament (such as a freehold estate) cannot be lost by mere abandonment, no matter how long continued,[345] but incorporeal hereditaments (such as easements) can be lost by abandonment.[346]

It is with reference to easements that the surveyor will most often be

343. O.C.G.A. §§ 9-13-140 to 9-13-178; Matthews, *Judicial Sales*, in 18 Encyclopedia of Georgia Law §§ 1–30 (1981).

344. 1 C.J.S. Abandonment § 1.

345. Aggregate Supply Co. v. Sewell, 217 Ga. 407 (1961).

346. Tietjen v. Meldrim, 169 Ga. 678 (1929).

concerned with the rules as to abandonment. If an easement has been acquired by mere use, its nonuse for the prescriptive period raises the presumption of abandonment; and if the owner of the easement does some unequivocal act to show an intent to abandon, then an even stronger case is made.[347] If an easement has been acquired by grant, then, in addition to nonuse, the owner *must* perform some unequivocal act evidencing his intention to abandon.[348] Such acts include an express statement by the owner of his intent to abandon,[349] allowing the area to grow up in trees and thick underbrush,[350] permanent removal of the physical installations for which the easement was conveyed and vacating the premises,[351] conveyance of the easement to someone who diverts the property to a use other than the one for which it was conveyed,[352] and allowing others to erect and maintain obstructions that prevent the use of the easement.[353]

The doctrine of loss by abandonment applies to government corporations as well as to individuals.[354] Thus, the fact that prescription does not run against a municipality by no means precludes the possibility that a municipality, by voluntary abandonment, may relinquish its rights to a street dedicated to the public.[355]

Accretion and Reliction

Accretion and reliction[356] as means of transferring real property differ from other methods in that they are caused by nature rather than by human resolution.

347. Gilbert v. Reynolds, 233 Ga. 488 (1975); O.C.G.A. § 44-9-6.

348. Walker v. Sims, 211 Ga. 472 (1955); Westbrook v. Comer, 197 Ga. 433 (1944); Atlanta, B. & A. Ry. Co. v. Coffee Co., 152 Ga. 432 (1921).

349. Atlanta, B. & A. Ry. Co. v. Coffee Co., id.

350. Ga. Real Est. Law § 8-31 (3d ed. 1986).

351. Id.

352. Annotation, 95 A.L.R.2d 498 (1964).

353. Ga. Real Est. Law § 8-31 (3d ed. 1986).

354. Mayor & Aldermen of Savannah v. Barnes, 148 Ga. 317 (1918).

355. Id.

356. Georgia v. South Carolina, 497 U.S.—, III L. Ed. 2d 309 (1990); 78 Am. Jur. 2d Waters §§ 406–15, 430, 432; 3 Am. Law of Prop. §§ 15.26–15.33; 93 C.J.S. Waters §§ 76, 78, 79; 5A G. Thompson, Commentaries on the Modern Law of Real Property §§ 2560–65 (1978).

Accretion is the process of gradual and imperceptible addition to riparian or littoral lands caused by the action of water in washing up sand, mud, gravel, and other materials. Land formed by this process belongs to the riparian owner on or against whose banks the matter is deposited. Reliction (or dereliction) is the gradual withdrawal or recession of waters so as to expose dry land. Land formed by this process belongs to the riparian owner from whose shore the water has receded.

A term that should be distinguished from accretion and reliction is avulsion. Avulsion is the sudden and perceptible gain or loss of riparian land. For example, a river may take a new course by cutting across a bend; or a considerable chunk of land may suddenly be washed from one bank of a river and deposited on the opposite bank. When avulsion occurs, the ownership of the land remains according to the former boundaries.

Generally the owner of the shore or bed of a stream or other body of water may reclaim submerged land by filling in as far as his ownership or as far as the point of navigability. An upland owner cannot, however, by merely filling in publicly owned submerged lands that abut his lands, extend his ownership over such lands.

Miscellaneous

A few other methods for transferring real property are either obsolete or so seldom used that they will be mentioned only in passing. The reader may refer to the citations for more detailed treatments.

Escheat, in feudal times, was the termination of a tenure either for failure of inheritable blood, or for "corruption" of the blood of the tenant by attainder of felony or treason. In such cases the tenant's lord resumed possession of the land.[357] Modern law prohibits the forfeiture or escheat of estates for crime or treason,[358] but escheat can result upon the failure of heirs or devisees. In such cases the estate reverts to the state.[359] If neither the grantee of land nor anyone claiming under him has been known or heard from for many years, it will be presumed that the land

357. H. Tiffany, A Treatise on the Modern Law of Real Property 19, 204, 821 (abr. ed. 1940).

358. Ga. Real Est. Law §§ 1-33, 15-13 (3d ed. 1986).

359. O.C.G.A. § 44-5-190.

has escheated.[360] The probate court is given jurisdiction in such matters. The escheated property is sold at public auction, and the proceeds are paid into the educational funds of the county.[361]

Although modern-day jurists generally frown upon the concept of forfeiture of property for treason, confiscation of Loyalists' estates was carried out by the state of Georgia during the Revolution. Titles to a large number of tracts in the colonial headright areas of the state can be traced back to these acts of confiscation.[362]

Despite very strong adherence to the terms of the statute of frauds, courts will sometimes allow the transfer of real property by parol gift or by parol contract if, otherwise, grave injustice would result on the part of the grantor toward the grantee. For land to pass by parol gift the donee must enter into possession of the land with the consent of the donor and make valuable improvements upon the faith of the gift.[363] For land to pass by parol contract there must be such part performance by the promisee of his obligations under the contract as to render it fraud by the promisor if he were not compelled to perform his part; or else the promisee must enter and make valuable improvements on the property.[364] A court decree is necessary to confirm legal title to the property in the donee or promisee.[365]

Formerly certain legal life estates such as dower and curtesy arose by operation of the law, but these are either obsolete or have been abolished in Georgia.[366]

The lease and release was the most commonly used method of transferring land in Georgia during colonial times because early American conveyancers followed forms given in English legal works. The reasons

360. Sutton v. McLeod, 29 Ga. 589 (1860).

361. Vickery v. Benson, 26 Ga. 582 (1858); O.C.G.A. §§ 44-5-191 to 44-5-195.

362. Cooper v. Telfair, 4 U.S. (4 Dall.) 14, 1 L. Ed. 721 (1800); Lambert, *The Confiscation of Loyalist Property in Georgia, 1782–1786,* 20 William and Mary Quarterly 80 (1963). The complete text of all acts of confiscation passed by Georgia during and after the Revolution can be found in 19 Col. Rec. Ga., pt. 2; and 1 Rev. Rec. Ga.

363. Parker v. Parker, 214 Ga. 509 (1958); Sharpton v. Givens, 209 Ga. 868 (1953); O.C.G.A. § 23-2-132.

364. Park v. Minton, 229 Ga. 765 (1972); Harris v. Underwood, 208 Ga. 247 (1951); Salmon v. McCrary, 197 Ga. 281 (1944); Wardlaw v. Wardlaw, 187 Ga. 467 (1939); Ga. Real Est. Law § 18-8 (3d ed. 1986).

365. Doe v. Newton, 171 Ga. 418 (1930).

366. 1969 Ga. Laws 123; 1 Am. Law of Prop. § 5.1; 2 Bl. Comm. 126–39.

for its application did not obtain here as in England, however, and the method fell into disuse at a very early date. The system is practically unknown today, even to title attorneys.[367]

Livery of seisin was occasionally used for transferring real property in America in colonial times but has never been used in Georgia. The statute of frauds renders it inefficacious.[368]

International law recognizes the right of ownership of land through discovery, occupancy, cession, or conquest, and these, of course, are the primal sources of all land titles.[369]

367. 26 C.J.S. Deeds § 7; H. Tiffany, A Treatise on the Modern Law of Real Property 156–57, 666–67, 840 (abr. ed. 1940). See F. Beckemeyer, Abstracts of Georgia Colonial Conveyance Book C-1 passim (1975).

368. See page 322.

369. 45 Am. Jur. 2d International Law §§ 27–29; 48 C.J.S. International Law §§ 21–22; E. Vattel, The Law of Nations 98–100, 118–19, 386–92 (Philadelphia 1817).

Chapter 14. Relative Importance of Conflicting Boundary Elements

The surveyor's most characteristic function is to remark and redefine the limits of a tract of land whose boundaries are already established on the ground or to lay out new boundaries on the ground according to instructions from the client. A thorough knowledge of boundary law is essential to both operations. In performing a resurvey, the surveyor acts in a quasi-judicial capacity, for in most instances the evidence as to the boundaries will be conflicting, and it will be his duty to gather and evaluate the evidence and then to apply the correct law to come up with a correct survey. In performing an original survey a knowledge of law is necessary because the surveyor has to know what evidence to leave behind so that property owners and later surveyors can correctly locate the boundaries. Also, original surveys must be tied to previously existing surveys. Surveyors can save landowners much trouble and sometimes litigation by accurately interpreting boundary descriptions and knowing their legal requirements and consequences.

Because boundary location is primarily a question of law and evidence, surveyors will at times disagree as to the correct location of a particular boundary. If judges and attorneys who are specially trained to interpret the law often disagree about the legal results of a given set of facts, then complete uniformity can hardly be expected in the opinions of surveyors. If certain general rules for locating boundaries can be devised and agreed upon, however, controversy can be reduced to a minimum. This and the following chapters will briefly state some of these rules and discuss how they are applied in actual practice.

General Rules of Construction

The construction of deeds is governed by rules applicable to contracts generally.[1] In the construction of deeds and other instruments, the para-

1. Ga. Real Est. Law § 19-133 (3d ed. 1986).

mount essential and controlling rule is to ascertain the intention of the parties as gathered from the language of the instrument and in the light of a body of case law and statutory rules.[2] In arriving at the intention, the whole instrument, not merely its disjointed parts, must be looked into.[3] If the intention is plain from the language of the instrument as a whole and contravenes no rule of law, it should be given effect regardless of all technical or arbitrary rules of construction and regardless of mere literal repugnancies in different clauses of the conveyance.[4] The intention may be ascertained by a consideration of the surrounding facts in connection with the recitals of the instrument.[5] Words should be construed according to their ordinary meaning and common acceptation unless they are shown to have a local, technical, or otherwise special meaning.[6] If an instrument raises no doubt or obscurity, there is no room for construction, and if the expressed meaning is clear on the face of the instrument, it will control.[7]

Since people ordinarily do not go to the trouble of drawing up conveyances unless they intend them to have an effective operation, a construction that renders an instrument valid is favored over one rendering it void.[8] Any description will suffice that identifies the land with such certainty that the specific parcel intended to be conveyed can be ascertained.[9] Effect should be given to the entire instrument if possible, and a construction that reconciles the different parts of a conveyance is favored over one that leads to a contradiction.[10]

What is most material and certain in a description shall prevail over that which is less material and less certain.[11] Thus monuments generally control over courses, distances, and area, because monuments, being visible and ascertainable objects, are more material and certain.[12]

2. Keith v. Chastain, 157 Ga. 1 (1923).
3. Jackson v. Rogers, 205 Ga. 581 (1949); Johnson v. Valdosta, Moultrie, & West. R.R. Co., 169 Ga. 559 (1929).
4. Keith v. Chastain, 157 Ga. 1 (1923); Harmon v. First Nat'l Bank, 50 Ga. App. 3 (1934).
5. Mitchell v. Spillers, 203 Ga. 565 (1948).
6. O.C.G.A. § 13-2-2; 23 Am. Jur. 2d Deeds §§ 286, 303; 26 C.J.S. Deeds § 86.
7. 26 C.J.S. Deeds § 86.
8. McCaskill v. Stearns, 138 Ga. 123 (1912); Payton v. McPhaul, 128 Ga. 510 (1907).
9. Moody v. Vondereau, 131 Ga. 521 (1908); O.C.G.A. § 44-5-33.
10. Simpson v. Brown, 162 Ga. 529 (1926).
11. Prescott v. Herring, 212 Ga. 571 (1956); Holder v. Jordan Rlty. Co., 163 Ga. 645 (1926).
12. White v. Spahr, 207 Ga. 10 (1950).

Formerly there was rigid adherence to the rule that, if two clauses in a deed are utterly inconsistent, the former prevails. Present-day authorities have greatly relaxed this rule, however, and the current trend is to look into the surrounding facts and construe the whole language of the deed so as to carry out the evident intent of the parties.[13]

Any calls inserted by mistake and obviously inconsistent with the true intentions of the parties may be rejected as false or mistaken and the deed construed according to other calls.[14] The courts will even go so far as to correct mistakes when necessary or to supply missing calls to make the instrument speak the true intentions of the parties. For example, if the courses and distances do not close because a bearing was inadvertently reversed or a distance incorrectly stated, it is permissible to change the bearing or distance so that the survey will close.[15] A bearing and distance accidentally omitted may be supplied if sufficient particulars exist to show they were intended.[16]

Under certain circumstances extrinsic evidence is admissible to aid the interpretation of descriptions because the office of a description is not to identify but to afford a means of identification.[17] Thus if a deed refers to other matter such as another deed or a plat for a description of the premises, this matter is admissible to aid in the description.[18] The deed itself must, however, point to the source from which evidence aliunde is sought to make the description of the land complete.[19]

All oral negotiations and agreements preceding the execution of a deed are merged in the final written instrument. Therefore, as a general rule, where the boundaries are definitely stated in the body of the instrument, parol evidence is inadmissible to add to or vary these terms.[20] Parol evi-

13. Prescott v. Herring, 212 Ga. 571 (1956); Collinsville Granite Co. v. Phillips, 123 Ga. 830 (1905); Harmon v. First Nat'l Bank, 50 Ga. App. 3 (1934).

14. Burson v. Shields, 160 Ga. 723 (1925); Shackelford v. Orris, 129 Ga. 791 (1907); Harris v. Hull, 70 Ga. 831 (1883).

15. See text at notes 119–21 below.

16. See text at notes 122–23 below.

17. King v. Brice, 145 Ga. 65 (1916); Williams v. Gilley, 133 Ga. 546 (1909); 3 Am. Law of Prop. § 12.98; 26 C.J.S. Deeds § 30; 6 G. Thompson, Commentaries on the Modern Law of Real Property § 3020 (1962).

18. See pages 459–62.

19. Stewart v. Cary, 220 N.C. 214, 17 S.E.2d 29, 144 A.L.R. 1287 (1941).

20. Feingold v. McDonald Mtge. & Rlty. Co., 166 Ga. 838 (1928); O'Neal v. Ward, 148 Ga. 62 (1918); Tippins v. Phillips, 123 Ga. 415 (1905); Hall v. Davis, 122 Ga. 252 (1905); Huntress v. Portwood, 116 Ga. 351 (1902). Compare Ga. Real Est. Law § 19-146 (3d ed. 1986).

dence is admissible, however, to explain and interpret the descriptive terms used in a conveyance for the purpose of applying the terms to their subject matter[21] and also to clear up an ambiguous description.[22] In fact, however detailed the description, there must always be resort to some parol evidence, either in the form of direct proof or of tacit admission or assumption by the parties.[23]

All the rules adopted for guidance in locating boundary lines have been to the end that the steps of the surveyor who originally projected the lines on the ground may be retraced as nearly as possible; further, in determining the location of a survey, the fundamental principle is that it is to be located where the surveyor ran it.[24] Errors of location cannot be corrected by the courts or by later surveyors.[25] Thus the primary duty of a surveyor in resurveying a tract of land is to relocate, on the best available evidence, the corners and lines at the same places where the first surveyor on the ground located them.[26]

Although the number of different types of evidence used to locate and define boundaries is almost endless, the major categories can be listed according to their approximate relative importance as follows: (1) lines established by adverse possession, agreement, or acquiescence; (2) lines fixed by a conveyance for which proper notice was given take precedence over lines fixed by later conveyances from the same grantor, in case of overlap; (3) natural monuments; (4) artificial monuments; (5) call for adjoining owners; (6) courses and distances; (7) coordinates; and (8) area.[27]

21. Haley v. Ray, 142 Ga. 390 (1914); Huntress v. Portwood, 116 Ga. 351 (1902); Summerlin v. Hesterly, 20 Ga. 689 (1856).

22. Prud. Ins. Co. v. Hill, 170 Ga. 600 (1930); Douglass v. Bunn, 110 Ga. 159 (1899). Where there was an ambiguity as to which of two lots bearing the same number a deed intended to convey, evidence was admissible to show that the grantor owned one of the lots and did not own the other. Head v. Lane, 153 Ga. 688 (1922).

23. Star Laundry Co. v. May Dry Cleaning Co., 176 Ga. 34 (1932).

24. 11 C.J.S. Boundaries § 3. "These rules of construction are designed to carry out the intention of the parties. The intention of the parties is considered to be essentially the same as that of the surveyor." United States v. Champion Papers, 361 F. Supp. 481 (S.D. Tex. 1973).

25. 12 Am. Jur. 2d Boundaries § 61; 9 C.J. Boundaries § 18.

26. Pereles v. Gross, 126 Wis. 122, 105 N.W. 217 (1905).

27. Broadsword v. Kauer, 161 Ohio St. 524, 120 N.E.2d 111, 46 A.L.R.2d 1309 (1954); Providence Properties v. United Virginia Bank, 219 Va. 735, 251 S.E.2d 474 (1979); 12 Am. Jur. 2d Boundaries § 65; C. Brown, Boundary Control and Legal Principles 2, 125–26 (2d ed. 1969); 6 G. Thompson, Commentaries on the Modern Law of Real Property § 3044 (1962).

The above sequence constitutes merely a broad summary of the rules of evidence that have been devised for determining the intentions of the parties. The sequence therefore is not meant to be comprehensive, absolute, or inflexible but should be adapted to the particular circumstances at hand. It will not be followed and the priorities may be reversed when its application will defeat the intention.[28]

Certain other rules have been adopted for guidance in construing descriptions. The courts presume that a person intends to describe what he owns rather that what he does not own.[29] If a deed calls for monuments marking the boundaries, it speaks as of the day of its execution.[30] When a deed contains two descriptions of the land conveyed, one general and the other particular, if there is any repugnance, the particular will usually prevail.[31] If a description is equally susceptible to differing constructions, the one most favorable to the grantee prevails.[32]

The courts will go a long way in upholding descriptions. There are times, however, when a description is considered too vague or ambiguous for construction, and the conveyance must be declared void.[33] Obviously, no description can be perfect. But a description may be so vague and imprecise that even the general location of the property cannot be determined. Between these two extremes the courts attempt (sometimes arbitrarily) to define the distinction between a valid and an invalid description. Generally a description will be sufficiently certain when from it the property can be identified either by a person familiar with the locality or by an actual survey.[34] The Georgia decisions state that the description must be sufficiently definite to enable the sheriff, in the event of recovery by the plaintiff in a court suit, to so identify the premises that he can deliver possession to the plaintiff.[35] Although the instrument need not embody a minute or perfectly accurate description of the land,

28. United States v. Champion Papers, 361 F. Supp. 481 (S.D. Tex. 1973); 3 Am. Law of Prop. § 12.115; 11 C.J.S. Boundaries § 47.

29. See page 466.

30. Brantly v. Huff, 62 Ga. 532 (1879).

31. See pages 457–58.

32. See pages 468–69.

33. For a metes and bounds description that the court stretched the rules of construction to the limit to uphold see Wisener v. Gulledge, 251 Ga. 419 (1983).

34. 1 R. Patton et al., Patton on Land Titles § 80 (2d ed. 1957); 6 G. Thompson, Commentaries on the Modern Law of Real Property § 3020 (1962).

35. Dodd v. Madaris, 206 Ga. 497 (1950).

it must furnish the key to the identification of the land.[36] A description that is sufficient as of the day the deed is executed generally cannot be rendered insufficient by any subsequent change in conditions.[37] Factors that may render a description void include failing to disclose the general location of the property, giving indefinite courses and distances, fixing an indefinite beginning point, and calling for adjoiners who cannot be identified or who had never owned any land in the area. These and other factors will be discussed in Chapter 16.

Overlaps and Gores

It is elementary that land not owned by the grantor cannot be conveyed by his deed.[38] As to the rights of third parties acting in good faith, however, a deed takes effect, not from the time of its execution, but only from the time notice is provided of its existence.[39] If an owner attempts to convey the same parcel of land to two different parties for valuable consideration, the party in the senior deed has the right of ownership if he provides either actual or constructive notice of his claim prior to the recording of the junior deed; the party in the junior deed has the right of ownership if he takes his deed without notice of the senior deed, and also records his deed.[40]

When an owner conveys a tract of land in separate parcels to suc-

36. Crawford v. Verner, 122 Ga. 814 (1905). "A description of land in a deed, in order to be valid, must identify the land or must contain a key by the use of which the description may be applied by extrinsic evidence. Possibly this idea of a 'key' has been overworked, and it has certainly been frequently misunderstood. There need not be confusion about this word, and no confusion will result if the word is given its true and literal meaning. A metallic bar is a key only when it serves the purpose of unlocking the door, and is not a key if it fails in its primary purpose, which is to unlock the door. Likewise any descriptive words in a contract for the sale of land, which will lead unerringly to the land in question, constitute the key which the law contemplates. But no amount of words in such a contract which fail to lead definitely to the land therein will constitute a key. If such words, when aided by extrinsic evidence, fail to locate and identify a certain tract of land, the description fails and the instrument is void." Blumberg v. Nathan, 190 Ga. 64 (1940).

37. 26 C.J.S. Deeds § 30a.

38. Ga. Real Est. Law § 19-20 (3d ed. 1986).

39. O.C.G.A. §§ 44-2-1 to 44-2-4.

40. Michael v. Poss, 209 Ga. 559 (1953); Nickerson v. Porter, 189 Ga. 671 (1940); Ga. Real Est. Law §§ 19-124, 19-124.1. Thus Georgia is a "race-notice" state, as opposed to the "race" states and the "notice" states. For an explanation of the three basic types of recording acts and the reasoning behind each, see 4 Am. Law of Prop. § 17.5.

cessive purchasers by any form of metes and bounds description or by definite quantities, and it later develops that there is an overlap in the descriptions, the overlaps are generally accorded to the respective grantees according to this rule.[41] The recording of a deed furnishes constructive notice, provided the deed contains a description sufficient to identify the lands so conveyed and is otherwise properly executed so as to entitle it to record.[42] Additionally, if the party in the senior deed goes into actual possession of the property under an unrecorded deed, his actual possession is notice to the world of his claim thereto; and any later purchasers will be charged with notice of the unrecorded deed.[43] If none of the deeds are recorded and notice is not provided by possession or otherwise, the oldest deed prevails.[44]

To clarify these rules, several examples are in order. In each of the following, A attempts to convey the same parcel of land to B and C.

Example: (1) A conveys to B. Then (2) B records or otherwise provides notice. Thereafter (3) A conveys to C. Subsequently (4) C records. *Result:* B acquires ownership of the land.

Example: (1) A conveys to B. Then (2) A conveys to C, who is without notice. Thereafter (3) B records or otherwise provides notice. Subsequently (4) C records. *Result:* B acquires ownership of the land.

Example: (1) A conveys to B. Then (2) A conveys to C, who is without notice. Thereafter (3) C records. Subsequently (4) B records or otherwise provides notice. *Result:* C acquires ownership of the land.

Example: (1) A conveys to B. Then (2) A conveys to C, who is with-

41. Hughes v. Yates, 228 Ark. 860, 311 S.W. 2d 179 (1958); Thurman v. Doss, 312 Ken. 604, 229 S.W. 2d 317 (1950); Flanagan v. Elder, 90 So. 2d 540 (La. App. 1956); Napoli v. Philbrick, 8 N.C. App. 9, 173 S.E. 2d 574 (1970); Merlino v. Eannotti, 177 Pa. Super. 307, 110 A. 2d 783 (1955). This illustrates why it is essential that the surveyor, prior to performing a survey, examine the deeds not only to the particular tract he is surveying but to all adjoining tracts as well. Many of the decisions dealing with overlapping descriptions from the same grantor fail to address the question of notice and, instead, simply give priority to the senior deed. See Pearson v. Shadix, 237 Ga. 817 (1976); Adams v. Powell, 87 Ga. 138 (1891); and cases cited at 12 Am. Jur. 2d Boundaries § 63 n. 8; 3 Am. Law of Prop. § 12.123 n. 3; and 11 C.J.S. Boundaries § 60 n. 24.

42. Lee v. O'Quin, 103 Ga. 355 (1897); Ga. Real Est. Law § 19-118 (3d ed. 1986). See pages 360–62.

43. See page 361.

44. Randall v. Hamilton, 156 Ga. 661 (1923); Wadley Lbr. Co. v. Lott, 130 Ga. 135 (1907); Turner v. Tyson, 49 Ga. 165 (1873).

out notice. Neither party records. *Result:* B acquires ownership of the land.

Sometimes an owner may sell off a tract in sections that he believes comprise the entire tract, but a later survey reveals strips or gores remaining between the tracts. For obvious reasons, court decisions dealing with this situation are almost nonexistent. Of the few cases that have been decided, some have held that the strips or gores remain the property of the grantor.[45] At least one decision, however, held that the strips or gores should be apportioned between the respective grantees.[46]

Monuments

General

Perhaps few principles of law are better established and more universally accepted than the principle that all other boundary elements, except prescriptive rights and lines fixed by prior conveyances from the same grantor for which proper notice was given, yield to monuments in determining the location of boundaries.[47] In comparison with courses and distances, it is said that "no rule in real estate law is more inflexible than that monuments control course and distance."[48] This rule is based on both justice and reason.

It is well known that no measurement is exact and that different surveys will yield different measurements for the same line.[49] Monuments,

45. Wells v. Lagorio, 112 Va. 522, 71 S.E. 713 (1911); 3 Am. Law of Prop. § 12.123.

46. Lincoln v. Edgecomb, 28 Me. 275 (1848). Compare text at notes 159–66 below.

47. United States v. Williams, 441 F.2d 637 (5th Cir. 1971); Lyons v. Bassford, 242 Ga. 466 (1978); Tift v. Gulf Oil Corp., 223 Ga. 83 (1967); White v. Spahr, 207 Ga. 10 (1950); Stewart v. Latimer, 197 Ga. 735 (1944); Thompson v. Hill, 137 Ga. 308 (1911); Harris v. Hull, 70 Ga. 831 (1883); Brantly v. Huff, 62 Ga. 532 (1879); Riley v. Griffin, 16 Ga. 141 (1854).

48. Diehl v. Zanger, 39 Mich. 601 (1878).

49. "Every observed or measured quantity contains errors of unknown magnitude due to a variety of causes, and hence a measurement is never exact." R. Davis, F. Foote, & J. Kelly, Surveying: Theory and Practice 72 (5th ed. 1966). "No matter how precise the measuring equipment or how skillful the observer, the determination of the value of a quantity can never be perfect." P. Kissam, Surveying for Civil Engineers, 659 (2d ed. 1981). Both of these books contain excellent discussions on the theory of errors in surveying and methods for their treatment in actual practice.

It was declared in Lively v. Lively, 206 Ga. 606 (1950) that surveying instruments are

however, are exact because they mark definite points in space. Courses and distances, on the other hand (although theoretically absolute), cannot be laid down on the ground in the precise same place each time the property is resurveyed because of the impossibility of making exact measurements. Illustrative of this is a California case in which an ordinance passed by the city of Santa Barbara declared certain points to be the initial points for all surveys in the city. All streets and lots (which had been laid out previously) were to be located from these initial points, according to their theoretical locations as shown in the plans of the city. The court, in declaring the ordinance void, noted: "No absolutely accurate survey ever was or ever will be made. In order to build cities and towns there must be some finality as to the location of blocks and streets. Under such an ordinance, if valid, there could be none. All lines would be forever subject to be revised and corrected or changed by new surveys not really more accurate, but only for the time deemed so."[50] In other words, monuments, as long as they remain undisturbed, will always mark the same lines, whereas courses and distances will not.

Perhaps an even more compelling reason for holding that monuments control courses and distances is that lay persons usually do not have the training in science, mathematics, or surveying to enable them to interpret courses and distances, but they generally are able to interpret monuments correctly. Property owners will usually depend on and construct improvements in accordance with the monuments that mark their boundaries. It would be unfair to alter these lines after such improvements are made, simply because the original surveyor failed to set his monuments in the measured position called for.[51] The courts have held that a landowner has the legal right to rely on the monuments that mark his boundaries, regardless of whether they were correctly located.[52] If a prospective land purchaser has any doubts concerning the true dimensions of a lot, he can always have the lot measured by a competent surveyor to determine the dimensions.

---

capable of laying down exact courses and distances, but the justice who made the remark shows a very lamentable lack of knowledge of the subject. Not only does the statement contradict the vast weight of legal authority, it also defies the laws of nature. We may hope that the Georgia Supreme Court will repudiate the doctrine.

50. Orena v. City of Santa Barbara, 91 Cal. 621, 28 P. 268 (1891).

51. C. Brown, Boundary Control and Legal Principles 157 (2d ed. 1969).

52. 9 C.J. Boundaries § 18; Id. § 42 n.9.

> There is an intrinsic justice and propriety in this rule, for the reason that the applicant for land, however unlearned he may be, needs no scientific education to identify and settle upon his land when the surveyor . . . authoritatively announces to him that certain well-known rivers, lakes, creeks, springs, marked corners, and lines constitute the boundaries of his land. But it would require some scientific knowledge and skill to know that the courses and distances called for are true and correct, and with the aid of the best scientific skill, mistakes and errors are often committed in respect to the calls for course and distance. . . . The unskilled are unable to detect them, and the learned surveyor often much confused.[53]

Another reason for the rule is that monuments, as long as they remain undisturbed, are always present on the land so that its boundaries can be readily ascertained at any time by inspection. Courses and distances are of value in determining the location of boundaries only when the property is being surveyed.

Finally, no matter how precise the courses and distances in a deed or plat, all surveys must begin at a known point on the ground—a monument. Even latitude and longitude are fixed by monuments: a spot in Greenwich, England, and the poles of the earth.[54] All the scientific advancements that have been made for locating property boundaries have not eliminated the monument as the ultimate means by which land is controlled.[55]

The often-cited case of *Riley v. Griffin*[56] involved a dispute over the location of the dividing line between Fractional Lots 2 and 3 in the Macon Reserve. The boundary between the two tracts, as marked by the original state surveyor, was not run in a straight line between the corners of the tracts. The subsequent owners of the two tracts acquiesced in and possessed up to this crooked line for more than eighteen years until the plaintiffs in the suit had the property resurveyed and tried to establish a straight line between the corners. The plaintiffs alleged that their deeds covered the premises and that the defendants had gone into possession of the disputed strip under a mistake as to the location of the boundary. The defendants, on the other hand, attempted to claim the strip in dispute by adverse possession.

53. Stafford v. King, 30 Tex. 257, 94 Am. Dec. 304 (1867).

54. C. Brown, Boundary Control and Legal Principles 4–5 (2d ed. 1969).

55. Id. 5.

56. 16 Ga. 141 (1854).

The trial court found for the plaintiffs, and the defendants appealed. The Georgia Supreme Court rejected the arguments of both the plaintiffs and defendants and reversed the trial court's judgment, giving the verdict to the defendants. In the opinion written by Justice Lumpkin, the court laid down the following rules which emphasize the importance of monuments:

> [1.] We recognize the doctrine, that a possession which is the result of ignorance, inadvertence, misapprehension, or, in other words, mistake, will not work a disseizin. . . . It is the intention to claim title which makes the possession of the holder adverse; and this is the doctrine upon which the decision in every case proceeds. If it be clear, therefore, that there is no such intention, there can be no pretence of an adverse possession. . . .
>
> Whether there be any evidence to justify the charge, that the defendant's occupancy . . . may have been the result of mistake, is somewhat questionable.
>
> But there is another portion of the charge, which requires more consideration. The Court instructed the Jury, that if Griffin's [one of the plaintiffs] grant covered the premises in dispute, then the verdict must be for the plaintiff.
>
> Is this proposition necessarily correct? We think not.
>
> Let us refer, for a moment, to the testimony of young Riley and Jonathan Wilder. Warren B. Riley swears, that he has known the premises in dispute since 1835 or 1836. . . . This witness testifies, that he knows the lines well, and that the line to which his father claimed, was the old original line. He has frequently seen the old blazes and trees which had all the appearance of the original Surveyor's marks. That the owners of the adjoining lands set it up, as the original line, and were governed by it, and there was no dispute about it until Adams bought the adjoining land. He further stated, that all the parties, that is, those residing on the contiguous tracts, acquiesced in the line as fixed, until Adams bought [the adjoining tract]. . . .
>
> Jonathan Wilder swore that . . . at the time it was sold, the blazes made by the Surveyor who ran the land, were plain on the trees; and that he followed the original marks. That Riley's fence is nearly on the line as run round by witness. . . . He knows he is not mistaken as to the lines, because he followed the original Surveyor's marks, then fresh and plain on the trees, which were then standing, very few, if any, having been cut down. . . . There never was any dispute about the boundary, while witness controlled it, as the agent of his uncle. Witness has known the place, ever since the original survey was made, . . . has often seen the original Surveyor's marks, and could trace the original lines by them, and did so.
>
> [2.] Now, it would seem, according to the proof, that when Lot No. 2 was

originally surveyed, the lines may not have been run straight, according to courses and distances. But still, if the Surveyor marked these as the true lines, it is quite clear that the owner of No. 2 will hold to these boundaries. Marked trees, or the line, as actually run, must control the line which courses and distances would indicate.

[3.] If nothing exists to control the call for course and distance, the land must be bounded by the courses and distances of the grant, according to the Magnetic Meridian; for it is the practice, undoubtedly, of Surveyors, to express, in their plots and certificates of survey, the courses which are designated by the needle. But it is a general principle, that the course and distance must yield to natural objects called for in the grant.

[4.] All lands are supposed to be actually surveyed; and the intention of the grant is, to convey the land, according to that actual survey.

[5.] Consequently, if marked trees and marked corners be found, distances must be lengthened or shortened, and courses varied, so as to conform to these objects. . . .

[6.] Where the calls of a deed or other instrument, are for natural, as well as known artificial objects both courses and distances, when inconsistent, must be disregarded. . . .

[7.] Whenever a natural boundary is called for in a grant or deed, the line is to determine at it, however wide of the course called for it may be, or however short, or beyond the distance specified.

[8.] And whenever it can be proved that there was a line actually run by the surveyor, was marked and a corner made, the party claiming under the grant or deed, shall hold accordingly, notwithstanding a mistaken description of the land in the grant or deed.

[9.] When the lines or courses of an adjoining tract, are called for in a deed or grant, the lines shall be extended to them, without regard to distance, provided these lines and courses be sufficiently established.

[10.] Where there are no natural boundaries called for—no marked trees or corners to be found, nor the places where they once stood, ascertained and identified by evidence; or where no lines or courses of an adjacent tract are called for—in all such cases, Courts are, of necessity, confined to the courses and distances prescribed in the grant or deed; for however fallacious such guides may be, there are none other left for the location. . . .

[11.] And thus, it will be seen that courses and distances occupy the lowest grade, instead of the highest, in the scale of evidence, as to the identity of land.

[12.] And it is reasonable that this should be so; for any natural object, when called for distinctly, and satisfactorily proved . . . becomes a landmark not to be rejected, because the certainty which it affords, excludes the probability of mistake;

[13.] While course and distance, depending, for their correctness, on a great variety of circumstances, are constantly liable to be incorrect. Difference in the instrument used, and in the care of Surveyors and their assistants, lead to different results. . . . This doctrine is found scattered . . . throughout the authorities; and I had supposed to be too well understood and established to require to be discussed at this day.

[14.] So, in *Brewer vs. Gay,* (3 *Greenleaf's R.* 126,) it was held, that in ascertaining the boundaries of lots of land . . . the locations of the original Surveyor, so far as they can be found, are to be resorted to; and where they vary from the proprietor's plan, the locations actually made will control the plan.

[15.] So, in *Dodge vs. Smith,* (2 *N. Hamp. R.* 303,) the Supreme Court say, "whenever, in a conveyance, the deed refers to monuments actually erected, as the boundaries of the land, it is well settled that those monuments must prevail, whatever mistakes the deed may contain, as to the courses and distances. . . ."

[16.] In *Doe vs. Paine & Sawyer,* (4 *Hawk's N. Rep.* 64.) the Court refer to courses and distances, as pointers or guides merely, to ascertain the natural objects of boundary.

[17.] So, also, it has been held, that where a given line is exceeded in a grant, according to the courses and distances, evidence may be received, of long occupation under it, to prove the boundaries. . . .

[18.] And, as landmarks are frequently formed of perishable materials, which pass away with the generation in which they are made; and are often destroyed . . . by the improvement of the country and other causes, the boundary and corners may be proved by hearsay, from the necessity of the case. . . .

[19.] Again: Suppose the line sworn to, is not that which was marked by the original Surveyor; still, if it were agreed on by the coterminous proprietors, and acquiesced in, and possession to it held for eighteen or twenty years, the parties, and those claiming under them, would be bound by it, no matter when nor by whom the line was run and chopped.

An important concept to understand is that a monument is not in itself a property corner but only marks the location of the property corner; that is, if the monument is moved or destroyed, the actual corner remains where it was if its location can be proved. The surveyor, in performing a resurvey, should therefore find the monuments of the original survey or, failing that, ascertain as nearly as possible their former locations and base his measurements accordingly.

Natural and Artificial Monuments

Formerly the courts placed greater reliance on natural monuments as opposed to artificial monuments in determining boundaries, but they seem less prone to do so today. Natural monuments are permanent objects that were placed by nature and remain undisturbed by people.[57] Examples include trees, rocks, rivers, natural lakes, and the ocean. Artificial monuments are objects put in place by people such as stakes or piles of rocks.[58] Natural monuments formerly were preferred over artificial ones because they usually were more permanent and notorious and less capable of counterfeit.[59] The advent and widespread use of permanent, identifiable, artificial monuments such as metal and concrete posts, coupled with an improved technology capable of altering and destroying natural monuments almost as easily as artificial ones, have inclined courts today to make less of a distinction between natural and artificial monuments.

Center of Monument Controls

As a general rule, when a monument marks a boundary, the legal terminus of the boundary is the center of the monument.[60] For example, if a tree is called for, the surveyor should measure to the center of the tree.[61] The rule, of course, will not apply if the conveyance clearly expresses a contrary intention or if the grantor does not own to the center of the object called for.[62]

Presumption of Straight Lines

If a line is described in a deed or plat as running from one point to another, it is presumed to be a straight line; but if the line is marked on the ground, this marked line will control even though it is crooked (assuming, of course, that no prior conveyance from the same grantor for which proper notice was given is interfered with).[63]

57. 28 Words and Phrases "Natural Monument."

58. 4 Words and Phrases "Artificial Monument."

59. Thompson v. Hill, 137 Ga. 308 (1911); O.C.G.A. § 44-4-5.

60. Coombs v. West, 115 Me. 489, 99 A. 445, 2 A.L.R. 1424 (1916); 3 Am. Law of Prop. § 12.105.

61. Coombs v. West, id.

62. Compare pages 450–51, 462–64.

63. Riley v. Griffin, 16 Ga. 141 (1854); O.C.G.A. § 44-4-5; 3 Am. Law of Prop. § 12.105;

Roads and Streets as Monuments

A call for a road as a boundary is a call for a monument, if the road is in existence at the time of the conveyance. Such a call will usually carry the boundary to the centerline of the right-of-way or to the edge of the right-of-way, depending on whether title to the underlying fee is vested in the adjacent owners or in a public body.[64] If title is vested in a public body, then such a call is technically a call for an adjoining owner. The intention of a conveyance calling for a road is to convey to the road as it is located at the time of the conveyance, and later changes in the route of the road will work no change in the boundary.[65] If a called-for street is not in existence at the time of the conveyance, resort must be had to the courses and distances specified in determining the boundaries.[66]

Boundary Improvements Held Controlling

If the monuments of the original survey have disappeared, a resurvey is for the purpose of determining where they were and not where they should have been.[67] In such cases a series of old fences, walls, hedgerows, turn rows, buildings, or other boundary improvements will become controlling as to the location of the boundaries, if they appear to have been constructed soon after the original survey and in accordance with the original survey and if the lines are otherwise uncertain.[68] Loosely speaking, such boundary improvements may be considered monuments. Such monuments are often better indicators of the original survey lines than

---

11 C.J.S. Boundaries § 104; 6 G. Thompson, Commentaries on the Modern Law of Real Property § 3054 (1962).

64. See pages 438–40.

65. Johnston v. Town of Palmetto, 139 Ga. 556 (1912); Brantley v. Huff, 62 Ga. 532 (1879).

66. Hammond v. George, 116 Ga. 792 (1902).

67. Pereles v. Gross, 126 Wis. 122, 105 N.W. 217 (1905); 9 C.J. Boundaries § 13 n. 80(c).

68. Wacker v. Price, 70 Ariz. 99, 216 P.2d 707 (1950); DeEscobar v. Isom, 112 Cal. App. 2d 172, 245 P.2d 1105 (1952); Westgate v. Ohlmacher, 251 Ill. 538, 96 N.E. 518 (1911); Mills v. Penny, 74 Ia. 172, 37 N.W. 135 (1888); Cutts v. King, 5 Me. (Greenleaf) 482 (1829); Breakey v. Woolsey, 149 Mich. 86, 112 N.W. 719 (1907); Diehl v. Zanger, 39 Mich. 601 (1878); Dittrich v. Ubl, 216 Minn. 396, 13 N.W.2d 384 (1944); Beck v. Brown, 8 N.J. Super. 103, 73 A.2d 594 (1950); Murray Hotel Co. v. Golding, 54 N.M. 149, 216 P.2d 364 (1950); Norcom v. Leary, 25 N.C. 49 (1842); Wilson v. Sidle, 4 Ohio N.P.(N.S.) 465, 17 Ohio Dec. 393 (1906); Wunnicke v. Dederich, 160 Wis. 462, 152 N.W. 139 (1915); Village of Galesville v. Parker, 107 Wis. 363, 83 N.W. 646 (1900).

the courses and distances given in the deed or plat.[69] In many cases, old possession lines are all that remain to indicate the location of boundaries.

If there is proof, however, that no lines have ever been actually surveyed out on the ground, then of course improvements cannot be regarded as marking the lines of an original survey.[70]

Monuments Held Controlling by Reputation and Tradition

Hearsay is evidence whose probative force depends on the competency and credibility of some person other than the one who produces it.[71] An example is testimony a witness gives regarding declarations made by someone no longer living.[72] Reputation and tradition are forms of hearsay that consist of matters of public interest that have been commonly accepted as true by the people in a community, but the source of the matter is lost in antiquity.[73]

Because of the secondhand nature of hearsay and the impossibility of cross-examining the original declarant, such evidence is usually inadmissible for proving facts.[74] Several exceptions are made, however, one in the case of boundaries. Generally it cannot absolutely be shown that ancient boundaries and corners (that is, ones established in the remote past) are the original boundaries and corners or that they occupy the same location as the originals because the eyewitnesses to their original placement have passed away and there is insufficient documentary evidence to prove them. In such cases the boundaries must be proved by the best available evidence—reputation and tradition.[75]

For a boundary to be proved in this manner the reputation and tradition must have originated at a time before the commencement of the boundary controversy and must have been of such interest in the neighborhood as to have provoked discussion and attracted general attention.

69. Village of Galesville v. Parker, id.; Harris v. Lewis, 156 Ia. 413, 136 N.W. 674 (1912); Diehl v. Zanger, 39 Mich. 601 (1878).

70. Pereless v. Gross, 126 Wis. 122, 105 N.W. 217 (1905); Wollman v. Ruehle, 104 Wis. 603, 80 N.W. 919 (1899).

71. O.C.G.A. § 24-3-1; 31A C.J.S. Evidence § 192.

72. Waddell v. Cole, 138 Ga. App. 15 (1976); O.C.G.A. § 24-3-9.

73. 31 C.J.S. Evidence §§ 196, 199.

74. O.C.G.A. § 24-3-1.

75. Deaton v. Swanson, 196 Ga. 833 (1943); McAfee v. Newberry, 144 Ga. 473 (1915); Martin v. Atkinson, 7 Ga. 228 (1849); O.C.G.A. §§ 24-3-13, 44-4-6.

The reputation and tradition must also attach itself to some monument or boundary and be fortified by evidence of occupation and acquiescence tending to give the land in question some fixed and definite location.[76] For a monument to be accepted by reputation and tradition, it must have been commonly accepted as correct, it must be in a position that could have been marked by the original surveyor, and there must be no evidence to disprove the monument.[77] Reputation and tradition are never admissible if the location of boundaries is ascertainable from deeds, field notes, or other authentic records.[78]

All surveyors at times accept monuments that cannot possibly be proved to mark the locations of the original monuments. Such acceptance must necessarily be based on reputation and tradition.[79]

When Monuments Do Not Control

In spite of the great importance placed on monuments, there are certain types that do not control in determining boundaries. As a general rule, to be controlling, a monument must be the original one[80] or a replacement set in the same location as the original monument;[81] it must be called for in the deed or plat[82] or otherwise be connected with the conveyance in such a way that one can presume the parties intended to adopt it;[83] and, if an original monument, it must be in its original loca-

76. 9 C.J. Boundaries § 311; 11 C.J.S. Boundaries § 106.

77. C. Brown, Boundary Control and Legal Principles 165 (2d ed. 1969).

78. 9 C.J. Boundaries § 311; 11 C.J.S. Boundaries § 106.

79. C. Brown & W. Eldridge, Evidence and Procedures for Boundary Location 54–55 (1962).

80. O.C.G.A. § 44-4-5.

81. Arnold v. Hanson, 91 Cal. App. 2d 15, 204 P.2d 96 (1949); Powell v. Reid, 519 S.W.2d 388 (Ky. 1975); Humble Oil & Refining Co. v. State, 162 S.W.2d 119 (Tex. 1942); 11 C.J.S. Boundaries § 5. A replacement monument set by proper authority is presumed to have been placed where the original monument was located, until the contrary is shown. Home Owners' Loan Corp. v. Dudley, 105 U. 208, 141 P.2d 160 (1943).

82. Stewart v. Latimer, 197 Ga. 735 (1944); Bell v. Redd, 133 Ga. 5 (1909); 9 C.J. Boundaries §§ 132, 135; 11 C.J.S. Boundaries § 49(e). As with all rules of law, however, there are exceptions to the general rule that a monument must be called for to be controlling. For example, monuments that were established or acquiesced in to settle an unknown or disputed boundary may become controlling. See pages 429–36. Other exceptions are noted in the text.

83. 12 Am. Jur. 2d Boundaries § 72. Occasionally the courts will hold that if lots are staked off prior to or at the time of a conveyance and the parties act with reference thereto,

tion.[84] Also, the monument must not interfere with prescriptive rights or a prior title.[85] Artificial monuments placed by a river or road, where the river or the road itself is the boundary, merely point the direction of the lines running to the river or road and do not otherwise control.[86] Artificial monuments erected along a body of water to mark a meander line do not control if the body of water is the boundary.[87]

Call for Adjoining Owners

A call for adjoining owners is a call for the boundary of an adjoining tract of land. An adjacent tract called for as a boundary may be designated by any form of expression definitely indicating its identity, such as the name by which it is commonly known, the name of its owner or former owner, or the place in the public records where a deed conveying such adjacent tract may be found.[88] A call for an adjoiner is a call for any monuments, deeds, plats, or other evidence that establish the lines of the adjoining tract.

When an adjoining tract is called for as a boundary, the lines shall be extended to it without regard to the specified distance, provided these

---

the stakes will control even though they are not mentioned in the deed or plat. Burke v. McCowan, 115 Cal. 481, 47 P. 367 (1896); Safret v. Hartman, 50 N.C. (5 Jones) 187 (1857); Reed v. Shenck, 13 N.C.(2 Dev.) 415 (1829); Bearden v. Schenecker, 240 S.W. 996 (Tex. 1922); Thiel v. Damrau, 268 Wis. 76, 66 N.W.2d 747 (1954); 9 C.J. Boundaries § 330 n. 37.

84. Brantley v. Huff, 62 Ga. 532 (1879); Ga. Real Est. Law § 13-2 (3d ed. 1986). Bodies of water are often an exception to this rule. See pages 391–92, 481, 487.

85. 9 C.J. Boundaries §§ 124, 167. See pages 374–88, 400–402.

86. State of Georgia v. Georgia Ry. & Pwr. Co., 141 Ga. 153 (1913); 12 Am. Jur. 2d Boundaries §§ 27, 71; 11 C.J.S. Boundaries § 31.

87. 12 Am. Jur. 2d Boundaries §§ 29 to 31; 3 Am. Law of Prop. § 12.114; 9 C.J. Boundaries § 69; 11 C.J.S. Boundaries § 30; 93 C.J.S. Waters § 71; 1 R. Patton et al., Patton on Land Titles § 117 (2d ed. 1957); 6 G. Thompson, Commentaries on the Modern Law of Real Property § 3075 (1962). In the public lands surveys of the United States, however, if a meander line is established at an excessive distance from the actual shore, leaving so great an excess of land between the meander line and the shore as to indicate fraud or mistake in the survey, the courts may—for equitable reasons—decree the meander line as the boundary. Albrecht v. United States, 529 F. Supp. 135 (D. Wyo. 1981); Snake River Ranch v. United States, 395 F. Supp. 886 (D. Wyo. 1975); 11 C.J.S. Boundaries § 30; 93 C.J.S. Waters § 107.

88. 12 Am. Jur. 2d Boundaries § 6.

lines are sufficiently established.[89] It is not necessary that the line of the adjoining tract be marked if its position can be ascertained with accuracy.[90]

When a deed describes land as bounded on one side by the land of another, the true boundary line between the two tracts must be taken as the boundary—not a conventional line orally agreed to by the parties at the time the deed was executed if there is a variance between such two lines.[91]

A deed by which a grantor conveys part of an irregularly shaped tract of land, describing the land conveyed as containing a definite number of acres and bounded on three sides by definite boundaries and on the fourth side by part of his own land, in the absence of evidence of a contrary intent, will be deemed as locating the divisional line between himself and the granted land to be run in a cardinal direction so as to give the grantee the quantity of land specified.[92] See Plate 3. Also, if the grantor and grantee orally agree on a line under such a conveyance and make a practical location of the line, this will suffice whether or not the line runs in a cardinal direction.[93]

Where the land of an adjoining owner is called for as the boundary and the adjoiner's title is defective, the description is still good if it can be shown that the maker of the deed recognized the adjoiner as the owner of the adjacent tract and the boundary of the adjacent tract had been established.[94]

Courses and Distances

General

As with other types of evidence, courses and distances should be used to reestablish property lines only when they constitute the best available evidence as to where the original lines were located on the ground. Only in the absence of all monuments and marks on the ground and in the

89. Ford v. Williams, 73 Ga. 106 (1884); Roberts v. Williamson, 116 Ga. App. 553 (1967).
90. 11 C.J.S. Boundaries § 53.
91. Morgan v. Godbee, 146 Ga. 352 (1916); Bell v. Redd, 133 Ga. 5 (1909).
92. Moody v. Vondereau, 131 Ga. 521 (1908); Walden v. Walden, 128 Ga. 126 (1907).
93. Moody v. Vondereau, 131 Ga. 521 (1908).
94. O'Farrell v. Vondereau, 134 Ga. 696 (1910).

total failure of evidence to supply them may recourse be had to calls for courses and distances in determining boundaries.[95] When the corners of a tract are marked by monuments, the courses and distances are considered approximate and are disregarded when the monuments are found. Sometimes land is described merely by courses and distances and without reference to an actual ground survey. If, however, the boundaries are later marked off on the ground and the adjoining owners either agree to these marked boundaries or acquiesce in them for seven years, the marked boundaries become the true boundaries whether or not they were correctly located.[96]

Distances in land measure are usually expressed in feet, yards, inches, rods (also called "poles" or "perches"), chains, links, and occasionally meters. The acre is also sometimes used as a lineal measurement,[97] corresponding to a length of a side of a square acre, or approximately 209 feet.[98] The Georgia courts, however, have held that such a call constitutes no definite measurement and, without more, renders the description void.[99]

Directions may be stated as points on a compass (based on true north, magnetic north, grid north, or some arbitrarily chosen direction assumed as north) or as interior angles, exterior angles, deflection angles, azimuths, and so on.

The Georgia courts will hold void for uncertainty of terms any description that, without otherwise fixing the boundaries, designates the distances as "about" or as "more or less," and/or gives directions such as "northerly," "westerly," or "southeasterly."[100] Many other jurisdictions, however, hold that a distance such as "85 feet more or less" or a direction such as "northerly," when not controlled by other descriptive averments, will be construed as meaning respectively "85 feet absolutely" and "due north" to prevent having to void the conveyance.[101] In one case, the Geor-

95. 12 Am. Jur. 2d Boundaries § 73.

96. See pages 429–36.

97. Shahan v. Watkins, 194 Ga. 164 (1942); Studstill v. Wilcox, 94 Ga. 690 (1894).

98. Ga. Real Est. Law § 13-24 (3d ed. 1986).

99. Odd Fellows v. City of Thomasville, 226 Ga. 4 (1970). Compare Rye v. Baumann, 231 Ark. 278, 329 S.W.2d 161 (1959) where such a call was upheld.

100. Smith v. Georgia Ind. Rlty. Co., 215 Ga. 431 (1959); Rogers v. Manning, 203 Ga. 771 (1948).

101. Ga. Real Est. Law § 13-61 n. 2 (3d ed. 1986); 12 Am. Jur. 2d Boundaries §§ 56, 57; 1 R. Patton et al., Patton on Land Titles § 125 (2d ed. 1957); 6 G. Thompson, Commentaries on the Modern Law of Real Property § 3056 (1962).

gia court even held that the descriptive call in a deed "thence northwest" was so indefinite as to make the deed inoperative either as a conveyance or as color of title.[102]

If there is an ambiguity as to which of two or more found monuments marks the true corner, courses and distances may be resorted to to resolve the ambiguity.[103] If a call in a deed is along a land lot line "to a rock corner," the land lot line is regarded as a mere course and must yield to the monument. In one such case, this resulted in setting up a line that did not coincide with the land lot line and giving the grantee land owned by the grantor in an adjoining land lot not mentioned in the general description of the premises being conveyed.[104]

When a lot bounded by a street is described by courses and distances only, the distances should be measured from the edge of the street right-of-way rather than from its center, even though the grantor owns to the center.[105] The conveyance, however, will generally be construed as passing title to the center of the street.[106]

Effect Given When Either Course or Distance, But Not Both, Must Prevail

Situations occur in which a conveyance gives a course and a distance for a line that cannot be reconciled, and the question arises as to which shall prevail—the course or the distance? On this subject the strong trend of authority in other jurisdictions is that course governs distance.[107] There seems to be little rhyme or reason for such rigid adherence to the rule,[108]

102. Hughes v. Heard, 215 Ga. 156 (1959). "The line so described could be a straight line from the beginning point . . . at any location . . . west of a due north course from such point." Id. But compare Holden v. Alexander, 82 S.C. 441, 62 S.E. 1108 (1908), where it was held that "the course 'southwest' called for in the deed without stating degrees is unusual, and possibly might not have been intended to mean an exact course, yet southwest means a course equally diverging from south and west, or south 45° west." Compare also Ga. Real Est. Law § 13-55 n. 7 (3d ed. 1986).

103. 9 C.J. Boundaries §§ 124, 130, 134; 11 C.J.S. Boundaries §§ 50, 51.

104. White v. Spahr, 207 Ga. 10 (1950).

105. Walker v. Hill, 253 Ga. 126 (1984); Bell v. Redd, 133 Ga. 5 (1909); Summerville Macadamized Graded or Plank Rd. Co. v. Baker, 68 Ga. 412 (1882).

106. Foute v. Elder, 109 Ga. 713 (1899).

107. 12 Am. Jur. 2d Boundaries § 74; 3 Am. Law of Prop. § 12.119; 9 C.J. Boundaries § 160; 11 C.J.S. Boundaries § 56; 1 R. Patton et al., Patton on Land Titles §§ 125, 154 (2d ed. 1957). There are apparently no Georgia decisions on the matter.

108. "Such a rule of construction has been maintained for generations, but for what reason

although surveyors are purportedly more likely to miscount tape lengths than to misread angles.[109] Perhaps the earliest cases that applied the doctrine based it on the rule that if two clauses in a conveyance are utterly inconsistent, the former prevails.[110] Since course is usually given before distance in descriptions, it wound up controlling over distance. Later decisions followed the early authorities based on the doctrine of stare decisis.[111]

The rule has not been universally followed, however, and situations exist in which one or the other may be preferred after minute examination of all the circumstances. For example, the facts may clearly show a mistake in course instead of in distance, the distance may be described with more certainty than the course, or other factors may occur in which distance is held controlling over course.[112]

Magnetic Meridian, True Meridian, and Grid Meridian

In the absence of monuments or other indications as to the direction of a line, there is sometimes ambiguity as to whether a bearing is to be based on magnetic, true, or grid meridian. The determination must be predi-

---

is not apparent to surveyors of to-day. With modern instruments and the skill with which the more enlightened surveyor of to-day handles them, there can be no just reason for a preference of the call for courses over a call for distances. If one is preferred to the other, it must be due to attendant circumstances. . . . The better law of to-day is that there is no fast and firm rule that obliges a court or surveyor to prefer the one or the other. . . . If any preference is given, it must depend upon the circumstances of each particular case. . . . It is well to remember, however, that there has been a rule that courses shall govern distances, and that courts are likely to adhere to such precedents. Whether due to their conservatism or to their ignorance of modern methods of surveying, they are likely to favor courses and give them the greater weight when in conflict with distances." J. Wait, The Law of Operations in Engineering and Architecture 403–4 (1901). Notice the foregoing was written more than ninety years ago.

109. Forest Preserve Dist. v. Lehmann Est., 388 Ill. 416, 58 N.E.2d 538 (1944); Stafford v. King, 30 Tex. 257, 94 Am. Dec. 304 (1867). This probably constituted a valid basis for the rule in the remote past. Using a compass and chain, it was a simple matter to double-check angles in the field but impractical to double-check distances. Lack of computation equipment and mathematical expertise frequently precluded discovering errors by office calculations. In addition, the person who operated the compass was usually an experienced surveyor, while chainmen were apprentices or laymen.

110. See page 397.

111. Morgan v. Renfro, 124 Ky. 314, 99 S.W. 311 (1907).

112. 11 C.J.S. Boundaries § 56.

cated upon the particular circumstances of each individual situation, and no hard and fast rules apply. Of course, if the deed or plat specifies the basis of the meridian, this will govern.[113] If the given bearings are based on an arbitrary bearing and computed from interior angles (as most modern surveys are), the meridian may be ignored—provided that at least two original monuments can be found on which to base the resurvey.[114]

Most of the earlier decisions held that, in the absence of other stipulations, magnetic bearings governed, because the magnetic compass was almost universally used and surveyors generally made no allowance for magnetic declination.[115] If magnetic bearings are used to reestablish property lines, the surveyor should take into account the variation in declination as of the time of the conveyance so as to reproduce the original line as closely as possible.[116]

The contention that a course described as running to one of the cardinal points of the compass, especially when qualified by the term *due* (that is, a course of north, due north, and the like), indicates a reference to the true meridian has usually been rejected and such calls may be based on true, magnetic, or grid meridian according to other circumstances.[117]

Surface and Horizontal Measurement of Distances

Because it is such a well-established practice among land surveyors to reduce slope distances to horizontal measure, it is almost inconceivable that surface distances would ever be used to describe land. Nevertheless, a few cases have held that where surface measurements have been customarily employed by the surveyors of a locality, the same should be adopted in relocating the lines.[118]

113. The acts for laying out the land lots, districts, and sections of Georgia specified that the compasses were to be adjusted for magnetic declination. Modern technical standards for surveys prohibit the use of magnetic readings as the basis for resurveys.

114. Henrie v. Hyer, 92 Utah 530, 70 P.2d 154 (1937).

115. Riley v. Griffin, 16 Ga. 141 (1854); Martin v. Tucker, 111 R.I. 192, 300 A.2d 480, 70 A.L.R.3d 1215 (1973); 1 R. Patton et al., Patton on Land Titles § 125 (2d ed. 1957).

116. Goodwin v. Greene, 237 N.C. 244, 74 S.E.2d 630 (1953).

117. Martin v. Tucker, 111 R.I. 192, 300 A.2d 480, 70 A.L.R.3d 1215 (1973).

118. Justice v. McCoy, 332 S.W.2d 846, 80 A.L.R.2d 1206 (Ky. 1960); 12 Am. Jur. 2d Boundaries § 58; 3 Am. Law of Prop. § 12.105; 11 C.J.S. Boundaries § 9(b).

Errors and Omissions

Clerical errors are often made in the courses and distances given in deeds and plats, but such errors alone will not void a description. A closure analysis may indicate where the error lies; and if so, the error may be corrected.[119] For example, it may be shown that the bearing "northeasterly" actually should read "northwesterly"[120] or the distance "1,270 feet" should have been "1,750 feet."[121] An omitted call may be supplied if there are sufficient particulars to show that it was intended.[122] For example, where a deed described the tract as "commencing at the northwest corner of said [land] lot, running south on Boring lot fifteen hundred feet; thence east six hundred and eight and a half feet to a stake corner; west six hundred and eight and a half feet to commencing corner, containing ten acres more or less," it was obvious that the east side had been mistakenly omitted from the description. The court held that enough information was available to supply the side so that the description would be construed as embracing a lot with the dimensions shown in Plate 1, even though this construction would give the grantee almost twenty-one acres instead of the ten acres called for.[123]

Closing Line of a Survey

If the lines of a survey are designated by courses and distances starting from a definite point of beginning but the courses and distances do not close back into the point of beginning, and there is no indication of an error or omission, then all of the error should be thrown in the last call where it states "thence . . . to the point of beginning."[124]

119. White v. Luning, 93 U.S. 514, 23 L. Ed. 938 (1876); Speed v. Creech, 307 Ky. 765, 212 S.W.2d 265 (1948); Combs v. Virginia Iron, Coal & Coke Co., 32 Ky. Law 601, 106 S.W. 815 (1908); Johnson v. Bowlware, 149 Mo. 451, 51 S.W. 109 (1899); 3 Am. Law of Prop. § 12.105; 1 R. Patton et al., Patton on Land Titles § 125 (2d ed. 1957).

120. Grant v. Fourth Nat'l Bank, 229 Ga. 855 (1972); Tanner v. Williamson, 199 Ga. 216 (1945); Leverett v. Bullard, 121 Ga. 534 (1904).

121. Clark v. Robinson, 162 Ga. 395 (1926).

122. 1 R. Patton et al., Patton on Land Titles § 125 (2d ed. 1957).

123. Ray v. Pease, 95 Ga. 153 (1894).

124. Simpkins' Adm'r v. Wells, 19 Ky. Law Rep. 881, 42 S.W. 348 (1897); Owings v. Freeman, 48 Minn. 483, 51 N.W. 476 (1892); Cowles v. Reavis, 109 N.C. 417, 13 S.E. 930 (1891); Fullam v. Foster, 68 Vt. 590, 35 A. 484 (1896).

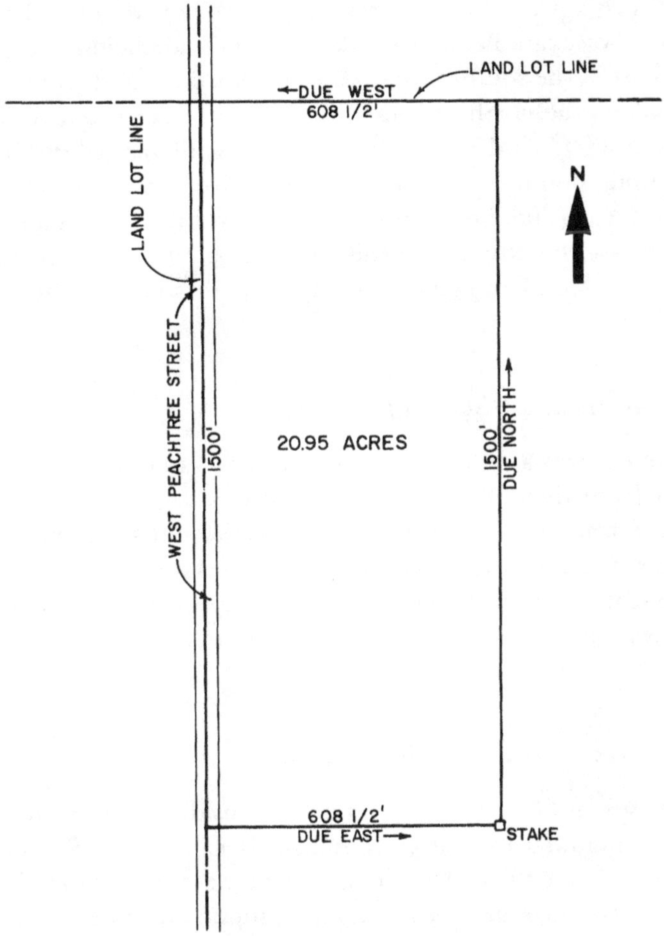

Plate 1. Diagram of 20.95 acres abutting West Peachtree Street.

Reversing Calls

In performing a resurvey, if difficulty is met in running the lines in the direction of the calls, it is generally permissible to reverse the direction of the resurvey.[125] But if all calls in the original survey cannot be harmo-

125. Duncan v. Frye, 236 Ga. 351 (1976); 12 Am. Jur. 2d Boundaries § 60; 3 Am. Law of Prop. § 12.105; 11 C.J.S. Boundaries § 9(3); 1 R. Patton et al., Patton on Land Titles § 125 (2d ed. 1957); 6 G. Thompson, Commentaries on the Modern Law of Real Property § 3059 (1962).

nized by such a procedure, the resurvey must be run in the direction of the calls.[126] For example, where a description read "beginning at a point 216 feet east of the southeast corner of lot No. 6 . . . and from said point running along the northerly side of Oakwood Street 100 feet to the east side of a culvert," and it was later found that the actual distance from the beginning point to the east side of the culvert was 118 feet, the court held that it would not be permissible to begin at the culvert and run 100 feet in reverse direction.[127] In this case the grantee received the entire frontage (118 feet) between the beginning point and the east side of the culvert.

Scaling Courses and Distances from Plats

At one time it was a practice of some surveyors to omit the courses and distances from their plats. The rationale was that since land formerly was often not surveyed with any greater precision than it could be scaled from plats, there was no need to state courses and distances on a plat drawn to scale.[128] In the absence of monuments or other indications as to boundaries, courses and distances scaled from plats may be used as a last resort.[129]

Effect of Excess or Deficiency in Distances

Since courses and distances are generally considered mere guides to aid in locating boundaries, the grantee usually cannot sue for damages if the distances come up short, nor can the grantor receive extra consideration if the distances are greater than he supposed.[130] But if the grantor is guilty of fraud in representing the boundary line or if the stated distances are of the essence of the contract, an award of damages is not precluded.[131]

126. Stewart v. Latimer, 197 Ga. 735 (1944).

127. Id.

128. Andrews v. Moss, 160 Ga. 793 (1925).

129. Grant v. Fourth Nat'l Bank, 229 Ga. 855 (1972); Anderson-Prichard Oil Corp. v. Key Okla Oil Co., 299 P. 850 (Okla. 1931).

130. Weathers Bros. Transfer Co. v. Loyd, 224 Ga. 157 (1968); Hanson v. McLendon, 170 Ga. 426 (1930); Lynburn Enterprises, Inc. v. Lawyers Title Insurance Corp., 191 Ga. App. 710 (1989); Land Trust Co. v. Morgan, 22 Ga. App. 388 (1918).

131. Foute v. Elder, 109 Ga. 713 (1899); Halliburton v. Collier, 75 Ga. App. 316 (1947).

Coordinates

A comparatively new method of describing land is by state-plane coordinates, though the theory behind the system has been known for centuries. The coordinate system as established by the U.S. Coast and Geodetic Survey (now the National Geodetic Survey) was adopted by Georgia in 1939 for use in Glynn County,[132] in 1941 for use in Bryan and Liberty counties,[133] and in 1945 for use in the state as a whole.[134] Nevertheless, statutory recognition of the use of coordinates in describing land is no more necessary than for other methods used to describe land.[135]

Undeniably, coordinates could greatly facilitate the relocation of boundaries, but as yet they have been little used in boundary surveys, primarily because of the increased costs involved and, sad to say, because few surveyors recognize their potential value.

In determining boundaries, coordinates should probably be considered inferior to courses and distances because they are computed from courses and distances, and superior to area because area is usually computed from coordinates.[136] There appear to be no decisions on the matter.[137]

Area

General

Area is usually considered the lowest indicia as to the location of land and the last to be resorted to.[138] A description may be so worded, how-

132. 1939 Ga. Laws 609.

133. 1941 Ga. Laws 788, 918.

134. 1945 Ga. Laws 218; O.C.G.A. §§ 44-4-20 to 44-4-26.

135. 3 Am. Law of Prop. § 12.109.

136. C. Brown, Boundary Control and Legal Principles 141–42 (2d ed. 1969); U.S. Dept. of Commerce, Coast and Geodetic Survey, Spec. Pub. No. 235, The State Coordinate Systems (A Manual for Surveyors) 6–7 (1945).

137. Although there are a number of books and articles dealing with coordinates from a scientific standpoint, almost nothing appears in the legal literature on the subject. Articles from this latter source, however, can be found in 3 Am. Law of Prop. § 12.109 and 1 R. Patton et al., Patton on Land Titles § 162 (2d ed. 1957). For a suitable deed format describing land by coordinates, see Ga. Real Est. Law § 28-8 (3d ed. 1986).

138. "In the description of land it is usual, after the description by metes and bounds or subdivisions, to add a clause stating that the land described contains so many acres. But

ever, as to make the stated area the primary criterion for determining the boundaries. Several examples presently will be given.

When land is conveyed by metes and bounds, whether there be more or less than the quantity named, the purchaser obtains the whole of it. For example, where a devise conveyed "25 acres land" and described the boundaries and there were actually about one hundred acres within the described boundaries, the devisee received the entire tract.[139]

If the quantity conveyed is specified as "more or less," it is generally understood that the parties assume the risk of a gain or loss in the quantity estimated.[140] For example, a certain metes and bounds description that purported to convey "forty acres, more or less" actually conveyed about ninety-two acres.[141] The rule works the other way, too. A metes and bounds description that described the tract as containing "two hundred and fifty (250) acres, more or less" actually conveyed only about 170 acres.[142]

When the area is given in precise terms, it may be accorded some weight in determining which of two disputed bounds is the correct one.[143] The fact, however, that the quantity of land called for in a deed is satisfied without including the bed of a stream which constitutes a boundary does not limit the title to the bank of the stream.[144]

---

unless there is an express covenant that there is the quantity of land mentioned, the clause as to quantity is considered simply as a part of the description, and will be rejected if it is inconsistent with the actual area, when the same is capable of being ascertained by monuments and boundaries. The mention of the quantity of land conveyed may aid in defining the premises, but it can not control the rest of the description." Collinsville Granite Co. v. Phillips, 123 Ga. 830 (1905), quoting 2 R. Devlin, The Law of Real Property and Deeds § 1044 (2d ed. 1897). See also Kendall v. Wells, 126 Ga. 343 (1906); and Ga. Real Est. Law §§ 18-41, 19-175 (3d ed. 1986). However, absence of title to part of the land described is distinguished from a deficiency in area. Holliday v. Ashford, 163 Ga. 505 (1926); Godwin v. Maxwell, 106 Ga. 194 (1898); Smith v. Eason, 46 Ga. 316 (1872); Lunsford v. King, 132 Ga. App. 749 (1974).

139. Rowland v. Matthews, 153 Ga. 849 (1922). See also Kendall v. Wells, 126 Ga. 343 (1906).

140. Baker v. Corbin, 148 Ga. 267 (1918); O.C.G.A. § 44-5-35.

141. May v. Sorrell, 153 Ga. 47 (1921). See also McDonald v. Taylor, 200 Ga. 445 (1946).

142. Baker v. Corbin, 148 Ga. 267 (1918). See also McCook v. Council, 202 Ga. 313 (1947); Currie v. Collins, 136 Ga. 473 (1911); Finney v. Morris, 116 Ga. 758 (1902); Walton v. Ramsey, 50 Ga. 618 (1874).

143. Allen v. Hiwassee Land Co., 162 Ga. App. 356 (1982); Mitchell v. Nicholson, 71 N.D. 521, 3 N.W.2d 83, 139 A.L.R. 1175 (1942); 11 C.J.S. Boundaries § 57.

144. 6 G. Thompson, Commentaries on the Modern Law of Real Property § 3075 (1962).

A deed that describes a tract as containing a definite area and that gives definite boundaries on three sides but an indefinite boundary on the fourth, in the absence of a manifest contrary intent, will be deemed as locating the fourth side parallel to the side opposite it so that the tract encloses the area specified (Plate 2).[145] If the tract is very irregularly shaped, however, such a description will locate the indefinite side as running in a cardinal direction (Plate 3).[146]

Where a land lot is in the shape of a square with its sides oriented in the general cardinal directions, a deed which describes a tract in that land lot as containing a definite area and being in the (north, south, east, west) portion of the land lot but the deed does not otherwise specify the boundaries, the tract will be deemed as embracing a parallelogram that would result from drawing a line across the land lot, parallel with its (north, south, east, west) boundary so as to cut off that definite area (Plate 4).[147]

Where a portion of a square-shaped land lot oriented in the general cardinal directions is conveyed and the tract is described as containing a definite area and lying in a certain corner of the land lot, the specified corner will be taken as a base point from which two sides of the tract shall extend equal distances and lines respectively parallel to these two sides will extend to enclose the quantity conveyed (Plate 5).[148]

A deed that conveys a tract "containing one hundred acres in the northwest corner of lot of land No. 18, the survey to start at the northwest corner of said lot and run the line south to [a certain point]" describes two corners and two sides so that the other two sides can be run forming a rectangular-shaped parallelogram that cuts out one hundred acres (Plate 6).[149]

The description "three acres of land lying in the northwest corner of the [northeast] one-fourth of the lot of land number 131" would convey a tract as shown in Plate 7.[150]

145. Sewell v. Sprayberry, 186 Ga. 1 (1938); Shackelford v. Orris, 129 Ga. 791 (1907); Walden v. Walden, 128 Ga. 126 (1907).

146. Clower v. Godwin, 140 Ga. 128 (1913); Moody v. Vondereau, 131 Ga. 521 (1908).

147. Bennett v. Rewis, 212 Ga. 800 (1957); Rowland v. Matthews, 153 Ga. 849 (1922); Collinsville Granite Co. v. Phillips, 123 Ga. 830 (1905).

148. Head v. Lee, 203 Ga. 191 (1947); Grace v. Roush, 202 Ga. 720 (1947); Harper v. Hesterlee, 152 Ga. 251 (1921); Osteen v. Wynn, 131 Ga. 209 (1908).

149. Smith v. Federal Land Bank, 181 Ga. 1 (1935).

150. Guess v. Morgan, 196 Ga. 265 (1943).

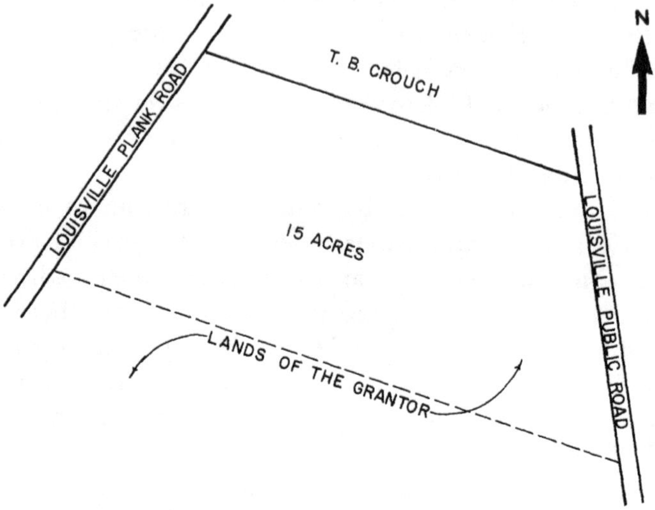

Plate 2. Diagram of "15 acres . . . bounded as follows: north by T. B. Crouch . . . ; south by [grantor]; east by the Louisville Public Road; and west by the Louisville Plank Road."

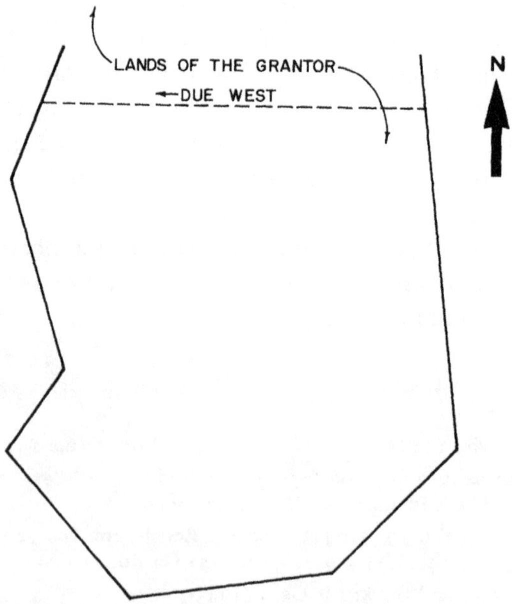

Plate 3. Diagram of an irregular tract bounded by "lands of" the grantor.

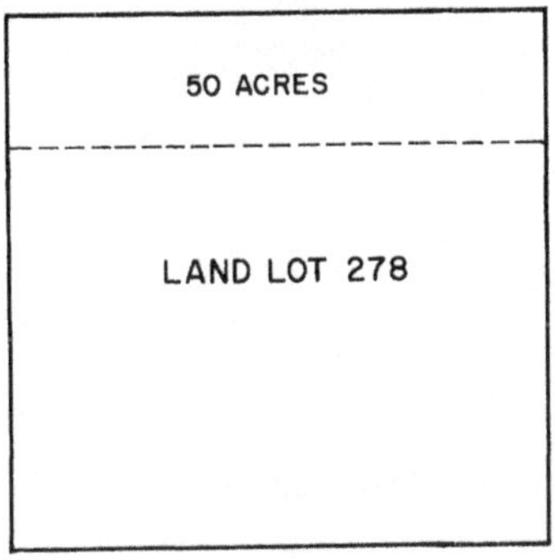

Plate 4. Diagram of "north 50 acres of Land Lot 278."

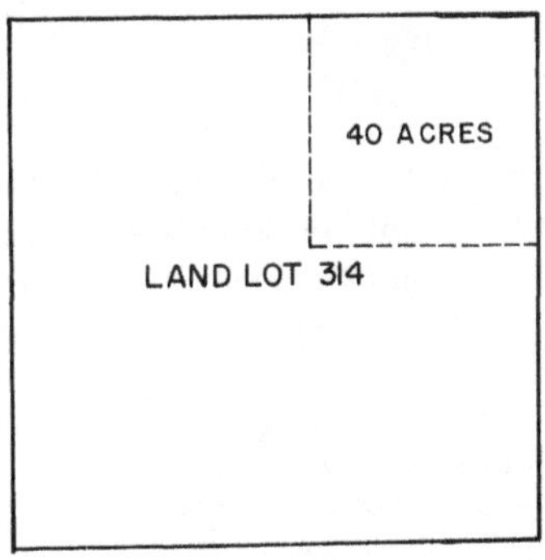

Plate 5. Diagram of "40 acres in the northeast corner of Land Lot 314."

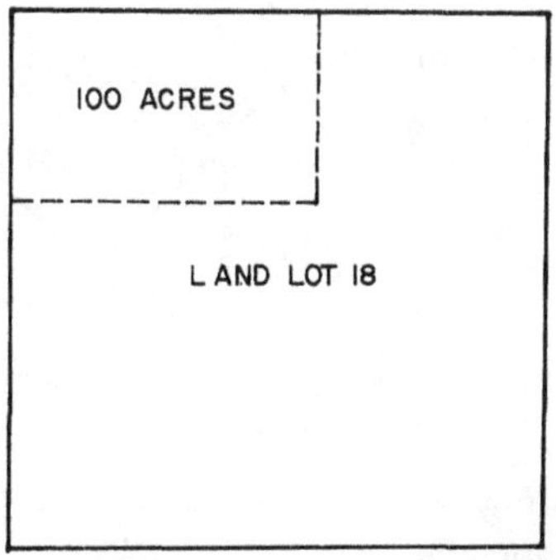

Plate 6. Diagram of "one hundred acres in the northwest corner of lot of land No. 18, the survey to start at the northwest corner of said lot and run the line south to [a certain point]."

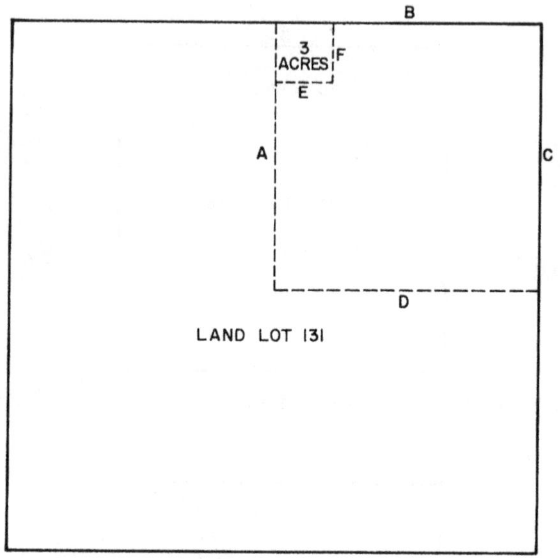

Plate 7. Diagram of "three acres of land lying in the northwest corner of the [northeast] one-fourth of the lot of land number 131." Sides D and E are parallel to side B, and sides A and F are parallel to side C.

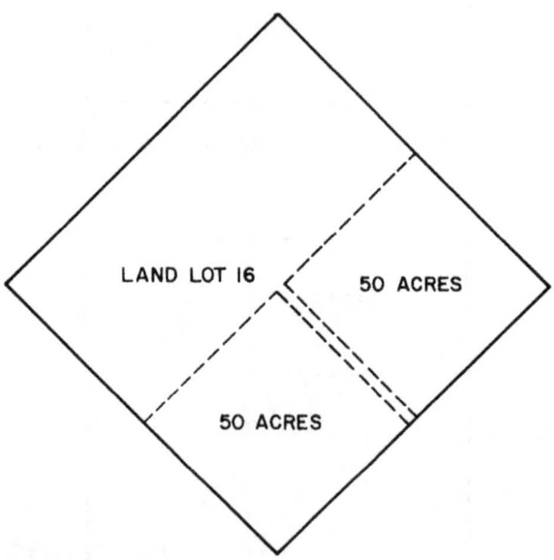

Plate 8. Diagram illustrating the ambiguity of "fifty acres in the southeast corner" of a diamond-shaped land lot.

If the land lot lines do not run in the general cardinal directions, a description such as "fifty acres . . . in the southeast corner" of such a land lot is void (Plate 8).[151]

The description "fourteen acres of land carved out of the north corner" of a diamond-shaped land lot, the tract "being in the shape of a triangle" would cut out a tract in the shape of an isosceles triangle and embracing the number of acres specified.[152]

If the parties to any of the above types of conveyances agree to or acquiesce in certain boundary lines and make a practical location of the lines, this will suffice whether or not the agreed-to lines run in the directions stated above.[153]

It should be reemphasized that in construing each of the above types of descriptions it is essential that a definite area be stated. If the area is indefinite and the description does not otherwise fix the boundaries, then the conveyance is void.[154]

The reason the courts place the above constructions on such descriptions is not because the parties stated these constructions in their writings but because the writings are silent on the matter. To prevent having to void the conveyance, the courts will presume that the parties intended to do what people ordinarily do under like circumstances.[155]

If a deed conveys a definite area out of a larger tract and the larger tract is identifiable but the part being conveyed is not identifiable, the grantee receives an interest in the whole tract, as tenant in common with

151. Darley v. Starr, 150 Ga. 88 (1920). But where an entire tract of land is conveyed and is described as being in the "south corner" of a land lot, the land lot being oriented in the general cardinal directions, the ambiguity could be resolved by extrinsic evidence in that the description named some of the adjoining owners and by the fact that the grantor owned a tract in the southwest corner but not in the southeast corner of the land lot. Crider v. Woodward, 162 Ga. 743 (1926).

152. Benton v. Munday, 183 Ga. 228 (1936).

153. Stone v. Jernigan, 214 Ga. 249 (1958); Moody v. Vondereau, 131 Ga. 521 (1908). See pages 429–36.

154. Matthews v. Logan, 242 Ga. 69 (1978); Brown v. Mitchell, 225 Ga. 115 (1969); Livingston v. Livingston, 210 Ga. 607 (1954). But the description "all of said lot of land except forty acres, more or less, on the northeast side of a certain branch known as Boggy Branch in the northeast corner of said lot" was held sufficient when it was shown that Boggy Branch extended through the northeast corner of the land lot. The exception was held to include all the land northeast of the branch. Sweat v. Mullis, 145 Ga. 450 (1916).

155. Payton v. McPhaul, 128 Ga. 510 (1907); Woods v. Selby Oil & Gas Co., 2 S.W.2d 895 (Tex. 1927) (concurring opinion), affirmed 12 S.W.2d 994 (Tex. 1929).

the grantor, in the proportion that the amount conveyed bears to the amount in the whole tract.[156]

Effect of Excess or Deficiency in Area

If land is described by metes and bounds or by monuments and the statement of area is merely part of the description, then the sale is by the tract and not by a specified area. Under such descriptions the buyer cannot recover for a deficiency, nor can the seller receive extra compensation in the event of an excess.[157] If, however, land is sold at an agreed price per unit area, or if the grantor is guilty of willful deception in representing the area, and a deficiency is later discovered, the grantee may either rescind the contract or demand a reduction of the price according to the relative value of the land so lost.[158]

Apportionment of Excess or Deficiency

For certain types of conveyances, the courts have generally held that the rules dealing with priority of notice do not apply in the case of excess or deficiency. This is because the titles under such conveyances are regarded as all passing simultaneously. Examples include divisions in partition proceedings, devises of a tract by two or more designated areas, conveyances all executed at the same time with the expressed or apparent purpose of conveying the entire tract, and conveyances of subdivision lots in which a plat is made showing all the lots so subdivided.[159]

When a tract of land is subdivided by one of these methods and is thereafter found to be larger or smaller in any dimension than the aggregate dimensions of the tracts comprising it, the excess or deficiency should generally be apportioned among the component tracts.[160] The

156. Harlowe v. Harlowe, 160 Ga. 822 (1925).

157. Keiley v. Citizens Savings Bank, 173 Ga. 11 (1931); Parks v. Norman, 108 Ga. 373 (1899); O.C.G.A. § 44-5-35.

158. Roberts v. Groover, 156 Ga. 386 (1923); Strickland v. Hutchinson, 123 Ga. 396 (1905); O.C.G.A. §§ 44-5-35, 44-5-36.

159. 3 Am. Law of Prop. § 12.123; 1 R. Patton et al., Patton on Land Titles § 158 (2d ed. 1957).

160. Hillside Cotton Mills v. Bartley, 156 Ga. 271 (1923); 12 Am. Jur. 2d Boundaries § 63; 3 Am. Law of Prop. § 12.123; 11 C.J.S. Boundaries § 124; 6 G. Thompson, Commentaries on the Modern Law of Real Property § 3066 (1962); Annotation, 97 A.L.R. 1227 (1935).

apportionment, however, should not be extended beyond any original monuments called for[161] or across any streets dedicated to the public.[162] In general, each street should be given its recorded right-of-way dimensions (unless, of course, the street is already staked off on the ground) and the error prorated within the blocks.

The underlying assumption in regard to apportionment is that the discrepancy arose from an imperfect measurement of the whole line and not of any particular part.[163] The excess or deficiency should not be apportioned across public roads because the rights of the traveling public are considered paramount to the rights of the abutting lot owners. Also, there is the convenience of having roads of uniform width rather than odd and varying widths. As always, where a line has been fixed by agreement or acquiescence,[164] or by adverse possession,[165] the rules with regard to apportionment will not apply.

The rules dealing with apportionment are looked upon with disfavor by the courts and should be applied only as a last resort.

Where a tract is subdivided into lots of specified dimensions, plus one undimensioned lot, the apparent intention of the proprietor and his grantees is carried out by giving each of the dimensioned lots their indicated measurements and throwing all the excess or deficiency in the undimensioned lot.[166]

Boundary Establishment by Agreement or Acquiescence

General

The law provides that where the boundary line between adjoining landowners is indefinite, unascertained, or in dispute it may be settled by the

161. 3 Am. Law of Prop. § 12.123; 1 R. Patton et al., Patton on Land Titles § 158 (2d ed. 1957).

162. Hillside Cotton Mills v. Bartley, 156 Ga. 271 (1923).

163. 3 Am. Law of Prop. § 12.123; 11 C.J.S. Boundaries § 124; 1 R. Patton et al., Patton on Land Titles § 158 (2d ed. 1957).

164. Gulf Oil Corp. v. Timms, 116 S.W.2d 940 (Tex. 1938).

165. Hillside Cotton Mills v. Bartley, 156 Ga. 271 (1923); 3 Am. Law of Prop. § 12.123; 6 G. Thompson, Commentaries on the Modern Law of Real Property § 3066 (1962).

166. 12 Am. Jur. 2d Boundaries § 63; 11 C.J.S. Boundaries § 124; 1 R. Patton et al., Patton on Land Titles § 158 (2d ed. 1957).

parties by an oral or written agreement if duly executed, or by acquiescence of a line coupled with possession up to the line for seven years. The line thus fixed will thereafter control the descriptions given in the parties' deeds.[167] A boundary settlement is distinguished from prescription in that the former is friendly, growing out of agreement, whereas the latter is a hostile relation unconnected with agreement.[168]

Purists have expressed some dismay at the proposition that boundaries can be settled by oral agreement or acquiescence, charging that the rule allows the oral transfer of real property in violation of the statute of frauds and that it is obnoxious to the rule prohibiting the introduction of parol evidence to contradict a written instrument. The courts have attempted to get around the first of these objections by stating that such boundary settlements are not considered within the statute of frauds because they do not extend to the title itself and do not have the operation of a conveyance so as to pass title from one to another.[169] They only determine the extent of what has already been conveyed.[170] As to the objection that boundary settlements are obnoxious to the rule prohibiting the introduction of oral evidence to contradict written instruments, it is said that a practical location of boundaries is but an actual designation by the parties on the ground of the monuments and bounds called for by their deeds.[171]

Whatever the theoretical objections to this rule, there is no doubt that some very sound practical reasons exist for allowing adjoining owners to fix their boundaries by agreement or acquiescence. Because the monuments of the original survey are often destroyed so that their exact locations cannot thereafter be determined, and because land is sometimes conveyed without reference to any physical evidence on the ground as to the boundaries—the descriptions merely giving courses and distances—it becomes imperative that a method be established to fix the practical location of boundaries. Otherwise lines described by mere courses and distances would not be fixed but would instead be subject to alteration by every resurvey. As was said in a California case,

167. Hickox v. Griffin, 205 Ga. 859 (1949); Tietjen v. Dobson, 170 Ga. 123 (1929); Swinson v. Jones, 66 Ga. App. 598 (1942); O.C.G.A. § 44-4-6.

168. 9 C.J. Boundaries § 171; 11 C.J.S. Boundaries § 63.

169. Farr v. Woolfolk, 118 Ga. 277 (1903).

170. Id.

171. 12 Am. Jur. 2d Boundaries §§ 78, 83; 11 C.J.S. Boundaries § 77.

> When the division line of adjoining owners is designated in their respective deeds as a line beginning at a specified distance from a fixed object, the only method of ascertaining the location of the line on the ground is by measuring the required distance from the object. Experience shows that such measurements, made at different times by different persons with different instruments, will usually vary somewhat. The position of the object or monument at which the course begins may also be changed and the change may not be known to the parties, or there may be no means of ascertaining its original position. If the position of the line always remained to be ascertained by measurement alone, the result would be that it would not be a fixed boundary, but would be subject to change with every new measurement. Such uncertainty and instability in the title to land would be intolerable. For these and other reasons the rule has been established that when such owners, being uncertain of the true position of the boundary so described, agree upon its true location, mark it upon the ground, or build up to it, occupy on each side up to the place thus fixed and acquiesce in such location for a period equal to the statute of limitations, or under such circumstances that substantial loss would be caused by a change of its position, such line becomes, in law, the true line called for by the respective descriptions, regardless of the accuracy of the agreed location, as it may appear by subsequent measurements.[172]

There are other reasons for allowing adjoining owners to settle their boundaries by agreement or acquiescence. The courts favor peaceable compromise rather than a legal or physical contest in settling landline disputes, and adjoining landowners will never be forced to litigate in order to fix a boundary which they are willing to determine voluntarily.[173] Frequently, to avoid delay and expense, adjoining owners resort to such a method in fixing their boundaries without attempting to determine the true line. After the lapse of many years it is difficult or impossible to determine where the true line was. Thus the rule, in keeping with the doctrine of laches, helps to prevent strife and disputes concerning boundaries and makes titles permanent and stable.[174] A Texas case noted:

> It is notorious that surveys have been hitherto very loosely made, and often by incompetent surveyors. Lines have been insufficiently marked, and corners designated by perishable objects. And the settling of the country

172. Young v. Blakeman, 153 Cal. 477, 95 P. 888 (1908).
173. Ga. Real Est. Law § 13-34 (3d ed. 1986).
174. Young v. Blakeman, 153 Cal. 477, 95 P. 888 (1908).

and the destroying ax of the settler, and time, have obliterated the path of the surveyor and destroyed the monuments he made. And so, when the lines he ran cannot now be run, and the boundaries he fixed have become of doubtful identity, and the parties to be affected by them have mutually agreed that here he fixed his lines and set their bounds, such agreement should be held conclusive; not subject to be set aside or reopened upon any subsequent discovery that possibly a mistake was made in that agreement as to the true locality.[175]

The courts often state that a boundary line, supposed to be the true boundary and long acquiesced in as such, affords better evidence of the location of the true line than any survey made after the original monuments have disappeared.[176]

The vast majority of boundary cases that have come before the courts have been settled on the doctrines of agreement or acquiescence. But the law dealing with the subject is somewhat chaotic, primarily because the courts often confuse the two doctrines by treating them as one, when in truth they should be treated as separate tenets that lead to the same result. The remainder of this discussion will attempt to sort out some of the particulars found in the Georgia law dealing with the matter.

A line may not be established by agreement or acquiescence unless there is some contention between the adjoining owners over the location of the line.[177] It is immaterial that the true line can be established by a competent survey. If, as between the parties, the line is unascertained or in dispute, then it may be settled by the parties.[178] Thus where a land lot line was the boundary between two tracts but the practical location of the land lot line was not known by the parties, an agreement as to the line fixed the boundary.[179]

Boundary lines cannot be established by agreement or acquiescence

175. Coleman v. Smith, 55 Tex. 254 (1881).

176. Hoffman v. City of Port Huron, 102 Mich. 417, 60 N.W. 831 (1894); Annotation, 110 Am. St. Rep. 685–86 (1906).

177. Cothran v. Burk, 234 Ga. 460 (1975); Seaboard Air Line R.R. Co. v. Taylor, 214 Ga. 212 (1958); Gee v. McDowell, 209 Ga. 265 (1952); Taylor v. Bd. of Trustees, 185 Ga. 61 (1937).

178. "The line is definite in so far as the words of the deed are concerned; but it had never . . . been located and marked upon the land itself. Despite the clarity of the deed, it remained necessary to ascertain by measurement, the location upon the land of the dividing line, and hence it must be held that the line was uncertain and unascertained." Warwick v. Ocean Pond Fishing Club, 206 Ga. 680 (1950).

179. Hethcock v. Padgett, 217 Ga. 328 (1961).

by parties who do not own the lands in question, unless the parties have authority from the true owners to do so.[180] Thus a tenant cannot agree on a boundary with the adjoining owner so as to bind his landlord,[181] and a partner cannot bind his copartner to an agreed-upon boundary without the consent of the copartner.[182] As to whether acquiescence can be used to fix boundaries between public lands and privately owned lands, such as a public highway and abutting privately owned lots, the decisions are in conflict.[183]

When a boundary line has been located by the coterminous owners as provided by law, the line thus located and established is binding on the grantees of the owners[184] unless a grantee takes without notice, either actual or constructive.[185]

Establishment by Oral or Written Agreement

An oral agreement to fix a boundary must have its basis in color of title under the deeds of the coterminous landowners and must be accompanied by possession up to the agreed-upon line, or it must be otherwise duly executed.[186] If the line is definite and ascertainable by the respective owners, an oral agreement to a new line violates the statute of frauds.[187] If, however, adjoining owners acquiesce in and possess up to an incorrect line for seven years, the incorrect line will become permanently binding on them and their successors in title.[188]

An agreement, to be "duly executed," must at least mark out or blaze

180. Kiker v. Anderson, 226 Ga. 121 (1970).

181. Cox v. Daugherty, 62 Ark. 629, 36 S.W. 184 (1896); Hunter v. Malone, 49 Tex. Civ. App. 116, 108 S.W. 709 (1908).

182. Godley v. Hinely, 165 Ga. 717 (1927).

183. 39 Am. Jur. 2d Highways, Streets, and Bridges § 156; 11 C.J.S. Boundaries § 85.

184. Brewer v. Head, 233 Ga. 585 (1975); Peacock v. Boatright, 221 Ga. 661 (1966); Wright v. Anthony, 205 Ga. 47 (1949); McGill v. Dowman, 195 Ga. 357 (1943); Swinson v. Jones, 66 Ga. App. 598 (1942).

185. McNeal v. Carter, 191 Ga. 441 (1940); Miller v. McGlaun, 63 Ga. 435 (1879).

186. United States v. Williams, 441 F.2d 637 (5th Cir. 1971); Tietjen v. Dobson, 170 Ga. 123 (1929).

187. United States v. Williams, 441 F.2d 637 (5th Cir. 1971); Holland v. Shackelford, 220 Ga. 104 (1964).

188. Lockwood v. Daniel, 193 Ga. 122 (1942); Henderson v. Walker, 157 Ga. 856 (1924); Osteen v. Wynn, 131 Ga. 209 (1908).

a boundary so that it can be physically identified by the parties.[189] Examples include cultivating up to the agreed-upon line,[190] erecting monuments or marking trees along the agreed-upon line,[191] driving stakes to mark both ends of the straight line,[192] erecting a fence along the agreed-upon line,[193] and orally agreeing to a line marked by processioners.[194] But putting down a single stake and agreeing that a district line shall be the boundary is insufficient.[195] If the agreed-upon line is physically marked on the ground, actual possession up to the line is not necessary to show acquiescence in the line as the true line.[196]

An oral or written agreement as to a boundary becomes binding as soon as it is entered into.[197] But if the parties agree on a line not as a compromise to an uncertain boundary but to reproduce the true line and it is later discovered that the agreed-upon line is not the true one, the agreement may be set aside. Even such an erroneously marked line will become permanently binding if the agreement is ratified by seven years' acquiescence or if the land has been occupied for a sufficient period of time to give title by adverse possession.[198]

If adjoining owners agree upon a temporary line with the stipulation that the actual line is to be run when a surveyor can be obtained, the agreed-upon temporary line will not become a permanently binding one unless the line is later ratified by agreement, acquiescence, or adverse possession.[199]

189. Myers v. Jackson, 87 Ga. App. 161 (1953); Payne v. Green, 84 Ga. App. 689 (1951).

190. Callaway v. Armour, 207 Ga. 229 (1950).

191. Wright v. Anthony, 205 Ga. 47 (1949); Buffington v. Carter, 199 Ga. 811 (1945); Freeman v. Nelson, 138 Ga. App. 697 (1976); Cagle v. Brady, 69 Ga. App. 162 (1943).

192. Barron v. Chamblee, 199 Ga. 591 (1945).

193. Allen v. Smith, 202 Ga. 363 (1947); Lockwood v. Daniel, 193 Ga. 122 (1941); Childers v. Dedman, 157 Ga. 632 (1924).

194. Huff v. Holley, 101 Ga. App. 292 (1960).

195. Payne v. Green, 84 Ga. App. 689 (1951).

196. Buffington v. Carter, 199 Ga. 811 (1945); Teitjen v. Dobson, 170 Ga. 123 (1930); Hart v. Carter, 150 Ga. 289 (1920).

197. Wright v. Anthony, 205 Ga. 47 (1949); Farr v. Woolfolk, 118 Ga. 277 (1903); Huff v. Holley, 101 Ga. App. 292 (1960).

198. Brunswick Pulp & Paper Co. v. Wilson, 211 Ga. 862 (1955); 12 Am. Jur. 2d Boundaries § 82; 11 C.J.S. Boundaries §§ 64b, 67b.

199. Williams v. Smith, 210 Ga. 325 (1954).

Establishment by Acquiescence

Acquiescence in such objects as fences,[200] ditches,[201] roads,[202] hedgerows,[203] or party walls[204] as boundaries for seven years will be deemed conclusive evidence of a boundary agreement even if, in fact, there was no express agreement between the parties. The acquiescence, to be effective, must be by the acts or declarations of *both* adjoining landowners.[205] Passive acquiescence without any specific acts or declarations is not sufficient to establish a line.[206] Thus failure to dispute the location of a fence as a boundary is not necessarily acquiescence because a fence may be placed for purposes other than fixing a boundary.[207]

Acquiescence cannot be founded on parol negotiations that fix a boundary line preceding the execution of a deed that fixes a different boundary. Such parol negotiations are merged in the deed and if the description in the deed fixes a different line, acquiescence will not operate to create title in the grantee up to the line attempted to be fixed by parol agreement.[208]

Surveyors' Duties with Respect to Boundaries Fixed by Agreement or Acquiescence

If adjoining owners have fixed an unknown or disputed boundary by a written agreement, no problem is posed for the land surveyor because it is his duty to locate lines from the writings of the parties involved. If the surveyor comes upon a situation in which coterminous owners have orally agreed to a line, an attempt should be made to get them to reduce their agreement to writing. If either party refuses, the surveyor should

200. Young v. Wiggins, 229 Ga. 392 (1972); Dye v. Dotson, 201 Ga. 1 (1946).

201. Croft v. Beverly, 202 Ga. 210 (1947).

202. Tyson v. Anderson, 164 Ga. 673 (1927); Chewning v. Bryson, 108 Ga. 750 (1899).

203. Brewer v. Head, 233 Ga. 585 (1975); Wood v. Fraker, 199 Ga. 190 (1945).

204. Collins v. Rebb, 174 Ga. 250 (1931); Liberty Nat'l Bank v. Merchant's & Manufacturer's Paint Co., 307 Ky. 184, 209 S.W.2d 828 (1948).

205. Robertson v. Abernathy, 192 Ga. 694 (1941); Banks v. Holt, 111 Ga. App. 473 (1965); Dixon v. Dixon, 97 Ga. App. 54 (1958); Sou. Timber Co. v. Bland, 32 Ga. App. 658 (1924).

206. Adair v. Atlanta Jewish Community, 228 Ga. 422 (1971); Banks v. Holt, 111 Ga. App. 473 (1965).

207. Cothran v. Burk, 234 Ga. 460 (1975); Carter v. Wyatt, 113 Ga. App. 235 (1966).

208. Smith v. Lanier, 202 Ga. 165 (1947); Taylor v. Bd. of Trustees, 185 Ga. 61 (1937).

not attempt to establish what their unwritten agreement was. Possession is then treated as an encroachment on deed lines.[209]

Surveyors have no right to determine, judicially, when lines have become fixed and binding upon adjoining landowners by acquiescence.[210] If the surveyor encounters a situation in which adjoining owners have acquiesced for many years in a line that differs from the deed line, an attempt should be made to have the parties execute quitclaim deeds to settle the matter or the possession should be treated as encroachments.

The object of a survey is frequently to establish an unknown boundary between coterminous properties. By understanding some of the legal principles of boundary agreements, the surveyor is in a unique position to perform a real service for landowners. When a boundary line is unknown, a simple written agreement upon any reasonable line as the boundary could prevent strife, ill feeling, and even litigation in the future. Property surveyors, as professionals, have a duty to assist the public in settling boundaries in a peaceful atmosphere.[211]

209. C. Brown & W. Eldridge, Evidence and Procedures for Boundary Location 152–53 (1962); J. Wait, The Law of Operations Preliminary to Construction in Engineering and Architecture 331 (1901).

210. Joyce v. Williams, 26 Mich. 332 (1873).

211. C. Brown & W. Eldridge, Evidence and Procedures for Boundary Location 153 (1962). For forms for boundary agreements see 8 R. Brown, Georgia Pleading, Practice and Legal Forms § 44-5-33 Form 39 (1989); and Ga. Real Est. Law § 28-9 (3d ed. 1986). The author has found it convenient to place boundary agreements directly on the plat of the property being surveyed rather than drawing up separate documents. Of course, it is advisable to have the agreements recorded. Consultation with an attorney is often helpful.

Chapter 15. Strip Conveyances

Land is often parceled off into tracts of extreme length and comparatively narrow width, which, for the purposes of this discussion, will be collectively termed strip conveyances. Such tracts are used for highways, streets, and roads; alleys; sidewalks; private ways; railroads and spurtracks; subways; utility and transmission lines; bridges; culverts; viaducts; overpasses and underpasses; tunnels; ditches; sewer lines; canals; and ferries. The title or interest a person or the public has in land set aside for such purposes may vary from fee simple down to a mere license, though the easement is the most common form and the fee simple and defeasible fee are fairly common.

Modern descriptions used in strip conveyances often define a line and convey all the land within a certain distance on each side of the line. The stationing system is often used—that of starting from an arbitrary point called 0+00 and assigning each point along the line a station that is dependent upon its distance from the starting point.[1]

Any description will suffice, however, that designates with reasonable certainty the land over which the conveyance extends. It is generally sufficient to identify the entire tract of the grantor through which the strip will run, without designating a particular path, but once the grantee enters and constructs facilities along a particular path, with the acquiescence of the grantor, he cannot thereafter appropriate a new path without the consent of the grantor.[2] If land is set aside for a public purpose by implied dedication and acceptance or by prescription, then generally only what is actually used passes hands.

Many of the legal technicalities pertaining to particular types of strip conveyances are common to all types. Therefore much of the following discussion may be regarded as applying to all strip conveyances in general rather than merely to the particular type under consideration.

1. C. Brown, Boundary Control and Legal Principles 10 (2d ed. 1969).

2. Ga. Real Est. Law § 8-18 (3d ed. 1986). A deed that conveyed a lot but reserved for the grantor the right to use a driveway "as presently located" across the lot was held sufficient when it was shown that such a driveway actually was in use at the time of the conveyance. Champion v. Neason, 220 Ga. 15 (1964).

Highways, Streets, and Roads

General

Although the courts sometimes attempt to draw a distinction between highways, streets, and roads,[3] it will suffice for the purposes of this discussion to use these words interchangeably.

Highways may come into being in a variety of ways. They may be created by deed,[4] by dedication,[5] by seven years' adverse possession under color of title,[6] or by twenty years' adverse possession without color of title.[7] In addition, in 1973 it was enacted that a prescriptive road would be created if public authorities incorporate into the road system a strip of land that has been used adversely as a road by the public for seven years or more.[8]

Before 1914 a condemning authority could acquire only "such interest . . . as . . . necessary" in land taken for road purposes, and the interest continued only as long as the use continued, with reversion to the owner upon cessation.[9] Many roads throughout the state were acquired on this basis.[10] The present high cost of building streets and highways, however, makes it undesirable that the improvements be made on property subject to reversionary rights. Cities, counties, and the state now usually insist on a fee simple title to the area sought.[11]

Conveyance of Abutting Land Usually Carries Grantee to the Centerline

In spite of the modern trend toward acquiring fee simple title to roadbeds, the underlying fee of most roads is vested in the abutting owners, usually to the centerline, and the public has only an easement in the

3. See City of Statesboro v. Dorman, 203 Ga. 25 (1947); Snow v. Johnston, 197 Ga. 146 (1943); and Ga. Real Est. Law § 5-2 (3d ed. 1986).

4. Ga. Real Est. Law § 5-7 (3d ed. 1986).

5. See pages 367–74.

6. Davis v. The State, 9 Ga. App. 430 (1911); O.C.G.A. § 44-5-164.

7. Jordan v. Way, 235 Ga. 496 (1975); Dunaway v. Windsor, 197 Ga. 705 (1944); Louisville & Nashville R.R. Co. v. Hames, 135 Ga. 67 (1910).

8. O.C.G.A. § 32-3-3(c).

9. Ga. Real Est. Law § 5-15 (3d ed. 1986).

10. Id.

11. Id. § 5-16.

road. A conveyance of land bordering a road will carry the grantee to the centerline, subject to the public right of easement, provided that the grantor owns so far and there are no provisions in the conveyance to the contrary.[12] This rule prevails even though the lot is described by courses and distances sufficient to carry it only to the edge of the road[13] or that monuments are called for which were placed along the edge of the right-of-way.[14] Several reasons exist for the rule: the unlikelihood that a grantor would desire to retain ownership of a small, odd-shaped parcel of land encumbered with an easement and of no apparent use to him; the public policy of discouraging separate ownership of narrow strips and gores of land which are a prolific source of disputes and litigation; and the public convenience, which is best served by having control of highways and streets in the owners of the adjacent properties rather than in distant owners who have no incentive to keep them up.[15]

> Perhaps the strongest reason in favor of this construction . . . is that there ought to be some general and settled rule applicable alike to all conveyances which bound the premises by a marginal road, and the best general rule in behalf of public policy, and the one most likely to conform to the real intention of the parties in most instances, is that which we have just indicated. It is favorable to the general public interest that the fee in all roads should be vested either exclusively in the owner of the adjacent land on one side of the road, or in him as to one half of the road, and as to the other half, in the proprietor of the land on the opposite side of the road. This is much better than that the fee in long and narrow strips or gores of land scattered all over the country and occupied or intended to be occupied with roads, should belong to persons other than the adjacent owners. In the main, the fee in such property under such detached ownership would be and forever continue unproductive and valueless. True it is that the fee in a road or in one half of the breadth of land occupied by a road is generally not of much value to an adjacent proprietor, but it goes to enlarge his holding and probably enhances somewhat the value of his estate; when a detached ownership would usually leave it of no value whatever. At all events, this much may be asserted confidently, that as the fee in roads has to reside somewhere, it is

12. R. G. Foster & Co. v. Fountain, 216 Ga. 113 (1960); Hannah v. Kenny, 210 Ga. 824 (1954); Mozley v. City of Marietta, 180 Ga. 590 (1935).

13. Long v. Faulkner, 151 Ga. 837 (1921).

14. 6 G. Thompson, Commentaries on the Modern Law of Real Property § 3068 (1962). See page 412.

15. 12 Am. Jur. 2d Boundaries § 38; 11 C.J.S. Boundaries § 35; 6 G. Thompson, Commentaries on the Modern Law of Real Property § 3068 (1962).

more desirable that it should be in the owners of the adjacent lands than elsewhere.[16]

Purchasers of lots in subdivisions acquire the fee all the way to the centerline of the streets abutting their lots unless the owner expressly reserves it.[17] After the owner of the subdivision has sold all the lots, he generally has no further rights in the streets thereof.[18]

Grants by the state or a municipality of lands abutting a street or road are usually construed to pass title to half the roadway in the same way that private conveyances do.[19] If a person owns the entire width of a roadway but the land on one side only, a conveyance by him of the tract will pass title to the entire width of the roadway.[20]

A person owning the fee to a portion of a roadway abutting his lands may generally use his land in the roadway in any manner not inconsistent with the rights of the public.[21] For example, trees growing within the right-of-way belong to the adjoining owner and he may take the fruit of the trees.[22] He owns any superfluous soil, rock, or minerals in the right-of-way and any grass or herbage growing therein.[23] He may also construct vaults or cellars under the street and may lay pipes or conduits beneath the surface. Moreover, he has rights to light and air and of lateral support.[24]

Width and Location of Right-of-Way

Except for roads expressly conveyed to or condemned by government agencies and roads established in accordance with a subdivision or town plat, most roads have no clearly defined right-of-way. This situation may pose serious problems for the land surveyor, especially inasmuch

16. Johnson v. Arnold, 91 Ga. 659 (1893).

17. Owens Hdwe. Co. v. Walters, 210 Ga. 321 (1954).

18. Harrison v. The Augusta Factory, 73 Ga. 447 (1884).

19. 12 Am. Jur. 2d Boundaries § 46. The fees to the original streets of Columbus, Macon, and probably Milledgeville are, however, vested in the state. See text at notes 40–44 below.

20. Johnson v. Arnold, 91 Ga. 659 (1893).

21. Seaboard Airline Ry. Co. v. Greenfield, 160 Ga. 407 (1925).

22. Long v. Faulkner, 151 Ga. 837 (1921). The governing authority may, however, remove the trees if they interfere with the public right of travel. Mozley v. City of Marietta, 180 Ga. 590 (1935).

23. 39 Am. Jur. 2d Highways, Streets, and Bridges §§ 164–68; 39A C.J.S. Highways § 138.

24. 39 Am. Jur. 2d Highways, Streets, and Bridges §§ 169–77; 39A C.J.S. Highways § 138.

as almost all parcels of land have some sort of public road leading to, through, or by them. Since the interest the public has in most roads is an easement, the roadbed itself is usually vested in the adjoining owners. Ordinarily, therefore, the question for consideration is where the road easement ends and full private ownership begins.

From 1799 until 1973, a statutory procedure was in effect for the establishment or alteration of roads by the counties.[25] This act provided that "public roads" be cleared to a width of thirty feet.[26] The courts construed this to raise a rebutable presumption that all roads maintained by the county are thirty feet wide.[27] The act was repealed, however, in 1973.[28] The end result seems to be that all roads maintained by the county up to 1973 are thirty feet wide unless otherwise shown, whereas the right-of-way of roads not county-maintained until after 1973 should be determined by what is actually maintained and used. The width of a prescriptive road that was never laid out by the county, registered, or classified must be determined by what is used.[29] It is often a question for the jury.[30]

If a strip of land is conveyed for road purposes but the conveyance does not define the width or location of the right-of-way, the width and location may subsequently be fixed by express agreement between the parties or by the grantee's use of a particular way coupled with acquiescence on the part of the grantor in the use of the route selected.[31]

Streets Created by Implication

When land is described as "bounded" by a street and the grantor owns the fee in the land thus represented as a street, a street is created whether or not it exists at the time of the conveyance.[32]

When the owner of a tract of land subdivides the tract into lots and makes a plat showing the tract as subdivided with designated streets, he

25. Ga. Code of 1933, Chapters 95-1 to 95-2.
26. Id. § 95-103.
27. Hyde v. Chappell, 194 Ga. 536 (1942); Buchanan v. James, 130 Ga. 546 (1908).
28. 1973 Ga. Laws 947.
29. Turner v. Brown, 234 Ga. 605 (1975).
30. Id.; Thrash v. Wood, 215 Ga. 609 (1960).
31. 28 C.J.S. Easements §§ 77, 82.
32. Martin v. First State Bank, 231 Ga. 511 (1973); Mills v. Smith, 203 Ga. 444 (1948); Hamil v. Pone, 160 Ga. 774 (1925).

is presumed to have irrevocably dedicated the streets to the public when he sells the lots.[33]

Limited-Access Highways

Generally the right of access to a street by the owner of land abutting the street is a property right which he cannot be deprived of without compensation.[34] In the special class of highways known as limited-access highways, freeways, or expressways, however, the state acquires not only fee simple title to the area designated for traffic but all express or implied rights of the adjoining owners to ingress or egress as well.[35] The conversion of a highway into a limited-access highway will require additional compensation to the abutting owners for their loss of access.[36]

Streets in Cities

By special statutory enactment, the city of Atlanta is authorized to permit the construction and maintenance of overhead and underground passageways across its streets and alleys as long as they do not interfere with present or future requirements for public purposes.[37] The city's right to construct viaducts over its streets without paying compensation to the abutting owners has also been upheld, even though the fee of the streets is in the abutting owners and such construction would impede those owners' access to their property.[38]

If a viaduct is constructed over a street in a city and the fee of the street is vested in the abutting owners, the surface of the street underneath the viaduct may be used by the owners of the fee in any manner not inconsistent with the rights of the public.[39]

33. See page 370.

34. Charles v. Cobb County, 231 Ga. 696 (1974); Mayor & Council of Macon v. Wing, 113 Ga. 90 (1901). The Department of Transportation, however, has the authority to regulate the location and number of private access driveways turning on to state highways. O.C.G.A. § 32-6-130 to 32-6-133.

35. Ga. Real Est. Law §§ 5-2, 5-20 (3d ed. 1986); 1973 Ga. Laws 961, 1087.

36. Clayton County v. Billups, 104 Ga. App. 778 (1961).

37. 1953 Ga. Laws (Jan.–Feb. sess.) 2258.

38. Brown v. City of Atlanta, 167 Ga. 416 (1928).

39. Adair v. City of Atlanta, 124 Ga. 288 (1905); Trustees Atlanta University v. City of Atlanta, 93 Ga. 468 (1893).

The original streets of Macon were laid out two hundred feet wide, and the state reserved title to the fee in them. By special legislative authority, however, the Macon city council is authorized to grant the adjoining owners the right to encroach into the streets up to the actual width in use.[40] These grants are regarded as appurtenant to the abutting lots, and a conveyance of the lots will automatically pass title to the portion of the street encroached upon, whether or not it is mentioned in the conveyance.[41]

By colonial act of May 1, 1760, the common, streets, lanes, and passageways of Savannah that are described in the plan of the town on file in the surveyor general's office are declared to be the common property of the lot holders in the town.[42] The fee to the original streets of Columbus, laid out under the act of December 24, 1827, is vested in the state.[43] By analogy to the rules regarding the fees to the streets of Macon and Columbus, it would seem that the fee to the original streets of Milledgeville is also vested in the state since that city was laid out under a legislative enactment similar to the ones for Macon and Columbus.[44]

Abandonment and Reversion Rights in Streets and Roads

Unless the evidence is clearly to the contrary, the presumption is that the fee to a street or road is in the adjacent owners and that the public has acquired only an easement over it.[45] The termination of such an easement may occur by abandonment, diversion to other uses, or official action by the municipal, county, or state authorities.[46]

When a street has been vacated, the owner of the fee again becomes entitled to use his property in such manner as he may see fit.[47] If the fee is in a public body, the vacating of the street leaves the public body in

40. 1857 Ga. Laws 182; Daly v. Georgia Sou. & Fla. R.R. Co., 80 Ga. 793 (1888); Kirtland v. Mayor etc. of Macon, 66 Ga. 385 (1881).

41. Acme Brewing Co. v. Central R.R. & Banking Co., 115 Ga. 494 (1902).

42. 18 Col. Rec. Ga. 455; Save the Bay Committee v. City of Savannah, 227 Ga. 436 (1971); Savannah, Albany & Gulf R.R. Co. v. Shiels, 33 Ga. 601 (1863); Mayor, etc., of Savannah v. The Steam Boat Co. of Georgia, R. M. Charlton, 342 (1830).

43. Dawson 470; Kavanaugh v. Mobile & Girard R.R. Co., 78 Ga. 271 (1886).

44. See page 284.

45. Marietta Chair Co. v. Henderson, 121 Ga. 399 (1904).

46. Ga. Real Est. Law § 5-26 (3d ed. 1986).

47. Marietta Chair Co. v. Henderson, 121 Ga. 399 (1904).

possession of the property, to use it for any purpose that the body deems proper, without reference to the property's former use.[48] If the fee to the street is in the adjacent landowners, then the street becomes subject to use by the abutting owners without reference to the former rights of the public.[49]

When the fee to a street is in the adjoining owners, each owner is presumed to own to the centerline. Therefore, upon abandonment, each takes full title to the center of the right-of-way.[50] This principle will not apply, however, if the right-of-way is shown to have been acquired originally in unequal portions. In such cases, the owners may claim to the original property lines.[51] If a person owns the entire width of a road but land on one side only, then the entire road reverts to him in case of abandonment.[52] For example, where a subdivision was laid out with a street running along the north edge of the property, but completely inside the property so subdivided, and later another subdivision was laid out on the adjoining property just to the north of the street, the owners of the lots on the north side of the street acquired no title to the soil of the street.[53]

In extending a boundary into a street, the lines between the centerline of the right-of-way and the edge of the right-of-way should, in most cases, be run at right angles to the centerline and not as prolongations of the lines running to the edge of the right-of-way. Gaps and overlaps are thus avoided.[54]

Alleys

An alley is a narrow street or passageway in common use, and it may be public or private.[55] A public alley is in most ways the same as a street, though it has been held that there is sufficient difference between the two that a city may not convert an alley into a public street without paying

48. Id.

49. Id.; Harbuck v. Richland Box Co., 207 Ga. 537 (1951); Harrison v. The Augusta Factory, 73 Ga. 447 (1884).

50. Marietta Chair Co. v. Henderson, 121 Ga. 399 (1904).

51. 29 C.J. Highways § 258; 39A C.J.S. Highways § 137.

52. Johnson v. Arnold, 91 Ga. 659 (1893).

53. Fessenden v. Parrigin, 228 Ga. 61 (1971).

54. Ga. Real Est. Law § 13-11 (3d ed. 1986).

55. Id. §§ 5-2, 8-38.

compensation to the adjoining owners.[56] A private alley is the same as a private way.[57]

In general, alleys are created in the same way as highways and other easements. A deed that conveys a lot "with the appurtenances" conveys to the grantee the right to use an existing alley adjoining the lot.[58]

Ordinarily a deed conveying land described as bounded by an alley carries with it the base fee to half the area of the alley, together with an easement over its full width—provided that the grantor is capable of conveying so much.[59] In one case, however, where a lot adjoining an alley was conveyed, the description did not expressly include the alley but granted an easement in the alley to the grantee. The grantor reserved to himself and his heirs the right to build over the alley so long as such structures did not interfere with the easement. The court construed this as an intent not to convey the alley itself, especially since the grantor also owned the lot on the other side of the alley.[60] A subsequent conveyance of the lot across the alley included the fee to the entire alley, subject to the rights in the alley which had been conveyed in the first deed.[61]

A passageway connecting a city lot with a public street, although originally laid out as a private alley, will become a public alley by its continuous use by the public and maintenance by the municipal authorities for many years.[62] On the other hand, mere use of a person's driveway by a portion of the public without its maintenance by public authorities will not convert the driveway into a public alley.[63]

Private Ways

General

A private way is a private right of ingress and egress over the lands of another.[64] The term *private way* is gradually falling into disuse. In its stead

56. Id.

57. Moon v. Jones, 101 Ga. App. 79 (1960). See next section.

58. Cantrell v. Kaylor, 201 Ga. 406 (1946).

59. Ga. Real Est. Law §§ 8-8, 8-38 (3d ed. 1986); 12 Am. Jur. 2d Boundaries § 53; 11 C.J.S. Boundaries § 39.

60. Murphey v. Harker, 115 Ga. 77 (1902).

61. Id.

62. East v. Mayor etc. of Wrightsville, 217 Ga. 846 (1962).

63. Tribble v. Mayor etc. of Forsyth, 225 Ga. 204 (1969).

64. Ga. Real Est. Law § 8-11 (3d ed. 1986).

terms such as *private roadway, private driveway,* and *private easement* are being used more and more by treatise writers and in ordinary speech. Because the term *private way* is used almost exclusively in the existing Georgia legal literature, however, that form will be generally adhered to here.

A private way may be acquired by express grant; by adverse possession for seven years' uninterrupted use through improved lands, or twenty years' use over wild lands; by implication of the law when such right is necessary to the enjoyment of lands granted by the same owner; or by compulsory purchase and sale through the superior court.[65]

Grant

Private ways can arise by outright conveyance in the same way any other rights or title to land are transferred.[66] For example, a grantor may reserve a private way across a lot to reach his other lands while selling off the lot itself;[67] and a reference to a plat showing a private way contiguous to the lot being conveyed is sufficient to create a private way.[68] A deed that describes the lot conveyed as bounded by a private way will give the grantee the right to use a way adjoining the lot.[69] On the other hand, where a deed described a lot as bounded on the north by a certain strip of land and attached to the deed was a plat which showed the words "Cement Driveway" written across the strip, the grantee received no rights to the driveway by implication.[70]

65. Jones v. Mauldin, 208 Ga. 14 (1951); O.C.G.A. § 44-9-1.

66. O.C.G.A. § 44-9-49.

67. Champion v. Neason, 220 Ga. 15 (1964).

68. Thompson v. Hutchins, 207 Ga. 226 (1950).

69. Schreck v. Blun, 131 Ga. 489 (1908).

70. Miller v. Slater, 182 Ga. 552 (1936). "The deed described the property sold as bounded on the north by the northern ten feet of the existing lot of the seller. This was not to describe the property sold as bounded by a street or way on the north side. Attached to the deed, however, was a plat of the property conveyed, and on the map the words 'Cement Driveway' appeared on the northern ten feet. In the circumstances, the word 'Driveway' indicated only the existence of a passage or driveway belonging solely to the seller, and did not show an intention between the parties that any right or use in regard to it was conveyed to the purchaser. The deed, in describing the property conveyed as being bounded by 'the northern ten feet' of the seller's tract, showed that the strip of ten feet was reserved to the same extent as other portions of the seller's tract, and the map itself contained nothing inconsistent with such description. Under the facts, the words 'Cement Driveway' were merely descriptive in part of the property not conveyed, and the map containing these

Prescription

To acquire prescriptive title to a private way over another's land, the prescriber must show that he has been in the uninterrupted use thereof for seven years over improved lands or twenty years over wild lands, that the way used does not exceed twenty feet in width, that it follows the same path originally appropriated, and that it has been kept open and in repair during such period.[71]

Improved lands are lands that people have entered into and left their mark by clearing, cultivation, enclosing, and so on.[72] The term refers to the whole tract in question. Thus if a person has woodland adjoining cultivated fields, the woodland is considered improved land.[73] Wild lands are lands located separate and apart from lands partly in cultivation. They remain in a state of nature, unenclosed, and with no indicia pointing to use by the owner.[74]

The purpose of requiring the prescriber to make repairs is to give the owner notice of adverse possession against him. It appears that notice in other ways, without repairs, is insufficient.[75] Regularly sweeping and on one occasion removing a tree limb from a concrete drive is not sufficient to show repairs.[76] Pulling brush and other debris out of the way which happen to fall into an alleged private way is not sufficient repairs.[77] And just because the alleged private way needs no repairs does not relieve the prescriber of the necessity of making them.[78]

The prescriber must use the same path throughout the period of adverse possession, for if he were allowed to wander about on another's land in different directions and upon different tracks the owner would

---

words had no greater significance toward enlarging the grant than if it had shown that a garage, servant's house, or some other object was situated upon such strip." Id.

71. Hunt v. Parker, 221 Ga. 484 (1965); O.C.G.A. §§ 44-9-1, 44-9-54.

72. Hopkins v. Roach, 127 Ga. 153 (1906); Sou. Bank. & Trust Co. v. Wilcox Lbr. Co., 119 Ga. 519 (1903).

73. Hopkins v. Roach, 127 Ga. 153 (1906); Kirkland v. Pitman, 122 Ga. 256 (1904); Henderson v. Cam Development Co., 190 Ga. App. 199 (1989).

74. Smith v. E. B. Burney Const. Co., 231 Ga. 772 (1974).

75. Ga. Real Est. Law § 8-11 n. 5 (3d ed. 1986).

76. First Christian Church v. Rlty. Inv. Co., 180 Ga. 35 (1934).

77. Aaron v. Gunnels, 68 Ga. 528 (1882).

78. First Christian Church v. Rlty. Inv. Co., 180 Ga. 35 (1934). "The gist of the requirement as to repairs is not so much the repairs as the notice which is given by the repairs." Id. See also Rothberg v. Peachtree Inv., 220 Ga. 776 (1965).

probably have no notice of his intent to claim.[79] Running around a bad spot for a few hours while the way is undergoing repairs, however, will not prevent the prescription from running.[80]

When the use of a private way originates by permission of the landowner, prescription does not begin to run until the user notifies the owner, by repairs or otherwise, that he has changed his position from a mere licensee to a prescriber.[81]

One year of uninterrupted use of a way through improved lands of another will give the users an inchoate right so as to entitle them to thirty days' written notice from the owner of his intention to close the way. The purpose is to afford them an opportunity to bring proceedings to prevent the closing.[82]

Implication

One who grants a thing is also deemed to grant that, within his ownership, without which the grant itself would be of no effect.[83] Therefore, when a person owns a tract and sells part of it so that the part sold is landlocked, an implication arises that he inadvertently omitted to convey a means of access. Under such circumstances the landlocked grantee is entitled to the use of a private way over the land of the grantor.[84] The existence of another shorter or better way to reach a public road or street over the land of a stranger will not destroy such an easement, nor will it require a person entitled thereto to acquire a new way from the stranger by paying him compensation under the method provided by statute.[85] If the landlocked grantee later purchases other adjoining land so that he can make a way of access to a highway over his own land, then his rights to a private way over the lands of his grantor cease.[86]

In other jurisdictions it is held that if a person sells off a portion of his land so that the part he retains is landlocked, he is entitled to a private

79. Collier v. Farr, 81 Ga. 749 (1888).

80. Everedge v. Alexander, 75 Ga. 858 (1885).

81. Hunt v. Parker, 221 Ga. 484 (1965); Nassar v. Salter, 213 Ga. 253 (1957).

82. Hall v. Browning, 195 Ga. 423 (1943); Barnes v. Holcomb, 35 Ga. App. 713 (1926); O.C.G.A. § 44-9-56.

83. Calhoun v. Ozburn, 186 Ga. 569 (1938).

84. Burk v. Tyrrell, 212 Ga. 239 (1956); O.C.G.A. § 44-9-1.

85. Calhoun v. Ozburn, 186 Ga. 569 (1938).

86. Russell v. Napier, 82 Ga. 770 (1889); O.C.G.A. § 44-9-5.

way over the lands of the grantee.[87] Seemingly this doctrine has been rejected in Georgia, though the case was not directly in point.[88]

Compulsory Purchase and Sale

The process of obtaining a private way by compulsory purchase and sale is an exception to the general rule that the power of condemnation cannot be used to acquire property solely for private use and gain.[89] The code provides that when a person owns land to which he has no means of access, he may petition the superior court for a private way across the lands of another.[90] Provision is made for notifying the owner of the property to be condemned, for advertising the condemnation in the local newspaper, and for naming assessors to fix the value of the property to be taken.[91] The strip condemned must not exceed twenty feet in width, and the condemnor and his successors in title must keep the way open and in repair.[92] Failure to comply with this provision for one year shall constitute abandonment with reversion to the original owner or his successor in title.[93]

Provision is also made for obtaining private ways for railroad, pipeline, power line, and other types of easements.[94]

A private way cannot be obtained by compulsory purchase and sale merely because such a way would be shorter or more convenient than one already in use or because the present way is obstructed.[95] Moreover, the applicant is not entitled to choose his route over the lands of another if the other landowner has tendered him a way reasonably convenient to both parties.[96]

87. 25 Am. Jur. 2d Easements & Licenses § 34.
88. Farris Const. Co., Inc. v. 3032 Briarcliff Road Asso. Ltd., 247 Ga. 578 (1981).
89. Ga. Const. of 1983, art. 1, § 3, para. 2.
90. O.C.G.A. § 44-9-40.
91. Id. §§ 44-9-41 to 44-9-44.
92. Id. § 44-9-45.
93. Id.
94. Id. §§ 44-9-70 to 44-9-92.
95. Moore v. Dooley, 240 Ga. 472 (1978); Gaines v. Lunsford, 120 Ga. 370 (1904); Charleston & West. Carolina Ry. Co. v. Fleming, 118 Ga. 699 (1903).
96. Wyatt v. Hendrix, 146 Ga. 143 (1916).

Private Ways as Boundaries

There are apparently no decisions in Georgia dealing with private ways as boundaries. Most other jurisdictions hold that a grantee bounded by a private way takes to the center thereof, unless of course the grantor does not own to the center or the conveyance clearly expresses a different intention.[97]

Railroad Rights-of-Way

As Boundaries

Generally, if land is bounded by a railroad the grantee takes to the center of the right-of-way unless the grantor does not own to the center.[98] This rule is based on the presumed intentions of the parties as analogous to the same rule applied to highways and rivers and on a sound public policy opposed to separate ownership of long, narrow, relatively useless strips of land.[99]

Particular language may be used so as to exclude the right-of-way from the conveyance. Thus a conveyance of land bounded by the "right-of-way" of a railroad should be construed as carrying the grantee only to the margin of the railroad right-of-way.[100] In one case a deed described a tract as bounded on the east by the "right-of-way" of a certain railroad and on the west by a street. The deed also referred to a plat that showed the distance between the street and the railroad as "500 feet," but the actual distance between the street and the edge of the railroad right-of-way was less than that distance. The grantee contended that the distance of five hundred feet on the plat evinced an intention to convey part of the right-of-way, especially since the grantor owned the right-of-way. The court held otherwise, however:

> The lot fronts on the right-of-way; it is bounded eastwardly by the right-of-way; and, as the deed stands, the title of defendant would seem to cover

97. 12 Am. Jur. 2d Boundaries § 53; 11 C.J.S. Boundaries § 43.

98. Fambro v. Davis, 256 Ga. 326 (1986).

99. 12 Am. Jur. 2d Boundaries § 54; 3 Am. Law of Prop. § 12.112; 11 C.J.S. Boundaries § 45; Annotation, 85 A.L.R. 404 (1933).

100. Georgia R.R. & Bank. Co. v. Hamilton, 59 Ga. 171 (1877); 9 C.J. Boundaries § 112; 11 C.J.S. Boundaries § 45; Annotation, 85 A.L.R. 409 (1933).

> the lot only to the right-of-way. The plat is like the deed, except that 500 feet is marked as the distance from the right-of-way to the opposite street; but it seems to us to be law, that where boundaries are defined distinctly in a deed and plat, distances must yield to them, and no matter what had been the distance marked on a plat between these boundaries, whether greater or smaller than the actual distance, the defendant would take title to the land between those boundaries only.[101]

A conveyance of land bounded by "the north line of the railroad track" carries the grantee to the north rail of the track, not to the north line of the right-of-way.[102] A deed conveying the land "that lies south of the railway cut" conveys only to the upper and outer edge of the cut and not to the lower and inner edge.[103]

Description of Right-of-Way

Often deeds to railroad rights-of-way, rather than attempting to describe the particular route over which the railroad is to be built, merely describe the grantor's entire tract through which the facility is to be constructed. Under such a description, the construction of the railroad along a particular route with the consent or acquiescence of the grantor will locate and define the right-of-way.[104] If a deed to a railroad right-of-way does not disclose its width, the amount of right-of-way taken by the railroad when it first constructs its tracks will define the width. The railroad company cannot at a later date acquire more right-of-way without the consent of the owner.

> The land through which the right of way was granted was located in the town of Lawrenceville. The whole tract was not granted, but only a strip of land through the tract. The width of the strip was not specified in feet, or other unit of measurement, but the fact that it was not does not require a construction of the deed that would authorize successive encroachments on the land of the grantor after the railroad company had selected the location for its road-bed and cuts and fills necessary to the maintenance of the road-bed through such land. So to construe the deed would authorize the grantee

101. Georgia R.R. & Bank. Co. v. Hamilton, 59 Ga. 171 (1877).

102. Reid v. Klein, 138 Ind. 484, 37 N.E. 967 (1894).

103. Newton v. Louisville & Nashville R.R. Co., 110 Ala. 474, 19 So. 19 (1895).

104. Gainesville Midland R.R. Co. v. Tyner, 204 Ga. 535 (1948); Johnson v. Valdosta, Moultrie & West. R.R. Co., 169 Ga. 559 (1929).

> to eventually take all of the grantor's land, if the necessities of the railroad company should in [the] future require it. In that event the grantee would get all, and not merely a strip through the land. While the width of the right of way was not expressed in feet, or the like, it was described as "extending each side of and at right angles to the center of the track or road-bed," and, "in case of high banks or deep cuts, such additional width as may be necessary." The clear meaning of this is that the strip of land should be of sufficient width to locate the road-bed of the railroad company; and inasmuch as in some places, on account of the necessity for high banks or deep cuts, a greater width might be necessary than where such would not exist, the right of way so given need not be of uniform width, but, though irregular, should not be wider than actually necessary for the construction and maintenance of the railroad through the grantor's land. That such was the intention of the parties is made the more plainly to appear by the further language of the deed, viz.: "The land hereby conveyed is the ground actually occupied by the road-bed, either fill or cut, and which may be necessary for the operation of the business of the said road in running its trains through the lands of the undersigned." The grant did not identify the land intended to be granted, but left it to the grantee to complete the identification by locating the road through the grantor's land. This is indicated by the language which refers to the road-bed intended to be located "as the same may be located and established by the party of the second part." Subsequently, when the grantee located its road-bed, that completed the identification of the land granted, and fixed the rights of the parties under the deed, thereby confining the grantee to the land so appropriated. . . . The deed, therefore, was no authority for the successors in title of the grantee, many years after the location of the road-bed, to encroach upon the remaining part of the land from which the strip was taken, for the purpose of building the additional side-track which it was sought to enjoin.[105]

The mere fact that a railroad company is authorized by its charter to acquire a right-of-way of a certain width raises no presumption that the railroad did actually acquire that width.[106]

Fee Simple Title or Easement

Because some railroads are constructed for only temporary use such as mining or lumbering operations, while others are built for more or less permanent use such as trunk lines for hauling passengers and freight,

105. Martin v. Seaboard Air-Line Ry., 139 Ga. 807 (1913).
106. Williams v. Sav., Fla. & West. Ry. Co., 94 Ga. 540 (1894).

controversies often occur as to whether the conveyance for a railroad right-of-way confers a permanent fee simple title or only an easement that is subject to forfeiture upon abandonment. This question could be of considerable import to the surveyor. For example, he may be called on to survey a tract formerly traversed by a railroad which has been taken up. The question arises as to whether the railroad company still has any rights or title in its former right-of-way. Also, if a tract of land is described as bounded by a railroad, the grantee could take title to part of the right-of-way if the railroad's title is an easement. If the railroad's title is fee simple, however, and the railroad company is not the grantor, then the grantee could take title no farther than the outer margin of the right-of-way.

From the many hairsplitting rulings that have been made in Georgia and in other jurisdictions on this issue, it would seemingly be hard to predict with any degree of certainty how a court would rule in a very close case. Should the situation arise, a careful reading of the cases along with close consultation with an attorney is advisable. Several general observations may be helpful, however. In construing deeds, the primary objective is to ascertain the intentions of the parties as gathered from the instrument. In arriving at the intention the whole instrument must be looked into, not merely its disjointed parts.[107] Where a deed conveys "land" for a railroad, it is usually construed as vesting fee simple title in the railroad.[108] But if the deed conveys a "right" or a "right-of-way" for a railroad, it is usually construed as an easement.[109] Nominal consideration usually implies that the grant is an easement,[110] but not always.[111] If the consideration of the conveyance is "benefits and advantages" accruing to the grantor, the implication is that only an easement is conveyed.[112] If the

107. Jackson v. Rogers, 205 Ga. 581 (1949); Johnson v. Valdosta, Moultrie & West. R.R. Co., 169 Ga. 559 (1929).

108. Jackson v. Rogers, 205 Ga. 581 (1949); Annotation, 6 A.L.R.3d 979 (1966). Compare Jackson v. Sorrells, 212 Ga. 333 (1956).

109. Duggan v. Dennard, 171 Ga. 622 (1930); Atlanta, B. & A. Ry. Co. v. Coffee County, 152 Ga. 432 (1921); Cent. of Georgia Ry. Co. v. Lawley, 33 Ga. App. 375 (1925); Annotation, 6 A.L.R.3d 983 (1966). Compare Johnson v. Valdosta, Moultrie & West. R.R. Co., 169 Ga. 559 (1929).

110. Jackson v. Sorrells, 212 Ga. 333 (1956); Askew v. Spence, 210 Ga. 279 (1954); Rogers v. Pitchford, 181 Ga. 845 (1935).

111. Jackson v. Rogers, 205 Ga. 581 (1949).

112. Rogers v. Pitchford, 181 Ga. 845 (1935). Compare Jackson v. Rogers, 205 Ga. 581 (1949).

deed states that the land is being conveyed for railroad purposes only, an easement is implied.[113] If no warranty of title is given, an easement is implied.[114] The use of such words as "fee simple" or "forever in fee simple" has not always been held to vest an absolute title in the railroad.[115]

Right-of-Way Held under a Parol License or by Acquiescence

It often happens that a property owner gives a railroad company oral permission to construct its tracks across his land. If such parol license has been given and the licensee, upon the faith of the parol license, expends money in constructing facilities, he obtains an easement in the property so occupied. The owner and those claiming under him with notice will be estopped from revoking the license.[116]

Also, if a railroad company builds tracks over someone's land and the owner, knowing of the construction, remains inactive and permits the railroad company to expend large sums of money on the work or permits the completion of the railroad, he will be regarded as having acquiesced in the construction. The owner cannot at a later date maintain an action for trespass or ejectment against the railroad company.[117]

A conveyance of land subject to such a burden does not free the land of the burden unless the grantee takes without notice, and there does not have to be any written record of the burden.[118] Physical evidence such as tracks running across the land is sufficient notice of an easement, and a purchaser of the land will take subject to it.[119]

Abandonment and Reversion Rights in Railroad Easements

Mere nonuse of an easement, even for a long period of time, is not in itself sufficient to show abandonment,[120] unless the conveyance specifi-

113. Jackson v. Sorrells, 212 Ga. 333 (1956); Askew v. Spence, 210 Ga. 279 (1954).

114. Askew v. Spence, id.

115. Atlanta, B. & A. Ry. Co. v. Coffee County, 152 Ga. 432 (1921); Georgia & Fla. Ry. v. Swain, 145 Ga. 817 (1916); Tompkins v. Atlantic Coast Line Ry. Co., 89 Ga. App. 171 (1953).

116. Atlanta, Knoxville & North. Ry. Co. v. Barker, 105 Ga. 534 (1898); Charleston & West. Carolina Ry. Co. v. Hughes, 105 Ga. 1 (1898); Cook v. Pridgen, 45 Ga. 331 (1872).

117. Central of Georgia Ry. Co. v. Stand. Fuel Supply Co., 144 Ga. 91 (1915).

118. Green v. South Bound R.R. Co., 112 Ga. 849 (1900); McLendon v. Atlanta & West Point R.R. Co., 54 Ga. 293 (1875).

119. Ga. Real Est. Law § 8-24 (3d ed. 1986).

120. Atlanta, B. & A. Ry. Co. v. Coffee County, 152 Ga. 432 (1921).

cally provides that nonuse shall constitute abandonment,[121] or unless the right-of-way was acquired by mere use.[122] Abandonment may be shown by an express statement of the railroad that it has abandoned its right-of-way,[123] by removal of the tracks and other structures and vacating the property,[124] by acquisition of a new route for the railroad together with nonuse of the old route,[125] and by conveyance of the right-of-way to another who diverts the property to some other use.[126] When a railroad abandons an easement-type right-of-way, the owner of the land adjacent and contiguous to the right-of-way becomes the absolute owner of the right-of-way if it was granted by the owner or his predecessor in title.[127]

Power Line, Pipeline, and Sewer Easements

A grant in general terms of the right to place a power line, pipeline, or sewer across the land of another without specifying the precise location or width of the right-of-way will be made certain by the act of the grantee with the acquiescence of the grantor in actually locating and constructing the facility. The grantee cannot at a later date change the easement by moving the facilities to another location or by adding facilities, without the consent of the grantor.[128] The grantee, however, obtains an implied easement to replace and maintain the original facilities.[129]

A parol license to construct such facilities becomes an easement running with the land when it has been executed by the licensee and in doing so he has incurred such expense that revocation would be harmful,[130] but if the terms of the license are deviated from substantially, no easement is created.[131]

121. Atlantic Coast Line R.R. Co. v. Sweat, 177 Ga. 698 (1933).

122. See page 391.

123. Atlanta, B. & A. Ry. Co. v. Coffee County, 152 Ga. 432 (1921); Annotation, 95 A.L.R.2d 492 (1964).

124. Jackson v. Crutchfield, 184 Ga. 412 (1937); Annotation, 95 A.L.R.2d 482 (1964).

125. Annotation, 95 A.L.R.2d 488 (1964).

126. Annotation, 95 A.L.R.2d 498 (1964).

127. Brown v. Weare, 348 Mo. 135, 152 S.W.2d 649, 136 A.L.R. 286 (1941).

128. 25 Am. Jur. 2d Easements & Licenses §§ 64–67, 69.

129. Bures v. Stephens, 122 Neb. 751, 241 N.W. 542 (1932).

130. Mayor etc. of Athens v. Gregory, 231 Ga. 710 (1974); Rome Gas-Light Co. v. Meyerhardt, 61 Ga. 287 (1878); Waters v. Pervis, 153 Ga. App. 71 (1980).

131. Mayor etc. of Athens v. Gregory, 231 Ga. 710 (1974).

Generally, adverse use of a strip of land for such purposes as power lines, pipelines, and sewers merely confers upon the prescriber an easement and not the fee to the land.[132]

Where a city attempted to acquire a sewer easement and the conveyance referred to an attached plat for a description of the easement but no plat was attached, the conveyance was held void.[133]

132. See page 387.

133. Mayor etc. of Athens v. Gregory, 231 Ga. 710 (1974).

Chapter 16. Construction of Various Types of Descriptions

This chapter will briefly mention a few specific property descriptions that have not already been discussed in detail and will cover some of the constructions given them by the courts. An extended discourse, drawn from the decisions of all jurisdictions, on the numerous types of descriptive matter used in conveyances would fill several large volumes, yet no pertinent Georgia decisions exist for a number of commonly used descriptions. The following discussion, though admittedly limited in scope, is based on the Georgia decisions with a few comments from decisions of outside jurisdictions.

General and Particular Descriptions

Deeds often give two complete and definite descriptions of the property being conveyed—one general and the other particular. Attempt should be made to harmonize both, but if there is any contradiction, the particular will usually prevail.[1] Thus in one case a tract was described by lot number and metes and bounds, and the deed also stated that "there exists against said property a loan in favor of the Jefferson Mortgage Company . . . ; that said property is presently rented to one Ralph R. Lockwood." It was shown that the description by lot number and metes and bounds covered one tract, whereas the description of the loan and the tenant covered another tract. The former description was held to prevail over the latter.[2] In another case a deed described a definite part of a land lot and concluded with the phrase, "Said lands known as the Wooldridge plantation." It was held that the general description of the tract by name yielded to the previous particular description with which it was inconsistent.[3] In yet another case a lot, fronting on a road and bounded

1. Reeves v. Whittle, 170 Ga. 408 (1930); McCaskill v. Stearns, 138 Ga. 123 (1912).
2. Harlan v. Ellis, 198 Ga. 678 (1944).
3. Osteen v. Wynn, 131 Ga. 209 (1908).

on the east and west by rows of marked trees, was severed by a deed conveying the western portion of the lot. The deed described the portion conveyed as measuring "7½ rods" on the road and stated that "this property is one-half" of the tree-bounded lot. It was later discovered that the tree-bounded lot actually measured 180 feet in frontage. The court held that the grantee received a frontage of 7½ rods (123.75 feet) rather than 90 feet.[4] The particular description of the frontage controlled over the general description that the lot being conveyed was one-half of the tree-bounded lot. If the description in a deed is clear and complete in itself, a general reference to another deed cannot enlarge or restrict the description.[5]

A particular description ordinarily prevails over a general one because the particular is less likely to be in error, usually having been more carefully considered by the parties than a general description written without detailed thought. The rule as stated is merely one of construction. It therefore has no application where the general description can be harmonized with the particular, or where the particular description is obviously erroneous, uncertain, ambiguous, or impossible.[6] One deed, for example, described a tract by courses and distances and concluded with the phrase "being the south end of a tract surveyed by virtue of a warrant in the name of Hugh Moore, being the remaining part of said tract hitherto unsold." It was found that the courses and distances did not correspond to the marks on the ground and that the survey did not and could not be closed by any reasonable construction. It was shown that the Hugh Moore tract originally covered 447 acres and that 250 acres had been sold off from the northern part of it in a previous conveyance. The court held that the description of the tract as the south portion of the Hugh Moore tract (general description) would control and that the description by courses and distances (particular description) would be rejected.[7]

4. Morse v. Kelley, 305 Mass. 501, 26 N.E.2d 326 (1940).

5. 6 G. Thompson, Commentaries on the Modern Law of Real Property § 3055 (1962); J. Wait, The Law of Operations in Engineering and Architecture 375–76 (1901); Annotation, 134 A.L.R. 1044 (1941).

6. Rowland v. Matthews, 153 Ga. 849 (1922); 9 C.J. Boundaries § 155; 18 C.J. Deeds § 254; 11 C.J.S. Boundaries § 20; 1 R. Patton et al., Patton on Land Titles § 156 (2d ed. 1957).

7. Duncan v. Madara, 106 Pa. St. (10 Out.) 562 (1884).

Reference to Plats

A deed will generally be deemed sufficient if it merely refers to a plat or map for a description of the land being conveyed. When a deed refers to a plat as furnishing the description of the land, the plat itself and the words and marks on it are as much a part of the deed and will control as if such descriptive features were written out on the face of the deed itself.[8] As in the case of other types of land descriptions, any monuments called for on a plat will generally prevail over other descriptive calls.[9] Generally a plat may not be admitted to supply descriptive data for a deed that contains no reference to the plat.[10]

When a deed fully describes the land conveyed by metes and bounds and refers to a plat representing them, the reference is not to enlarge or diminish the effect of the descriptive words but to give them efficacy. The operative words are found in the deed itself, and the metes and bounds description contained in the deed will prevail over the description shown by the plat if there is a difference in the two.[11]

But when the descriptive clause in a deed contains particular words of description and also refers to a plat which is more definite but not inconsistent with the descriptive words, the description by plat prevails.[12] For example, a certain deed bounded a lot by a street, a railroad right-of-way, and another lot, but these boundaries did not completely circumscribe the lot being conveyed. The deed also referred to a plat that was complete and definite and did not contradict the deed. The plat was held to control.[13]

To make a plat a part of a conveyance, there must be something in

8. Kilby v. Sawtell, 203 Ga. 256 (1948); Westbrook v. Comer, 197 Ga. 433 (1944); O.C.G.A. § 44-2-28.

9. Sellman v. Schaaf, 26 O. App. 2d 35, 269 N.E.2d 60 (1971); 3 Am. Law of Prop. § 12.116; 11 C.J.S. Boundaries §§ 49c, 50c; 6 G. Thompson, Commentaries on the Modern Law of Real Property § 3052 (1962).

10. Arrendale v. Dockins, 166 Ga. 62 (1927); Calhoun v. Ryals, 159 Ga. 35 (1924) (dissenting opinion); 11 C.J.S. Boundaries § 113; 26 C.J.S. Deeds § 101b; 6 G. Thompson, Commentaries on the Modern Law of Real Property § 3052 (1962). See comment at note 14 below.

11. United States v. Williams, 441 F.2d 637 (5th Cir. 1971); Johnson v. Willingham, 212 Ga. 310 (1956).

12. Patrick v. Sheppard, 182 Ga. 788 (1936).

13. Id.

the conveyance definitely connecting the plat to the conveyance,[14] but the conveyance need not describe the plat with minute detail. Thus a deed to a tract in Macon "known in the plan of said city as lot no. four of subdivision of lot no. 7, square 53, southwest commons," was sufficient to make the plan of the town a part of the deed without specifically describing the plat.[15] Where a grantor sold a lot described in a deed as bounded by a street not expressly defined in the deed but shown on a plat attached to it, the grantor was estopped from denying the existence of the street as shown on the plat.[16] A description by land lots and district makes the plats of the land lots and district deposited in the office of the secretary of state admissible as evidence.[17]

Persons dealing with descriptions calling for a plat become chargeable with notice thereto whether or not the plat is recorded.[18]

If a deed refers to a plat that does not conform to statute and is not duly certified by a surveyor, the plat may still be used to identify the lands referred to in the deed.[19] Even a plat or sketch compiled from deeds, aerial photographs, and other plats and referred to in a conveyance may be sufficient to identify the lands.[20] Tax maps, although helpful to title examiners, contain many errors and conflicts and should not be relied on as evidence of boundaries.[21]

If a deed description is too vague to be enforced but refers to a plat that is also too vague, the conveyance as a whole is unenforceable.[22] Thus

14. 26 C.J.S. Deeds § 101b. Despite much lip service to this rule, there are a number of cases in which the courts have held plats admissible to aid defective descriptions when there was no reference to the plat in the deed. See, for example, Arrendale v. Dockins, 166 Ga. 62 (1927) and Calhoun v. Ryals, 159 Ga. 35 (1924) (one dissent).

15. Holder v. Jordan Rlty. Co., 170 Ga. 764 (1930). See also Acme Brewing Co. v. Cent. R.R. & Bank. Co., 115 Ga. 494 (1902).

16. Schreck v. Blun, 131 Ga. 489 (1908).

17. Bridges v. Brackett, 205 Ga. 637 (1949).

18. Buckner v. Mallett, 245 Ga. 245 (1980); Hardigree v. Hardigree, 244 Ga. 830 (1979); Talmadge Bros. & Co. v. Interstate Bldg. & Loan Asso., 105 Ga. 550 (1898); Wiggins v. Gillette, 93 Ga. 20 (1893); Shantha v. West Georgia Nat'l Bank, 145 Ga. App. 712 (1978).

19. 26 C.J.S. Deeds § 101b; 6 G. Thompson, Commentaries on the Modern Law of Real Property § 3052 (1962).

20. Hanners v. Woodruff, 257 Ga. 73 (1987); Furney v. Dukes, 226 Ga. 804 (1970).

21. Ga. Real Est. Law § 13-31 (3d ed. 1986). Compare Sackett v. Wilson, 258 Ga. 612 (1988) and Johnson v. Sackett, 256 Ga. 552 (1986).

22. Plantation Land Co. v. Bradshaw, 232 Ga. 435 (1974); Smith v. Georgia Ind. Rlty. Co., 215 Ga. 431 (1959).

in one instance a deed purported to convey several lots as shown by a certain plat. The plat was not drawn to scale, no courses or distances were given, all the streets shown except one were unnamed, no land lot lines or starting and ending points were indicated, and several engineers testified to the difficulty of locating the lots from the plat. The court held the conveyance void.[23] In another instance a sales contract referred to a plat but the plat failed to designate the dimensions as feet, yards, chains, inches, or any other unit of measure. In addition, the dimension numbers were designated as "±." The contract was held void.[24]

If a deed refers to a plat that is lost, the grant has to be construed from the terms of the deed only.[25]

A recorded plat constitutes constructive notice to the world of its contents.[26]

Reference to Other Instruments

If a deed refers to another instrument for a description of the land, that instrument will be considered as incorporated in the deed itself. Such a reference is sufficient to make the deed operative if the instrument referred to contains a valid description.[27] For example, a reference to the page of a given deed book in the courthouse in which a prior deed to the same tract of land can be found is sufficient to identify the premises.[28] A description that tells the approximate location of the land and contains a further description such as "being the same lands that W. H. Rolls bought from J. S. Easterling" is sufficient to allow introduction of the deed of the former conveyance of the property, though the former deed is not called for directly.[29]

An instrument referred to for a description of the premises does not

23. Herrington v. Rose, 225 Ga. 452 (1969).

24. Lansky v. Brannon, 225 Ga. 378 (1969).

25. Oemler v. Green, 134 Ga. 198 (1910).

26. Conyers v. Fulton County, 117 Ga. App. 649 (1968).

27. Deaton v. Swanson, 196 Ga. 833 (1944). For reference to plats see the preceding section.

28. Turner v. Helton, 212 Ga. 208 (1956); Turner v. Boyd, 212 Ga. 191 (1956); Cannon v. Whiddon, 194 Ga. 417 (1942).

29. Burns v. Tootle, 166 Ga. 226 (1928). See also Williams v. Gilly, 133 Ga. 546 (1909) and Derrick v. Sams, 98 Ga. 397 (1896).

have to be recorded to make that instrument admissible for identifying the lands being conveyed.[30] Thus a reference to another deed is valid even if the cited deed is not properly executed so as to entitle it to record.[31] Even a reference to an application for a loan[32] or that the deed is subject to the dower of a named widow[33] is sufficient to make the loan application or the proceedings by which the dower was assigned to the widow a part of the deed.

A reference to another deed made merely to show from what source title was derived and not to provide an identification or description of the property conveyed will not restrict the description given in a deed that incorporates such a reference.[34]

Of course, if a deed depends upon a reference to another instrument that contains a void description, the deed is void.[35]

Buildings

A deed to land generally includes all buildings and other things permanently attached to the land.[36] The converse, however, is not necessarily true: a deed to a building may or may not include all the land lying under the building, depending on the intentions of the parties.[37] If a grantor conveys land with a building standing on it that encroaches on adjoining land not owned by the grantor, of course no title can pass as to the area encroached upon.[38] If the chief object of the conveyance appears to be the building, then generally all the land underneath the building and so much adjoining land owned by the grantor as is incident to or necessary to the use of the building will pass.[39] In a number of cases, however, where a

30. Deaton v. Swanson, 196 Ga. 833 (1944); Williams v. Gilly, 133 Ga. 546 (1909); Talmadge Bros. & Co. v. Interstate Bldg. & Loan Asso., 105 Ga. 550 (1898).

31. Deaton v. Swanson, 196 Ga. 833 (1944).

32. Talmadge Bros. & Co. v. Interstate Bldg. & Loan Asso., 105 Ga. 550 (1898).

33. Parler v. Johnson, 81 Ga. 254 (1888).

34. Talmadge v. Adams, 240 Ga. 193 (1977).

35. Rogers v. Manning, 203 Ga. 771 (1948).

36. Sawyer Ice & Coal Co. v. Kinnett-Odom Co., 192 Ga. 166 (1941); O.C.G.A. §§ 44-1-2, 44-1-6.

37. Sawyer Ice & Coal Co. v. Kinnett-Odom Co., 192 Ga. 166 (1941).

38. Id.

39. Irwin v. Dailey, 216 Ga. 630 (1961); Faulkner v. McKelvey, 207 Ga. 354 (1950); Stanley v.

tract was described by metes and bounds and the conveyance incidentally mentioned buildings standing on the land, the conveyance was held not to include adjoining land owned by the grantor upon which the buildings encroached.[40] In some decisions, if a building on the property conveyed encroaches on or projects over other property of the grantor, the grantee takes an implied easement for the property encroached upon; and if a building on a lot retained by the grantor encroaches on or projects over adjoining land that he has sold, the grantor reserves an implied easement as to the property encroached upon.[41]

When land is "bounded" by a building, unless a contrary intent appears, the actual boundary is the farthest edge of the building (such as the edge of the eaves or the edge of the footing).[42] The reason is that the parties are presumed to intend that the property line shall be wholly on one side of every portion of the building.[43] Likewise, if a line is to be measured "to" or "from" a building, the line should be measured to or from the outermost edge of the building.[44] This is not in conflict with the rule that the center of a monument controls in marking a boundary.[45] The distinction is stated as follows:

> Whenever land is described as bounded by other land, or by a building or structure, the name of which, according to its legal and ordinary meaning, includes the title in the land of which it has been made part, as a house, a mill, a wharf, or the like, the side of the land or structure referred to as a boundary is the limit of the grant; but when the boundary line is simply by an object, whether natural or artificial, the name of which is used in ordinary speech as defining a boundary, and not as describing a title in fee, and which does not in its description or nature include the earth as far down as the grantor owns, and yet which has width, as in the case of a way, a river, a ditch, a wall, a fence, a tree, or a stake and stones, then the centre of the

---

City of Glennville, 140 Ga. 306 (1913); 26 C.J.S. Deeds § 106b; H. Tiffany, The Modern Law of Real Property 683–84 (abr. ed. 1940).

40. Winters v. Bloom, 96 Kan. 443, 151 P. 1109 (1915); Carville v. Hutchins, 73 Me. 227 (1882); 26 C.J.S. Deeds § 106b.

41. Annotation, 53 A.L.R. 910 (1928).

42. 9 C.J. Boundaries § 111; 11 C.J.S. Boundaries § 41.

43. Kendall v. Green, 67 N.H. 557, 42 A. 178 (1894); 9 C.J. Boundaries § 111.

44. 9 C.J. Boundaries § 5; 11 C.J.S. Boundaries § 4.

45. See page 408.

thing so running over or standing on the land is the boundary of the lot granted.[46]

House Number

In city limits and metropolitan areas, house numbers are sometimes used as makeshift land descriptions.[47] Since house numbers are usually simply a designation for the building, such descriptions leave open the question of how much land is included.[48] Very little judicial construction is found as to the use of house numbers in describing land, although the description "#401 Spring known as the Cob Home 50 x 160 more or less" was held to be a valid description when the town and state were also given.[49] A lease and option that described the premises as "Beach Cottage, known as 1310 Butler Ave." was held to include the cottage and the entire lot it stood on, after the lessor had allowed the lessee to go into possession of the whole premises and make valuable improvements thereon.[50]

A devise in a will conveying "the house and land at 70 Clark Street" was held to include the house and the lot it stood on, as well as an adjoining vacant lot owned by the testatrix.[51] The court's reasoning was that the use of the word *land* and not *lot* indicated that the testatrix intended to devise not only the lot the house stood on but all of her land fronting on Clark Street. The court also applied the rule that intestacies are not favored in construing wills. The will under consideration, although devising other lands owned by the testatrix, made no specific mention of the vacant lot and contained no residuary clause. Therefore, to hold that the devise did not include the vacant lot would have resulted in a partial intestacy.

46. City of Boston v. Richardson, 95 Mass. (8 Allen) 146 (1866).

47. Ga. Real Est. Law § 13-59 (3d ed. 1986).

48. Id.

49. Bush v. Black, 142 Ga. 157 (1914). See also Greenspan v. Caison, 101 Ga. App. 632 (1910) and Carter v. Ray, 70 Ga. App. 419 (1943).

50. Deich v. Reeves, 203 Ga. 596 (1948).

51. Matthews v. Loftin, 224 Ga. 98 (1968).

Local or Descriptive Name

When property has a descriptive name, it may be conveyed by that name.[52] Extrinsic evidence in such cases is receivable to apply the description to its subject matter.[53] In determining what land is included under a descriptive name, the language of former deeds coupling such name with more particular descriptions should be given weight.[54] A description by name will prevail over a further description which is uncertain and imperfect.[55] Thus when the general location of the property was given or could be inferred from the language in the instrument, the following descriptions were held valid: "the Pollett old mill tract of land";[56] "the 'Humphrey Place,' containing 330 acres more or less";[57] "one house and lot in Georgetown, known as the Oatis place";[58] "the Anderson place on Rt. No. 1, 12 miles out of Augusta";[59] and "that tract of land . . . known as Litchfield, containing one thousand and eighty-four (1084) acres, more or less."[60] But the description "Known as Wilkinson Pond consisting of 25 acres, more or less; with the exception of 3 acres, more or less, to be set aside as a home place containing the Wilkinson home" was held void because the "3 acres, more or less" were not identified. The court conceded that the description of the overall tract would have otherwise been valid.[61] The description "fifty (50) acres of land, more or less . . . known as the Bunk Grace Home Place" was held insufficient after it was shown that the Bunk Grace home place contained several hundred acres.[62]

52. Bunger v. Grimm, 142 Ga. 448 (1914).
53. Id.; Reeves v. Whittle, 170 Ga. 408 (1930).
54. Ga. Real Est. Law § 19-154 n. 22 (3d ed. 1986).
55. Bunger v. Grimm, 142 Ga. 448 (1914).
56. Nasworthy v. James, 152 Ga. 368 (1921).
57. Clark v. Cagle, 141 Ga. 703 (1914).
58. Oatis v. Brown, 59 Ga. 711 (1877).
59. Marsh v. Baird, 203 Ga. 819 (1948).
60. Bunger v. Grimm, 142 Ga. 448 (1914).
61. Smith v. Wilkinson, 208 Ga. 489 (1951).
62. Grace v. Grace, 92 Ga. App. 693 (1955).

Reference to Ownership or Residency

If the general locality of the property is also disclosed, the courts will usually uphold descriptions such as "my place adjoining W. T. Mosher S.E., including house and everything on said place";[63] "my half interest in the property corner of Second and Cherry Streets";[64] "that certain tract or parcel of land belonging to the Atlanta Loan and Trust Company, located on the northeast corner of Auburn Avenue and Bell Street";[65] "two hundred and forty-five (245) acres . . . known as the place whereon the said C. A. Lindsey now resides";[66] and "my property consisting of 85 acres, more or less, in DeKalb county, being in Redan District, and more particularly described as follows: Fronting the Rock Bridge Road, with a five-room house thereon, now occupied by tenant by name of Wilson."[67]

If a description is equally applicable to two or more tracts of land, only one of which the grantor owns, it will be presumed that he intends to convey the tract to which he has title.[68]

Omnibus Descriptions

Deeds conveying "all" of the grantor's property or "all" of his property in a certain locality are known as omnibus deeds, blanket deeds, or Mother Hubbard deeds. Other jurisdictions generally hold such descriptions valid,[69] but the Georgia decisions are in conflict—some holding them valid[70] and some holding them not valid.[71] It is certainly question-

63. Mosher v. Rogers, 141 Ga. 557 (1914).

64. Pearson v. Horne, 139 Ga. 453 (1912).

65. Petretes v. Atlanta Loan & Trust Co., 161 Ga. 468 (1925).

66. Allen v. Lindsey, 139 Ga. 648 (1913); See also Brice v. Sheffield, 118 Ga. 128 (1903).

67. Lyle v. Phillips, 141 Ga. 618 (1914).

68. Crider v. Woodward, 162 Ga. 743 (1926); Head v. Lane, 153 Ga. 688 (1922); Tolbert v. Short, 150 Ga. 413 (1920); Romanik v. Buitrago, 153 Ga. App. 886 (1980).

69. 23 Am. Jur. 2d Deeds § 59; 3 Am. Law of Prop. § 12.104; 26 C.J.S. Deeds § 30b; 1 R. Patton et al., Patton on Land Titles § 123 (2d ed. 1957); 6 G. Thompson, Commentaries on the Modern Law of Real Property § 3023 (1962); Annotation, 55 A.L.R. 162 (1928).

70. Copeland v. Carpenter, 206 Ga. 822 (1950); Bennett v. Green, 156 Ga. 572 (1923); Lanier, Hamilton & Co. v. Hebard, 123 Ga. 626 (1905).

71. Newton v. Allen, 220 Ga. 681 (1965); Blue Ridge Apt. Co. Inc. v. Telfair Stockton & Co. Inc., 205 Ga. 552 (1949); Johnson v. Shook, 156 Ga. App. 878 (1981).

able as to whether such descriptions afford sufficient record notice to bona fide purchasers.[72]

Despite the conflict of authority, it seems to be settled that such descriptions are valid in this state when used in wills because greater liberality is allowed in the construction of wills than deeds and other instruments.[73]

Inaccuracies and Omissions

A mere inaccuracy in a description will not in itself void the description.[74] If the thing to be granted appears clearly and satisfactorily from any part of the description and other circumstances of description are mentioned which do not apply to that thing, the grant will not be defeated, but those circumstances will be rejected as false or mistaken.[75] It has already been mentioned that a description is not necessarily void if it contains incorrect courses and distances,[76] an inaccurate area,[77] or other erroneous descriptive terms.[78] Likewise, stating the wrong county,[79] land section,[80] land district,[81] land lot,[82] militia district,[83] or subdivision lot number;[84] stating the wrong location of the point of beginning;[85] or calling for an adjoining owner on the east when the adjoiner is actually on the west[86]

72. Ga. Real Est. Law § 19-14 (3d ed. 1986).

73. Harriss v. Howard, 126 Ga. 325 (1906). It should be pointed out that most of the cases upholding omnibus descriptions involve wills, powers of attorney, and conveyances of personal property.

74. Pressley v. Jennings, 227 Ga. 366 (1971); McCaskill v. Stearns, 138 Ga. 123 (1912); Shantha v. West Georgia Nat'l Bank, 145 Ga. App. 712 (1978).

75. Burson v. Shields, 160 Ga. 723 (1925); Thompson v. Hill, 137 Ga. 308 (1911).

76. See page 418.

77. See page 422.

78. See pages 457–58.

79. McCaskill v. Stearns, 138 Ga. 123 (1912); Stanford v. Bailey, 122 Ga. 404 (1905).

80. Morgan v. Lester, 215 Ga. 570 (1959).

81. Smith v. Federal Land Bank, 181 Ga. 1 (1935); Boggess v. Lowrey, 78 Ga. 539 (1887).

82. Johnson v. McKay, 119 Ga. 196 (1903); Rogers v. Rogers, 78 Ga. 688 (1887); Summerlin v. Hesterly, 20 Ga. 689 (1856); Harmon v. First Nat'l Bank, 50 Ga. App. 3 (1934).

83. Burson v. Shields, 160 Ga. 723 (1925).

84. Smith v. Smith, 112 Ga. 351 (1900).

85. Everett v. Culberson, 215 Ga. 577 (1959).

86. Pressley v. Jennings, 227 Ga. 366 (1971).

will not void a description if the other stated particulars serve to locate the precise land intended.

Inadvertently omitting a course and distance from a description that describes a tract by courses and distances is not a fatal defect if a closure analysis would indicate the omission.[87] If a large, irregularly shaped tract is described by calls for adjoiners, and all the adjoiners are named except one and the grantor himself is not named as an adjoiner, nor is any division of the tract into parts indicated, the description is sufficient to convey the tract.[88] If land is described as being in a certain land lot in a certain county but the district number is left out, the number of the district may be supplied by extrinsic evidence showing that there is only one land lot of that number in the county.[89] Omitting the name of the county where the land lies does not invalidate a description, if the name of the town it lies in is given.[90]

Construed Most Favorably to the Grantee

When all other means of ascertaining the true construction of a deed fail, and a doubt still remains, the construction most favorable to the grantee is preferred.[91] This rule is one of last resort and should not be applied if an ambiguity can be resolved by any other means of construction.[92]

In a case in which the rule was applied, a grantor owned a lot fronting 364.06 feet on the western side of Cedar Street. He sold off the northern portion of the lot. The deed described the portion sold as "Bounded easterly by Cedar street, one hundred and twenty-two feet; southerly . . . by my own land, from which it is separated by a straight line drawn westerly from the west line of said street, at right angles thereto, from a point in said west line of said street two hundred and forty seven feet distant northerly [from the south line of the grantor's property]." The court, in

87. See page 418.

88. Broach v. O'Neal, 94 Ga. 474 (1894). Compare text at note 105 below.

89. Copeland v. Carpenter, 206 Ga. 822 (1950); Moore v. McAfee, 151 Ga. 270 (1921).

90. Essuon v. Raynor, 231 Ga. 297 (1973); King v. Brice, 145 Ga. 65 (1916); Horton v. Murden, 117 Ga. 72 (1902).

91. Holder v. Jordan Rlty. Co., 163 Ga. 645 (1926); Shackelford v. Orris, 129 Ga. 791 (1907); Harmon v. First Nat'l Bank, 50 Ga. App. 3 (1934); Colley v. Atlanta & W. P. R.R. Co., 32 Ga. App. 711 (1924).

92. 23 Am. Jur. 2d Deeds §§ 229, 322; 26 C.J.S. Deeds § 82e.

applying the rule, held that the grantee received a frontage of 122 feet rather than 117.06 feet—the amount that would have been obtained if the separation line had been drawn at a point 247 feet northerly of the grantor's south line.[93]

One exception to the rule that a conveyance is construed most favorably to the grantee is that grants by the state are construed in favor of the state and against the grantee.[94]

Void Descriptions

General

Many factors affect the validity of deeds and other conveyances: identification of the parties, capacity of the parties, attestation, consideration, delivery, recording and notice, covenants and encumbrances, the presence or absence of fraud or mistake, and the description of the property conveyed. Consideraton of most of these factors is beyond the scope of the land surveyor's duties and, consequently, they are not treated in this book. Indeed, a full discussion of all such factors would fill many volumes. Nevertheless, because one of the most important functions of land surveyors is the ability to interpret property descriptions properly and because surveyors are frequently employed to write descriptions of property, a brief summary will be given of a few factors that may void a description.

No prescribed form is necessary to effect a valid description.[95] Furthermore, the courts are loath to hold any instrument void and thus render the work of the maker futile.[96] Some intended conveyances are so imprecise in their descriptions, however, that, rather than run the risk of putting a construction on them that neither party envisioned or contemplated, the courts will declare the instrument void.[97]

Several examples of void descriptions have already been discussed in this and the preceding two chapters. Other examples follow.

93. Clark v. Beloff, 71 Conn. 237, 41 A. 801 (1898).

94. State of Georgia v. Ashmore, 236 Ga. 401 (1976).

95. O.C.G.A. § 44-5-33.

96. McCaskill v. Stearns, 138 Ga. 123 (1912).

97. Williams v. Manchester Bldg. Sup. Co., 213 Ga. 99 (1957).

State and County

The conveyance must disclose the state and county in which the land lies or otherwise tell its approximate location.[98] If the body of a deed fails to mention these elements, the land will be presumed to lie in the state and county denoted in the caption of the deed.[99]

Monuments

Calls such as "thence . . . to near a white oak tree,"[100] "a point, approximately two (2) feet from Earl Ballew's garage,"[101] "a stake on the Highway right of way line,"[102] and "an agreed point,"[103] without more, are insufficient. But the description "a certain designated pine tree on the Sparta and Milledgeville public road, just to the right of the road leading from said public road to the church house" was held valid. The name of the church was given in the description.[104]

Call for Adjoining Owners

A description depending on references to adjoiners may be rendered invalid if all the adjoiners are not given.[105] If one of the called-for adjoiners is the grantor, the deed will usually be void even if the grantor and grantee orally agreed on a division line.[106] In one case, however, the owner of a tract of land caused it to be subdivided into several smaller tracts with the lines of each surveyed and marked by a surveyor. The

98. Callaway v. White, 222 Ga. 371 (1966); Glover v. Newsome, 132 Ga. 797 (1909); Gatins v. Angier, 104 Ga. 386 (1898).

99. Callaway v. White, 222 Ga. 371 (1966); Copeland v. Carpenter, 206 Ga. 822 (1950).

100. Mull v. Allen, 202 Ga. 176 (1947).

101. Id.

102. Hedden v. Hilton, 236 Ga. 641 (1976).

103. Swindle v. Curry, 218 Ga. 552 (1962).

104. Bass v. African M.E. Church, 150 Ga. 452 (1920).

105. Conyers v. West, 210 Ga. 190 (1953); Holt v. Tate, 193 Ga. 256 (1941). Compare text at note 88 above.

106. Livingston v. Livingston, 210 Ga. 607 (1954); Oglesby v. Volunteer State Life Ins. Co., 195 Ga. 65 (1942). "Even if [the grantor and grantee] had agreed upon a dividing line and understood perfectly how the same should run, this would not suffice; for a deed must itself contain descriptive words with respect to its subject-matter, such as will enable a third person to apply the same to the locus in quo without resorting to any secret and undisclosed intention on the part of the parties thereto." Huntress v. Portwood, 116 Ga. 351 (1902).

owner then executed a deed conveying one of the tracts. The deed described this tract by giving definite boundaries on the east, south, and west and described the north boundary as "lands of" the grantor. The conveyance was held sufficient in that extraneous evidence could be introduced to show that the line had been actually surveyed and marked on the ground prior to the conveyance.[107]

A description calling for adjoiners who had never owned any land in the area is void,[108] as is a description naming adjoiners who cannot be identified.[109] A tax levy that described the land as bounded on the west "by the Central railroad" was held void because, among other reasons, there were two branches of the railroad in the area and the levy did not identify which branch was intended.[110]

Courses and Distances

Courses such as "southerly," "easterly," and even "northwest";[111] distances designated as "±," "about," or as "more or less";[112] and distances designated merely by numbers without disclosing the unit of measure[113] will void a conveyance if no further descriptive calls are given to indicate the land to be conveyed. A description that located a four-sided lot at a designated corner of two named streets, gave the lengths of the respective four sides, but did not disclose the street on which any of the sides abutted or give other data by which the precise land could be located was ruled too vague.[114] A deed that attempted to convey a twenty-foot strip of land along a road for widening purposes was held void because it failed to state on which side of the road the twenty-foot strip was located and the grantor owned land on both sides of the road.[115]

107. Calhoun v. Ryals, 159 Ga. 35 (1924) (one dissent). See also Prudential Ins. Co. v. Hill, 170 Ga. 600 (1930).

108. Cross v. Nicholson, 211 Ga. 769 (1955).

109. Conyers v. West, 210 Ga. 190 (1953). A sales contract that described the property as "lot of land in W. R. Roberts Property, Perkins, Ga., bordered by R.R. track, State Park, and field belonging to widow" was held void because neither the railroad nor the widow was identified. Carswell v. Roberts, 215 Ga. 283 (1959).

110. Brinson v. Lassiter, 81 Ga. 40 (1888).

111. See pages 414–15.

112. See page 414. Compare Grant v. Fourth Nat'l Bank, 229 Ga. 855 (1972).

113. See page 461.

114. Nettles v. Charles P. Glover Rlty. Co., 141 Ga. 126 (1913).

115. City of Atlanta v. Atlanta Trailer City, 213 Ga. 825 (1958).

Area

Descriptions such as the following are deemed insufficient: "424 acres in Tattnall county";[116] "Two hundred and sixty-one acres of land off of lots numbers five, twenty-seven and twenty-eight, in the ninth district of Randolph county";[117] "2 acres of land, of lot No. 1101 (eleven hundred one), on the east side of Orange and Roswell Road, near the center of said lot";[118] and "fifty acres of land where dwelling house now stands."[119]

Plats

A reference to a plat not drawn to scale that gives indefinite courses, distances, and boundaries is insufficient to save an already defective description.[120] A reference to a plat showing several lots or parcels is insufficient when the deed, by its terms, attempts to convey only one of the parcels but does not specify which parcel is intended.[121] If a conveyance refers to an attached plat for a description of the premises and there is no plat attached, the conveyance is void.[122]

Point of Beginning

A general description by metes and bounds is fatally defective if it has no definite starting point and if the property is not otherwise described so that it can be identified.[123] Beginning points such as "a stake on the west side of the road,"[124] "beginning on the North line . . . at an agreed point,"[125] "an iron pin located on the easterly side of Church Street,"[126]

116. Tippins v. Phillips, 123 Ga. 415 (1905).

117. Atkins v. Paul, 67 Ga. 97 (1881). See also Nolan v. Cook, 231 Ga. 737 (1974); Chattahoochee Fertilizer Co. v. Quinn, 169 Ga. 801 (1929); Allen v. Smith, 169 Ga. 395 (1929); Clayton v. Newberry, 138 Ga. 735 (1912); Estes v. Winn, 136 Ga. 344 (1911); and Clarke Bros. v. Stowe, 132 Ga. 621 (1909).

118. Bruce v. Strickland, 201 Ga. 526 (1946).

119. Lee v. English Co., 107 Ga. 152 (1899).

120. See page 461.

121. Delfosse v. Coleman, 211 Ga. 888 (1955); King v. Sears, 91 Ga. 577 (1893).

122. Mayor etc. of Athens v. Gregory, 231 Ga. 710 (1974).

123. Floyd v. Carswell, 211 Ga. 36 (1954).

124. Hedden v. Hilton, 236 Ga. 641 (1976); Dodd v. Madaris, 206 Ga. 497 (1950).

125. Swindle v. Curry, 218 Ga. 552 (1962).

126. Martin v. Sunset Hill Mem. Gardens, 212 Ga. 159 (1956).

and "starting at an oak tree at a point above Ebow Landing on Dunbars Creek"[127] have all been held too vague. A description that read "beginning at a stake or point in the middle line of present Washington-Lexington Highway" was held void because "middle line of . . . Highway" does not indicate whether the survey started in the middle of the paved surface or the middle of the right-of-way.[128] A beginning point described as seventy-five feet "more or less" east of an intersection, without more, is insufficient.[129]

Survey "to be made"

Descriptions such as "the north 100 acres, more or less, of the said Land Lot and which 100 acres will be more fully described in a plat of survey which is to be made";[130] "ten acres of land which is hereafter to be surveyed, leaving said schoolhouse in the center";[131] "fifty acres of land to include the home place . . . lines to be agreed upon . . . as soon as survey and plat can be made";[132] and "Residential Dwelling occupied by seller together with a tract of property not exceeding 10 acres selected by seller"[133] are insufficient.

127. Gould v. Gould, 194 Ga. 132 (1942).

128. Callaway v. Armour, 208 Ga. 136 (1951).

129. Malone v. Klaer, 203 Ga. 291 (1948).

130. Brown v. Mitchell, 225 Ga. 115 (1969). See also McCumbers v. Trans-Columbia, Inc., 172 Ga. App. 275 (1984).

131. Laurens Co. Bd. of Education v. Stanley, 187 Ga. 389 (1938). Compare Kauka Farms, Inc. v. Scott, 256 Ga. 642 (1987).

132. Wardlaw v. Wardlaw, 182 Ga. 209 (1935).

133. Plantation Land Co. v. Bradshaw, 232 Ga. 435 (1974). See also McMichael Rlty. & Ins. Agency v. Tysinger, 155 Ga. App. 131 (1980).

Chapter 17. Riparian and Littoral Boundaries

An owner of land bounded by a watercourse or lake, or through which a stream flows, is known as a riparian owner.[1] Those whose lands border on the ocean are called littoral owners.[2] Bodies of water provide economic advantages so different from those provided by dry land that a distinctive body of law has evolved with regard to their use. The privileges and rights conferred upon riparian and littoral owners may include the right to construct levees, dams, docks, and mills; the right to the bed of the stream, lake, or sea; the right to land formed by accretion or reliction; the right of reclamation; the right of navigation and boomage; the right to take fish, minerals, ice, and stranded or floating objects; the right to pollute; and the right to appropriate water for power, irrigation, and water supply. To the surveyor the laws governing riparian and littoral rights often play a primary role in determining boundary lines between adjacent properties.

Although a tremendous body of law exists dealing with the general subject of riparian and littoral rights, there is not a great deal of Georgia statutory or case law concerning the more specific topic of riparian and littoral boundaries. This is probably because the state has few large seaports and few large natural inland lakes. Furthermore, its lands along rivers, being primarily swampland, have generally been considered of relatively low value. Because of expanding development of waterfront property and the growing environmental concerns of the public, however, litigation over these matters should become increasingly prevalent in the future.

1. Rome Ry. & Light Co. v. Loeb, 141 Ga. 202 (1913); Stoner v. Patten, 132 Ga. 178 (1908).

2. Some definitions classify land bordered by a lake as "littoral" rather than as "riparian." See 25A Words and Phrases "Littoral" and 37A Words and Phrases "Riparian Owner."

Rivers and Streams

Before the Georgia Code of 1863 was adopted, the Georgia Supreme Court (like most American jurisdictions at that time) followed the strict, literal interpretation of the British common law classification of rivers. This doctrine, in a suit involving a portion of the Chattahoochee River in Randolph (later Quitman) County, was stated by Justice Lumpkin as follows:

> Rivers are of three kinds: 1st. Such as are wholly and absolutely private property. 2d. Such as are private property, subject to the servitude of the public interest, by a passage upon them. The distinguishing test between these two is, whether they are susceptible or not of use for a common passage. 3d. Rivers where the tide ebbs and flows, which are called arms of the sea. . . . It is not pretended that the Chattahoochee is at this point a navigable river, where the tide ebbs and flows; such is notoriously not the fact; it belongs to the second class or division in the foregoing enumeration. Now, it is well settled in England, and the doctrine is pretty uniform in this country, that the proprietor of the land on the margin, owns the bed over which the river passes; and though it be nominally and in terms, bounded on the margin, it extends, by construction of law, to the middle of the stream.[3]

Thus all rivers above the point of ebb and flow of the tides were classified as nonnavigable, and property rights in such places, subject to easement rights of the public where the river was capable of being used as a common passageway, extended to the center or thread of the river. The purported reason that matters of tide and navigability were so tightly interwoven under British law was that, because of the limited area of the country, very few streams in the British Isles are, in fact, navigable until tidewater is reached.[4]

The Code of 1863 fundamentally altered the definition of a navigable river. It defined a navigable river as one capable of bearing upon its bosom, either for the whole or part of the year, boats loaded with freight in regular course of trade. It added that the mere rafting of timber or transporting of wood on small boats did not make a stream navigable.[5] The code went on to state that "the right of the owner of lands adja-

3. Young v. Harrison, 6 Ga. 130 (1849). See also Jones v. Water Lot Co., 18 Ga. 539 (1855).

4. 1970 Op. Atty. Gen. 279; 45 C.J. Navigable Waters §§ 3, 4, nn. 14, 22; 65 C.J.S. Navigable Waters § 4 n. 27.

5. Ga. Code of 1863 § 2208.

cent to navigable streams, extends to low water mark in the bed of the stream."[6] This latter provision was presumably an attempt to limit private ownership of land to the low-water mark and leave title to the bed of the stream itself in the state. The same definition and rule are in the present code.[7]

In a 1930 decision, also involving the Chattahoochee River, the Supreme Court held that the 1863 code provision fixing the right of landowners adjacent to navigable streams at the low-water mark would not be construed to apply to lands granted by the state before the adoption of that code.[8] The court, in effect, ruled that the common law classification of rivers would remain in force, at least as far as property rights are concerned, for all lands granted by the state before 1863. Though the question was not specifically raised in this case, there are doubts as to whether the 1863 code provision would stand up to the state and federal constitutional prohibitions against the taking of private property without payment of just and adequate compensation for any land granted before 1863.[9]

Thus the base fee of most beds of navigable,[10] nontidal rivers, subject

6. Id. § 2209.

7. O.C.G.A. § 44-8-5.

8. Fla. Gravel Co. v. Capital City Sand & Gravel Co., 170 Ga. 855 (1930). See also Parker v. Durham, 258 Ga. 140 (1988). The Code of 1863 took effect on January 1, 1863. 1861 Ga. Laws 28.

9. Ga. Const. of 1983, art. 1, § 1, paragraphs 1 & 2; U.S. Const. amendments V & XIV. There seems to be some question, however, as to whether the Code of 1863 even attempted to change the common law rule with respect to property boundaries. It will be noticed that the 1863 code provision stated that the *right* (not ownership) of the owner of land adjacent to navigable streams extends to the low-water mark. An obiter dictum by the Supreme Court suggests that, rather than attempting to change the common law rule, the purpose of this section was simply to ensure to riparian proprietors the right to use the river bottoms on their lands for agricultural purposes. (Under common law the riparian proprietor could not use his land in the bed of the river in any manner inconsistent with the public right to easement.) See State of Georgia v. Ashmore, 236 Ga. 401 (1976) at 411–12. Compare this code section with the code section on nonnavigable streams, which states, "The beds of streams not navigable *belong* to the owner of the adjacent land" (emphasis added). Ga. Code of 1863 § 2207; O.C.G.A. § 44-8-2.

As yet there is no precise judicial interpretation of the code provision, though the implication in *Fla. Gravel Co., etc.* (note 8 above) seems to be that it did attempt to limit private ownership to the low-water mark along factually navigable streams. For other similar implications see Parker v. Durham, 258 Ga. 140 (1988); Maddox v. Threatt, 225 Ga. 730 (1960); and Hicks v. Seaboard Coast Line R.R. Co., 123 Ga. App. 95 (1970).

10. From here on, unless otherwise noted, the terms *navigable* and *nonnavigable* shall mean navigable in fact and nonnavigable in fact, rather than navigable and nonnavigable in the old sense with reference to the ebb and flow of tides.

to the public right of passage, is vested in the adjoining owners because most of the state-owned lands had been granted by 1863.[11] The only probable exceptions would be lands under rivers abutting uplands which were granted on or after January 1, 1863, possible grants prior to 1863 in which the state expressly reserved the riverbed, and grants in which the upland owner has somehow become divested of his title to the bed subsequent to the original grant.

Since the adoption of the 1863 code, interest in property rights in navigable rivers has apparently been low in Georgia, as evidenced by the lack of judicial construction in the area. This is probably because the use of rivers as highways of commerce and as sources of power by individual proprietors has declined.

The thread of a stream is the line midway between the opposite shore lines, when the water is in its natural and ordinary stage, at medium height, and neither swollen by freshets nor shrunken by droughts.[12] Along the Chattahoochee River, where it marks the boundary between Georgia and Alabama, property rights extend not to the centerline but to the west bank of the river, which is the boundary between Georgia and Alabama.[13] Where the river reaches the Florida line (the thirty-first

11. In reference to the Ocmulgee River see Shannon, *It's Not Easy to Build Your Own Bridge*, Atlanta Journal & Constitution Magazine, Jan. 18, 1976, at 9.

12. 12 Am. Jur. 2d Boundaries § 28; 11 C.J.S. Boundaries § 33.

13. Fla. Gravel Co. v. Capital City Sand & Gravel Co., 170 Ga. 855 (1930); Young v. Harrison, 6 Ga. 130 (1849). By the Articles of Agreement and Cession of 1802 Georgia ceded to the federal government its lands "west of a line beginning on the western bank of the Chattahoochee river, where the same crosses the boundary line between the United States and Spain, running thence up said river Chattahoochee, and along the western bank thereof, to the great bend thereof, and next above the place where a certain creek or river called Uchee . . . empties into the said Chattahoochee river. . . ." See page 114. The rule that grantees from the state along the river take to the west bank thereof is apparently based on the principle that, when a grantor owns the entire bed of a river but the land on one side only, a conveyance by him of the land will pass title to the whole riverbed unless there are provisions to the contrary. The presumption is that it is unlikely he would desire to retain a portion of the riverbed which would be of no apparent use to him. Compare page 440. Some state grants, however, expressly conveyed to the west bank of the river. In 1840 the Georgia General Assembly passed an act empowering the mayor and council of Columbus to sell certain "water lots" along the waterfront in that city. The act described the western boundary of the lots as "high water mark on the Western bank of the Chattahoochee River." 1840 Ga. Laws 187. See also 1845 Ga. Laws 123; 1897 Ga. Laws 586; 1910 Ga. Laws 484; and Muscogee Man. Co. v. Eagle & Phenix Mills, 126 Ga. 210 (1906).

Sometime before 1842 Stephen M. Ingersoll, a citizen of Russell County, Alabama, received patents to certain lands in that county from the U.S. government, and on the land he built a water mill. The mill was located on the west bank of the Chattahoochee River between high-water and low-water marks and opposite Columbus. Ingersoll believed his

parallel), however, the boundary becomes the center of the river, that being the boundary between Georgia and Florida.[14] Along the Savannah, Tugaloo, and Chattooga rivers, the line is midway between the banks with the water at its ordinary stage; where there are islands, the line is midway between the island bank and the South Carolina shore, with the water at its ordinary stage.[15] A change in the location of the channel of the Savannah River brought about by its improvement for navigation by the United States government does not work a change in boundaries.[16]

There may be a few cases in which property rights would stop at the low-water mark along navigable streams, under the 1863 code provi-

---

mill was in Alabama (property rights extend to the low-water mark along navigable streams in the state of Alabama).

In 1845 John H. Howard, relying on grants from the city of Columbus under the 1840 and 1845 acts, built a mill on the west bank of the Chattahoochee between high-water and low-water marks and three hundred yards below Ingersoll's mill. Howard also built a dam across the river which backed water up and overflowed Ingersoll's mill. Ingersoll sued Howard for damages. The legal question for determination was whether Ingersoll's mill was in Alabama or Georgia, since an action for damages to land has to be brought in the state where the land lies. Ingersoll brought suit in Alabama, and the circuit court ruled that the boundary between Georgia and Alabama along the Chattahoochee was ordinary low-water mark on the west bank. Ingersoll was awarded $4,000 in damages, and the decision was affirmed by the Alabama Supreme Court. Howard v. Ingersoll, 17 Ala. 780 (1850). Howard then took the case to the U.S. Supreme Court, which reversed the Alabama decision. Howard v. Ingersoll, 54 U.S. (13 How.) 381, 14 L. Ed. 189 (1851); Howard v. Ingersoll, 23 Ala. 673 (1853). The state of Alabama, charging that the definition of the boundary between Georgia and Alabama as given in the opinion of the U.S. Supreme Court was too vague for either state to fix the boundaries of its jurisdiction, then instigated an action of its own against Georgia in the U.S. Supreme Court to have the line defined with precision. The court this time declared that Georgia owns the entire bed and banks of the Chattahoochee River to the high-water mark on the western side where the banks are steep and well-defined, and that where the western banks are low and flat Georgia owns to the line traced by the western edge of the river when it is in its average and mean stage. Alabama v. Georgia, 64 U.S. (23 How.) 505, 16 L. Ed. 556 (1859).

By statute Georgia purportedly reserves to itself title to the west bank of the Chattahoochee River. O.C.G.A. § 50-16-1. This provision, of course, will not apply where the state has expressly granted away such lands, as in the case of the lands along the waterfront in Columbus. In fact, in view of the rulings made by Georgia courts, it seems doubtful that the provision would have any application at all because it did not go into effect until the adoption of the 1863 code.

14. Fla. Gravel Co. v. Capital City Sand & Gravel Co., 170 Ga. 855 (1930). O.C.G.A. § 50-2-1 defines the boundary between Georgia and Florida along the Chattahoochee as the west bank of the river, but this is obviously an oversight.

15. State of Georgia v. State of South Carolina, 257 U.S. 516; 66 L. Ed. 347 (1922).

16. James v. The State, 10 Ga. App. 13 (1911).

sions.[17] The problem then would be to determine whether the river, or portion of it under consideration, is in fact navigable under the present code definition. In many cases the navigability of a particular stream cannot be known with certainty until there has been a full judicial inquiry into all the evidence.[18]

As stated above, the code defines a navigable river as one capable of bearing upon its bosom boats loaded with freight in regular course of trade for at least part of the year. Therefore, the stream does not have to be used for commercial navigation if its capability for such use can be shown;[19] nor does it have to have a public terminus at each end ("a terminus ad quem and a terminus a quo") as required in some jurisdictions.[20]

In an 1884 case the Supreme Court held that a stream "running into the Savannah river, from seventy-five to two-hundred feet in width and fourteen feet in depth . . . and in which the tide rises and falls three and a half feet, and which is open so as to allow free passage to all water craft running thereon, is a navigable stream."[21]

There are several state and federal court decisions regarding the navigability of particular rivers (though many of these pronouncements are more or less obiter). The publication *Water Resources Development by the U.S. Army Corps of Engineers in Georgia,* published intermittently by the government, tells what rivers the Corps of Engineers considers navigable.[22] Appendix B summarizes these findings.

17. Low-water mark is the point to which the water recedes at its lowest ordinary stage and not that of an unusually dry season or period of drought. 9 C.J. Boundaries § 75; 11 C.J.S. Boundaries § 32.

18. Ga. Real Est. Law § 6-8 (3d ed. 1986).

19. United States v. Lewis, 355 F. Supp. 1132 (S.D. Ga. 1973); Ga. Real Est. Law § 6-7 (3d ed. 1986); 78 Am. Jur. 2d Waters § 62; 65 C.J.S. Navigable Waters § 6.

20. Rauers v. Persons, 144 Ga. 23 (1915). See 65 C.J.S. Navigable Waters § 5(4).

21. Charleston & Sav. Ry. v. Johnson, 73 Ga. 306 (1884). Although the stream under consideration in this case (Knoxboro Creek in Chatham and Effingham counties) is tidal, the court classified it as a navigable stream rather than as an arm of the sea, apparently because it flows into the Savannah River and not directly into the ocean. Also, the question of boundaries was not involved in this case. For comment see Johnson v. The State, 114 Ga. 790 (1901).

22. The U.S. Supreme Court originally followed the common law classification of rivers as to navigability but eventually adopted the "navigable in fact" test and laid down a definition very similar to the one in the Georgia code. It stated, "Those rivers must be regarded as public navigable rivers in law which are navigable in fact. And they are navigable in fact when they are used, or are susceptible of being used, in their ordinary condition, as highways for commerce, over which trade and travel are or may be conducted in the customary modes of public trade and travel on water." The Daniel Ball v. United States, 77 U.S. (10

If there is any question as to whether the bed of a particular stream is vested in the upland owner or in the state, perhaps the best thing for the surveyor to do when called on to survey an abutting tract is to show two lines on the plat—one at the low-water mark and the other at the centerline of the stream—and to note that there is doubt as to which line property rights extend. The matter can then be left to the client and his attorney.[23]

In the foregoing discussion very little has been said about nonnavigable rivers and streams. Under common law all streams above the point of ebb and flow of tides were classified as nonnavigable. The present view is that a nonnavigable stream is a watercourse above the ebb and flow of tides that flows between more or less well-defined banks during at least part of the year and does not meet the requirements of navigability as stated in the code.[24] Thus a watercourse may be wholly dry at times without losing the character of a stream,[25] and a swampy or marshy area may be held to constitute a stream if it has a slight flow or run to it.[26]

The code provides that the beds of nonnavigable streams belong to the abutting landowners to the center or thread of the stream,[27] but "inde-

---

Wall.) 557, 19 L. Ed. 999 (1871). The definition was broadened, however, in a 1940 decision in which the court held that "the navigability of a stream is to be determined on the basis, not only of its natural condition, but also of its possible availability for navigation after the making of reasonable improvements; and it is not necessary that such improvements be actually completed or even authorized." United States v. Appalachian Elec. Pwr. Co., 311 U.S. 377, 85 L. Ed. 243 (1940).

It should be emphasized that the Georgia courts are in no way bound to follow the Corps of Engineers' classification of streams as to navigability, though it may be of some evidentiary value. At least one authority has seen fit to make special note of it. See Ga. Real Est. Law § 6-8 (3d ed. 1986). Nor are the Georgia courts bound by federal decisions about property rights in rivers because state law, rather than federal law, governs the acquisition, ownership, and disposition of real property. Ga. Real Est. Law § 1-4 (3d ed. 1986); Oregon v. Corvallis Sand & Gravel Co., 429 U.S. 363, 50 L. Ed. 2d 550 (1977); Archer v. Greenville Sand & Gravel Co., 233 U.S. 60, 58 L. Ed. 850 (1914); Hardin v. Jordan, 140 U.S. 371, 35 L. Ed. 428 (1890); Barney v. City of Keokuk, 94 U.S. 324, 24 L. Ed. 224 (1876).

23. For general discussions on tests of navigability in other jurisdictions see 78 Am. Jur. 2d Waters; and 65 C.J.S. Navigable Waters.

24. Pelham Phosphate Co. v. Daniels, 21 Ga. App. 547 (1917); Johnson v. The State, 114 Ga. 790 (1901).

25. O'Connell v. East Tenn. Va. & Ga. Ry. Co., 87 Ga. 246 (1891); Parker v. Adamson, 109 Ga. App. 172 (1964); Pelham Phosphate Co. v. Daniels, 21 Ga. App. 547 (1917).

26. Outlaw v. Outlaw, 225 Ga. 100 (1969).

27. O.C.G.A. § 44-8-2.

pendently of the statute, 'such ever has been the law in this state.'"[28]

It should be pointed out that all of the above discussion on property rights in rivers and streams is based on mere presumptions of law. The general rule is that conveyances of land bounded by rivers and streams, where the expressions designating the water line are general, confer proprietorship to the thread (or low-water mark) of the stream.[29] If the conveyance, however, specifically states that the line will be somewhere else, such a stipulation will control; but the intent has to be clearly manifested.[30] Where a grantor conveyed all of a certain tract of land lying "north" of a named nonnavigable stream, the court construed the description as carrying the property line to the center of the stream, not to the north edge of the bank.[31]

In 1913 the state of Georgia brought suit against a power company to recover land located in a deep gorge through which the Tallulah River flows. The original grants referred to plats that showed lines extending only to the edge of the gorge but called for the river as the boundary. The state contended that the grants stopped at the edge of the gorge and that the land in the gorge had never been surveyed and was still part of the public domain. The court ruled that the boundary extended to the center of the stream.[32]

If, by natural forces, the channel of a stream changes gradually, then the property line follows the changes; but if the stream suddenly takes on a new channel because of natural or man-made causes, and the old line is still capable of identification, the latter remains the boundary.[33] Accretions accrue to the owner of the bed of the stream.[34]

Islands in a river or stream generally belong to whoever owns the bed of the stream.[35] Therefore, the ownership of an island will usually de-

28. State of Georgia v. Georgia Ry. & Pwr. Co., 141 Ga. 153 (1913). See also Spivey v. Barwick, 157 Ga. 853 (1924); Thompson v. Tennyson, 148 Ga. 701 (1919); and Hendrick v. Cook, 4 Ga. 241 (1848).

29. 12 Am. Jur. 2d Boundaries § 20; 65 C.J.S. Navigable Waters § 120; 93 C.J.S. Waters § 84.

30. Westmoreland v. Beutell, 153 Ga. App. 558 (1980).

31. Outlaw v. Outlaw, 225 Ga. 100 (1969).

32. State of Georgia v. Georgia Ry. & Pwr. Co., 141 Ga. 153 (1913).

33. Johnson v. Hume, 163 Ga. 867 (1927); James v. State, 10 Ga. App. 13 (1911); O.C.G.A. § 44-8-2.

34. Johnson v. Watson, 157 Ga. 349 (1923).

35. 78 Am. Jur. 2d Waters § 437; 65 C.J.S. Navigable Waters § 115; 1 R. Patton et al., Patton on Land Titles § 142 (2d ed. 1957).

pend on which side of the island the thread of the stream lies.[36] If an island straddles the thread, the owner on each side of the stream owns the part of the island that lies on his side of the thread.[37] All islands in the Savannah, Tugaloo, and Chattooga Rivers[38] and all islands in the Chattahoochee River along the Alabama line[39] are within the boundaries of the state of Georgia and therefore would be included in the original grants by the state to lands abutting these rivers. It is generally held that no land can be called an island unless it is surrounded by water at low-water mark. Thus it is sometimes difficult to classify an area as an island or a peninsula.[40]

Lakes and Ponds

A lake is a large inland body of standing water in a depression of land or expanded part of a river.[41] A pond is an inland body of water, with or without an outlet, and smaller than a lake but larger than a puddle.[42] As seen from these definitions, there is no clear distinction between a lake and a pond, and the two words are often used interchangeably. Lakes and ponds are distinguished from marshes in that they are definite bodies of standing water, rather than areas of soft, low-lying, waterlogged land that may or may not have water standing in places on the surface.[43]

Impounding the water of a stream will cause a spread of the upper stream into the adjoining land to form a lake or pond. Such an act will not in itself affect title to the submerged land involved, and it remains vested as before.[44] Developing a subdivision around a lake and showing the lake

36. Johnson v. Watson, 157 Ga. 350 (1923); Stanford v. Mongin, 32 Ga. 625 (1861); Stanford v. Mangin, 30 Ga. 355 (1860).

37. 3 J. Kent, Commentaries on American Law 428 (12th ed. 1896); 1 R. Patton et al., Patton on Land Titles § 142 (2d ed. 1957); 6 G. Thompson, Commentaries on the Modern Law of Real Property § 3075 (1962).

38. State of Georgia v. State of South Carolina, 257 U.S. 516, 66 L. Ed. 347 (1922); Georgia Ry. & Pwr. Co. v. Wright, 146 Ga. 29 (1916).

39. Fla. Gravel Co. v. Capital City Sand & Gravel Co., 170 Ga. 855 (1930); Cook v. Winter, 68 Ga. 259 (1881).

40. Ga. Real Est. Law § 6-11 (3d ed. 1986); 78 Am. Jur. 2d Waters § 436.

41. Ga. Real Est. Law § 6-2 (3d ed. 1986).

42. Id.

43. R. Kates, Georgia Water Law 8 n. 23 (1969).

44. Ga. Real Est. Law § 6-16 (3d ed. 1986); 12 Am. Jur. 2d Boundaries § 18.

on the subdivision plat will generally give the subsequent lot owners in the subdivision an easement in the lake.[45]

A deed that described the property being conveyed as all of the tract of land in the northeast corner of a certain land lot "covered by what is known as the Bradford Mill Pond" was held to convey all the land covered by the pond even though the pond extended down into the south portion of the land lot.[46]

If land is described as "bounded" by an artificial lake created by the damming of a stream, the prevailing view in other jurisdictions is that the grantee takes to the center or thread of the original stream.[47] The basis of this presumption is the assumption that the impoundment is merely temporary in nature, existing only so long as the dam is maintained.[48] There seems to be no reason why this general rule would not also apply in Georgia. In one of the very few Georgia cases dealing with ownership of lake beds, however, the Georgia Supreme Court held the contrary view because of the peculiar circumstances involved.[49] In that case the plaintiff owned all the land adjacent to and covered by a certain artificial pond known as "McCall's mill-pond" except for a small tract owned by the defendant, whose deed described his tract as bounded on the south by "McCall's mill-pond." A dispute arose over this southern boundary.

In construing this description, the court rejected the centerline rule generally applicable to rivers and artificial lakes by deciding that the pond had become a permanent body of water on the basis of its existence for more than forty years under a contract with a railroad to keep the dam in good repair—"thus insuring its stability and permanency."

The court also touched on the "pie" method of apportionment[50] and rejected this rule because of the inequities involved. Quoting from an Indiana case the court stated, "It would be unfair and unjust to allow a party to claim and hold against his grantor the bed of a lake containing thousands of acres, solely on the ground that he had bought and paid for the small surrounding fractional tracts—the mere rim."

45. Higgins v. Odom, 246 Ga. 309 (1980); Sawyer v. Bush, 244 Ga. 785 (1979).

46. Prescott v. Herring, 212 Ga. 571 (1956).

47. R. Kates, Georgia Water Law 185–86 (1969); 12 Am. Jur. 2d Boundaries § 17; 11 C.J.S. Boundaries § 27; 1 R. Patton et al., Patton on Land Titles § 140 (2d ed. 1957); 6 G. Thompson, Commentaries on the Modern Law of Real Property § 3081 (1962).

48. R. Kates, Georgia Water Law 185 (1969).

49. Boardman v. Scott, 102 Ga. 404 (1897).

50. That is, projecting lines radially from the center of the pond to where they intersect the property lines on the shore.

In deciding that the edge of the pond as of the date of the conveyance would control as the boundary, the court reasoned that in this case the grantor used the lake to operate his mill and it was unlikely that he would intend to part with a portion of the bed. In addition, the designation of "McCall's mill-pond" as the boundary was deemed analogous to a call for an adjoiner, thus indicating that the edge of the lake was intended. It was decided that the low-water mark, rather than any other level, would control because the use of the water was of great importance to the adjoining tracts. Unless rebutted by some proof, the parties were presumed to have intended that the tract should be bounded by the pond during all seasons of the year.[51]

In another case[52] a deed conveyed all the land covered by a certain pond and contained the provision "that the water in said pond shall at all times be maintained at a height not exceeding 7½ feet from the mill sheeting of bottom floors in the millhouse." Though the deed did not expressly say so, the court held that the obvious intent of this provision was to fix the boundaries of the tract being conveyed.

Tidewaters and the Ocean

Under common law, in the absence of a special title by grant or prescription, the boundary of landowners abutting on the sea or upon any estuary, tidal stream, or arm of the sea, where there was a regular rise and fall of the tide, extended only to the high-water mark.[53] The soil between high-water mark and low-water mark was the property of the crown, and upon gaining independence, the states succeeded to this ownership.[54]

51. For a criticism of this decision see 3 H. Farnham, The Law of Waters and Water Rights 2534 (1904) and for rebuttal see R. Kates, Georgia Water Law 185–86 (1969). Similar rulings have been made in other jurisdictions. See 12 Am. Jur. 2d Boundaries § 17; 9 C.J. Boundaries § 72; and 6 G. Thompson, Commentaries on the Modern Law of Real Property § 3081 (1962). There are a large number of decisions in other jurisdictions regarding the apportionment of pond and lake beds. Because of the great difficulty, if not utter impossibility, of making such apportionments equitable, very few general rules can be laid down. It is not advisable for the surveyor to attempt to make such apportionments without the prior consent of all parties involved. For a discussion on the subject of apportioning pond and lake beds see 3 Am. Law of Prop. § 15.31. For annotations of cases on the subject see 74 A.L.R. 618 (1931) and 14 A.L.R. 4th 1028 (1982).

52. Prescott v. Herring, 212 Ga. 571 (1956).

53. Johnson v. The State, 114 Ga. 790 (1901); Jones v. Oemler, 110 Ga. 202 (1899).

54. 65 C.J.S. Navigable Waters § 94. American and English cases dealing with ownership rights in navigable and nonnavigable waters often allude to the idea that the English crown,

In 1901 criminal charges were brought against certain persons for taking oysters from allegedly privately owned oyster beds which had been planted on the beach by the owner of the abutting uplands. The case was dismissed on the grounds that the upland owner had no title to the land over which the tide flowed.[55] The court ruled that the 1863 code provision fixing rights of landowners along navigable streams at the low-water mark did not apply to tidal waters.[56]

This decision greatly disturbed the oyster industry because oyster beds are expensive to plant and care for. Under this ruling, oystermen had no private rights to oyster beds they may have located on public land. The legislature remedied the situation in 1902 by passing an act providing, among other things, that "the boundaries and rights of owners of land adjacent to or covered in whole or in part by navigable tide-waters . . . shall extend to low water mark in the bed of the water" and that the title to the beds of nonnavigable tidewaters "shall vest in the owner of the adjacent land," to the main thread or channel of the water.[57] The act imposed certain restrictions on the landowners and gave certain rights to the public in such areas.

---

from the earliest times, had prima facie held title to all beds of navigable (tidal) waters to the high-water mark. It is of historical interest to note that the facts do not seem to support this conclusion. A search into early English law indicates that the idea was unheard-of until the time of Queen Elizabeth I. In the beginning no distinction was made between dry and submerged lands, at least as far as ownership rights were concerned. The king could part with any of the lands of the realm at will. It is said that by the end of the reign of King John (1199–1216) the crown had, by grants to its subjects, parted with title to almost all the seacoast.

The theory of the crown's presumptive ownership of lands under tidal waters seems to have arisen during the reign of Queen Elizabeth I and then gradually to have developed over the next two hundred years until it came to be commonly accepted that the king was presumed to own all tidelands and that he had rarely parted with title to any of them. For an interesting discussion of this see S. Moore, A History of the Foreshore and the Law Relating Thereto passim (1888); and MacGrady, *The Navigability Concept in the Civil and Common Law: Historical Development, Current Importance, and Some Doctrines That Don't Hold Water,* 3 Florida State University Law Review 511 (1975).

55. Johnson v. The State, 114 Ga. 790 (1901).

56. Some had contended that the 1863 code provisions defining navigable streams and adjacent owners' rights therein (text at notes 5–6 above) had conferred some sort of title to the abutting owners in the beaches down to low-tide mark. See, for example, Shively v. Bowlby, 152 U.S. 1, 25; 38 L. Ed. 331, 341 (1894): "In Georgia . . . the rules of the common law would seem to be in force as to tide waters, except as affected by statutes of the state providing that 'the right of the owner of lands adjacent to navigable streams extends to low water mark in the bed of the stream.'"

57. 1902 Ga. Laws 108 (O.C.G.A. §§ 44-8-6 to 44-8-8).

Because of the wording of this act and the circumstances under which it was passed, there was controversy for many years over its exact intent. Did the act merely give oystermen the exclusive right to their oyster beds or was it a grant to the abutting owners of all the marshes and tidelands of the state? The provision in the Constitution of 1945 stating, "The act of the General Assembly approved December 16, 1902, which extends the title of ownership of lands abutting on tidal water to low water mark is hereby ratified and confirmed"[58] did not clear the matter up. The attorney general took the position that the original act was unconstitutional and therefore this attempted confirmation was void.[59]

The Supreme Court touched on the question in 1904[60] and 1974[61] and in two other decisions,[62] the court seemed to imply that the act had conferred title to the upland owners along the coast. No definite ruling on the matter was made in any of the cases.

In a 1976 landmark decision[63] the court, with two dissents, ruled that title to lands below the high-water mark is vested in the state; that the 1902 act only conferred certain rights (not title) to the abutting landowners; and that the provision in the Constitution of 1945 only confirmed the 1902 act and did not confer any greater rights to the adjoining owners.

So it will be seen that private ownership along tidewaters and the ocean does and always has, presumptively, extended only to the ordinary high-water mark.[64] Ordinary high-water mark is the line marked by the periodical flow of the tide, excluding the advance of the water above that line, occasioned by winds and storms or other unusual conditions.[65]

The possibility of specific grants to individuals of the beds of tide-

58. Ga. Const. of 1945 art. I, § VI, para. 1; Ga. Const. of 1983, art. I, § III, para. 3.

59. 1970 Op. Atty. Gen. 279.

60. Prey v. Oemler, 120 Ga. 223 (1904).

61. State of Georgia v. Bruce, 231 Ga. 783 (1974).

62. Rauers v. Persons, 144 Ga. 23 (1915); Aiken v. Wallace, 134 Ga. 873 (1910), at 878–79.

63. State of Georgia v. Ashmore, 236 Ga. 401 (1976).

64. Technically *State of Georgia v. Ashmore* only affirmed the state's title to beds of navigable tidewaters, but most observers have treated it as effectively dealing with nonnavigable tidewaters also. The practical problem is that a large portion of the marshes and beaches in Georgia are claimed by private individuals, some of whom have built costly improvements on them. How this dilemma will be solved poses an interesting problem for the future.

65. 12 Am. Jur. 2d Boundaries § 13; 3 Am. Law of Prop. § 12.27; 11 C.J.S. Boundaries § 32; 65 C.J.S. Navigable Waters § 88; 19A Words and Phrases "High-water Mark." See 5 Ga. L. Rev. 578–82.

waters is not absolutely precluded, however. Colonial grants of wharf lots in Savannah, Brunswick, and other coastal towns commonly conveyed to low-water mark or to some designated distance below high-water mark, by express terms. An 1802 act of the General Assembly, providing for the laying out of certain lots in the town of Frederica on St. Simons Island, specified that the lots along the beach were to extend to low-tide mark.[66] There are a number of other original grants of tidelands and beaches to private proprietors.[67] However, a grant of an island by King George II, which bounded the island on the "northwesterly by Large Marshes," was construed not to include the marshes themselves.[68]

Gradual accretions of land along the oceans accrue to the adjacent landowner.[69]

When water encroaches so far on land that a tract that was formerly littoral is completely submerged or washed away and land that was formerly nonlittoral becomes littoral, subsequent accretions belong to the owner of the tract newly made littoral even though the accretions in time extend into or over the area formerly owned by the original littoral owner.[70]

Under common law the right of the littoral owner of access to the ocean was not denied, and the public had the right to bathe, fish, and navigate in such waters, subject to lawful regulation by the state.[71] By special statutory enactment, littoral owners in Georgia are given an exclusive license to plant, cultivate, and harvest oysters, clams, and other shellfish in tidal waters adjacent to their land.[72] The littoral owner, by common law right, can also, with the consent or acquiescence of the state, erect wharves or piers in front of his littoral holdings to facilitate access to

66. Clayton 63.

67. See, for example, Aiken v. Wallace, 134 Ga. 873 (1910); 1870 Ga. Laws 400; 1935 Ga. Laws 408; 1945 Ga. Laws 1231. An examination by the author of several thousand original headright grants in Georgia's six coastal counties indicates that conveyances of tidal marshes by the colonial and state governments were fairly common in the eighteenth and early nineteenth centuries. Such areas were highly prized for rice culture. In addition, access to river frontage was deemed important for transportation purposes because the few existing roads were usually in deplorable condition.

68. Oemler v. Green, 134 Ga. 198 (1910). See 1912 Ga. Laws 1605.

69. Cherry v. Hopkins, 254 Ga. 260 (1985); State of Georgia v. Ashmore, 236 Ga. 401 (1976); O.C.G.A. § 44-8-6.

70. Smith v. Bruce, 241 Ga. 133 (1978).

71. West v. Baumgartner, 124 Ga. App. 318 (1971), cert. to Supreme Court 228 Ga. 671 (1972); 78 Am. Jur. 2d Waters § 266; 6A Am. Law of Prop. § 28.59.

72. State of Georgia v. Ashmore, 236 Ga. 401 (1976).

navigable waters.[73] If a landowner constructs a pier that extends over the ocean beyond the low-water mark, he can enjoin others from using the pier even though the pier lies over publicly owned lands.[74]

Federal Rights in Waters

Until 1947 the states assumed that they held title to the lands under the seas along their coasts.[75] Some states leased oil rights and granted other privileges to individuals in such lands. In three decisions beginning in 1947 the U.S. Supreme Court ruled that the United States possessed paramount rights in, and full dominion and power over, the lands, minerals, and other things underlying the oceans between ordinary low-water mark and the outer edge of the continental shelf.[76] The rulings were construed to apply to all the states, including the thirteen original ones.

These decisions were much criticized. As a result, Congress, in 1953, quitclaimed to the states or to individual owners all rights, title, or interest the United States had in the beds of navigable and tidal waters within three geographical miles of the coastline, or, in the case of the Great Lakes, to the international boundary. The act reserved title to the United States, however, to the bed beyond the three-geographical-mile limit, along with the right to regulate the beds within the three-geographical-mile limit for commerce, navigation, national defense, and so on.[77]

Even though the federal government no longer has title to submerged lands within three geographical miles of the coast, it still has very broad powers over these areas, as well as over navigable streams, through the following specific powers: the commerce clause, the property clause, the general welfare clause, the war power clause, the power of taxation, and

73. 3 Am. Law of Prop. § 15.35; 2 R. Patton et al., Patton on Land Titles § 298 (2d ed. 1957).

74. Coney v. Brunswick & Florida Steamboat Co., 116 Ga. 222 (1902).

75. Silver v. The State, 147 Ga. 162 (1917); 1916 Ga. Laws 29; 45 C.J. Navigable Waters §§ 213, 220; H. Tiffany, The Modern Law of Real Property 452, 648 (abr. ed. 1940).

76. United States v. California, 332 U.S. 19, 91 L. Ed. 1889 (1947); United States v. Louisiana, 339 U.S. 699, 94 L. Ed. 1216 (1950); United States v. Texas, 339 U.S. 707, 94 L. Ed. 1221 (1950).

77. Submerged Lands Act, Pub. L. No. 83-31, 67 Stat. 29 (43 U.S.C.A. §§ 1311 to 1315). Georgia subsequently declared its seaward boundary to be three geographical miles from the mean low-water mark of the foreshore. Dept. of Natural Resources v. Joyner, 241 Ga. 390 (1978); 1969 Ga. Laws 678.

the jurisdiction of the federal judiciary in disputes between states.[78] With the ever-broadening judicial interpretation of the extent of such powers, along with the trend toward the growth of federal powers in general, it may be said that all waters of the state, with the possible theoretical exception of a factually nonnavigable, small stream that is entirely intrastate in its boundaries, are subject to the superior federal power of use.[79]

78. R. Kates, Georgia Water Law 158-83 (1969); Ga. Real Est. Law § 6-3 (3d ed. 1986).

79. R. Kates, Georgia Water Law 158 (1969). So rigid have both federal and state restrictions become regarding the use and development of wetlands that, in many cases, the owner is denied the right of making any practical use of his property.

Chapter 18. Other Topics

Restrictive Covenants

In its broadest context a covenant is an agreement to do or not to do a particular act.[1] A real covenant, or a covenant running with the land, is one so closely connected with land that either the right to enforce it or the liability for its performance passes to later owners of the land.[2] For a covenant to run with the land it must relate to an interest or estate granted therein; its performance or nonperformance must affect the nature, quality, or value of the property or must affect the mode of enjoyment of the property; and there must be privity between the contracting parties.[3] Examples of covenants running with the land include covenants of title in deeds,[4] agreements to build and maintain dams or levees,[5] agreements to furnish water or water power,[6] party wall agreements,[7] and restrictive covenants such as building lines, building restrictions, use restrictions, and billboard and sign restrictions. The present discussion will be limited to restrictive covenants because the broader topic of covenants running with the land is beyond the scope of this book.

In general, the law favors the full, free, and unlimited use of property, resolving doubts and ambiguities in favor thereof.[8] Restrictive covenants, though not favored, will be enforced when reasonable and within the policy of the law. Any claim as to restrictions, however, must be clearly established.[9] Restrictions will not be enlarged or extended by construction, and doubts will be construed in favor of the grantee.[10]

1. 21 C.J.S. Covenants § 1.

2. Copelan v. Acree Oil Co., 249 Ga. 276 (1982); Atlanta Consol. Ry. Co. v. Jackson, 108 Ga. 634 (1899).

3. Copelan v. Acree Oil Co., 249 Ga. 276 (1982); Grant-Jeter Co. v. American Real Est. Co., 159 Ga. 80 (1924).

4. See page 358.

5. 20 Am. Jur. 2d Covenants, Conditions, Etc. § 42.

6. Id.; Atlanta, K. & N. Ry. Co. v. McKinney, 124 Ga. 929 (1905).

7. See next section.

8. Ga. Real Est. Law § 19-184 (3d ed. 1986); 20 Am. Jur. 2d Covenants, Conditions, Etc. § 187; 21 C.J.S. Covenants § 19.

9. Voyles v. Knight, 220 Ga. 305 (1964); David v. Bowen, 191 Ga. 467 (1941).

No precise form or technical words are necessary to create covenants.[11] They may be created by their incorporation into deeds,[12] by an agreement apart from a conveyance,[13] by the selling of lots with reference to a plat showing restrictions,[14] by representations made in sales brochures and signs placed on the property,[15] or by oral statements relied on by a purchaser in making his purchase.[16] When a grantee accepts a deed he is bound by any covenants contained therein, even though the deed is not signed by him.[17] Successors in title to the grantee, claiming through him, take the land subject to such covenants whether or not their immediate deeds make reference to the covenants.[18] Of course, a purchaser without actual or constructive notice of covenants takes the land free therefrom; and this includes covenants set forth in instruments outside the purchaser's chain of title.[19] Where a subdivision with restrictive covenants was laid out and the subdivision plat outlined other adjoining property owned by the grantor but not subdivided, the restrictive covenants were held not to apply to the adjoining lands.[20] On the other hand, the sale of lots by reference to a plat that designated some of the lots as "reserved" did not relieve the lots so designated from restrictive covenants that applied to the subdivision.[21]

A valid restriction that applies to property may be enforced by the parties thereto and by their successors in title.[22] Restrictive covenants for

10. Copelan v. Acree Oil Co., 249 Ga. 276 (1982); Randall v. Atlanta Advertising Service, 159 Ga. 217 (1924).

11. Taylor v. Smith, 221 Ga. 55 (1965).

12. 20 Am. Jur. 2d Covenants, Conditions, Etc. § 169; 26 C.J.S. Deeds § 162.

13. 20 Am. Jur. 2d Covenants, Conditions, Etc. § 170.

14. Kilby v. Sawtell, 203 Ga. 256 (1948).

15. 20 Am. Jur. 2d Covenants, Conditions, Etc. § 175. As a case in point, subdivision lots were sold with no restrictions in the deeds. There existed on the property at the time of the sales, however, an advertising sign that read "This is Morningside, exclusive residential section, with adequate restrictions." Deeming this sufficient to create a restrictive covenant limiting the lots to residential purposes, the court enjoined the erection of a filling station on one of the lots. Raven v. Laurens, 164 Ga. 868 (1927).

16. Westhampton, Inc. v. Kehoe, 227 Ga. 642 (1971).

17. Antill v. Sigman, 240 Ga. 511 (1978); Lawson v. Lewis, 205 Ga. 227 (1949).

18. Ga. Real Est. Law § 19-195 (3d ed. 1986).

19. Id.

20. Kitchens v. Noland, 172 Ga. 684 (1931).

21. Regency Const. Co., Inc. v. Gwinnett County, 227 Ga. 798 (1971); Atlanta Assoc. of Baptist Churches v. Cowan, 186 Ga. 10 (1938) (two dissents).

22. 20 Am. Jur. 2d Covenants, Conditions, Etc. §§ 20, 288–303.

a subdivision may be enforced by the lot owners of the subdivision.[23] If an owner sells a portion of his land and imposes restrictive covenants on the portion sold, it is presumed that the restrictions are imposed for the benefit of the land retained. If the person later sells other portions of the land, owners thereof may enforce the restrictions against the owner of the portion first sold.[24]

Restrictions limiting property to "residential purposes" have been held to be violated if the land is used for a cow pasture,[25] chicken farm,[26] beauty shop,[27] fraternity house,[28] or church[29] but not violated if it is used for apartments,[30] a boardinghouse,[31] kindergarten,[32] or tennis courts.[33] A restriction setting a certain minimum number of square feet and a certain minimum cost for residences is violated by the erection of a garage containing temporary living quarters, the area and cost of which are less than the set minimum.[34] A restriction prohibiting the use of land for "business purposes" is not violated by the erection of a church.[35] The use by a church of a portion of its land for a cemetery does not violate a covenant limiting the property to a "religious purpose."[36] A covenant requiring the grantee to submit his building plans to the grantor for approval as to whether such plans are in "conformity and harmony of external design and general quality with the existing standards of the neighborhood" is not so vague and indefinite as to be void.[37] When the grantee has submitted the plans and they comply with such conditions, the grantor will not be allowed arbitrarily to withhold approval.[38]

23. Phillips v. Blackwell, 164 Ga. 856 (1927).

24. Cawthon v. Anderson, 211 Ga. 77 (1954); Lawson v. Lewis, 205 Ga. 227 (1949).

25. Grove Lakes Sub., Inc. v. Hollingsworth, 218 Ga. 443 (1962). Compare Holbrook v. Davison, 258 Ga. 844 (1989).

26. Kilby v. Sawtell, 203 Ga. 256 (1948).

27. Taylor v. Smith, 221 Ga. 55 (1965).

28. Mu Chapter Bldg. Fund Inc. v. Henry, 204 Ga. 846 (1949).

29. Renfroe v. Alden, 164 Ga. 77 (1927).

30. Lewenstein v. Brown, 200 Ga. 433 (1946); Hamm v. Wilson, 169 Ga. 570 (1929).

31. David v. Bowen, 191 Ga. 467 (1941); John Hancock Mut. Life Ins. Co. v. Davis, 173 Ga. 443 (1931) (one dissent).

32. Shoaf v. Bland, 208 Ga. 709 (1952).

33. Elder v. Watts, 252 Ga. 212 (1984).

34. Barber v. Winter, 208 Ga. 712 (1952).

35. Wiggins v. Young, 206 Ga. 440 (1950).

36. Williams v. Ramey, 201 Ga. 737 (1947).

37. Winslette v. Keeler, 220 Ga. 100 (1964).

38. Id.

Building line restrictions have generally been held to prohibit the encroachment of only the main portion of the building into the restricted area. Therefore, the encroachment by verandas, porches, and steps is not considered a violation.[39] Nonetheless, many deeds now prohibit the encroachment of the "building or any part thereof."[40] A building line specified in a deed will prevail over one shown on a plat referred to if there is a conflict.[41] In one instance subdivision lots were sold by reference to a plat that showed building lines on some of the lots but not on others. The deeds to all the lots stated, "No building shall be placed upon said property . . . nearer to the street line than such building limits as are shown on said plat." The restriction was held to apply only to those lots for which building lines were depicted on the plat.[42] The discovery of an accidental encroachment of four and one-half inches across a building line after the building had been commenced and partly completed at great expense was held not to justify an injuction against the completion of the building.[43]

Restrictive covenants may be extinguished by terms incorporated therein providing for their extinguishment after a certain length of time or upon the occurrence of a certain event,[44] by agreement or release between the parties thereto,[45] by merger of the rights of the covenantor and covenantee in one person,[46] by acquisition of the land by a governing body under the power of eminent domain,[47] or by radical changes in the neighborhood which render their enforcement oppressive and inequitable.[48] If violations of restrictive covenants in a neighborhood have become so general and widespread as to indicate an intention on the part of the residents to abandon them, they will not be enforced.[49] Georgia law limits the duration of restrictive covenants to twenty years in cities and counties where zoning regulations have been adopted.[50] Covenants

39. Seawright v. Blount, 139 Ga. 323 (1912).
40. Ga. Real Est. Law § 19-189 (3d ed. 1986).
41. Stanley v. Greenfield, 207 Ga. 390 (1950).
42. Real Est. Operators v. McMahon, 171 Ga. 454 (1930).
43. Aspinwall v. Enterprise Dev. Co., 165 Ga. 83 (1927) (two dissents).
44. Ga. Real Est. Law §§ 19-196, 22-12 (3d ed. 1986).
45. 20 Am. Jur. 2d Covenants, Conditions, Etc. § 270; 21 C.J.S. Covenants § 76.
46. Muscogee Mfg. Co. v. Eagle & Phenix Mills, 126 Ga. 210 (1906).
47. Anderson v. Lynch, 188 Ga. 154 (1939).
48. 20 Am. Jur. 2d Covenants, Conditions, Etc. § 281; 21 C.J.S. Covenants § 75.
49. 20 Am. Jur. 2d Covenants, Conditions, Etc. § 272.
50. O.C.G.A. § 44-5-60.

which by their terms are to run for a longer period are not invalidated thereby but are merely limited to the statutory period.[51] A subdivision developer may, by express terms, retain the right to waive restrictions imposed upon the subdivision. If the developer does so for some of the lots, the other lot owners lose the right to enforce the restrictions on the lots for which they have been waived.[52] Restrictive covenants will not be set aside merely on the grounds that their annulment will enhance the value of the property so as to yield more taxes to the government;[53] and permits issued by zoning authorities to use property in a particular way do not abrogate restrictive covenants as to the use of the property.[54]

When one who has the right to enforce restrictions has notice of violations of the restrictions, he must move promptly to enjoin the violation or he will forfeit his right to demand cessation of the violation.[55]

Party Walls

A party wall is one built partly on the land of one owner and partly on the land of another for the mutual benefit of both in supporting contiguous buildings; or a wall standing entirely on the property of one owner but subject to an easement or right in the adjoining property to have it maintained as a dividing wall between the two tenements.[56] When a party wall is located on the boundary between two adjoining tracts, the owners are not tenants in common of the wall. Instead, each owns in severalty the part of the wall that rests on his side of the line, together with an easement of support from the other.[57]

Party walls may be created by an oral agreement duly executed by the parties,[58] by a written agreement,[59] by deed,[60] by acquiescence,[61] by pre-

51. Goddard v. Irby, 255 Ga. 47 (1985); Antill v. Sigman, 240 Ga. 511 (1978).

52. Davis v. Miller, 212 Ga. 836 (1957). The developer loses the right to waive restrictions after selling all the lots in the subdivision. Armstrong v. Roberts, 254 Ga. 15 (1985).

53. Cawthon v. Anderson, 211 Ga. 77 (1954).

54. Arlington Cemetery Corp. v. Hoffman, 216 Ga. 735 (1961).

55. Ga. Real Est. Law § 19-198 (3d ed. 1986).

56. Id. §§ 8-35, 14-8; 69 C.J.S. Party Walls § 1.

57. Wilensky v. Robinson, 203 Ga. 423 (1948).

58. Rawson v. Bell, 46 Ga. 19 (1872).

59. Simmons v. Hall, 180 Ga. 492 (1935); Horne v. Macon Tel. Pub. Co., 142 Ga. 489 (1914).

60. Forsyth Corp. v. Rich's Inc., 215 Ga. 333 (1959).

61. 69 C.J.S. Party Walls § 8.

scription,[62] or by implication (as, for example, when the owner of a tract with buildings erected on it, conveys the tract in separate parcels with the dividing line described as running through a common wall between the buildings).[63] In the latter case, each grantee takes an implied easement to as much of the wall on the adjoining land as is necessary for the support of his building.[64] The mere construction of a building flush with a property line confers no interest or easement in the wall of the building to the adjoining owner.[65]

Generally party wall rights are extinguished in the same way rights to other types of easements are extinguished.[66] The accidental destruction of a party wall will terminate the easement thereto, and under such circumstances each owner may assert his unqualified title to his land up to the division line.[67] The destruction of a building, leaving a party wall standing, does not terminate the rights of the adjoining owner to the wall.[68] An owner may tear down an insufficient or dangerous party wall and replace it with a better one, provided he does so promptly and gives the adjoining owner the same rights in the new wall as he had in the old wall.[69]

A party wall agreement may constitute a covenant running with the land.[70] A purchaser who can readily discover by measurements or otherwise that a party wall lies partly or wholly on his land and that it is a supporting wall of an adjoining building is charged with notice of an easement for the use of the wall as a party wall.[71]

When a party wall is called for as marking a boundary, it is regarded as a monument and will generally prevail over specified courses and distances.[72]

62. Joel v. Publix-Lucas Theatres Inc., 193 Ga. 531 (1942); Levinson v. Goode, 164 Ga. 361 (1927).

63. 60 Am. Jur. 2d Party Walls § 8; 69 C.J.S. Party Walls § 10.

64. 60 Am. Jur. 2d Party Walls § 8; 69 C.J.S. Party Walls § 10.

65. 60 Am. Jur. 2d Party Walls § 2.

66. See page 351.

67. 60 Am. Jur. 2d Party Walls § 15; 69 C.J.S. Party Walls § 11.

68. Forsyth Corp. v. Rich's Inc., 215 Ga. 333 (1959); Simmons v. Hall, 180 Ga. 492 (1935).

69. Forsyth Corp. v. Rich's Inc., 215 Ga. 333 (1959).

70. Horne v. Macon Tel. Pub. Co., 142 Ga. 489 (1914).

71. 69 C.J.S. Party Walls § 22.

72. Darling v. Alexander, 130 App. Div. 85, 114 N.Y.S. 334 (1909); Ross v. Golden, 146 Pa. Super. Ct. 417, 22 A.2d 310 (1941), affirmed 344 Pa. 487, 25 A.2d 700 (1942); Dunlap v. Reardon, 24 Pa. Super. Ct. 35 (1903). Contra: Ehrenreich v. Froment, 73 App. Div. 213, 76 N.Y.S. 861 (1902).

Cemeteries

As a general rule, one who purchases a lot in a public cemetery does not acquire the fee to the soil but only the easement or license of burial therein.[73] This is so even though the deed is absolute in form.[74] A cemetery lot conveyed for burial purposes cannot be devoted to any other use, whatever the form of conveyance.[75]

A person who is permitted to bury his dead in a cemetery by the express or implied consent of those in control of the cemetery acquires an easement of possession in the spot in which the body is buried.[76] Title to an easement for burial in a cemetery can also be obtained by prescription.[77] In such cases the burial of a dead body and the erection of gravestones are sufficient to make the possession actual, notorious, and adverse.[78]

Although a cemetery constitutes an encumbrance on the land whereon it lies, if a family burial plot has been so neglected for many years that it has lost all identity and appearance as a place of burial, a purchaser of the land takes the land free of the encumbrance of the cemetery and may plow over the area.[79]

Since cemeteries constitute an encumbrance on the land, the surveyor should show them in his surveys when they are present.

Georgia Militia Districts

Georgia militia districts[80] had their origin in the colonial act of January 24, 1755,[81] although the militia itself had existed since the founding

73. Mayor, etc., of Savannah v. Colding, 181 Ga. 260 (1935); Jacobus v. Cong. of the Children of Israel, 107 Ga. 518 (1899); Walker v. Georgia Power Co., 177 Ga. App. 493 (1986).

74. Stewart v. Garrett, 119 Ga. 386 (1903).

75. Id.

76. Hale v. Hale, 199 Ga. 150 (1945).

77. Roumillot v. Gardner, 113 Ga. 60 (1901).

78. Jacobus v. Cong. of the Children of Israel, 107 Ga. 518 (1899).

79. Mayes v. Simmons, 189 Ga. 845 (1940).

80. Hitz, *Georgia Militia Districts*, 18 Ga. B.J. 283 (1956); F. Hudson, An 1800 Census for Lincoln County, Georgia 3–37 (1977).

81. 18 Col. Rec. Ga. 7.

of the colony.[82] The militia served as a source of reserve manpower during military emergencies, especially Indian uprisings. In succession, the royal governor, the state governor, the militia officers, the inferior court, and the ordinary (now the probate judge) were empowered to create militia districts and define their boundaries. Today in some counties this power has been conferred on the county commissioners.[83] Originally militia district lines were drawn so that each district had enough qualified males to raise one company.[84]

The practice of numbering the districts apparently began about 1804 as a regulation of the state adjutant general, but the regulation was written into law in 1807. Before this date militia districts had been designated by name, for example "Captain Smith's District" (Burke County) and "Captain Wiggins' District" (Washington County). Such names often continued in use along with the number, and many of these names survive even today. Militia districts were numbered in the order of their creation. As of the 1990s there were approximately nineteen hundred districts in the state. The active, organized Georgia militia no longer exists,[85] but in headright counties militia districts are employed in conveyancing, to define the general location of property.

Current law provides for the laying out of militia districts and the altering of militia district lines by three commissioners appointed by the probate judge or county commissioners, whichever has jurisdiction.[86] The three commissioners have the authority to engage the services of a surveyor.

Specific information regarding militia district boundaries is scattered. County maps published by the department of transportation show militia district lines, but this information often comes from dubious sources and is frequently inaccurate. The minutes of the inferior court in each county contain data on militia district lines. After 1868, the year the in-

82. 1 Col. Rec. Ga. 24; L. Ivers, British Drums on the Southern Frontier passim (1974); 1 W. Stevens, A History of Georgia 250 (New York 1847).

83. 1982 Op. Atty. Gen. 203; 1968 Op. Atty. Gen. 218.

84. Sixty-three men by act of February 26, 1784. 19 Col. Rec. Ga. (pt. 2) 348. Currently the number is fixed at two hundred. O.C.G.A. § 36-2-2. Though technically still law, this quaint criterion for fixing militia district lines has been ignored for well over a century.

85. Under present law, all able-bodied citizens of the state between the ages of seventeen and forty-five years, who are not members of some organized military force, are members of the unorganized Georgia militia. O.C.G.A. § 38-2-3.

86. O.C.G.A. §§ 36-2-1 to 36-2-7.

ferior courts were abolished, such information would have been recorded in the minutes of the court of ordinary or in the minutes of the county commissioners. These records are usually poorly indexed. In addition, because of frequent burnings of courthouses, carelessness of clerks, and general apathy, they are often incomplete or not to be found. The *Georgia Official and Statistical Register* for 1931 contains information on militia district lines for thirty-eight counties. Much of this information is now out-of-date. The surveyor general section of the Georgia Department of Archives and History has some information relative to militia district boundaries.

In a case in which an attempt was made to alter militia district lines as provided by law, the description of the new lines did not call for a point of beginning, nor did the lines close into any of the old lines. Furthermore, the county commissioners continued to recognize the district lines as they existed prior to the proceeding. The court held that there was no valid change in the lines and that the original lines would remain in effect.[87] Where a militia district boundary is described as the city limits of a town, a change in the city limits will not change the militia district lines.[88] The law does not contemplate the laying out of militia districts so that a district is made up of noncontiguous parts.[89]

The determination of militia district lines by the probate judge or county commissioners is final and not subject to review or appeal unless void by reason of fraud or abuse of discretion.[90]

Processioning

Processioning is a method for having the boundaries between adjacent tracts of land marked anew to help prevent controversies between the respective owners.[91] It is a summary proceeding whose primary object is the settling of boundary line disputes between coterminous owners, and it is not designed to be a substitute for an action to try title to land.[92]

87. Camp v. Trapp, 209 Ga. 298 (1952).
88. 1968 Op. Atty. Gen. 321.
89. Howell v. Kinney, 99 Ga. 544 (1896).
90. Id.; Johnson v. Chappell, 198 Ga. 162 (1944); O.C.G.A. § 36-5-1.
91. Watson v. Bishop, 69 Ga. 51 (1882).
92. Georgia Talc Co. v. Cohutta Talc Co., 140 Ga. 245 (1913).

The history of processioning is of great antiquity. Very likely it has existed in some form for as long as people have felt a need to demarcate any sort of boundaries on the landscape, though the concept appears to be outdated today. Originally it was a method of periodically restoring and renewing boundary markers and of perpetuating the memory of their locations at a time when almost all land transfers were oral and few, if any, written records were kept as to the locations of boundaries.[93] It was therefore somewhat analogous to livery of seisen in that, while livery of seisen preserved the memory of the title to property, processioning preserved the memory of the location of its boundaries. Unlike livery of seisen, however, processioning did not die out but instead has been embalmed by statutes and invigorated by court decisions so as to persist, despite its apparent incongruity in modern society, to this day.

No doubt meticulous research into some very obscure sources would reveal many fascinating details about the history of processioning. A few extracts will be given.

The ancient Romans had a god of the landmark called Terminus, meaning "end, limit, or boundary." Terminus was supposed to reside in the stones or posts that marked land boundaries. A Roman author on land surveying, Siculus Flaccus, gives the following account of the Roman method of placing boundary stones:

> When they were putting out the boundary stones, they would place the stones themselves upright on solid ground next to the places where they were going to set them once they had dug holes, and they would top them with oil, veils, and garlands. As for the holes in which they were going to set them, after performing a sacrifice and slaughtering the victim and burning it with flaming wood, they would, with covered heads, let the blood drip into the holes, and they would throw in incense and grain. They would also throw honeycombs and wine, and other things which custom holds sacred to Terminus, into the holes. When all the food had been burned up, they would set the stones on top of the warm remains. After putting some broken bits of rock next to the boundary stones, they would stamp around them to make them stand more securely. Such a sacrifice did the landowners between whom boundaries were being marked out make. And if boundary stones were being put at a *trifinium,* that is, at a point which three landholders

93. Townsend v. Radcliffe, 63 Ill. 9 (1872); Pitman v. Town of Albany, 34 N.H. 577 (1857). Scholarly studies on the history of processioning include Stilgoe, *Jack-o'-lanterns to Surveyors: The Secularization of Landscape Boundaries,* 1 Environmental Review 14 (1976); and Seiler, *Land Processioning in Colonial Virginia,* 6 William and Mary Quarterly 416 (1949).

had in common, all three would perform the sacrifice. And everyone else who had a common boundary with a landowner would all erect boundary stones as was mutually agreeable, and would perform the sacrifice. Mutual agreement among the landholders would confirm the boundary stones.[94]

Though the sacrifice was an atonement to Terminus for protecting the landmark, it undoubtedly served the practical purpose of impressing upon the minds of those present the location of the boundary marker. Rather than holding the ceremony just once, the Romans held an annual festival called the Terminalia. In this festival adjoining landowners met and sacrificed to the sacred boundary marker to prevent its shifting and to maintain peace and good relations between neighbors.[95]

In England processioning has been variously known as perambulation, circumambulation, beating the bounds, or riding the boundaries. It was, and still is to some extent, used as a method of preserving parish and manorial boundaries. Commonly carried out once each year during Rogation week, it was presided over by the minister and churchwardens. The custom had apparently fallen into some disuse when George Wither published the following verses in 1634:

That, ev'ry man might keep his owne possessions,
Our fathers, us'd in reverent Processions
(With zealous prayers, and with praiseful cheere)
To walke their Parish-limits, once a yeare:
And, well knowne markes (which sacrilegious Hands
Now cut or breake), so bord'red out their Lands,
That, ev'ry one distinctly knew his owne;
And, many brawles, now rife, were then unknowne.[96]

The following descriptions of English processionings are also of interest:

In the Middle Ages there was one duty of the parish and town officers that was of paramount importance and had to be carried out annually, the

94. Siculus Flaccus, De condicionibus agrorum, in Gromatici Veteres 140–41 (C. Lachmanni ed. 1848); *quoted in* 2 Publius Ovidius Naso, Fastorum libri sex; the Fasti of Ovid 483 n. 1 (J. Frazer ed. & trans. 1929) (as translated by Philip G. Holt, professor of classics, University of Georgia, Athens, Ga.).

95. 2 Publius Ovidius Naso, Fastorum libri sex; the Fasti of Ovid 481–89 (J. Frazer ed. & trans. 1929).

96. G. Wither, A Collection of Emblemes, Ancient and Moderne 161 (London 1634) (reprinted by the University of South Carolina Press, Columbia 1975).

"beating of the bounds." When map-making and map-drawing were unknown mysteries, the bounds of a district had to be remembered to prevent encroachment; so, once a year, the civic or parochial officials perambulated the boundaries accompanied by the boys of the district, and the beating of certain landmarks with long canes was intended to impress these marks on the boys' memories. Any additional ceremonies observed at the same time, or any accompanying ritual was also intended to help the younger generation to memorize the bounds. What was once a universal matter of great importance has now become an isolated custom, but several districts keep up the old practice of impressing the boundaries on the minds of the rising generation.

In London, the Liberty of the Savoy, the Tower of London, and that arch preserver of old customs, the City, still keep up this ceremony. In the case of the Savoy, permission has to be obtained from the Benchers of the Temple for the officials of the Liberty to enter the Temple grounds in search of one of their marks. They must also enter the vaults of a bank to find another. The City officials have also to enter the Temple for the same purpose.

At Helston, a clod of earth and a slip of hawthorn are placed on each boundary stone and duly beaten by the boys; but in some cases it is not only the bounds but also the boys and the civic officials who are beaten, or bumped, on the boundary marks. At Newbiggin in Northumberland two new freeholders are bumped on a stone on the moor called the "dunting stone," and in the parish of St. Mary's, Leicester, it was the custom to place any newly appointed parish official head downwards in a hole, which had previously been dug on the boundary line, and then thump him with a spade. Marlborough in Devonshire has a water boundary, as has Oxford, and once in each of them, when beating these water boundaries, the mayor has been reminded of the mark in a way he is not likely to forget. The Mayor of Marlborough was soundly ducked some years ago to impress the limit on his memory, and at Oxford the boat containing the beating party was upset in the stream and the whole boatload received a very wet reminder.[97]

A *Perambulation,* or, as it might be more correctly called, a circumambulation, is the custom of going round the boundaries of a manor or parish, with witnesses, to determine and preserve [the] recollection of its extent, and to see that no encroachments have been made upon it, and that the landmarks have not been taken away. It is a proceeding commonly regulated by the steward, who takes with him a few men and several boys who are required to particularly observe the boundary lines traced out, and thereby qualify themselves for witnesses, in the event of any dispute about the landmarks

97. F. Drake-Carnell, Old English Customs and Ceremonies 77–78 (Charles Scribner's Sons, New York 1938) (quoted with permission).

> or extent of the manor, at a future day. In order that they may not forget the lines and marks of separation, they "take pains" at almost every turning. For instance, if the boundary be a stream, one of the boys is tossed into it; if a broad ditch, the boys are offered money to jump over it, in which they of course fail, and pitch into the mud, where they stick as firmly as if they had been rooted there for the season; if a hedge, a sapling is cut out of it, and used in afflicting that part of their bodies upon which they rest in the posture between standing and lying; if a wall, they are to have a race on the top of it, when, in trying to pass each other, they fall over on each side,—some descending, perhaps, into the still stygian waters of a ditch, and others thrusting the "human face divine" into a bed of nettles; if the boundary be a sunny bank, they sit down upon it, and get a treat of beer, and bread and cheese, and, perhaps, a glass of spirits. When these boys grow up to be men, if it happens that one of them should be asked if a particular stream were the boundary of the manor he had perambulated, he would be sure to say, in the manner of Sancho Pança, "Ees, that 'tis, I'm sure o't, by the same token that I were tossed into't, and paddled about there lik a water-rot, till I wor hafe dead." If he should be asked whether the aforesaid pleasant bank were a boundary,—"O, ees it be," he would say, "that's where we squat down, and tucked in a skinvull of vittles and drink." With regard to any boundary perambulated after that, he would most likely declare, "I won't be sartin; I got zo muddled up top o' the banks, that don' know where we ambulated arter that."[98]

Processioning was adopted in several of the American colonies. The following is an account of the method used in Virginia in the 1600s:

> While these laws were highly useful, they did not reach all the evils springing from the imperfections of the first surveys. Casting about for a remedy for the confusion and harassments created by their defects, the Assembly adopted what was known as the law of processioning. According to this law, the people in every neighborhood were required once in the course of every four years to assemble at a designated spot, and from that point march in a body to examine and, if necessary, to renew the terminal marks of every plantation in their precinct. If the mark in any instance consisted of blazes on a tree, they were to be recut to their former depth. If it consisted of a pile of stones, the original number were to be restored in case many had been removed. If the determining line were the bed of a small stream or a public road, then, as soon as this was reached, it was to be proclaimed as the legal boundary in the presence of the entire company, who were to

98. W. Hone, The Year Book of Daily Recreation and Information 1178–79 (London 1832).

serve as witnesses of the fact in the next processioning. When the bounds of every plantation had been recognized and acknowledged by all participating in the procession, the owners of the land at the time, or those afterwards claiming their title from them, were shut out from making any alteration in the terminal marks, monuments, or lines. In case an altercation arose between two neighbors in the course of the processioning, as to the boundaries of their estates, the two surveyors who always accompanied the procession were required, in sight of the assembled people, to draw again the lines in dispute, and if a conflict in the original surveys were shown, the difference was to be equitably adjusted and the bounds thus laid off were to be accepted as permanent. The area which was to be covered by a procession was designated by the vestry of each parish for the domain in their special jurisdiction. The time first assigned for the performance of the duty of processioning was in the interval between Easter and Whitsuntide, but in the latter part of the seventeenth century, it was advanced to the interval between the 7th of September and the 31st of March, a period during which the planters were most at leisure, as their crops had been gathered in, and in most cases sold. Moreover, during this section of the year, the physical obstacles were less serious, as the trees were now devoid of leaves, and the thickets and undergrowth were more easily penetrated in consequence. The supreme importance of processioning in the view of the law is revealed in the requirement that all persons who, without having a sound excuse to offer, failed to take part in it, were exposed to a fine of three hundred and fifty pounds of tobacco.[99]

One of the earliest processioning acts passed in Georgia provided that all lands in the state were to be processioned within twelve months after June 1, 1798, and every ten years thereafter.[100] The captain of each militia district was to hold an election for selecting three commissioners to preside over the processionings. At the time of the processioning, each landowner was required to procession the boundaries of his land with his adjoining neighbor and at least one disinterested person. They were to chop out or mark the lines plainly. Anyone refusing was subject to a fine of one hundred dollars. This act seemingly was passed to deal with problems that had arisen out of the Pine Barrens Fraud.[101]

99. 1 P. Bruce, Economic History of Virginia in the Seventeenth Century 543–44 (1896) (footnotes omitted).

100. Marbury & Crawford 339.

101. S. McLendon, History of the Public Domain of Georgia 56–57 (1924). The earliest processioning act passed in Georgia is not printed in any of the statute books. It was, however, printed on loose sheets by Alexander McMillan, state printer, at Augusta in 1793. A

The present law of processioning provides that the probate judge shall appoint "three suitable persons" in each militia district to serve two-year terms as processioners.[102] Any landowner who desires to have his boundaries marked anew can apply to the processioners of the district in which the tract of land or portion of it is located.[103] Ten days' written notice of the running and marking are to be given to all adjoining owners residing within the state.[104] On the appointed day the processioners and county surveyor mark out the boundaries. If bad weather or other adverse conditions intervene, the marking may be postponed. Written or verbal notice must be given to the adjoining landowner under such circumstances.[105] If there is no county surveyor, the processioners may engage any competent person residing in the county to perform the duties of the county surveyor, provided the person is sworn.[106] The surveyor is required to make a certified plat to the proceeding. One copy is given to the applicant and another copy is deposited in the office of the probate judge, along with the return of the processioners, within thirty days.[107] The applicant pays the costs. The return of the processioners, where no objections are filed and no trial held, constitutes merely prima facie evidence of the correctness of the lines surveyed and marked anew.[108]

Processioners have no authority to establish new boundary lines or to establish lines merely drawn on paper if those lines vary from lines marked on the ground.[109] Neither is processioning a method for trying conflicting claims of title.[110] The only duty of processioners is to trace and mark anew old lines or those actually located and designated on the

---

copy is in the DeRenne Collection, University of Georgia Libraries, Athens, Ga. The date of the act—December 17, 1793—may be significant as the Pine Barrens Fraud was then in full swing and the legislature was in the process of passing other acts to deal with the fraud. See pages 96–97.

102. O.C.G.A. § 44-4-1.

103. O.C.G.A. § 44-4-2.

104. Id.

105. Garrett v. Masse & Felton Lbr. Co., 134 Ga. 442 (1910); Reynolds v. Kinsey, 50 Ga. App. 385 (1934).

106. Tisinger v. Ellerbee, 37 Ga. App. 391 (1927).

107. Moore v. Hood, 131 Ga. 479 (1908); O.C.G.A. §§ 44-4-3, 44-4-4.

108. Bradley v. Chattanooga Iron & Coal Co., 144 Ga. 478 (1915); Woodcock v. Rayonier, Inc., 97 Ga. App. 133 (1958); O.C.G.A. § 44-4-3.

109. Minor v. Ray, 118 Ga. App. 164 (1968); Cosby v. Reid, 21 Ga. App. 604 (1917).

110. Darnell v. Betty's Creek Baptist Church, 230 Ga. 461 (1973); Minor v. Ray, 118 Ga. App. 164 (1968).

ground and not necessarily to locate them as they were originally or as they should have been.[111] Processioners should respect all acts of possession of the adjoining landowners under a claim of right for more than seven years, or of adverse possession which has ripened into title, or of previous acquiescence or agreement between them. The lines are to be marked so as not to interfere therewith, even though they are not in harmony with the recitals given in the parties' deeds.[112]

Although it is the duty of processioners to mark anew established lines, and not necessarily to locate the lines as they were originally laid out, this does not mean that where established boundaries are not physically marked in their entirety, processioners cannot function.[113] If such were the case, their office would be useless.[114] Thus even though the course and extent of the lines may not be physically marked out in their entirety upon the earth's surface, if there is a sufficient number of physically established corners or landmarks, the connecting of which by straight lines will complete the boundaries; or if there is an established line of demarcation (such as an abandoned railroad bed) to which parallel boundaries can be determined by courses and distances with respect thereto, it is the duty of processioners to so ascertain and establish the same, always respecting any rights had under actual possession.[115] In the absence of monuments or marks on the ground, processioners may resort to courses and distances given in the parties' deeds to set up the boundaries.[116]

Formerly it was necessary for the processioners and county surveyor

111. Goodson v. Pope, 112 Ga. App. 71 (1965); Palmer v. Jackson, 82 Ga. App. 702 (1950); Ledford v. Hill, 82 Ga. App. 299 (1950); Hall v. Browning, 71 Ga. App. 694 (1944). "The evidence here—being sufficient to authorize a finding that the processioners . . . took into consideration the contentions of both parties as to possession, and evidence of possession as shown by fences and cultivated fields, and also that they observed natural landmarks such as a hedgerow, an old fence line, and a branch line—was sufficient to show that they were not running a new line, but were marking anew an established line." Milligan v. Hale, 88 Ga. App. 70 (1953) (headnote). Where the processioners and surveyor, after making certain trial surveys, decided to establish a "compromise" line between adjoining tracts, such a solution was no attempt to run and mark the boundaries anew. Therefore, the resulting line could not be sustained. Dodson v. Knox, 89 Ga. App. 760 (1954).

112. Lansford v. Gatliff, 119 Ga. App. 145 (1969); Milligan v. Hale, 88 Ga. App. 70 (1953); O.C.G.A. § 44-4-7. Although processioners are required to mark prescriptive boundaries, they do not, by this action, determine title to the land within the prescriptive boundary. Byrd v. McLucas, 194 Ga. 40 (1942).

113. Byrd v. McLucas, 194 Ga. 40 (1942).

114. Id.

115. Id.; Cosby v. Reid, 21 Ga. App. 604 (1917).

116. Wood v. Hamilton, 109 Ga. App. 608 (1964).

to trace and mark anew the lines around the applicant's entire tract before the proceeding could be considered valid. In 1901, however, the law was changed so that now only the line or lines to be renewed need to be run.[117]

Any owner dissatisfied with the lines as run and marked may file a protest with the probate judge within thirty days after the processioners have made their return. The protester specifies the lines to which he objects and the true lines he claims.[118] The probate judge then sends the papers, the plat, and the protest to the clerk of the superior court who enters them on the docket to be tried as other causes.[119] The issue in such cases is not necessarily confined to the question of sustaining or overruling the lines marked by the processioners but may include ascertainment of the true lines.[120] Whatever the final verdict of the court, it is conclusive as to the location of the lines.[121]

Processioning does not apply to lands within the corporate limits of cities and towns.[122]

Although processioning has been defended as a quick and inexpensive method of settling boundary disputes,[123] it is actually an anachronism that should be abolished. Its chief problem is that if it is to constitute a sound and justifiable method of restoring property boundaries, it must be performed with enough frequency and regularity so that the memory of the physical evidences of such boundaries does not lapse in the minds of the laypersons who carry it out. The present law ignores this fundamental principle, and attempts to initiate it in these busy times probably would not meet with much success. Landowners, attorneys, and others, without understanding the legal and historical limitations of processioning, often employ it in attempts to resolve boundary disputes. As a consequence, they usually create more problems than they solve.

117. Russell v. Radford, 76 Ga. App. 302 (1947); Groover v. Durrence, 36 Ga. App. 543 (1926); O.C.G.A. § 44-4-9.

118. O.C.G.A. § 44-4-9.

119. Id.

120. Reynolds v. Kinsey, 50 Ga. App. 385 (1934).

121. Edenfield v. Lanier, 206 Ga. 696 (1950); Stovall v. Caverly, 139 Ga. 243 (1912).

122. Smith v. Bailey, 183 Ga. 869 (1937).

123. Ga. Real Est. Law § 13-51 (3d ed. 1986).

Appendix A. Order of Creation of Georgia's Counties

Name	*Date of Creation*	*Created From*
Wilkes	Feb. 5, 1777	Original county created from Ceded Lands (ceded by Treaty of June 1, 1773; ratified by Treaties of Aug. 7, 1790 and July 2, 1791).
Richmond	Feb. 5, 1777	Original county created from St. Paul Parish.
Burke	Feb. 5, 1777	Original county created from St. George Parish.
Effingham	Feb. 5, 1777	Original county created from St. Matthew Parish and part of St. Philip Parish.
Chatham	Feb. 5, 1777	Original county created from Christ Church Parish and part of St. Philip Parish.
Liberty	Feb. 5, 1777	Original county created from St. John Parish, St. Andrew Parish and St. James Parish.
Glynn	Feb. 5, 1777	Original county created from St. David Parish and St. Patrick Parish.
Camden	Feb. 5, 1777	Original county created from St. Thomas Parish and St. Mary Parish.
Franklin	Feb. 25, 1784	Original county created from land ceded by Treaties of May 31, 1783 and Nov. 1, 1783; ratified by Treaties of Aug. 7, 1790 and July 2, 1791.

Arranged chronologically from the *Georgia Official and Statistical Register, 1977–1978*, ed. Marian B. Holmes and Alice C. Cormany (Atlanta: Department of Archives and History), pp. 1089–93.

Name	*Date of Creation*	*Created From*
Washington	Feb. 25, 1784	Original county created from land ceded by Treaties of May 31, 1783 and Nov. 1, 1783; ratified by Treaties of Aug. 7, 1790 and July 2, 1791.
Greene	Feb. 3, 1786	Washington
Columbia	Dec. 10, 1790	Richmond
Elbert	Dec. 10, 1790	Wilkes
Screven	Dec. 14, 1793	Burke, Effingham
Hancock	Dec. 17, 1793	Washington, Greene
Warren	Dec. 19, 1793	Wilkes, Columbia, Burke, Richmond
Oglethorpe	Dec. 19, 1793	Wilkes
McIntosh	Dec. 19, 1793	Liberty
Bryan	Dec. 19, 1793	Chatham
Montgomery	Dec. 19, 1793	Washington
Bulloch	Feb. 8, 1796	Screven, Bryan
Jackson	Feb. 11, 1796	Franklin
Jefferson	Feb. 20, 1796	Burke, Warren
Lincoln	Feb. 20, 1796	Wilkes
Tattnall	Dec. 5, 1801	Montgomery
Clarke	Dec. 5, 1801	Jackson
Wayne	May 11, 1803	Original county created from land ceded by Treaty of June 16, 1802.
Wilkinson	May 11, 1803	Original county created from land ceded by Treaties of June 16, 1802 and Nov. 14, 1805.
Baldwin	May 11, 1803	Original county created from land ceded by Treaties of June 16, 1802 and Nov. 14, 1805.
Morgan	Dec. 10, 1807	Baldwin
Jasper	Created on Dec. 10, 1807 as Randolph; name changed to Jasper by Act of Dec. 10, 1812	Baldwin
Jones	Dec. 10, 1807	Baldwin
Putnam	Dec. 10, 1807	Baldwin
Laurens	Dec. 10, 1807	Wilkinson

Name	Date of Creation	Created From
Telfair	Dec. 10, 1807	Wilkinson
Pulaski	Dec. 13, 1808	Laurens
Twiggs	Dec. 14, 1809	Wilkinson
Madison	Dec. 5, 1811	Oglethorpe, Clarke, Jackson, Franklin, Elbert
Emanuel	Dec. 10, 1812	Bulloch, Montgomery
Early	Dec. 15, 1818	Original county created from land ceded by Treaty of Aug. 9, 1814.
Irwin	Dec. 15, 1818	Original county created from land ceded by Treaties of Aug. 9, 1814 and Jan. 22, 1818.
Appling	Dec. 15, 1818	Original county created from land ceded by Treaties of Aug. 9, 1814 and Jan. 22, 1818.
Walton	Dec. 15, 1818	Original county created from land ceded by Treaty of Jan. 22, 1818.
Gwinnett	Dec. 15, 1818	Original county created from land ceded by Treaties of July 8, 1817 and Jan. 22, 1818.
Hall	Dec. 15, 1818	Original county created from land ceded by Treaties of July 8, 1817 and Feb. 27, 1819.
Habersham	Dec. 15, 1818	Original county created from land ceded by Treaties of July 8, 1817 and Feb. 27, 1819.
Rabun	Dec. 21, 1819	Original county created from land ceded by Treaty of Feb. 27, 1819.
Dooly	May 15, 1821	Original county created from land ceded by Treaty of Jan. 8, 1821.
Houston	May 15, 1821	Original county created from land ceded by Treaty of Jan. 8, 1821.
Monroe	May 15, 1821	Original county created from land ceded by Treaty of Jan. 8, 1821.
Fayette	May 15, 1821	Original county created from land ceded by Treaty of Jan. 8, 1821.
Henry	May 15, 1821	Original county created from land ceded by Treaty of Jan. 8, 1821.
Newton	Dec. 24, 1821	Jasper, Walton, Henry
DeKalb	Dec. 9, 1822	Fayette, Gwinnett, Henry

Name	*Date of Creation*	*Created From*
Bibb	Dec. 9, 1822	Jones, Monroe, Twiggs
Pike	Dec. 9, 1822	Monroe
Crawford	Dec. 9, 1822	Houston
Decatur	Dec. 8, 1823	Early
Upson	Dec. 15, 1824	Crawford, Pike
Ware	Dec. 15, 1824	Appling
Lee	June 9, 1825 (Named by Act of Dec. 11, 1826)	Original county created from land ceded by Treaties of Feb. 12, 1825 (declared null and void), Jan. 24, 1826 and March 31, 1826.
Muscogee	June 9, 1825 (Named by Act of Dec. 11, 1826)	Original county created from land ceded by Treaties of Feb. 12, 1825 (declared null and void), Jan. 24, 1826 and March 31, 1826.
Troup	June 9, 1825 (Named by Act of Dec. 11, 1826)	Original county created from land ceded by Treaties of Feb. 12, 1825 (declared null and void), Jan. 24, 1826 and March 31, 1826.
Coweta	June 9, 1825 (Named by Act of Dec. 11, 1826)	Original county created from land ceded by Treaties of Feb. 12, 1825 (declared null and void), Jan. 24, 1826, March 31, 1826 and Nov. 15, 1827.
Carroll	June 9, 1825 (Named by Act of Dec. 11, 1826)	Original county created from land ceded by Treaties of Feb. 12, 1825 (declared null and void), Jan. 24, 1826, March 31, 1826 and Nov. 15, 1827.
Baker	Dec. 12, 1825	Early
Thomas	Dec. 23, 1825	Decatur, Irwin
Lowndes	Dec. 23, 1825	Irwin
Taliaferro	Dec. 24, 1825	Wilkes, Warren, Hancock, Greene, Oglethorpe
Butts	Dec. 24, 1825	Henry, Monroe
Meriwether	Dec. 14, 1827	Troup
Harris	Dec. 14, 1827	Troup, Muscogee
Talbot	Dec. 14, 1827	Muscogee
Marion	Dec. 14, 1827	Lee, Muscogee
Randolph	Dec. 20, 1828	Lee

Name	*Date of Creation*	*Created From*
Campbell	Dec. 20, 1828	Coweta, Carroll, DeKalb, Fayette (Abolished and merged with Fulton on Jan. 1, 1932).
Cherokee	Dec. 21, 1830 (Named by Acts of Dec. 26, 1831 and Dec. 3, 1832)	Original county created from land ceded by Treaty of Dec. 29, 1835.
Heard	Dec. 22, 1830	Troup, Coweta, Carroll
Stewart	Dec. 23, 1830	Randolph
Sumter	Dec. 26, 1831	Lee
Forsyth	Dec. 3, 1832	Cherokee
Lumpkin	Dec. 3, 1832	Cherokee
Union	Dec. 3, 1832	Cherokee
Cobb	Dec. 3, 1832	Cherokee
Gilmer	Dec. 3, 1832	Cherokee
Murray	Dec. 3, 1832	Cherokee
Bartow	Created on Dec. 3, 1832 as Cass; name changed to Bartow by Act of Dec. 6, 1861.	Cherokee
Floyd	Dec. 3, 1832	Cherokee
Paulding	Dec. 3, 1832	Cherokee
Walker	Dec. 18, 1833	Murray
Macon	Dec. 14, 1837	Houston, Marion
Dade	Dec. 25, 1837	Walker
Chattooga	Dec. 28, 1838	Floyd, Walker
Gordon	Feb. 13, 1850	Floyd, Cass (now Bartow)
Clinch	Feb. 14, 1850	Ware, Lowndes
Polk	Dec. 20, 1851	Paulding, Floyd
Spalding	Dec. 20, 1851	Pike, Henry, Fayette
Whitfield	Dec. 30, 1851	Murray
Taylor	Jan. 15, 1852	Talbot, Macon, Marion
Catoosa	Dec. 5, 1853	Walker, Whitfield
Pickens	Dec. 5, 1853	Cherokee, Gilmer

Name	*Date of Creation*	*Created From*
Hart	Dec. 7, 1853	Franklin, Elbert
Dougherty	Dec. 15, 1853	Baker
Webster	Created on Dec. 16, 1853, as Kinchafoonee; name changed to Webster by Act of Feb. 21, 1856	Stewart
Fulton	Dec. 20, 1853	DeKalb
Worth	Dec. 20, 1853	Dooly, Irwin
Fannin	Jan. 21, 1854	Gilmer, Union
Coffee	Feb. 9, 1854	Clinch, Ware, Telfair, Irwin
Chattahoochee	Feb. 13, 1854	Muscogee, Marion
Clay	Feb. 16, 1854	Early, Randolph
Charlton	Feb. 18, 1854	Camden
Calhoun	Feb. 20, 1854	Baker, Early
Haralson	Jan. 26, 1856	Polk, Carroll
Terrell	Feb. 16, 1856	Lee, Randolph
Colquitt	Feb. 25, 1856	Thomas, Lowndes
Berrien	Feb. 25, 1856	Lowndes, Irwin, Coffee
Miller	Feb. 26, 1856	Baker, Early
Towns	Mar. 6, 1856	Union, Rabun
Dawson	Dec. 3, 1857	Lumpkin, Gilmer
Pierce	Dec. 18, 1857	Ware, Appling
Milton	Dec. 18, 1857	Cherokee, Cobb, Forsyth (Abolished and merged with Fulton on Jan. 1, 1932).
Glascock	Dec. 19, 1857	Warren
Mitchell	Dec. 21, 1857	Baker
Schley	Dec. 22, 1857	Marion, Sumter
White	Dec. 22, 1857	Habersham
Wilcox	Dec. 22, 1857	Irwin, Dooly, Pulaski
Clayton	Nov. 30, 1858	Fayette, Henry
Quitman	Dec. 10, 1858	Randolph, Stewart
Banks	Dec. 11, 1858	Habersham, Franklin
Johnson	Dec. 11, 1858	Washington, Emanuel, Laurens
Brooks	Dec. 11, 1858	Lowndes, Thomas

Name	*Date of Creation*	*Created From*
Echols	Dec. 13, 1858	Lowndes, Clinch
Douglas	Oct. 17, 1870	Campbell, Carroll
McDuffie	Oct. 18, 1870	Warren, Columbia
Rockdale	Oct. 18, 1870	Newton, Henry
Dodge	Oct. 26, 1870	Pulaski, Telfair, Montgomery
Oconee	Feb. 25, 1875	Clarke
Crisp	Aug. 17, 1905	Dooly
Grady	Aug. 17, 1905	Decatur, Thomas
Jenkins	Aug. 17, 1905	Bulloch, Burke, Emanuel, Screven
Tift	Aug. 17, 1905	Berrien, Irwin, Worth
Jeff Davis	Aug. 18, 1905	Appling, Coffee
Stephens	Aug. 18, 1905	Franklin, Habersham
Toombs	Aug. 18, 1905	Emanuel, Montgomery, Tattnall
Turner	Aug. 18, 1905	Dooly, Irwin, Wilcox, Worth
Ben Hill	Nov. 6, 1906	Irwin, Wilcox
Bleckley	Oct. 2, 1912	Pulaski
Wheeler	Nov. 5, 1912	Montgomery
Barrow	Nov. 3, 1914	Gwinnett, Walton, Jackson
Candler	Nov. 3, 1914	Bulloch, Emanuel, Tattnall
Bacon	Nov. 3, 1914	Appling, Pierce, Ware
Evans	Nov. 3, 1914	Bulloch, Tattnall
Atkinson	Nov. 5, 1918	Coffee, Clinch
Treutlen	Nov. 5, 1918	Emanuel, Montgomery
Cook	Nov. 5, 1918	Berrien
Seminole	Nov. 2, 1920	Decatur, Early
Lanier	Nov. 2, 1920	Berrien, Clinch, Lowndes
Brantley	Nov. 2, 1920	Charlton, Pierce, Wayne
Long	Nov. 2, 1920	Liberty
Lamar	Nov. 2, 1920	Monroe, Pike
Peach	Nov. 4, 1924	Houston, Macon

Appendix B. Navigability of Particular Georgia Rivers and Streams

River or Stream	*Legal Decisions*	*Status**
Altamaha River	Parker v. Adamson, 109 Ga. App. 172 (1964); Georgia Pwr. Co. v. Federal Pwr. Com., 152 F.2d 908 (1946)—navigable.	Navigable for its entire length up to the junction of the Oconee and Ocmulgee rivers.
Back River	City of Brunswick v. King, 65 Ga. App. 44 (1941)—navigable.	
Canoochee River	Seaboard Air-Line Ry. v. Sikes, 4 Ga. App. 7 (1908)—nonnavigable.	
Chattahoochee River	Howard v. Ingersol, 54 U.S. (13 How.) 381, 421; 14 L. Ed. 189, 206 (1851)—navigable for steamboats to Columbus and navigable for small craft for more than 100 miles above Columbus.	Navigable to Columbus.
Chestatee River	Chestatee Pyrites Co. v. Cavenders Creek Gold Mining Co., 119 Ga. 354 (1903); Whelchel v. Gainesville & Dahlonega Elec. Ry. Co., 116 Ga. 431 (1902)—nonnavigable.	

*According to *Water Resources Development by the U.S. Army Corps of Engineers in Georgia* (Jan. 1981).

River or Stream	Legal Decisions	Status*
Coosa River	Ala. Pwr. Co. v. Gulf Pwr. Co., 283 F. 606 (1922)—navigable in part.	
Etowah River	Johnson v. Watson, 157 Ga. 349 (1923); 1954–56 Op. Atty. Gen. 555—nonnavigable. See 1985 Op. Atty. Gen. 156.	
Fancy Bluff Creek		Navigable to Fancy Bluff.
Flint River	Mitchell County v. Hudspeth, 151 Ga. 767 (1921); 1963–65 Op. Atty. Gen. 748—navigable. 1920 Ga. Laws 1776—nonnavigable above the south side of Broad Street in Albany.	Navigable to Bainbridge.
Kirklands Creek	Pyle v. Gilbert, 245 Ga. 403 (1980)—nonnavigable.	
Knoxboro Creek	Charleston & Savannah Ry. v. Johnson, 73 Ga. 306 (1884)—navigable.	
McQueen's Inlet	Rauers v. Persons, 144 Ga. 23 (1915)—navigable.	
Ochlochnee River	Robinson v. State, 12 Ga. App. 683 (1913)—undetermined.	
Ocmulgee River	State Highway Bd. v. Wilcox, 168 Ga. 883 (1929)—navigable. Central Georgia Pwr. Co. v. Ham, 139 Ga. 569 (1912)—nonnavigable where it runs between Butts and Jasper counties. By legislative act of 1873 the town of	Navigable to Macon.

River or Stream	*Legal Decisions*	*Status**
	Hawkinsville was declared to be the head of navigation of this river. 1873 Ga. Laws 148. The act was repealed, however, in 1889. 1888–89 Ga. Laws 153.	
Oconee River	Georgia Pwr. Co. v. Federal Pwr. Com., 152 F.2d 908 (1946)—navigable to Milledgeville.	Navigable to Milledgeville.
Ogeechee River	Brantley v. Lee, 139 Ga. 600 (1913)—nonnavigable where it runs between Bulloch and Screven counties.	
Oostanaula River	Gordon County v. Pyron, 120 Ga. 101 (1904)—navigable.	
St. Marys and North River		Navigable to Trader Hill.
Satilla River		Navigable to Waycross.
Savannah River	Georgia Ry. & Pwr. Co. v. Wright, 146 Ga. 29 (1916); Levy's Son & Co. v. Gibson Line of Steamers, 130 Ga. 581 (1908); South Carolina R.R. Co. v. Moore, 28 Ga. 398 (1859); South Carolina v. Georgia, 93 U.S. (3 Otto) 4, 23 L. Ed. 782 (1876); Mills v. United States, 46 F. 738, 12 L.R.A. 673 (1891); Lawton v. Comer, 40 F. 480	Navigable to Augusta.

River or Stream	Legal Decisions	Status*
	(1889)—navigable. See 1927 Ga. Laws 313.	
Tallulah River	State of Georgia v. Georgia Ry. & Pwr. Co., 141 Ga. 153 (1913)—nonnavigable.	
Tugaloo River	See 1925 Ga. Laws 267.	

Appendix C. Surveyors General of Georgia

For the colonial period, the following table is compiled from sources cited in Chapters 1 and 2 of this book. For the state period, the information comes from the Executive Minutes and Oath Books on file in the Georgia Department of Archives and History. In cases for which the record is not more specific, for the state period, it is assumed that each surveyor general's term ended with the beginning of the next surveyor general's commission. The records are sketchy for the Revolution. Some years ago Alex M. Hitz, then head of the Georgia Surveyor General Department, compiled a list of the state surveyors general, which he published in the Georgia Bar Journal (Hitz, *The Surveyor General of Georgia,* 19 Ga. B.J. 191 [1956]). The terms of office he gave vary slightly from those shown in this table. Apparently Hitz used the dates of election, whereas the dates of commission, deemed by the author to be more accurate, are given here.

Since 1861 the office of surveyor general has been exercised by the secretary of state. This table takes the office down to the period of the depletion of the public domain, though the secretary of state continues to exercise a few residual functions of the surveyor general. The author is indebted to Robert S. Davis, Jr., of Jasper for assistance in compiling this table.

Public Surveyors under the Trustees

Noble Jones	1733?–October 21, 1738
Joseph Avery	October 1, 1742–October 16, 1744
Thomas Ellis	November 11, 1745–June 23, 1752[1]
Henry Yonge	March 29, 1749–June 23, 1752[2]

Surveyors General under the Royal Government

Thomas Ellis	June 23, 1752–1754
Henry Yonge	June 23, 1752–May 1778
William Gerard DeBrahm	December 20, 1754–June 1764

1. Date of the surrender of the charter by the Trustees.
2. Date of the surrender of the charter by the Trustees.

Philip Yonge	May 13, 1778–January 19, 1782
George McKenzie	January 1782–July 11, 1782[3]

Surveyors General under the State Government

Thomas Chisholm	May 22, 1777–?
William Few	1778?
Richard Call	October 1, 1782–October 14, 1785
Thomas McCall	October 14, 1785–February 22, 1796
Samuel Hammond	February 22, 1796–January 26, 1797
Daniel Sturges	January 26, 1797–November 24, 1809
Edmund B. Jenkins	November 24, 1809–February 6, 1817
Daniel Sturges	February 14, 1817–September 17, 1823
Benjamin Hicks Sturges	September 19, 1823–November 22, 1823
Thomas Mitchell	November 22, 1823–November 25, 1825
Carlton Wellborn	November 25, 1825–November 24, 1827
John McBride	November 24, 1827–June 28, 1828
Everett H. Pierce	June 30, 1828–November 10, 1828
John Bethune	November 10, 1828–November 24, 1835
John Brewster	November 24, 1835–October 13, 1837
Obadiah Echols	October 13, 1837–November 15, 1837
Jesse Cox	November 15, 1837–November 25, 1839
Alfred M. Horton	November 25, 1839–November 24, 1843
Pleasant M. Compton	November 24, 1843–November 26, 1849
James Rogers Butts	November 26, 1849–November 24, 1851
Pleasant M. Compton	November 24, 1851–November 24, 1855
James Addison Green	November 24, 1855–November 25, 1859
Ahas J. Boggess	November 25, 1859–August 15, 1861
Archibald Alexander Gaulding	August 17, 1861–November 22, 1861

Secretaries of State

Nathan Crawford Barnett	November 22, 1861–January 28, 1868
Charles Wheaton	January 28, 1868–August 5, 1868
David G. Cotting	August 7, 1868–January 13, 1873
Nathan Crawford Barnett	January 13, 1873–February 2, 1890
Philip Cook, Sr.	February 5, 1890–May 22, 1894
Allen Daniel Candler	May 24, 1894–March 3, 1898
William C. Clifton	March 31, 1898–October 29, 1898
Philip Cook, Jr.	October 29, 1898–March 19, 1918

3. Date of the British evacuation of Georgia.

A Note about Sources

This text discusses the applicable decisions of the Supreme Court of Georgia through volume 259 of the Georgia Reports (388 S.E.2d) and the applicable decisions of the Court of Appeals of Georgia through volume 193 of the Georgia Appeals Reports (389 S.E.2d). Full bibliographic information about most of the nonlegal published sources cited can be found in Azalea Clizbee, Catalogue of the Wymberley Jones DeRenne Georgia Library . . . (Wormsloe: Privately printed, 1931) and Arthur R. Rowland & James E. Dorsey, A Bibliography of the Writings on Georgia History 1900–1970 (Spartanburg, S.C.: Reprint Co., 1978). Abbreviations of depositories of manuscripts cited are as follows:

GDAH	Georgia Department of Archives and History, Atlanta
GSGD	Georgia Surveyor General Department, Atlanta
UGL	University of Georgia Libraries, Athens

The meanings of most other abbreviations used can be found in the back of Black's Law Dictionary. Abbreviations and shortened citation forms not listed there, with their meanings, are as follows:

Ala. Law	The Alabama Lawyer
Am. Jur. 2d	American Jurisprudence 2d
Am. Law of Prop.	American Law of Property
Am. St. Papers	American State Papers
Annals of Cong.	Annals of the Congress of the United States
A.O.	Audit Office
art.	article
c.	chapter
ca.	circa
cert.	certiorari
C.I.S.	Congressional Information Service
Clayton	Augustin S. Clayton, A Compilation of the Laws . . . of Georgia, 1801–10
Cobb	Thomas R. R. Cobb, A Digest of the Statute Laws of . . . Georgia . . . to . . . 1851
Coll. Ga. Hist. Soc.	Collections of the Georgia Historical Society
Col. Rec. Ga.	Allen D. Candler et al. eds., The Colonial Records of the State of Georgia

Col. Rec. N.C.	William L. Saunders et al. eds., The Colonial Records of North Carolina
Cong. Deb.	The Debates and Proceedings in the Congress of the United States
Dawson	William C. Dawson, A Compilation of the Laws of . . . Georgia, 1820–29
ed.	edition; editor
Emory U.Q.	Emory University Quarterly
et seq.	et sequitur, and as follows
Ga. Const.	Georgia Constitution
Ga. Hist. Q.	Georgia Historical Quarterly
Ga. L.	Georgia Laws (Acts of the General Assembly)
Ga. Real Est. Law	George A. Pindar, Georgia Real Estate Law and Procedure
Ga. St. B.J.	Georgia State Bar Journal
GLB	Governor's Letter Book
H.R.	U.S. House of Representatives Resolution
H.R. Doc. No.	U.S. House of Representatives Document Number
H.R. Ex. Doc. No.	U.S. House of Representatives Executive Document Number
H.R. Rep. No.	U.S. House of Representatives Report Number
Id.	idem, the same
Lamar	Lucius Q. C. Lamar, A Compilation of the Laws of . . Georgia, 1811–19
Marbury & Crawford	Horatio Marbury & William H. Crawford, Digest of the Laws of . . . Georgia, 1755–1800
ms.	manuscript
n.	note
n.d.	no date
n.p.	no place
O.C.G.A.	Official Code of Georgia Annotated
Op. Atty. Gen.	Opinions of the Attorney General of Georgia
Papers Am. Hist. Assoc.	Papers of the American Historical Association
para.	paragraph
pt.	part
Pub. L. No.	Public Law Number
rev. ed.	revised edition
Rev. Rec. Ga.	Allen D. Candler, The Revolutionary Records of the State of Georgia
S. Doc. No.	U.S. Senate Document Number
S. Misc. Doc. No.	U.S. Senate Miscellaneous Document Number
S. Rep. No.	U.S. Senate Report Number
Ser. No.	Serial Number

St. Rec. N.C.	William L. Saunders et al. eds., The State Records of North Carolina
trans.	translator
unpub. ms.	unpublished manuscript
U.S. L. Rev.	United States Law Review
v.	versus
Watkins	Robert Watkins & George Watkins, A Digest of the Laws of . . . Georgia, 1755–98 (1st ed. 1800)
§	section

Table of Cases Cited

Table of Sections Cited to the Official Code of Georgia Annotated

Index

Numbers in italics refer to illustrations.